IT'S ABOUT TIME

ELEMENTARY MATHEMATICAL ASPECTS OF RELATIVITY

ROGER COOKE

AMERICAN MATHEMATICAL SOCIETY
Providence, Rhode Island

2010 *Mathematics Subject Classification.* Primary 89-01, 01-01.

For additional information and updates on this book, visit
www.ams.org/bookpages/mbk-102

Library of Congress Cataloging-in-Publication Data

Names: Cooke, Roger, 1942–
Title: It's about time : elementary mathematical aspects of relativity / Roger Cooke.
Description: Providence, Rhode Island : American Mathematical Society, [2017] | Includes bibliographical references and index.
Identifiers: LCCN 2016038198 | ISBN 9781470434830 (alk. paper)
Subjects: LCSH: Relativity (Physics)–Mathematics–Textbooks. | Mathematical physics–Textbooks. | Space and time–Mathematics–Textbooks. | AMS: Relativity and gravitational theory – Instructional exposition (textbooks, tutorial papers, etc.). msc | History and biography – Instructional exposition (textbooks, tutorial papers, etc.). msc
Classification: LCC QC173.55 .C66 2017 | DDC 530.1101/51–dc23 LC record available at https://lccn.loc.gov/2016038198

Contents

Preface

*And the same year I began to think of gravity extending to ye Orb
of the Moon & (having found out how to estimate the force w^{ch} [a]
globe revolving within a sphere presses the surface of the sphere)
from Kepler's rule of the periodical times of the planets being in
sesquilateral proportion to their distance from the center of their
Orbs, I deduced that the forces wch keep the planets in their Orbs
must [be] reciprocally as the square of their distances from the cen-
ters about which they revolve: & thereby compared the force requisite to keep the Moon in her Orb with the force of gravity at the
surface of the Earth, & found them answer [agree] pretty nearly.
All this was in the plague years 1665–1666.*

Isaac Newton, 1718. Quoted by Westfall ([**84**], p. 109).

The present book has grown from what was originally the very small project of
writing an article highlighting three mutually unrelated areas of special and general
relativity, namely the twin paradox (§ 6 of Chapter 1), the relativistic Maxwell
equations (Chapter 3), and the precession of the orbit of Mercury (§ 7 of Chapter 4).
As the writing proceeded, I realized that, while I was keeping the mathematical
details to a minimum and omitting all the historical context of the physics, this
material would still not be accessible to the audience I had in mind, consisting
of people with mid-level undergraduate preparation in mathematics. One would
still need to know at least the rudiments of calculus of variations and differential
geometry. One thing led to another, and the filling in of those details required
two years of work and expanded this work to its present Brobdingnagian size of
some 400 pages, plus two additional volumes (posted online) of ancillary material.
My vision of the core of the work (Volume 1) remains: it is intended to be a
random set of commentaries on certain aspects of relativity. This book is neither a
technical introduction to relativity, nor a systematic history of its development, nor
yet a professional-quality examination of philosophical issues. The physicists and
historians who vetted it for publication pointed out to me a number of areas where
my ignorance and brashness led me to make throwaway comments that describe
research already performed or in progress. As I shall keep reminding the reader, I
am not a specialist in any of these areas; if I make some suggestions that expose
my innocence, that is not the worst fate than can befall an author.

I have had two purposes in mind in writing the present work, one pedagogical,
the other humanistic. The pedagogical purpose is to present some highlights of
the special and general theories of relativity with full mathematical details in a
form accessible to advanced undergraduate mathematics, science, and engineering
majors (and, of course, any interested person who knows a little university-level

mathematics). The humanistic purpose is to reflect on what these theories have meant for the human understanding of the physical world. These two purposes are intertwined, and the material that follows was selected so that both purposes could be pursued together. The book constitutes a meandering journey from one fundamental equation of mechanics to another, from Newton's $\boldsymbol{F} = m\boldsymbol{r}''$ to Einstein's $G_{\mu\nu} + \Lambda g_{\mu\nu} = (8\pi G/c^4)T_{\mu\nu}$. There are very many digressions along the way. The core material in Chapters 5–7 is preceded by three chapters discussing the need for the Lorentz transformation and its properties (Chapter 1), a standard discussion of the mechanics of special relativity (Chapter 2), a side excursion to highlight the fruitful interaction between special relativity and the Maxwell equations (Chapter 3), and a mostly computational chapter (Chapter 4) that shows how general relativity produced its first two major successes, all mathematical justification being postponed. These are followed by the three chapters (Chapters 5–7) of exposition of general relativity, Chapter 5 and Chapter 6 being devoted entirely to the requisite differential geometry. Chapter 7 contains a very brief discussion of nonrotating black holes, followed by an attempt to make the Einstein field equations appear natural. Chapter 8, which brings this work to a close, consists of a chronology of landmark works on physics in general and some reflections on metaphysics from the point of view of an interested nonspecialist.

I hope that physicists will be tolerant of my inexperience in their specialty. In lieu of expertise, what I offer the reader is a selection of topics that I have found fascinating and wish to share with others. To get all of the apologies out of the way at the beginning, let me state again that this is not a systematic exposition of relativity theory, which is an enterprise I leave to the experts. The experts have not been found wanting in this regard. The best example is the 2013 book *Einstein Gravity* by Anthony Zee (see my comments below). Other recent specimens are the books of Narlikar [62] and Dray ([13], [14]). The purely mathematical heart of the present book—Chapters 5 and 6 on curvature—is treated in more detail and more systematically in the recent book of Sternberg [78]. The reader is encouraged to look at these books to see how many topics have been omitted here. Generally, these other sources contain fewer computational proofs and more conceptual ones than the present book and offer the additional advantage of having been written by people who *do* have expertise in this area.

The manuscript has been vetted by anonymous reviewers, at least two of whom were physicists. These two were both very kind in their criticism but dissatisfied at many of my omissions, for example the absence of any discussion of the 1923–24 work of Elie Cartan ([6], [7]), which subsumes my amateur attempt to demonstrate in Chapter 4 that Newtonian mechanics cannot be geometrized in such a way as to account for the precession of a planetary orbit, and the similar absence of any mention of the 1911 work of Willem de Sitter (1872–1934) [10], which subsumes my attempt to adapt special relativity to account for that precession. They added that I have also omitted the work of Alexander Friedmann (1888–1925) on curved spaces ([30], [31]) and that of Georges Lemaître (1894–1966) [52] on the so-called Hubble expanding universe. I have also not mentioned the original paper[1] [45] of Edwin Hubble (1889–1953) himself. These omissions are, one and all, due to my lack of familiarity with, and/or complete unawareness of, the topics mentioned. I would

---

[1] Hubble's paper does not mention the earlier work of Lemaître; as Hubble was not careless about citation, one must assume he had not heard of it at the time he published his own work.

not venture to write about subjects in which I am so ill-informed. I emphasize that this book is, as its subtitle indicates, a set of reflections on relativity by a mathematician who hopes someone else might find them as interesting as I do. Nevertheless, I think advanced undergraduates can learn something of value about relativity from reading it.

My title is one I remember seeing on a televised documentary on relativity some decades ago. Since completing the work, I have also chanced upon the 1995 book ([9]) by physicist Paul Davies whose title is similar and which discusses relativity and other issues of modern physics. Indeed, a search through almost any online catalog for this title will turn up literally thousands of books whose titles contain the phrase "About Time". Some of them use it in the sense I intend. For others, the word "time" is a synonym for a prison term. The reader will recognize that the title bears a double meaning in all cases. In ordinary speech, "It's about time" expresses vexation that a desired result has been delayed and satisfaction that it has finally arrived. That is *not* the meaning I intend for it here, since I do not flatter myself that the world has been waiting impatiently for the present book. Rather, I have in mind the literal sense that time is precisely what this book is about and even (with some exaggeration) what the theory of relativity itself is about. Poincaré recognized this over a century ago, when he heaped lavish praise on the introduction of what Henrik Lorentz called *local time* and what is now called (following Hermann Minkowski) *proper time*. Even in everyday life, if we want to know the distance to a remote location, we are not usually concerned with the distance itself, but more often with the amount of time it will take us to go there. Our language is full of expressions that describe distance in terms of travel time. In physics, we can do the same thing by prescribing a fixed velocity v_0 and using the equation $d = v_0 t$ to convert a distance d to a time t and vice versa. That technique leads to some interesting geometrized Newtonian astronomy in Chapter 4 and is at the heart of the relativistic approach to planetary motion. The central fact on which relativity is based is that there is an absolute unit of speed that can be used to carry out this conversion, namely $v_0 = c$, the speed of light in a vacuum, a speed that is the same for all observers. All of the equations of special relativity depend on the entanglement of time with a spatial axis of relative motion between two observers and the resulting distinction between observed time and proper time for a moving particle. "It" is literally *about time*.

So much for the title. The subtitle expresses my pedagogical purpose, to which I now turn.

Pedagogical Aims

One of the hardest questions asked by students studying high-school and college-level algebra is, "What is algebra good for?" They know perfectly well that people *never* need to solve even quadratic equations in everyday life, and no one is ever faced with the problem of computing where two trains will meet setting out in opposite directions from Chicago and New York at different times. One can tell them that algebra is the language in which science is written, but the unfortunate truth is that science seldom needs to solve only algebraic equations; much more often, the problems involve differential equations. It is true that one needs to know algebra very well in order to understand differential equations, but that explanation is generally lost on people just beginning the study of the subject. It needs to be pointed out that algebra plays a vital role in the *discovery* of scientific laws. Here are two examples:

- The rule that "distance = speed × time" at constant speed has the same form as the rule that, for a rectangle of constant width and variable length, "area = width × length". It is easy to see geometrically—as the scholars at Merton College, Oxford, seem to have reasoned in the thirteenth century—that if the width (w) happens to be directly proportional to the length (l), that is, $w = kl$ for a constant k, the graph of this relation provides a family of right triangles with legs varying in proportion to each other, and areas A given by the relation we nowadays write as the formula $A = \frac{1}{2}kl^2$. By the analogy that exists between the two relations just cited, the distance (s) fallen when speed is directly proportional to time (t) ought to be $s = \frac{1}{2}gt^2$, where g is the constant of proportionality between speed v and and time t ($v = gt$).[2] Assuming that g is constant near the surface of the Earth, it is the *gravitational acceleration*, which by observation is 9.81 meters per second-squared.

  To extend this example, consider the case of a particle moving around a circle of radius r at uniform speed v. Since it is not moving in a straight line, it has some acceleration, always directed toward the center of the circle. It is intuitively obvious that the magnitude of this acceleration is constant. What is its value? In the absence of any force, the particle would fly off the circle along a tangent line at speed v, and after time t, it would be on a circle of radius $\sqrt{r^2 + (vt)^2}$. In order to stay on the circle, it must "fall" a distance[3]

  $$s = \sqrt{r^2 + (vt)^2} - r = r\left(\sqrt{1 + (vt/r)^2} - 1\right).$$

  Over a small interval of time t, the standard "differential" arguments of calculus show that the right-hand side of this equation is closely approximated by the expression

  $$r \cdot \frac{1}{2}\frac{(vt)^2}{r^2} = \frac{1}{2}\frac{v^2}{r}t^2.$$

  The relative error in this approximation tends to zero as t tends to 0. Since it is the instantaneous acceleration we are interested in (essentially, the relation that results when $t = 0$), by comparing this expression with the Merton rule, we see that the factor g in the relation $s = \frac{1}{2}gt^2$ corresponds to v^2/r, which is therefore the magnitude of the acceleration. This is what Newton in the quotation above called "the force [with] which a globe revolving within a sphere presses the surface of the sphere".

  To extend the example still more, given the formula v^2/r for the acceleration of a body in motion at speed v around a circle of radius r, Kepler's third law implies, as Newton noted, an inverse square law of gravitational attraction.[4] The implication also goes in the opposite

---

[2]The analogy between the geometric and mechanical formulas here was explicitly noted and illustrated by the fourteenth-century Bishop of Lisieux Nicole d'Oresme (1323–1382), who laid the groundwork for the analytic geometry of Fermat and Descartes.

[3]Only the radial distance fallen is involved here, since the acceleration has no tangential component.

[4]From Newton's own words, quoted above, this appears to be the reason he believed in an inverse-square law. One can easily imagine other considerations that point to the same conclusion.

direction. An inverse-square law of attraction, together with Newton's second law of motion, implies Kepler's third law, as Newton later proved. (See Chapter 4 below.) To take the simplest case, for a body in circular orbit, the derivation of the inverse-square law is merely a short string of equations:

$$T = k\,r^{\frac{3}{2}} \quad \text{(Kepler's third law, } T = \text{period, } r = \text{radius of the orbit)},$$

$$v = \frac{2\pi r}{T} = \frac{2\pi}{k} r^{-\frac{1}{2}} \quad \text{(speed = distance/time)},$$

$$\frac{v^2}{r} = \left(\frac{2\pi}{k}\right)^2 r^{-2} \quad \text{(the inverse-square law)}.$$

Finally, to reap the harvest of this simple mathematics, consider, as Newton claimed to have done, the orbit of the Moon. Its sidereal period T is approximately $27\frac{1}{3}$ days,[5] which amounts to 2.361×10^6 seconds. The radius r of the Moon's orbit (approximating it by a circle) is 3.85×10^8 meters. By the formula given above, its acceleration v^2/r is $4\pi^2 r/T^2$, which amounts to 0.002725 meters per second-squared. Since the radius r is almost exactly 60 times the radius of the Earth, this ought to be about equal to the acceleration of gravity at the Earth's surface (9.81 meters per second-squared) divided by $60^2 = 3600$. And indeed, $9.81/3600 = 0.002725$. As Newton said, the two figures "answer pretty nearly"! (Newton wrote these words half a century after the alleged events they describe. Historians do not believe Newton had the law of universal gravitation in 1665. If he had possessed it at that time, the computation he claimed to have made would not have worked, since he believed one degree of a great circle on the Earth's surface was about 60 miles. In fact, it is about 70 miles (111 km). He really did make the computation later, with accurate figures on the size of the Earth provided in a posthumously published work of Jean-Félix Picard (1620–1682).) This astonishingly close agreement with observation is an awe-inspiring example of the power of simple mathematics to reveal the mysteries of the universe.

To summarize, in three brief, elementary arguments using simple geometry, simple algebra, and a tiny bit of calculus, we have set up a plausible physical law that can be applied and tested with astronomical observations and have performed such a test. Our proposed law has passed the test of observation amazingly well.

That is the kind of connection I enjoy making and will be making in these pages. As a mathematician, I am most interested in the contribution that symbolic algebra makes in the process just described. It was the analogy between the relations "length × width = area" and "speed × time = distance", expressed by two symbolic equations of exactly the

---

For example, if the total "gravitational attraction" remains constant as it spreads out from the center of attraction, its "density", which is the attraction on a particle at a given point at distance r, would be "diluted" in proportion to the area of the sphere of that radius, which is to say, in proportion to the square of the distance.

[5]Because the Earth moves about 26 degrees around the Sun during that sidereal period, the Moon must revolve about 390° degrees from one full moon to the next. Hence the synodic period of the Moon is about $29\frac{1}{2}$ days.

same algebraic form, that led us (not the Merton scholars) to the Merton rule, and ultimately to the law of falling bodies. It was the algebraic similarity of the formula connecting the distance a heavy body falls in time t with the distance a body in circular motion falls toward the center of the circle in time t that produced the law of acceleration for uniform circular motion.

My second example shows how algebra enables us to reason about the geometry of multi-dimensional spaces too complex to visualize.

- It will be shown in the discussion of curvature in Chapters 5 and 6 that the intuitive geometric idea of projecting the derivative of a tangent vector to a surface in three-dimensional real space $\mathbb{R}^3$ into the tangent plane of the surface leads to the Christoffel symbols, for which an explicit algebraic formula can be given. This algebraic formula can be trivially extrapolated to any finite number of variables by the simple *algebraic* process of extending the range of summation on the indices. In that way, the whole panoply of covariant derivatives, parallel transport, and curvature can be defined in a completely abstract way, without the need for any embedding in an absolute Euclidean space. That is a crucial point, since space-time is the only universe we know, and we have no direct evidence that it is embedded in a space of higher dimension. The geometric, intuitive origin of the Christoffel symbols lies in the familiar territory of surfaces in three-dimensional space. But, by a simple algebraic generalization, replacing 3 by n, we get Christoffel symbols that describe the curvature of manifolds of any dimension.

Humanistic Aims

The special and general theories of relativity are milestones in the progress of theoretical and experimental physics in the late nineteenth and early twentieth centuries. At the same time, these theories have produced a profound paradigm shift in the modes of thought by which scientists order the universe. They were necessary on purely theoretical grounds due to asymmetries in the equations of electromagnetism and on experimental grounds due to the failure of attempts to detect the hypothetical "luminiferous ether" in which light was believed to be an elastic wave. In that respect, they resemble the paradigm shift in chemistry in the eighteenth century, which coincided with the failure of the hypothetical "phlogiston" to show itself to experiment and its replacement by a quantitive theory of oxidation.

After presenting the mathematical skeleton of the theories in Chapters 1–7, I take some time in the final chapter to explore these issues and reflect on the evolution of physics all the way from Aristotle through Einstein. In this chapter, not being a specialist in philosophy, I am more concerned to raise questions for people who may not have thought of them than to propose answers that professional philosophers of science may have thought of already and perhaps even rejected as inadequate.

In order to fulfill my humanistic aims, I have sought breadth and generality rather than profundity and detail. But to avoid sacrificing my pedagogical goals, I have not sought *ultimate* generality. The whole book reflects the tension between these two goals. For example, a minimal exposition of special relativity is given in the first two chapters. That is pedagogically necessary for what follows, but to keep the student's imagination active, I replace some of the standard material on special relativity in the first three chapters with topics that I happen to find interesting. These sections are marked with asterisks to indicate that they can be omitted without loss of continuity. They are the kinds of questions that mathematicians ask—Is the composition of two Lorentz transformations a Lorentz transformation? If so, how can I get its matrix in standard form? Given that this standard form is obtained by "sandwiching" the actual coordinate transformation between two rotations on the spatial portions of the observers' space-times, can we be sure this composition is associative? Can relativistic velocities be made into a group?— and those who are not fascinated by such questions would be well advised to skip these starred sections. I do think, however, that the deduction of the Maxwell curl equations from the divergence equations in Chapter 3 by use of the relativistic transformation of electric and magnetic fields between observers will be interesting to nonmathematicians.

Special Features of This Book

The following is a list of aspects of relativity that are not part of standard physics textbooks, an exception being the recent book of Dray [**13**]. I believe they would be of both pedagogical and humanistic value to university professors teaching this material.

- Let c denote the speed of light.[6] Consider three observers moving away from a common origin, the three pairs having relative speeds u, v, and w (all less than c). With them, we associate three lines in a plane with lengths U, V, and W via transformations $U = k\operatorname{arctanh}(u/c)$, $V = k\operatorname{arctanh}(v/c)$, and $W = k\operatorname{arctanh}(w/c)$, where k is a fixed unit of length. The inverse transformations are $u = c\tanh(U/k)$, $v = c\tanh(V/k)$, and $w = c\tanh(W/k)$. As a consequence of the Lorentz transformation, the lengths U, V, and W will be such that they form a triangle, that is, the sum of the two smaller ones will exceed the largest. If we assign angles to that triangle equal to the angles that each of the three observers will measure between the lines of sight to the other two, its metric properties will be those given by the trigonometric relations in a hyperbolic plane of curvature $-1/k^2$. This fact has been well known for over a century, having been first noted by the Croatian mathematician (of Serbian descent) Vladimir Varićak (1865–1942), who wrote a series of articles on it (see, for example, [**82**]). It became better known after it was discussed in a book [**3**] by the eminent mathematician Emile Borel (1871–1956). It has been discussed in a number of places, most recently in the book of Dray [**13**]. One consequence of this fact is that the simple, very computable, Lorentz transformation can be used to derive all of hyperbolic plane trigonometry, including the formula for the angle of parallelism, horocycles, and equidistant lines, in just a few pages. That derivation is carried out in Appendix 1 (in Volume 2).

  The connection between relativistic velocities and hyperbolic trigonometry is even richer than just indicated. The set of all observable velocities in a fixed plane of reference can naturally be associated

---

[6]This is now the universal symbol for the speed of light, probably derived from the Latin word for speed, *celeritas*. In his 1905 paper on special relativity, Einstein used the symbol V.

with the points in a disk of radius c. If the relativistic law of cosines is applied to define the first fundamental form on that disk, the result is precisely the Beltrami–Klein model of the hyperbolic plane, named after Eugenio Beltrami (1835–1900) and Felix Klein (1849–1925). The connection between the geometry of relativistic velocities and the geometry of the hyperbolic plane is explained in detail in Appendix 1 (of Volume 2).

In all the discussions of this connection that I have seen, the authors have explained the Lorentz transformation in terms of the geometry of the hyperbolic plane. My aim in the present work is to do the converse: to show how easily the trigonometry of the hyperbolic plane follows from the Lorentz transformation. It seems to me that this would be an excellent way to present non-Euclidean geometry to students, avoiding all the hard computations involved in using the Poincaré disk model or the even more arduous, classical, "bare-fisted" derivation using Saccheri and Lambert quadrilaterals and the fact that the hyperbolic horosphere is a Euclidean plane.

In addition to the pedagogical considerations just mentioned, this connection between relativity and hyperbolic geometry is for me a source of historical wonder: The trigonometry of the hyperbolic plane was worked out by Gauss, Lobachevskii, János Bólyai, and others in the early nineteenth century. There seems to be no logical or physical reason why this purely mathematical creation should have any connection with modern physics, and yet—astonishingly—the Lorentz transformation automatically produces relations among the speeds and directions of three observers in uniform relative motion that mirror *exactly* the relations among the parts of a triangle in the hyperbolic plane. To me, this is an excellent example of what Felix Klein (1849–1925) must have had in mind when, lecturing on the theory of a spinning top at the 1893 World's Columbian Exposition, he referred to a "pre-established harmony" between the physical world and the world of mathematical concepts. Klein was referring to the excellent "fit" between the assumptions of the physical model and what we now call the Cayley–Klein parameters of a rotation, which he introduced in this lecture. He remarked, incidentally, that these parameters allowed the motion to be regarded as occurring in a non-Euclidean space, or as occurring in a Euclidean space, but combined with a strain.

The pre-established harmony described by Klein is nowadays better known as "the unreasonable effectiveness of mathematics" since the appearance of an article [**85**] bearing that title by the late Nobel Prize-winning physicist Eugene Wigner (1902–1995). This article has become famous and is accessible online, for example, at the following website:

http://www.maths.ed.ac.uk/~aar/papers/wigner.pdf

For an opposing point of view, see the article by Grattan-Guinness [**35**]. In Chapter 8, we attempt to make this effectiveness appear a bit less unreasonable.
- The relativistic transformation of electric and magnetic fields eliminates an asymmetry in the classical transformation of electric and magnetic fields between two observers in relative motion; and when the relativistic

correction is made, it becomes possible to derive Maxwell's two curl equations from the two divergence equations. Although this result is hardly new—I first read about it some 30 years ago—I find that few mathematicians seem to know about it. From the historical point of view, it helps to clarify the main achievement of Einstein's first (1905) paper on the subject. This material forms the subject matter of the optional (starred) Chapter 3.

- In Chapter 4, I compute the relativistic orbit of Mercury and use real data to show that the relativistic portion of the precession of its perihelion is indeed 43 seconds per century. Moreover, if the right-ascension angle $\theta(t)$ at time t is replaced by an angle $\varphi(\tau) = \theta(t)$, where τ is the proper time on Mercury corresponding to time t in the Sun-based coordinates, the orbit is an ellipse in a set of unobservable polar coordinates with φ as the polar angle. (In other words, there is no precession in these coordinates.) This fact illustrates a more general principle: Some classical laws of physics, such as conservation of angular momentum, can be transferred to relativity if observer time is replaced by proper time on the moving object being discussed.

- As already mentioned, my discussion of the precession of the orbit of Mercury is prefaced by two sections aimed at clearing away possible approaches that might have occurred—in fact, did occur—to people reflecting on the problem in the early twentieth century. In both cases (also as already mentioned) I have learned from anonymous reviewers that my thoughts were anticipated and developed in much more detail than I could have managed myself by two prominent mathematical physicists, namely Willem de Sitter and Elie Cartan. Let my explanation of these approaches in my own words serve as a rough approximation to parts of their work. These two explorations, both ultimately unsuccessful as an explanation of planetary motion, nevertheless produce differential equations surprisingly similar to those of the Schwarzschild solution. The "Newtonian geometric" approach, in particular, naturally introduces both the Schwarzschild radius and the elliptic integrals that provide an exact expression of relativistic orbits.

- I have noticed that physics texts tend to avoid the use of elliptic functions, even though their mathematical theory is well known and they have had applications in physics since the eighteenth century—in the study of pendulum motion and the rotation of rigid bodies, for example. There are several places where these functions arise in relativity theory. I use them in Chapter 1 to study a relativistic model of a geocentric universe—essentially giving the mathematical details of the relativistic geometry of a rotating plane, a problem discussed qualitatively by Einstein [21] and in Chapter 4 to give the exact (Schwarzschild) solution of the gravitational equation in the two-body problem.

- In order to put the Lorentz transformation into perspective, I point out that each individual observer uses his own proper time and his own proper space, and these have the same absolute qualities that Newton assigned to a *universal* time and space common to all observers, except that no material particle can move faster than light relative to that "personal absolute" space. (Just to be clear, two particles in the space *can* move faster than light relative to each other, as judged by an observer at rest in the framework; but neither of them will judge the relative speed to be larger than c.) It is the entangling of the time axis with the spatial axis along the common line of motion when two observers reconcile their coordinates that leads to the bizarre, yet real, phenomena of length and time contraction. A discussion of this transformation provides a healthy caution on the use of vector operations to avoid coordinates. In special relativity theory, vector notation is applicable to the space used by each observer, but is not transferrable between two observers unless a particular set of coordinates is used, since *the Lorentz transformation does not preserve dot and cross products on the spatial portion of space-time.* If Y has velocity $\boldsymbol{u}$ relative to X, and X has velocity $-\boldsymbol{u}$ relative to Y—that is, Y assigns to the velocity of X coordinates that are the negatives of those that X assigns to the velocity of Y—while Z has velocity $\boldsymbol{v}$ relative to Y, vector notation can be used to compute the velocity $\boldsymbol{w}$ that Z has relative to X. But unless $\boldsymbol{u}$ and $\boldsymbol{v}$ are parallel—that is, when expressed in terms of the coordinates used by Y, each is a scalar multiple of the other—this computation cannot be reversed to give the velocity that X has relative to Z through the simple replacement of $\boldsymbol{u}$ by $-\boldsymbol{v}$ and $\boldsymbol{v}$ by $-\boldsymbol{u}$ and computation of the dot products contained in the formula. When X and Y use arbitrary orthonormal coordinate systems, Y will *not* generally assign to the velocity of X the coordinates that are the negatives of those that X assigns to Y. Consequently, these velocities cannot be passed back and forth between different observers using the familiar dot and cross products that each individual observer can use for his own purposes. The observers do not have a common space that they can talk about. Vector notation remains useful for *recording* physical relations, but can be passed from one observer to another only when they both agree to use their common line of motion as the first spatial axis. In other words, one is forced to derive physical relations coordinate-wise, just as Einstein did in 1905, or else adopt a completely new approach to the subject. As this book aims to be elementary, we choose the first of these options.

Other Works on the Subject

As I am not the first person to write a book of this type, let me point out here the main respects in which the present book differs from or resembles others on the subject. This book was begun in 2012, before the publication of the recent book by Anthony Zee ([**88**]), and it was completed before I became aware of the existence of Professor Zee's book. Zee and I have similar pedagogical and humanistic aims. One significant difference is that Zee's book discusses the subject from the standpoint of late twentieth-century physics, a task that would be utterly beyond my ability. Zee's book covers many more topics and in much greater depth than I have done. Except for the Gödel metric, which dates to the 1940s, the present book takes the development of both differential geometry and general relativity only as far as the mid-1920s, and it does not attempt to apply the Einstein field equations to anything outside the simple case of a constant matter density. The most complicated example considered is the Gödel metric. I do explore certain special questions that are more of interest to mathematicians than to physicists.

Outside of Zee's book, perhaps the closest in its pedagogical aim of explaining relativity theory in plain language accessible to undergraduates is the book of the late Richard L. Faber [27]. Besides the works of Narlikar, Dray, and Sternberg mentioned above—and, no doubt, dozens of others unknown to me—the subject matter of this book is discussed in the book of Torretti [81]. Like Zee's book, Torretti's erudite book is many times more complete and detailed than the present one and contains, as well, detailed historical information on the evolution of Einstein's thought in the years 1913–1915. My pedagogical aim differs from what I perceive to be Torretti's, which appears to be aimed at specialists in the philosophy of science. My goal is to simplify *selected parts* of the theory of relativity, to make them accessible to a person who has studied only the basic three semesters of calculus and two semesters of linear algebra.

Among older works, Bertrand Russell devoted considerable space to an exposition of relativity theory, for example, in [72]. As often happens with such books, however, he explained the mechanics of manipulating the formulas without giving much information on the grounds for accepting them as laws of nature. His intended audience also appears to be people well versed in the philosophy of science, and he explores only philosophical issues. He is definitely not writing a textbook for undergraduates.

Given the sheer quantity of writings on this subject—many dozens of works that one can find in almost any library and which are not mentioned here—and my consequent unfamiliarity with most of them, I would not venture to say that anything I have written here is appearing for the first time. Like many other mathematicians, I have too often found that my ideas have also occurred to others, sometimes much earlier. For any passage in this book containing no citation of a source, there are two possible explanations: (1) the ideas contained in the passage are well known and can be found in many standard sources; (2) I have never found the ideas in any source, but have thought them up on my own. In neither case should the absence of a citation be interpreted as a claim of originality on my part.

Background Necessary to Read This Book

The reader is assumed to know the rudiments of advanced calculus,[7] a few techniques for solving differential equations, some linear algebra, and at least the nomenclature of set theory and groups. That is the reason for the word *elementary* in the title. A person with that mathematical background belongs to the audience to whom the present book is addressed. Of course, in order to understand general relativity it is necessary to know something about manifolds and geodesics, and that involves the very simplest parts of the calculus of variations and the existence and uniqueness theorem for the initial-value problem in ordinary differential equations. I have put this material into Volume 2 as Appendices 2, 4, and 5. Appendix 4 grew out of control during the writing, and I was forced to move its topological

---

[7]This is the name that used to be given to what is nowadays more frequently called elementary real analysis. We actually do not use much of it outside the appendices in Volume 2. Still, it will be helpful if the reader knows the implicit function theorem, Stokes's theorem, the divergence theorem, and the standard criteria for term-wise differentiation and (Riemann) integration of a uniformly convergent sequence.

foundations into Appendix 3. In addition, Appendix 3 contains a large assortment of inessential facts from point-set topology, which may be of intrinsic interest and accustom the reader to thinking in terms of abstract spaces. As for physics, I am assuming that the reader knows the concepts of mass, acceleration, force, momentum, work, and energy in mechanics, along with the basic facts of electricity and magnetism. Many of these concepts, however, are explained further in Chapters 2 and 4 and in Appendices 2 and 5.

Plan of the Work

As this work has grown, I have found it necessary to break it into three volumes.

The first volume consists of eight chapters of material on the main subject of the work, namely special and general relativity. This volume is divided into three parts, as follows: Part 1, which consists of Chapters 1–3, contains the rudiments of the special theory of relativity, explained from a particular point of view, and with some optional excursions into topics that especially intrigue me. Part 2, consisting of Chapters 4–7, is devoted to the general theory of relativity. Part 3, consisting of just the final Chapter 8, is an excursion through the historical and philosophical context of the theory. In it, I attempt to answer metaphysical questions as to the nature of arcane objects such as gravitational fields and explain how, even though we cannot directly experience these things, we can know facts about them and can be justified in regarding them as real. Since I am not a professional philosopher (as will be apparent to those who specialize in this area), my explanations are offered as suggestions to a generally educated reader. Those explanations, elaborated with many examples in this chapter, invoke the harmony between mathematics and the physical world to explain how facts come to be known about unknowable things and use an analogy with the theory of manifolds to explain that the agreement between different methods used to measure physical quantities reinforces the credibility of the theories on which they are based. What I suggest may be a way for broadly educated people to understand approximately what is meant by existence and reality. A high level of precision in this enterprise is, I think, unachievable.

At the end of each chapter, the reader will find a set of problems and questions to help fix the ideas from that chapter. These are almost without exception routine applications of what was discussed in the text, and few of them require any ingenuity to solve. I hope that the reader nevertheless finds them interesting to work. Some of them are intended to make a mathematical or philosophical point.

Sections that represent digressions or material that is not needed for reading what comes after it are marked with an asterisk. The reader who is looking for a minimal introduction to the problem of gravity in free space can omit these sections without losing the thread of the narrative.

Volume 2 arose because the mathematical background involved in Volume 1 is rather large. It consists of six appendices made up of background mathematical material that would have forced digressions if presented in the main narrative of this work. For example, in Chapter 4, on the orbit of Mercury, I have temporarily omitted the mathematical theory that justifies the use of tensor calculus to express the curvature of space-time in order to focus on the computation. The omitted theoretical tensor analysis can be found in Chapters 5 and 6 and Appendices 4 and 6. Taking Volumes 1 and 2 together, the reader should be able to find full mathematical details on all the material. In addition to using these appendices as a way to avoid digressions in the main narrative, I took the opportunity to include among them some topics that I find irresistible. Examples are (1) Euler's wonderful result that a particle moving along a surface but free of tangential acceleration will describe a geodesic on the surface (Appendix 2); (2) Jacobi's elegant last-multiplier principle (Appendix 5), which shows that a system of n ordinary first-order linear differential equations with algebraic coefficients whose divergence vanishes has solutions that can be expressed as quotients of theta functions, provided one can find $n - 1$ independent algebraic integrals (functions not identically constant that are constant on the trajectories of a solution); and (3) the whole subject of point-set topology (Appendix 3), which I include as a way for the reader to acquire some practice in visualizing completely abstract spaces, along with the basic facts of the theory, which are used in both real analysis and differential geometry.

Volume 3 contains all the *Mathematica* notebooks that I used to lighten the labor of some of the more complex computations in the book and suggested answers to the exercises in the first two volumes. Eleven *Mathematica* notebooks are referenced in the first volume and one more in the second volume. They are collected as a unit at the beginning of Volume 3. For the convenience of the reader, Volumes 2 and 3 and all twelve *Mathematica* notebooks can be downloaded from my website at the University of Vermont:

http://www.cems.uvm.edu/~rlcooke/RELATIVITY/

Acknowledgments

I am grateful to Stephen Wolfram for inventing *Mathematica* and thereby putting enormous computing and graphing power and accuracy in computation into the hands of people of limited patience, who would otherwise face the dreary prospect of spending days or weeks floundering in an attempt to carry out a computation that *Mathematica* enables us to do flawlessly in a fraction of a second. That it will also render a beautiful perspective drawing of the graph of a function of two variables is a delightful bonus.

I wish to thank the anonymous reviewers who vetted the manuscript for publication and made some extremely valuable suggestions, pointing out places where my writing was misleading or revealed too much of my innocence of the full history of this subject. Their advice to tone down the occasional references to controversial political issues was also sound, and I have heeded it.

Special thanks go to my lifelong friend Charles Gillard, with whom I shared philosophical reflections as a teenager when we were both delivering newspapers, at whose wedding I met my wife, and who, now that we are both retired, has been kind enough to read the entire manuscript of the first volume and send me detailed comments, which I have taken into account in the proofreading.

Very special thanks go to my wife Catherine, who endured my getting up at 5:00 AM every morning for two years to work on this project.

Roger Cooke
August 2016

Part 1

The Special Theory

CHAPTER 1

Time, Space, and Space-Time

I have seen many books which have objected to [Euclid's fifth postulate], among the earlier ones Heron and Autocus (Autolycus),and the later ones Al-Khazen, Al-Sheni, Al-Neyrizi, etc. None has given a proof. Then I have seen the book of Ibn Haytham, God bless his soul, called the solution of doubt in Chapter One. This postulate among other things was accepted without proof. There are many other things which are foreign to this field, such as:

If a straight line segment moves so that it remains perpendicular to a given line, and one end of it remains on the given line, then the other end of it draws a parallel.

There are many things wrong here. How could a proof be based on this idea? How could geometry and motion be connected?

Omar Khayyám (1048–1131). See [**1**], p. 277.

1. Simultaneity and Sequentiality

Omar Khayyám's criticism of the principle behind an attempted proof of the parallel postulate by ibn al-Haytham has some features in common with a basic principle of the special theory of relativity. In the attempted proof, one line segment is pictured as sliding along another one so as to be always perpendicular to it. Omar Khayyám merely asked how a line segment[1] could move. He was willing to grant that a point could move, and that all the points in a collection could move individually. But, he thought, if they move, they are no longer the same points they were before. How can we be sure that the relations among them will be preserved in such a way that they continue to form a straight line, much less a straight line that is perpendicular to a given line? In fact, the special theory of relativity says very explicitly that two observers in motion relative to each other, each watching the line move as described by ibn al-Haytham, will *probably not* agree that it remains perpendicular to the given fixed line. The difficulty, it turns out, concerns the notions of simultaneity and temporal order of events. When the base point moving along the line is at point Q, where is the rest of the moving line? The presence of each of these points at one place or another is a different event, and two observers will not generally agree as to which of those events is simultaneous with the passage of the moving line through point Q.

The reconsideration of the notion of simultaneity required by special relativity reinforces Omar Khayyám's point. The event that is occurring at a given location

---

[1]Until the advent of projective geometry, all geometers thought of a line as what we call a closed line segment. Euclid says that the extremities of a line are points. We shall use the word *line* in this traditional sense in order to save writing the word *segment* excessively many times.

at a given time (the presence or absence of a given particle, for example) depends on who is observing. Perpendicularity of two lines requires that a certain configuration among three variable points remain constant at all times. One of the points, say Q, is the moving point of intersection of the two lines; the second point, say P, is fixed on the moving line; and the third, say R, lies on the fixed line. As we are about to discuss, when special relativity is admitted, it is impossible to state in an observer-independent way what the relative positions of three particles are at a given instant of time. It follows from the relativistic equations of transformation between observers that the temporal order of two events occurring in different locations may depend on the observer, and it is easy to see that lines regarded as mutually perpendicular by one observer are normally not mutually perpendicular as measured by a second observer. In terms of spatial coordinates, this disagreement comes about because of the *FitzGerald–Lorentz contraction*,[2] whereby the length one observer ascribes to a line in the direction of the other observer is found to be shortened by the factor $\sqrt{1 - u^2/c^2}$ when measured by the other. Here u is the mutual speed with which the two are moving relative to each other, and c is the speed of light, approximately 300,000 km/sec. This inconsistency of spatial measurement, in turn, arises from the fact that the observers disagree as to which of two events occurred earlier: They are unable to agree as to where one end of the line is in relation to the other at a given instant of time, since simultaneity of two events occurring in different places is not observer-independent. The details of this phenomenon will be explored below. We are now going to flesh out these abstract ideas with a scenario taken from the modern world.

1.1. The car wash puzzle. Imagine a very long limousine parked in a car wash, the limousine being exactly as long as the car wash. It would just fit inside, and the attendant would be able to close the doors at both ends of the car wash. Now imagine that same limousine driving through the car wash (at any speed you like, but imagine it to be a very high speed). Is there an instant of time when the limousine is entirely inside the car wash? The limousine driver will say no: Due to the FitzGerald–Lorentz contraction, the car wash shrank in length, and the limousine wouldn't fit inside. The car wash attendant will say yes: Due to the FitzGerald–Lorentz contraction, the limousine shrank in length and fitted inside with room to spare. Who is right here?

The explanation involves the observer-dependence of the concept of simultaneity. We are looking at two events here that take place in different locations. One event is the rear end of the limousine entering the car wash. The other is the front end of the limousine leaving the car wash. Those events occur in the opposite order for the two observers. While they share the same four-dimensional space-time and agree about the four-dimensional "proper-time" interval between the two events—taking one of the events to have occurred at time zero and at the origin of the spatial coordinates in both systems, while the other occurred at time t and at a point (x, y, z), that interval is $t^2 - (x^2 + y^2 + z^2)/c^2$ and is the same for both of them—they do not agree about the *scales* on either the line in space along the common direction of motion or on the time axis. In other words, t and at least one of the spatial coordinates, say x, are different for the two observers even though the four-dimensional interval is the same for both of them. The answer to the question

---

[2]Named after George Francis FitzGerald (1851–1901) and Hendrik Antoon Lorentz (1853–1928).

is that the question does not make sense. The individual time order of two events A and B will be the same for all observers only if a ray of light could set out from the location of one of them, say A, at the time A occurred and arrive at the location of event B before B occurs there. That is to say, the four-dimensional interval between A and B is positive—"timelike," as physicists refer to it. Because this four-dimensional interval is the same for all observers, any two will agree whether this is the case or not.

Another way of saying that the interval between events A and B is timelike is to say that an observer could physically be present at both events A and B. The proper time interval[3] between the two events (for any observer) is obtained by parameterizing a path from A to B, say $r \mapsto \big(t(r), x(r), y(r), z(r)\big)$, where A corresponds to r_0 and B to r_1, and then integrating the infinitesimal proper time interval ds, which is given by

$$ds^2 = dt^2 - \frac{1}{c^2} dx^2 - \frac{1}{c^2} dy^2 - \frac{1}{c^2} dz^2 .$$

Thus we find that the proper time interval is

$$\Delta s = \int ds = \int_{r_0}^{r_1} \sqrt{\big(t'(r)\big)^2 - \left(\frac{x'(r)}{c}\right)^2 - \left(\frac{y'(r)}{c}\right)^2 - \left(\frac{z'(r)}{c}\right)^2} \, dr .$$

If an observer is present at all the events $\big(t(r), x(r), y(r), z(r)\big)$, then in that observer's coordinate system, $x(r) = y(r) = z(r) = 0$ at time $t(r)$ for all r. Hence $x'(r) = y'(r) = z'(r) = 0$ for that observer, and (assuming B is later than A for that observer), $t'(r) \geq 0$. Therefore the time interval recorded by that observer is

$$\int_{r_0}^{r_1} \sqrt{\big(t'(r)\big)^2} \, dr = \int_{r_0}^{r_1} t'(r) \, dr = t(r_1) - t(r_0) .$$

In short, *the proper time interval between two events is the time interval recorded by an observer who was present at both.* More generally, if an observer assigns the same spatial coordinates to two events, then the proper time interval between them is just the time difference recorded by that observer.

If there were any observer who perceived B as occurring before A, then paradoxes might result if the interval is timelike—that is, a ray of light could leave the location of A at the time A occurred and arrive at the location of B before B occurred (which is equivalent to saying that $\Delta s^2 > 0$). A second observer at the site of event B could transmit information about event A before event B occurs and thus the observer who perceives B as occurring first could get historical information about event A *before it occurred.* That observer would be "remembering the future."

If the space-time interval between A and B is not positive—it is "spacelike," as physicists say—there is no observer-independent temporal ordering between two events occurring in different locations, and there is no absolute sense in which one

---

[3]The terminology is due to Hermann Minkowski (1864–1909), who gave it this name in a paper [**60**] published in the last year of his life. Minkowski had read this paper at a meeting in Köln on 21 September 1908. He actually gave the name "proper time" (*Eigenzeit*) to a quantity that has the physical dimension of length, namely $-dx^2 - dy^2 - dz^2 - ds^2$, where s is time made into an *imaginary length* by means of what he called (p. 86), the "mystical formula" $3 \cdot 10^5$ km $= \sqrt{-1}$ sec. The concept of proper time had been introduced earlier, however, by Lorentz, who called it *local time.*

event occurred before the other. A geometric explanation of this puzzle can be found by working Problem 1.19 below.

The car wash puzzle throws some doubt on ibn al-Haytham's intuition. The points P, Q, and R are in different locations, and different observers will not agree as to their relative configuration at a given time. For two observers O and O' in relative motion, both watching the two lines slide along each other, there will not in general be any agreement as to the angle between the two lines. Unless the fixed line is the line joining O and O', the two will not both agree that the moving line is perpendicular to it. In that respect, Omar Khayyám's objection gains force when special relativity enters the picture. Nevertheless, it remains true that both observers are carrying out their measurements using Euclidean principles, and they agree that the end of the moving line *not* on the fixed line is describing a line parallel to the fixed line. Thus, one can retain ibn al-Haytham's conclusion while rejecting the considerations that led him to it.

In the following sections, we shall investigate the relations among the speeds and directions of three or more observers, assuming that the space-time coordinates of any pair are reconciled using the equations of special relativity. As we shall learn, there is an inherent difficulty in regarding the velocity spaces used by two different observers as having any vectors in common, even though the equations of transformation from the coordinates of one to those of the other assume that the vector spaces representing locations (not velocities) do share a line, namely the line joining the origins. A unifying theory, which unfortunately lies beyond the scope of the present book, is the *Lorentz group*, the six-dimensional Lie group of transformations of $\mathbb{R}^4$ that preserve the four-dimensional interval between events.

2. Synchronization in Newtonian Mechanics

I. Absolute, true, and mathematical time, of itself, and from its own nature, flows equably without relation to anything external, and by another name is called duration: relative, apparent, and common time, is some sensible and external (whether accurate or unequable) measure of duration by the means of motion, which is commonly used instead of true time; such as an hour, a day, a month, a year.

II. Absolute space, in its own nature, without relation to anything external, remains always similar and immovable. Relative space is some movable dimension or measure of the absolute spaces; which our senses determine by its position to bodies; and which is commonly taken for an immovable space; such is the dimension of a subterraneous, an aerial, or celestial space, determined by its position in respect of the Earth. Absolute and relative space are the same in figure and magnitude; but they do not remain always numerically the same. For if the Earth, for instance, moves, a space of our air, which relatively and in the respect of the Earth remains always the same, will at one time be another part of the same, and so, absolutely understood, it will be continually changed.

Isaac Newton, *Mathematical Principles of Natural Philosophy*, Scholium to Definitions I–VIII ([**63**], pp. 8–9).

To the ancient Greeks, most of whom adhered to a geocentric cosmology, there was a perfect geometric representation of the universe: It was a series of concentric spheres, whose center was at the center of the Earth. The absolute geometric objects corresponded perfectly to what they believed was the nature of physical space. Later, when heliocentric astronomy replaced the geocentric model, the sphere was banished as the embodiment of physical space and was replaced by the unbounded three-dimensional flat Euclidean space of solid geometry, soon algebraized by the use of Cartesian coordinates.

In Newtonian mechanics, all observers share this common three-dimensional Euclidean space and can agree about the scale on a time axis independent of space. The location and the time of an event are separate issues, and all observers agree about both of them. Underlying this view, as expressed by Newton in the quotations above, is an unconscious metaphysical idea, an intuitive notion that there is a well-defined "now" that is common to all locations in the universe. Newton referred to it as "absolute time," that is, the pure article, undiluted by any of the observable "events" such as the hands on a clock assuming a certain position by which we actually determine the time. That inferior, diluted form of time is what Newton called "relative time." Many modern philosophers hold the view—one congenial to the present author—that what Newton called absolute time is a pure mental creation and is unobservable and unknowable, analogous to the Aristotelian distinction between "substance" and "accidents." These philosophers hold that we can talk meaningfully only about Newton's relative time (or, in the Aristotelian context, only about the accidents). While these philosophers have very likely adopted an unassailable position, they may be underestimating the value of human imagination in the creation of physical theories. Other concepts of mathematical physics—force, for example, which Bertrand Russell called "a convenient fiction"—can be similarly challenged by an obstinately practical-minded philosopher. Yet, as a guide to thought, leading to insights about the properties of observable objects, the ability to hypostatize[4] abstract concepts, picturing them as if they were observable physical objects, can be of enormous value, provided the hypothesis of their reality leads to the prediction of the result of a physical measurement. The concept of gravity is a good example of such a mental creation.[5] It may be only a convenient fiction, but it is a *very* convenient one.

Newton's attempt to define absolute space without reference to any observable objects is subject to the same metaphysical objections as his definition of time. Mathematicians and physicists find it useful to picture space as endowed with imaginary "mileposts" labeling each of its points with three numbers giving the distances of that point in front of, to the right of, and above, a fixed imaginary reference point (the origin). That is absolute space, and it "exists" only as a mental construct. For the mathematician, the space $\mathbb{R}^3$ *actually is* just the set of ordered triples of real numbers (x, y, z). This is "absolute Euclidean space." Such a picture

---

[4]From the Merriam–Webster on-line dictionary. "*hypostatize:* To attribute real identity to (a concept)." The word comes from the Greek word *hypostatos* ($\dot{\upsilon}\pi\dot{o}\sigma\tau\alpha\tau o\varsigma$), which has the same two roots as the Latin-derived word *substance*.

[5]Originally, *gravity* (*gravitas*) was just the Latin word for heaviness. It was a property possessed by such things as earth, air, and water, but not by fire, which had the opposite property of *levity* (*levitas*) or lightness. The mental picture of gravity as a *force*, a *thing* "existing" independently of bodies having the *property* of gravity, even though no one has any clear ideas as to *what* that thing is, remains useful.

cannot be used experimentally, however. As Newton suggested, we need *observable physical* objects in order to define a three-dimensional frame of reference in which the locations of all objects can be given in terms of three coordinates. He said that there might be, in the remote depths of interstellar space, a body perfectly at rest, but that it is impossible for us to infer from the positions of the bodies we observe which, if any, of them is at rest relative to absolute space. Nevertheless, he thought people writing philosophical treatises such as his *Principia* ought to "abstract from our senses, and consider things themselves, distinct from what are only sensible measures of them." In such statements, he sounds very much like Plato, who argued (in Book 7 of the *Republic*) that astronomy becomes useful only after one makes it a subject to be perceived by the mind, ignoring what the senses perceive in the heavens.

Although it was eventually forced to yield to hard stubborn facts, such as the absence of any detectable medium that conducted light waves, the mental habit of hypostatizing absolute time and space is still with us, and has a great deal of appeal for minds that evolved in a world of slow velocities and distances that lie within the range of perception of human senses, things that we can grasp intuitively. One feels, for example, that there really is a difference between saying the Sun orbits the Earth once a day and saying that the Earth rotates on its axis once a day. Newton himself (still in the section of definitions at the beginning of the *Principia*) argued that we could distinguish absolute motion from relative motion by looking at the case of circular motion about an axis. He pointed out that if a bucket is hung on a twisted cord and the cord allowed to unwind, the bucket spins, eventually causes the water to spin, and then the water rises up the sides of the bucket. Surely that suggests some absolute frame, at least for rotational motion, since the relative motion of the bucket and the water was greatest at the beginning, when there was no tendency for the water to rise; when the relative motion was least (the water had attained its maximal rotational velocity, equal to that of the bucket), the tendency to rise was at a maximum. That picture is a familiar one, and leads naturally to the idea that there is an absolute set of invisible axes relative to which rotation occurs. Such a picture is at the basis of a scene from C. S. Lewis' 1944 science-fiction novel *Perelandra*.

> *It was born in upon him that the creatures were really moving, though not moving in relation to him. This planet which inevitably seemed to him while he was in it an unmoving world—the world, in fact—was to them a thing moving through the heavens. In relation to their own celestial frame of reference they were rushing forward to keep abreast of the mountain valley. Had they stood still, they would have flashed past him too quickly for him to see, doubly dropped behind by the planet's spin on its own axis and by its onward march around the Sun.* [*Perelandra*, Chapter 16.]

The idea of absolute time and space still lingers, and even general relativity appears at first sight to require some replacement for it.[6] Beyond doubt, a great deal has been achieved with this concept through Newtonian mechanics. By the twilight years of this concept, at the end of the nineteenth century, it had

---

[6]One such replacement is known as *Mach's principle*, after Ernst Mach (1838–1916). We shall return to that discussion in Chapter 8.

been brilliantly supplemented by vector analysis, an outgrowth of William Rowan Hamilton's quaternions, leading to an elegant, compact mathematical notation for physical laws that could easily be turned into a set of coordinates at any time when numerical results were needed. In this cozy, familiar-seeming picture of the universe, two observers[7] O and O' in uniform relative motion have little difficulty reconciling their observations. It is not difficult to convert the coordinates that O uses for the time and place of an event with those used by O', whether the event is something that occurs in daily life, or a celestial phenomenon. When the conversion of the time and space coordinates of an event cannot be separated into a conversion of the time coordinate independent of the conversion of the space coordinates, however, as we now know to be necessary, it becomes more difficult for O' to translate the coordinates used by O into his own coordinates. Let us now make a more detailed investigation of this problem.

2.1. The speed of information and Newtonian synchronization. To picture the Newtonian scheme in a more concrete manner, let us imagine a pair of mid-eighteenth-century twins. To make the backstory more colorful, let's assume that they are English and are named John and Mary. As a further part of the backstory, let us assume that Mary has emigrated to Massachusetts with her husband, one James Foster, while her brother John has remained in London. To keep up with each other's lives, each sends a local newspaper to the other along with a letter every week.

In that case, perhaps in the winter of 1763, when both of them would have been eager for news of the end of the Seven Years' War in Europe (known as the French and Indian War in the United States), Mary might receive the *London Chronicle or Universal Evening Post*, Vol. 13, No. 944, and on p. 33 (Saturday, 8 January to Tuesday, 11 January 1763), read the following story:

> *On Saturday the river Thames was frozen so hard at Isleworth that a fair was kept on it all day. A large booth was erected in which was sold beer and other liquors, and in which a leg of mutton was boiled for the company. There was a round-about for children to ride in and all sorts of toys sold, as at other fairs. Great numbers of people came from the adjacent parts to see it.*

And perhaps John would, about the same time, read the following sad note—confirmed by an accompanying letter from Mary—on p. 3 of the *Boston Evening Post* of January 10, 1763:

> *Last Saturday died at Dorchester, after a few Days Illness, James Foster, Esq; of that Town.*

We have made it easy for Mary and John to synchronize their time-keeping, which in their case involved calendars rather than clocks. Since the two would have been using a common calendar—the Gregorian, recently (1752) adopted throughout the British Empire—the reference to Saturday in both stories would cause both

---

[7]For convenience, two observers are just two people in motion relative to each other, and to avoid clumsy he/she constructions, we shall simply assume they are both male. Each is pictured as located at the origin of a set of three spatial coordinates and carrying a clock. Reconciling the spatial coordinates and the clocks between the two observers is the central problem of the present chapter.

siblings to conclude that these two events occurred simultaneously. (We shall ignore the five-hour difference in solar time between London and Boston and assume that events that occurred on a given date in Massachusetts also occurred on that same date by London reckoning, even though this is often not the case.) That is, Mary would conclude that the fair her brother attended on the frozen Thames was going on at the same time that her husband was on his deathbed. And John would agree that, unaware of this sad event, he had been enjoying beer and leg of mutton while his brother-in-law was breathing his last. The two events were simultaneous; and even though neither sibling observed both of them, they would have been able to reconcile their timekeeping by using a common calendar.

Even without that calendar, however, each of them could have computed the time of the event at which the other was present, provided they knew (1) the time t_0 elapsed between those events and the dispatch of the report aboard the ship, (2) the average speed v_0 at which the ship sailed, and (3) the distance d_0 by sea from London to Boston.[8] Dividing the distance d_0 by the speed v_0 would yield the time that the newspaper was en route. Subtracting the sum $t_0 + d_0/v_0$ from the time that the news arrived would give the time of the event on the *common* calendar both were using. To make the analogy with physics, the point of Newtonian mechanics is that there *is* a universal method of measuring time, usable by everyone, and it is not affected by any state of motion of any observer. Likewise, there is a universal method of specifying locations through a system of three rectangular coordinates (x, y, z), and everyone can use this system, thereby always agreeing on the absolute location of any particle at any absolute time.

This small anecdote—James Foster was a real person who died in Dorchester on 8 January 1763, but his having emigrated from London with a wife named Mary is fiction—illustrates two points of importance to physics. First, information travels at a finite speed.[9] Second, if an event in one place is the cause of a second event in another place, it must be possible for information about the first event to reach the second place before the second event occurs: John cannot write to his sister to express condolences and perhaps suggest that she consider returning to London until *after* he receives news of his brother-in-law's death. Thus, travel at eighteenth-century speeds, with the eighteenth-century "speed of information"[10] being what it was, has rather weak effects on the lives of ordinary people. If, anachronistically, they both had accurate calendar-watches that they synchronized when Mary departed from London, they would find when she returned that their watches were still synchronized, as far as they could tell. In particular they would find that they had both aged by the same amount, and they would agree as to the total distance Mary had traveled. We shall bring these siblings back to the stage later in the present chapter, moving their stories forward in time by some three centuries and making Mary's travel a high-speed journey to a planet orbiting

---

[8] As John and Mary were citizens of the British Empire in the eighteenth century, that distance would have been given as roughly 3300 miles (say 5300 kilometers).

[9] The claims that paranormalists make on behalf of precognition and clairvoyance do not hold up well under scrutiny.

[10] We shall make formal use of the concept of speed of information starting in Chapter 4, reserving the symbol v_0 for the arbitrary hypothetical value we assign to it in the Newtonian world. In Newtonian theory, it could have any positive value, but of course had a rather small one in the eighteenth century. Once communication by radio was invented, v_0 became coincident with the speed of light, for which we use the symbol c. Further increases in its value are not expected.

a nearby star. At that time, as we shall see, travel has rather more noticeable—relativistic—effects on a person, and their maps and timetables for the journey will not agree. For their final bow, in Chapter 7, we'll move them ahead by yet another three centuries and ask Mary to travel to a black hole. The discrepancies in their measurements of time and space will then be still more noticeable.

Physics is much more abstract than everyday life. The kind of events reported in news media are described by the "five W's": what, who, where, when, and why. That is, they tell what happened (picnic on the frozen Thames, death of James Foster), who was involved (large numbers of people, James Foster), when (8 January 1763), and (in reporting complicated issues) a context that makes the event comprehensible to the reader or viewer. We have confidence in the accuracy of reports if all journalists agree on these five points. Physics, in contrast, pays no attention to individual people and their motives. Accordingly, in physics we have observers, not journalists, and they are concerned only with what happened, where, and when. Now if two observers are to reconcile their observations at all, they surely have to agree as to what happened. (Say, one particle collided with another.) In Newtonian physics, two observers can reconcile their observations so as to agree separately about the time and location of the event. In special relativity, in contrast, they cannot. Instead of having a where and a when to reconcile, each observer has a "where-when," and one where-when is reconciled with another through the Lorentz transformation that will be introduced below.

From the concept of events, we get the notion of a space-time in which an event is simply a set of four coordinates $(t; x, y, z)$, t being the absolute Newtonian time of the event, and (x, y, z) its spatial location in terms of a conventional three-dimensional coordinate system. In one sense, it can be argued that the speed of information in Newtonian physics is infinite. In such equations as Laplace's equation, the heat (diffusion) equation, and Newton's law of gravity, any perturbation of the controlling initial/boundary conditions is propagated instantly to the solution at all points of space for all later times. The only exception is the wave equation, in which disturbances of the initial condition propagate at a finite speed. This unrealistic feature of the Newtonian equations is one reason for preferring the relativistic ones. The one exception—the wave equation—which governs electromagnetic radiation, lies at the very heart of the special theory of relativity.

It is hoped that the reader finds none of this subsection difficult to understand. Most readers will, more likely, be impatient at being patronized by such a detailed discussion of what is, after all, only common sense. The trouble is that Common Sense is Newtonian, but the physical universe isn't. We have included this subsection in order to lay down a detailed background of concepts that can be modified in intuitively reasonable ways, one step at a time, all the way to the general theory of relativity.

We are now in the realm of abstract mathematics, and to tie it to the physical world, we need an interpretation of the mathematical object $(t; x, y, z)$. In the Newtonian scheme, we can think of (x, y, z) as the location of an identifiable particle at time t, perhaps a proton or electron. Although these physical bodies do occupy some volume, we can idealize them as having all three of their geometric dimensions equal to zero. If the particle moves, we can identify its position at time t with a vector $\boldsymbol{r}(t) = \big(x(t), y(t), z(t)\big)$. Along with that position, the particle has other numbers associated with it, such as its mass m, possibly its electrical charge q, its velocity $r'(t)$, its acceleration $r''(t)$, its momentum $mr'(t)$, the force acting on it $mr''(t)$, its kinetic energy $(1/2)m|r'(t)|^2$, and, if the forces that are acting on it are conservative, its potential energy $V\big(r(t)\big)$, which depends on its location. But everything is defined in terms of the mass m and the four coordinates $(t; x, y, z)$, with t having the physical dimension of time, and the other three having the physical dimension of length. We are now going to explore the concept of ordered time in Newtonian mechanics from this more abstract point of view.

2.2. Four kinds of time. In the story told above, we see two related events occurring. One is a primary event, but secondary to it is another event, namely the *observation* of the primary event. The second event occurs later because information requires time to travel from the location of the primary event to the location of the secondary event. To "abstractify" these considerations and fit them into mechanics, yet at the same time present a simple model for understanding, we find it useful to consider a moving particle that is being viewed by an observer. For the time being, let us assume that the observer is using an orthogonal system in which the Pythagorean theorem holds, and we think of the observer as sitting at the origin of this coordinate system. Since Newtonian time is absolute, we can assume that the all clocks used by all observers are properly synchronized with one another. Finally, we shall assume that information travels at some fixed speed v_0 in any direction. If, at time t an observer at rest at the origin of Newton's universal space receives information that something interesting is happening to a particle at time t that is located at the point with universal coordinates $r = (x, y, z)$ at time t, then that observer, taking account of the finite speed of information, will conclude that that event *really* happened at an earlier time s given by

$$s = t - \sqrt{x^2 + y^2 + z^2}/v_0 = t - |r|/v_0.$$

The time s is the *Newtonian universal time* showing on a clock at the point (x, y, z) when the event occurred. It will be the same for any observer viewing the event. The *observation* time t, which differs from one observer to another, is the value that universal time has when information about the event reaches the observer. For two events that occur at locations r_1 and r_2 and are observed at the origin at times t_1 and t_2 respectively, the time interval between the events themselves is

$$\Delta s = s_2 - s_1 = \left(t_2 - \frac{|r_2|}{v_0}\right) - \left(t_1 - \frac{|r_1|}{v_0}\right) = \Delta t - \frac{\Delta|r|}{v_0}.$$

This expression will be the same for any two observers viewing the events, even though the information about the events will reach them at different times and the coordinates they assign to the locations of the events will generally be different. In order to think clearly about the four-dimensional relativistic world whose points are "events" $(t; x, y, z)$, we need to keep this "speed of information" v_0 in the background. In relativity, it will be the speed of light in free space, but in Newtonian mechanics, it can be any convenient positive speed. In Newtonian mechanics, the difference between observation time and universal time is merely the time required for information to travel from the site of the event to the observer. In relativity, that Newtonian adjustment is taken for granted as having been made by any given observer O, who therefore has a clear concept of simultaneity throughout his own personal Euclidean space. But communication with another observer O' in motion relative to O is complicated, since neither the similar synchronization O' has carried out nor his spatial measurements agree with those of O. The discrepancy all hinges on one issue, which is precisely time-keeping. Since we no longer have a universal time, we shall distinguish between Newtonian and relativistic measurements by calling the observer's time *laboratory time*. The best replacement we can get to help our two observers reconcile the order of events is yet another kind of time we shall call *proper time*, which we use as the replacement for universal time.

2.3. Proper time. The distinction we have just made between observation time and universal time in Newtonian mechanics becomes much more important in relativity, where the interval between the time s shown by a clock attached to a moving particle and the time t recorded by someone observing the particle is given by "enlarging" the Pythagorean theorem so as to include a time dimension. On the infinitesimal level, as mentioned above,

$$ds^2 = dt^2 - \frac{1}{c^2}\left(dx^2 + dy^2 + dz^2\right).$$

Here, c is the speed of light, the fastest speed at which information can be transmitted. The analogy is not perfect, as will be seen, since in relativity it is *impossible* for two observers in relative motion, each having an accurate clock, to synchronize those clocks. In relativistic mechanics, we need to imagine *two* clocks associated with any moving object that is being observed. The time t that the observer records, which we just agreed to call *laboratory time*, comes from a clock synchronized with the observer's clock and attached at the point of the observer's space that the moving object is passing through at a given instant; it is, from the point of view of someone riding on the moving object, not keeping accurate time. The proper time s is what is shown on a clock attached to the object. Although the first clock is synchronized with his own local clock, the observer will still have to make the correction for the time it takes for a signal from that clock to reach him in order to assign a time to this event. But even after that correction is made, the time the observer records for an event will not agree with the time broadcast at the same instant by the second clock, attached to the moving object.

All that will be explained in detail below. In the relativistic equations relating the space-time coordinates of two observers, the correction for time lag is already taken into account in the variable t. *The laboratory time t of an event is not the time at which the observation was recorded (what we just called "observation time"), but rather the time at which the event actually happened in the observer's personal space-time, taking the Newtonian adjustment into account.* The Newtonian correction amounts to taking account of the *time delay* involved in transmitting information about the event, and any two observers who have made that correction will agree on the time of the event. In the relativistic model, however, even after correcting for the time lag due to a finite speed of information, different observers will not in general ascribe the same space-time coordinates to an event.

There is in fact only the one invariant across all these coordinate systems, namely the infinitesimal squared-increment of proper time ds^2, which is expressed as a quadratic form in the differentials of the four coordinates. It is the same for any two observers in relative motion along a straight line at a constant speed. All of this falls into the domain of the special theory of relativity, which will be discussed in the next chapter. In this theory, two observers using different coordinate systems can "talk to" each other, if their coordinates are related by a *Lorentz* transformation, which is a linear transformation of $\mathbb{R}^4$ that preserves the quadratic form ds^2 just written.

2.4. Homogeneous coordinates*. The use of $\mathbb{R}^4$ to represent the events in the Newtonian universe has the seeming disadvantage that the physical dimensions of the four coordinates are different. Really, what we have is a three-dimensional space of distances and a one-dimensional space of time, and they are completely independent of each other. Nevertheless, we would like to be able to talk about the *interval* separating two events $(t_1; x_1, y_1, z_1)$ and $(t_2; x_2, y_2, z_2)$. That will involve somehow adding the time interval to the spatial interval so as to get a measure of the full interval, as we just did in the previous subsection. It is a well-known principle of all applied mathematics that concrete numbers of different types cannot be added. The way to solve this problem is to assign some equivalence to time and distance. We used the "speed of information" v_0 for that purpose in the example above.

We do this constantly in everyday speech, reporting distances as if they were times. Thus we say that it is "a twenty-minute walk" from one's home to the grocery store, or that a certain city is "two hours away by car." In military language, a scout at one time might have reported that the enemy camp is "three days' march from here." And astronomers tell us that the nearest star is approximately "four light-years" away. In all these cases, we specify two independent quantities: (1) a unit of time (minutes, hours, days, years) and (2) a speed (walking speed, average driving speed, marching speed, the speed of light) to convert distance into time. The important fact to be kept in mind is that in Newtonian mechanics, *both* of these standards are pure conventions, and any unit of time or speed will do. Thus there are *two mutually independent* arbitrary units in Newtonian mechanics. Once they are chosen, actual times and distances may (theoretically) be represented by any real numbers.

One consequence of this Newtonian independence of space and time is that any two events occur in a definite temporal order, one that will be agreed upon by any two observers. To reach agreement, each of them has only to compute the universal time of the two events, which will be the same for both observers, provided they synchronize their clocks at any given time. Then Event 1, occurring at universal time s_1 precedes Event 2, occurring at universal time s_2 if $s_1 < s_2$, and any two observers will agree whether this is the case or not. (In practice, the two observers may not be able to measure time precisely enough to say which of two events was earlier, but the theoretical ordering remains.) This intuitive ordering, as we saw in the car wash puzzle, gets shattered by the special theory of relativity, in which the ordering assigned to two events by one observer may be the opposite of the order assigned by another observer.

2.5. The Galilean transformation. Before introducing the special theory of relativity formally, we give the coordinate transformations in Newtonian mechanics linking two observers that are both using the same x-axis and have clocks that are synchronized. If Observer O_2 is moving with speed u along the x-axis shared with Observer O_1 and the axes of the two systems coincide at time $t_1 = t_2 = 0$, then the coordinates the two will assign to an event are related by

$$
\begin{aligned}
t_2 &= t_1, \\
x_2 &= x_1 - ut_1, \\
y_2 &= y_1, \\
z_2 &= z_1.
\end{aligned}
$$

This transformation is called the *Galilean* transformation, after the pioneering scientist Galileo Galilei (1564–1642).

REMARK 1.1. Newtonian space is Euclidean and each point in it, regarded as a vector $\boldsymbol{\xi} = (x, y, z)$, has a squared-distance from the origin given as its dot product with itself:

$$
|\boldsymbol{\xi}|^2 = \boldsymbol{\xi} \cdot \boldsymbol{\xi} = x^2 + y^2 + z^2.
$$

The vector notation for the dot product (invented in the late nineteenth century) is very useful because of the compact expression it gives to many physically important quantities. The Euclidean structure of space singles out certain coordinate systems as "preferred," namely those that are *orthonormal*, meaning that the basis coordinates $\boldsymbol{\xi}_1 = (1, 0, 0)$, $\boldsymbol{\xi}_2 = (0, 1, 0)$ and $\boldsymbol{\xi}_3 = (0, 0, 1)$ satisfy $\boldsymbol{\xi}_i \cdot \boldsymbol{\xi}_j = 0$ if $i \neq j$ and $\boldsymbol{\xi}_i \cdot \boldsymbol{\xi}_i = 1$. If we confine ourselves to orthonormal coordinate systems, it does not matter which particular one we choose, since the square-distance is given by the same expression: $|\boldsymbol{\xi}|^2 = x^2 + y^2 + z^2$ whenever $\boldsymbol{\xi} = x\boldsymbol{\xi}_1 + y\boldsymbol{\xi}_2 + z\boldsymbol{\xi}_3$. That is one huge advantage of orthogonal systems: The dot product is invariant under orthogonal transformations, which are defined by that property: $T\boldsymbol{\xi} \cdot T\boldsymbol{\eta} = \boldsymbol{\xi} \cdot \boldsymbol{\eta}$.[11]

This seems an appropriate point to foreshadow certain other aspects of physics that are affected by the use of alternative systems of coordinates. We have in mind particularly the concept of kinetic energy. Assume that O_1 and O_2 are both fixed at the same origin ($u = 0$), and O_2 continues to use an orthonormal coordinate system but that O_1 is using a general coordinate system. For what we want to do, we need to be slightly more formal and systematic about our labeling. Henceforth, we let $x_i = x_i^1$, $y_i = x_i^2$ and $z_i = x_i^3$, $i = 1, 2$. Then for some constants g_{ij}, $i, j = 1, 2, 3$, we have

$$
\begin{aligned}
x_2^1 &= g_{11}x_1^1 + g_{12}x_1^2 + g_{13}x_1^3, \\
x_2^2 &= g_{21}x_1^1 + g_{22}x_1^2 + g_{23}x_1^3, \\
x_2^3 &= g_{31}x_1^1 + g_{32}x_1^2 + g_{33}x_1^3.
\end{aligned}
$$

---

[11] The cross product $\boldsymbol{u} \times \boldsymbol{v}$ is invariant under a rotation, that is, an orthogonal transformation whose determinant is 1. For that reason, physicists sometimes refer to the cross product as a *pseudo-vector* reserving the term *vector* for vectors that are invariant under all orthogonal transformations.

Then the kinetic energy T of a particle of mass m that O_2 observes to be at position $\boldsymbol{r}_2(t)$ at time t, is

$$
\begin{aligned}
T = \frac{1}{2}m|\boldsymbol{r}_2'(t)|^2 &= \frac{1}{2}m\left((x_2^{1'})^2 + (x_2^{2'})^2 + (x_2^{3'})^2\right) \\
&= \frac{1}{2}m\sum_{j=1}^{3}\left(g_{j1}x_1^{1'} + g_{j2}x_1^{2'} + g_{j3}x_1^{3'}\right)^2 \\
&= \sum_{j=1}^{3}\sum_{i=1}^{3} t_{ij}(x_1^{i'})(x_1^{j'})\,,
\end{aligned}
$$

where

$$
t_{ij} = m\sum_{k=1}^{3} g_{ki}g_{kj}
$$

if $i \neq j$, and

$$
t_{ii} = \frac{1}{2}m\sum_{k=1}^{3} g_{ki}^2\,.
$$

In Newtonian mechanics, it is perfectly legitimate—though one might think it foolish—to use general coordinates. Still, one might be studying the crystalline structure of a body and prefer axes that follow the lines of symmetry of the crystals. In that case, we might actually use oblique coordinates. If we do so, we see that we need more than the simple scalar equation $T = (1/2)m|\boldsymbol{r}'(t)|^2$ to keep track of kinetic energy. In its place we need what is called a *tensor*, which in this case is a bilinear mapping of pairs of velocity vectors $\boldsymbol{u} = (u^1, u^2, u^3)$ and $\boldsymbol{v} = (v^1, v^2, v^3)$:

$$
T(\boldsymbol{u}, \boldsymbol{v}) = \sum_{i,j=1}^{3} t_{ij}u^i v^j\,.
$$

This same tensor, with the factor of $m/2$ divided out of every entry, gives the square of the infinitesimal element of arc length $ds = \sqrt{dx_1^2 + dy_1^2 + dz_1^2}$ in coordinates $x = x_2^1$, $y = x_2^2$, $z = x_2^3$:

$$
ds^2 = \sum_{i,j=1}^{3} g_{ij}\, dx_2^i\, dx_2^j\,,
$$

where $g_{ij} = 2t_{ij}/m$. This last tensor is of basic importance throughout differential geometry. It was the starting point for modern abstract differential geometry, introduced by Bernhard Riemann (1826–1866) in his 1854 inaugural lecture. It soon came to be called the *fundamental tensor* by Einstein and others. We shall call it the *metric tensor*, since it gives the metric by which intervals are measured on an abstract manifold. Here we have our first hint of the intimate connection between differential geometry and mechanics: the tensor that defines the geometry of a manifold also serves to convert velocities into kinetic energy, through exactly the same bilinear operation on a pair of vectors in both cases. As we progress from Newtonian mechanics to general relativity, that metric-energy connection will serve as a guide. We will think of the metric coefficients (the g_{ij}) as potential energy functions. Notice that all of this insight comes about because we attempted to free ourselves from dependence on particular coordinate systems. As long as we confined ourselves to orthonormal coordinates, where the metric is $ds^2 = dx^2 + dy^2 + dz^2$ and the kinetic energy is $m|\boldsymbol{r'}|^2/2$, we wouldn't necessarily think in terms of a matrix.

While we would not normally bother with this tensor in Newtonian mechanics, the analogous four-dimensional concept in relativity will turn out to be very useful to us. We shall see this at the end of Chapter 2 and again in Chapters 6 and 7. Certain bilinear mappings of pairs of velocity vectors occur quite naturally in the equations of geodesics, which lie at the heart of relativity theory. By putting this example here, we are foreshadowing some important results that will come later and preparing the reader to adjust to a new way of thinking about mechanics.

REMARK 1.2. The reader may also be wondering why we chose to measure distance as time, converting it via a conventional standard velocity v_0 (taken to be c, the speed of light, in relativity). Why not instead convert time to distance by defining $\tilde{t} = v_0 t$? After all, we generally find it easy to measure distance. We can, in the simplest case, carry a ruler around with us calibrated in standard units of length, such as millimeters, and determine the distance between two nearby points. Time, on the other hand, is a rather mysterious, mystical thing (see the discussion in Chapter 8). In order to measure it, we have to select some process that we accept as proceeding at a uniform rate—the dripping of water through a hole, or sand in an hour-glass, or the swing of a pendulum, or the unwinding of a watch spring, or the vibration of the crystal in a digital watch or the right ascension of a star—and use that process as a measure of time. It is not intuitively obvious that all these ways of measuring time are even mutually consistent. The now old-fashioned clock with hour and minute hands goes in exactly the opposite direction from the conversion we made, measuring time by the lengths of the arcs traversed by the tips of the two rotating hands. And we have all learned geometry by thinking of lines as lengths. Why this nonintuitive, seemingly needless complication? It is certainly possible to express time as a length in this way, in which case one specifies the conversion by giving the standard speed and a standard unit of length.

In fact, we do exactly that any time we draw a trend line on a piece of paper. The horizontal axis represents time, and each horizontal distance a certain amount of elapsed time. We shall even do so below on occasion. For the purposes of theoretical physics, however, we wish to invoke the least-time principle that simplifies so much of classical physics: A physical process evolves in such a way that the integral of the difference between kinetic and potential energy *with respect to time* is "stationary" (usually a minimum). That is why we shall generally homogenize dimensions and express them all as time. From Chapter 4 on, this aspect of the theory moves to center stage, and we shall then exclusively write intervals as time intervals.

Actually, we can measure the "distance" from one place to another in many different ways. For an astronaut, the distance between two points of space might be most practically measured as the amount of fuel required to get from one to the other. For an economy-minded ordinary citizen, the distance from, say Boston to Chicago, might be measured by the cost of the airline fare for a round-trip journey (in which case, many mutually inconsistent measures of the distance would exist).

The important aspect of Newtonian space-time to be kept in mind is that it involved *two* conventional standard units. There is no "natural" unit of time, and there is no "natural" unit of speed. The choice of each is arbitrary. We can think of the standard v_0 as the maximum rate at which information can be transmitted, calling it the "speed of information." This fact is in complete accord with the well-known fact that there is no natural unit of length in a Euclidean space. It is this "flatness" of Euclidean space that makes it possible to build scale models of vintage automobiles, ships, airplanes, and shopping malls, in which all lengths are shrunk in the same proportion and all angles are the same in the model as in the original. This is not possible, for example, when one tries to draw a map of a large portion of the Earth's surface. Changing the lengths also causes the angles to change. In fact, a sphere *does* have a natural unit of length, namely its radius. Similarly, the curved plane of hyperbolic geometry has a natural unit of length, which might be (for example) the distance at which the angle of parallelism is half of a right angle. (See Appendix 1 for details. Angles have an absolute meaning in all geometries, a right angle, for example, being exactly one-fourth of a complete rotation.)

In the special theory of relativity, by way of contrast, the constancy of the speed of light provides a natural link between space and time that is absent from the Newtonian model. It has profound—and observable—consequences for physics.

2.6. Absolute spaces and parameter spaces. Finally, we introduce one more foreshadowing of the world of general relativity, in the form of completely general coordinate systems. Even in pure mathematics a point x in n-dimensional Euclidean space $\mathbb{R}^n$ can often be usefully represented by *some other* n-tuple of real numbers different from its components. For example, in linear algebra it is often useful to change the basis of a vector space so that linear operator will have a simpler matrix (diagonal if possible). When we make a change of basis, the *coordinates* of a point in the absolute space are no longer equal to its *components*. The point *actually is* its components, but we may prefer to work with its coordinates. Thus, the space of components is the absolute Euclidean space $\mathbb{R}^n$ of n dimensions, and the space of coordinates in a given basis is a parameter space of n dimensions. The components coincide with the coordinates in the natural basis of the space, which consists of the vectors $(1, 0, 0 \ldots, 0, 0)$, $(0, 1, 0, \ldots, 0, 0)$,..., $(0, 0, 0, \ldots, 1, 0)$, $(0, 0, 0, \ldots, 0, 1)$.

Given that observers use parameter spaces, the problem of communication between them becomes important. How do they know when they are talking about the same physical quantity? What does it even mean for it to be "the same quantity" if it has no absolute definition? If the quantity is the star Sirius or the Atlantic Ocean, no real problem arises. But what if the two observers are talking about the intensity of a magnetic field? How do they know which measurements by one observer correspond to given measurements by the other? Indeed, using Newtonian mechanics, two observers in uniform relative motion do agree about magnetic fields, but not about electric fields. The "necessary evil" of parameter spaces, which Carl Friedrich Gauss (1777–1855) made into a convenience, brings with it the problem of reconciling parameters between observers. This is the problem of invariance, which has to be taken into account in general relativity (Chapters 5 and 6) and will be discussed in detail in Appendix 6.

Each observer uses an individual set of parameters to compute various important geometric objects, such as length, area, and curvature. Once we agree on a way for two observers to identify the parameter values corresponding to points, functions, and vectors, we are still left with the problem of reconciling the *processes* by which these objects are computed. If Observer O combines objects a, b, and c according to some algorithm to produce an object d, then Observer O' can interpret these objects as a', b', c', and d'. The question that naturally arises is: Suppose O' combines a', b', and c' following what is verbally the same algorithm that O followed? Will the result be the corresponding d'? The general answer is affirmative, provided all the objects that are combined are tensors. Detailed discussion of the problem is given in Appendix 6. Just to make this matter seem a bit less abstract, we note that the objects we are interested in are all obtained from the coordinate parameters through algebraic operations and differentiation. The criterion for an object to be a tensor is, in informal language, that when space-time coordinates are changed, its coordinates transform according to the chain rule for differentiation.

3. An Asymmetry in Newtonian Mechanics: Electromagnetic Forces

> *It is recognized that Maxwell's electrodynamics, as currently interpreted, leads to asymmetries that do not appear to be intrinsic to the phenomena when applied to moving bodies. Consider, for example, the electrodynamic interaction between a magnet and a conductor. Here the observable phenomenon depends only on the relative motion of the conductor and magnet, while in the usual interpretation a distinction is made as to which of them is moving. In particular, if the magnet moves and the conductor remains at rest, an electric field having a definite quantity of energy arises around the magnet, producing a current in regions where the conductor is present. But if the magnet is at rest and the conductor moves, no electric field arises around the magnet; on the other hand, an electromotive force arises in the conductor, not corresponding to any energy, but which nevertheless—assuming the equality of the relative motion in the two cases—produces electric currents of the same magnitude as those generated by the electric field in the first case.*

Einstein ([**17**], p. 891).

When vector analysis was invented in the late nineteenth century, it seemed that a very powerful mathematical language had been created, one ideally suited to the purpose of getting compact expressions of physical laws. Now vector analysis is indeed a powerful tool, but in the first decade of the twentieth century it was a recent creation and by no means universally used. Einstein did not use the vector operations of curl and divergence in his 1905 paper on special relativity, although he did use the word *vector*. But he wrote out the Maxwell equations connecting electric and magnetic fields componentwise. His use of this seemingly more cumbersome notation may have been caused by the fact that two observers in relative motion do not agree that they are both using Euclidean geometry, in which the curl and divergence have a coordinate-free meaning.

Because Newton's second law of motion asserts that forces are directly proportional to acceleration, it follows that two observers in relative motion at constant velocity (zero relative acceleration) must agree about the magnitude and direction of all forces. This principle does not appear to have raised any doubts until the discovery of Maxwell's four laws of electromagnetism, when an asymmetry arose for two such observers looking at a charged particle moving in a pair of electric and magnetic fields. The two observers, it turned out, would agree about the magnetic field, but not about the electric field. They agreed about the *magnitude and direction* of the forces on the particle, but not about the *physical nature* of those forces. It was this asymmetry that Einstein remarked upon in his fundamental 1905 paper on special relativity. He made only a casual allusion to the famous Michelson–Morley experiment[12] of 1887 that had failed to detect any dependence of the speed of light on its direction of motion in a hypothetical absolute space. Einstein did, later on, discuss an 1851 experiment by Armand Hippolyte Louis Fizeau (1819–1896) of which the Michelson–Morley experiment was an improved reconstruction.

In the Newtonian system, there is a time axis common to all observers and a three-dimensional Euclidean space also common to all observers. When two observers wish to communicate their observations to each other, it is only necessary for one of them to say what coordinates are assigned to three points in space relative to that observer's origin and what event marks the epoch (time 0) of the time axis at that origin. The physical anomalies mentioned above, however, forced a reformulation of mechanics, in which time and space could not be separated. What Observer O takes to be an orthonormal coordinate system in space is not orthonormal as seen by Observer O'. As a result of that stark difference, each individual observer might at first sight seem to be isolated in a set of time and space coordinates, which are for that observer just like the old Newtonian ones, but does not agree with the equally valid system of time and space coordinates used by another observer. Two observers in relative motion almost appear to be inhabiting parallel universes. How can they determine whether an event occurring at point x at time t in the coordinates used by O is to be regarded as the event occurring at point x' at time t' in the coordinates used by O'? What common observations will enable them to make such an identification?

The key to solving this problem is their common line of motion, assuming that each believes the other is moving along a straight line at constant speed. We assume that each can at the very least observe the origin used by the other—it is convenient to think of the observers as "sitting" at their respective origins at all times—and assign a location to the origin of that other observer at any given time. We can identify the line in O-coordinates joining the two origins with the line in O'-coordinates joining the two origins.

After these lengthy preliminaries, we are at last ready to tackle the problem posed by the constancy of the speed of light and thereby explain the impossibility of getting 100% agreement on the order of events between two observers in relative motion.

4. The Lorentz Transformation

The starting point for the reformulation of mechanics is the assumption that the speed of light is a universal constant for all observers, independent of the motion of the observer or the source of the light. This assumption requires us to revise a number of intuitive notions about the order in which events occur.

An *event* is a point $(t; x, y, z)$ in $\mathbb{R}^4$, written with a semicolon to distinguish the time coordinate t from the three spatial coordinates (x, y, z).

---

[12]Named after Albert Abraham Michelson (1852–1931) and Edward Williams Morley (1838–1923). For details, see Chapter 8.

When we discuss events, we must keep in mind that their temporal order may depend on the observer. If the observers are not in relative motion, however, that is, the spatial coordinates of all points are constant in both frames of reference and the origins always coincide, then we can assume that the time coordinate of any event is also the same for both, and coordinates can be converted as in classical mechanics, that is, by a linear transformation $(t; x, y, z) \mapsto (t'; x', y', z')$ given by

$$
\begin{aligned}
t' &= t\,, \\
x' &= a_{11}x + a_{12}y + a_{13}z\,, \\
y' &= a_{21}x + a_{22}y + a_{23}z\,, \\
z' &= a_{31}x + a_{32}y + a_{33}z\,,
\end{aligned}
$$

where the matrix

$$
\begin{pmatrix}
a_{11} & a_{12} & a_{13} \\
a_{21} & a_{22} & a_{23} \\
a_{31} & a_{32} & a_{33}
\end{pmatrix}
$$

is the invertible matrix that transforms coordinates in one fixed basis of $\mathbb{R}^3$ to another. In our case, it will always be a rotation matrix, since we are going to assume that our observers use only right-handed orthonormal bases in their coordinates.[13]

To see what these coordinate changes look like in relativity, imagine two observers, O and O', whose spatial frames of reference are moving relative to each other in a fixed direction at a constant speed u. Each of our observers is imagined to have a clock measuring time in the Newtonian way through some physical process that is, by definition, said to be proceeding at a uniform rate, and each has measuring instruments that measure distances and angles in such a way that triangles obey the trigonometric laws of Euclidean geometry. These are the *proper time* and *proper space* for that observer, and they have the properties of Newton's absolute time and space, including the property that time and space measurements are independent variables. The difference from Newton's system is that these times and spaces are not in agreement with the proper times and spaces of other observers. When two observers try to reconcile their measurements, each finds the space and time measurements of the other are *entangled* and so no longer appear to be independent variables when compared with his own. In particular, distances (x) along the common line of motion are mixed up with time (t), and only the difference $(ct)^2 - x^2$ (where c is the speed of light) is agreed upon by both observers.

For simplicity, we assume that at some instant of time, given the value $t = 0 = t'$ by both observers, the three mutually perpendicular coordinate axes used by O coincide with those used by O', and that the relative motion is a translation along the direction of the common x-axis (x'-axis) at constant speed. We assume that O''s origin is moving in the positive direction of the common axis at speed u (from O's point of view). Of course, from O''s point of view O's origin is moving along the x'-axis in the negative direction, that is, at speed $-u$. Because of the assumption

---

[13]As Einstein remarked in one of his popular expositions of relativity theory, it is impossible to define what a right-handed system is intrinsically. One can divide systems of orthonormal bases into two equivalence classes and always say whether two bases belong to the same class— the determinant of the transition matrix between coordinates in the two bases is positive—but otherwise, there is nothing intrinsic to either system that marks it as being "right-handed." Thus, what we are really saying is that we assume the coordinate transformation between any two observers has a positive determinant.

that the speed of light must be the same for both observers, we cannot now assume that they are using the same time coordinate, or that simultaneity means the same thing for both of them. The best we can assert is a kind of homogeneity in events, expressed by assertions like *if event P took place twice as far away from O's origin as event Q, and after an elapsed time (measured from the instant when the two origins coincided) twice as large as the time elapsed when Q occurred, as seen by O, then the same should be true from O''s point of view.* That is, if all four of the space-time coordinates of event P are twice those of event Q in O's system, the same should be true in O''s system.

Einstein must have had something like this in mind when he asserted that, because of our beliefs about the nature of time and space, it seems clear that the coordinates of an event in one frame of reference must be linear functions of those in the other frame. In other words, we are assuming that there is a linear transformation such that

$$
\begin{aligned}
t' &= a_{11}t + a_{12}x + a_{13}y + a_{14}z, \\
x' &= a_{21}t + a_{22}x + a_{23}y + a_{24}z, \\
y' &= a_{31}t + a_{32}x + a_{33}y + a_{34}z, \\
z' &= a_{41}t + a_{42}x + a_{43}y + a_{44}z.
\end{aligned}
$$

Our first assumption is that the yz-plane coincides point by point with the $y'z'$-plane at a time we shall take as the epoch (time 0) for both observers. That is, if $t = 0 = t'$ and $x = 0 = x'$, then $y' = y$ and $z' = z$. This assumption yields the equalities

$$
\begin{aligned}
0 &= a_{13}y + a_{14}z, \\
0 &= a_{23}y + a_{24}z, \\
y &= a_{33}y + a_{34}z, \\
z &= a_{43}y + a_{44}z.
\end{aligned}
$$

If these equalities are to hold for all y and z, then we must have $a_{13} = 0 = a_{14}$, $a_{23} = 0 = a_{24}$, $a_{33} = 1$, $a_{34} = 0$, $a_{43} = 0$, $a_{44} = 1$. The equations now read

$$
\begin{aligned}
t' &= a_{11}t + a_{12}x, \\
x' &= a_{21}t + a_{22}x, \\
y' &= a_{31}t + a_{32}x + y, \\
z' &= a_{41}t + a_{42}x + z.
\end{aligned}
$$

The assumption that the motion is along the x-axis in both systems implies that this axis is the same for both at all times. In other words, if $y = 0 = z$, then $y' = 0 = z'$ also. Putting these values in the last two equations, we find

$$
\begin{aligned}
0 &= a_{31}t + a_{32}x, \\
0 &= a_{41}t + a_{42}x.
\end{aligned}
$$

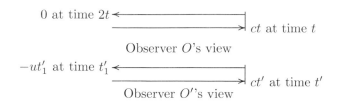

FIGURE 1.1. Top: A light ray traveling from the origin of Observer O's coordinate system to a mirror on the axis of common motion and then back to the origin, as described by O. Bottom: The same process as described by O'.

Since these equations hold for all t and x, we must have $a_{31} = 0 = a_{32}$ and $a_{41} = 0 = a_{42}$. Our transformation equations now read

$$\begin{aligned} t' &= a_{11}t + a_{12}x\,, \\ x' &= a_{21}t + a_{22}x\,, \\ y' &= y\,, \\ z' &= z\,. \end{aligned}$$

These determinations are not particular to the theory of relativity; they are independent of any assumptions about the speed of light c. In order to determine the four remaining coefficients a_{ij}, $i, j = 1, 2$, we need to introduce the assumption that c is the same for all observers. With that assumption, we first consider coordinates assigned to events on the axis of relative motion. Consider a light ray that leaves the common origin at time $t = 0 = t'$, travels to a mirror on the positive x-axis (which is also the positive x'-axis), arriving at time t according to Observer O, then is reflected straight back to the origin of O's system, necessarily arriving there at time $t_1 = 2t$ according to O. To Observer O', the light ray arrives at the mirror at some time t', then returns to O's origin at some later time t'_1, when that origin has coordinates $(-ut'_1, 0, 0)$. Thus we have two events, the arrival of the light ray at the mirror, to which the two observers assign coordinates $(t, ct, 0, 0)$ and $(t', ct', 0, 0)$ respectively, and its return to O's origin, to which the two observers assign coordinates $(2t, 0, 0, 0)$ and $(t'_1, -ut'_1, 0, 0)$, so that

$$\begin{aligned} t' &= a_{11}t + a_{12}ct\,, \\ ct' &= a_{21}t + a_{22}ct\,, \\ t'_1 &- 2a_{11}t\,, \\ -ut'_1 &- 2a_{21}t\,. \end{aligned}$$

The last two equations here imply that $a_{21} = -ua_{11}$.

Now let us look more closely at the return portion of this trip in the "light" of the fact that the speed of light c is the same for both observers. According to O', after leaving the mirror, the light traveled a distance $c(t'_1 - t') = ct' + ut'_1$. (Here the left-hand side represents the distance traveled as speed times time elapsed. The right-hand side represents it directly as the difference of the two distances from O''s origin to the starting point ct' and ending point $-ut'_1$.) Solving this relation for t', we find

$$t' = \frac{c-u}{2c}t'_1 \, .$$

Substituting this value of t' into the first two equations of the coordinate transformation, we find

$$\frac{c-u}{2c}t'_1 = a_{11}t + ca_{12}t \, ,$$
$$\frac{c-u}{2}t'_1 = -ua_{11}t + ca_{22}t \, .$$

Since $\frac{1}{2}t'_1 = a_{11}t$, we can cancel this factor; and we get, upon dividing the second equation by c,

$$\frac{c-u}{c} = 1 + \frac{ca_{12}}{a_{11}} \, ,$$
$$\frac{c-u}{c} = -\frac{u}{c} + \frac{a_{22}}{a_{11}} \, .$$

We now rewrite these equations as

$$\frac{a_{12}}{a_{11}} = \frac{c-u}{c^2} - \frac{1}{c} = -\frac{u}{c^2} \, ,$$
$$\frac{a_{22}}{a_{11}} = 1 \, .$$

Now, letting $\alpha = a_{11}$, we have $a_{21} = -u\alpha$, $a_{12} = -\frac{u}{c^2}\alpha$, and $a_{22} = \alpha$. Thus, we have the transformation

$$t' = \alpha\left(t - \frac{u}{c^2}x\right) \, ,$$
$$x' = \alpha(-ut + x) \, .$$

It remains to determine the factor α. This time we imagine a point fixed on O's y-axis. A light ray again leaves the common origin at time 0 (in both coordinate systems) and travels to this point, arriving at time t. Then O assigns to this arrival the coordinates $(t, 0, ct, 0)$ and, by what we know of the transformation so far, O' assigns coordinates, $(\alpha t, -\alpha ut, ct, 0)$ to this same event. For O', the event is the arrival at the point $(-\alpha ut, ct, 0)$ of a light ray that left O''s origin at time 0, and this arrival occurs at time αt. Since the proper space of each observer is Euclidean, we deduce that

$$c\alpha t = \sqrt{(\alpha ut)^2 + (ct)^2} = t\sqrt{(\alpha u)^2 + c^2} \, .$$

Canceling t, squaring the equation, and then solving for α, we find

$$\alpha = \frac{c}{\sqrt{c^2 - u^2}} = \frac{1}{\sqrt{1 - \frac{u^2}{c^2}}} \, .$$

The complete set of transformation equations is now best stated as a theorem.

THEOREM 1.1. *For two systems of measuring time t and t' and rectangular space coordinates (x, y, z) and (x', y', z') for which (1) the x-axis and the x'-axis coincide at all times and (2), the origin of the primed system is moving along the* *along the x-axis at speed u, the four coordinates in the two systems are related by the following system of equations:*

(1.1) $$t' = \alpha\left(t - \frac{u}{c^2}x\right),$$

(1.2) $$x' = \alpha(-ut + x),$$

(1.3) $$y' = y,$$

(1.4) $$z' = z.$$

We shall refer to this transformation as the *Lorentz transformation* and say that it corresponds to a velocity vector $\boldsymbol{u} = u\boldsymbol{i}$ of O' relative to O. The reader can easily compute that the *space-time interval* between two events is the same for both observers:

$$\Delta s^2 = \Delta t'^2 - \frac{1}{c^2}\Delta x'^2 \quad \frac{1}{c^2}\Delta y'^2 - \frac{1}{c^2}\Delta z'^2 = \Delta t^2 - \frac{1}{c^2}\Delta x^2 - \frac{1}{c^2}\Delta y^2 - \frac{1}{c^2}\Delta z^2.$$

The interpretation of Eqs. 1.1–1.4, is that the event recorded by observer O as $(t; x, y, z)$ is the same event that is recorded by O' as $(t'; x', y', z')$, provided the two sets of coordinates are related by these equations (and the x-axis/x'-axis is the line of mutual motion). These equations make it possible for two observers to agree on what happens, say, to a particle that both observe moving along trajectories $(x(t), y(t), z(t))$ and $(x'(t'), y'(t'), z'(t'))$ (where of course, the primes do *not* mean differentiation).

REMARK 1.3. It is sometimes useful to have a four-dimensional space-time in which all the coordinates have the physical dimension of length, (see Problem 1.19 below). For that reason, we shall occasionally replace the time coordinates t and t' by the "spatialized" times $\tau = ct$ and $\tau' = ct'$. When that is done, the matrix of the Lorentz transformation takes on a more symmetric appearance:

$$\tau' = \alpha\left(\tau - \frac{ux}{c}\right),$$

$$x' = \alpha\left(-\frac{u\tau}{c} + x\right),$$

$$y' = y,$$

$$z' = z.$$

Thus, the Lorentz transformation imposes two significant modifications on Newtonian space-time:

(1) It provides a natural absolute velocity—the speed of light, denoted c—by which we can "temporize" spatial coordinates through Minkowski's "mystical formula." (See the note on p. 5.) The arbitrary velocity v_0 we used earlier for this purpose is no longer arbitrary, being replaced by c. The unit of time remains arbitrary, however.

(2) It does not preserve the Euclidean metric on $\mathbb{R}^4$, in which a point $(t; x, y, z)$ has the square-norm $t^2 + x^2 + y^2 + z^2$; instead, it preserves the pseudo-metric given by the bilinear form $t^2 - (x^2 + y^2 + z^2)/c^2$, which assumes both positive and negative values. This bilinear form, used as a metric, creates a *pseudo-Euclidean* space. This space is still flat, since its metric coefficients (the coefficients in the bilinear form) are still constant, but the square of the space-time interval between two events can be negative. Such intervals are called *spacelike* because the spatial portion of the proper

time interval is larger than the time portion; intervals in which the time portion is larger are called *timelike*.

Those who appreciate the power and beauty of vectors will want to state it in vector form, and this can be done, by simply decomposing the spatial portion of the vectors in $\mathbb{R}^4$ into vectors parallel to $\boldsymbol{u}$ and perpendicular to $\boldsymbol{u}$, that is, writing

$$\boldsymbol{x}' = \frac{\boldsymbol{x}' \cdot \boldsymbol{u}}{\boldsymbol{u} \cdot \boldsymbol{u}} \boldsymbol{u} + \left(\boldsymbol{x}' - \frac{\boldsymbol{x}' \cdot \boldsymbol{u}}{\boldsymbol{u} \cdot \boldsymbol{u}} \boldsymbol{u} \right).$$

Since the time coordinate is easily written in terms of the vector dot product, the resulting vector equation is:

$$(t'; \boldsymbol{x}') = \left(\alpha \left(t - \frac{\boldsymbol{u} \cdot \boldsymbol{x}}{c^2} \right); \boldsymbol{x} + \left((\alpha - 1) \frac{\boldsymbol{x} \cdot \boldsymbol{u}}{\boldsymbol{u} \cdot \boldsymbol{u}} - \alpha t \right) \boldsymbol{u} \right),$$

where $\alpha = c/\sqrt{c^2 - \boldsymbol{u} \cdot \boldsymbol{u}}$. The inverse relation is

$$(t, \boldsymbol{x}) = \left(\alpha \left(t' + \frac{\boldsymbol{u} \cdot \boldsymbol{x}'}{c^2} \right); \boldsymbol{x}' + \left((\alpha - 1) \frac{\boldsymbol{x}' \cdot \boldsymbol{u}}{\boldsymbol{u} \cdot \boldsymbol{u}} + \alpha t' \right) \boldsymbol{u} \right),$$

Here, for the first but not the last time, we emphasize that both observers *must* be using $\boldsymbol{i} = \boldsymbol{u}/u$ as the first vector of an orthonormal basis in order for the vector equation to be valid. These must be such that the coordinates of the velocity vector $\boldsymbol{u}$ that O ascribes to O', are the negatives of the coordinates of the velocity that O' ascribes to O. Only in that sense is it helpful to say that the velocity of O relative to O' is $-\boldsymbol{u}$. The Lorentz transformation was derived assuming the two observers are using such coordinates. If either of them rotates the spatial axes, this vector equation becomes false when written out in terms of components in the two systems. When we introduce a third observer O'' having a velocity $\boldsymbol{v}$ relative to O'—we assume its components are assigned by O'—that is not parallel to $\boldsymbol{u}$, we shall find that the relative velocity vectors $\pm\boldsymbol{w}$ between O and O'' do not have this property unless all three of the observers rotate their spatial coordinates. Composing two Lorentz transformations is thus not just a simple matter of multiplying the matrices of the two transformations in fixed coordinate systems used by the three observers. If the three observers are using arbitrary orthonormal coordinate systems, the complete composition requires multiplying five matrices rather than two.

Thus, although the vector notation is compact, it must be used carefully. We remark that the quantities $\boldsymbol{u} \cdot \boldsymbol{u} = u^2$ and $\alpha = c/\sqrt{c^2 - u^2}$ determine each other ($\boldsymbol{u} \cdot \boldsymbol{u} = u^2 = c^2(\alpha^2 - 1)/\alpha^2$), so that we could eliminate one of the two from the Lorentz transformation, replacing $(\alpha - 1)/(\boldsymbol{u} \cdot \boldsymbol{u})$ by $\alpha^2 c^2/(\alpha + 1)$.

5. Contraction of Length and Time

Now that we have introduced the Lorentz transformation and the pseudo-Euclidean space-time that it operates on, we would like to analyze the effect of this transformation on time and space separately. For this purpose, we can suppress the y and z-coordinates of our two observers, since they are the same. We imagine two situations, in one of which O' observes the clock used by O over a period of time and in the other of which O' observes the distances that O assigns to points along their common line of motion. In each case, O' wants to compare O's measurements with his own. To that end, it is convenient to imagine that each observer has set up a post at every point along the common line of motion with a

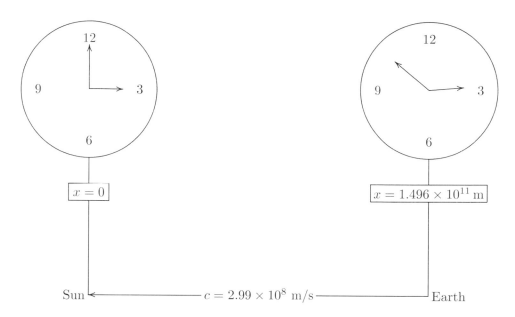

FIGURE 1.2. An observer located at the Sun uses a telescope to
observe a clock on the Earth synchronized with his own. Left:
The clock on the Sun. Right: The image of the clock on Earth
observed at the same instant on the Sun. This image requires
about 8.5 minutes to reach the Sun.

sign on the post giving its distance from the origin and a clock showing the time.
Since each observer is using the familiar Newtonian ideas of space and time, it is
possible for each to synchronize all the clocks in the system. The way to do so is
to imagine, say O seated at his own origin holding a clock, and looking through a
telescope at one of these sign posts, say corresponding to coordinate x. If O's clock
reads time t, then, looking through the telescope at the signpost located at x, O
should see the clock showing time $t - |x|/c$, to account for the fact that the light
revealing the clock to O left the sign post earlier and required time $|x|/c$ to reach
the origin. If the two clocks show times related in this way, then the clock at x
can be regarded as synchronized with the clock at the origin. In that way, O gets
a concept of "now" that applies all over the universe, although what is happening
"now" at a distant location cannot be known "now." That information must await
the arrival of a signal from the point with coordinate x, and cannot occur until
time $|x|/c$ after "now." Observer O' can perform a similar synchronization, and we
shall always assume that both observers have already equipped every point of space
with such a sign post bearing an accurate clock. The synchronization procedure is
illustrated in Fig. 1.2.

The problem is that, while the x-axis is the same for both observers, those sign
posts are moving past each other. Although posts with negative x-values are in
locations coinciding with posts bearing negative x'-values, a post with a negative
x' value will eventually reach O's origin and move to a position coinciding with a
post bearing a positive x-value. As it will turn out, the two observers will assign
different x-coordinates and different clock times to all points on the x-axis except

the origin itself, even at the time both call $t = 0 = t'$. We need to see what the discrepancy amounts to. We shall see that, at the very least, it amounts to the impossibility of agreement on what is happening "now" at time $t = 0 = t'$ anywhere except at the common origin of the two systems, which coincide at that time—or should we say, those times, since time is something they don't generally agree on.

5.1. Time contraction. Imagine O' reading the time t on the clock located at O's origin and comparing its reading with his own clock. Since O's origin at what O' calls time t' is located at $(-ut', 0, 0)$—that is, the event $(t; 0, 0, 0)$ to O is the event $(t'; x', 0, 0) = (\alpha t, -\alpha u t, 0, 0)$ to O'—we have the equality $t' = \alpha t$. Since $\alpha > 1$, O' perceives O's clock as "running slow." That is, O', reading the clock at O's origin and comparing it with his own, says that it assigns too small a value to the interval between events occurring at O's origin. There is perfect reciprocity here: O observes a clock located at O''s origin as running slow by the same factor $1/\alpha = \sqrt{1 - u^2/c^2}$. If these two facts appear to contradict each other, note that they refer to different sequences of "events." In the first case, we are looking at the event that O describes as $(t; 0, 0, 0)$ and computing the time coordinate assigned to it by O', which is $t' = \alpha t$. In the second case, we are looking at the event that O' describes as $(t'; 0, 0, 0)$. Thus we are describing events at a location regarded as fixed by one of the observers but not the other in each case.

It should not be thought that the apparent slowness each observes in the other's clock is caused by the fact that each observer is moving away from the other. The same time contraction is observed even when they move toward each other. That is, it holds for negative times t as well as positive ones. True, given that the two clocks agree when the origins coincide, O' will observe a later time on O's clock than on his own as the two observers approach each other and an earlier time as they recede from each other. The time elapsed between two *events* occurring at O's origin—say, for example, the time between two flashes of a beacon located at that origin—will be shorter as read by O' from O's clock than as read from O''s own clock, whether the two are getting closer to each other or farther away.

Nor is the difference due to the fact that the light from O's origin takes some time to arrive at O''s origin. To correct for that time lag, we apply the Newtonian synchronization discussed above. The relativistic computation assumes that O' has *already corrected for that time lapse* by keeping a clock synchronized with his own at every point of space and recording the time of an event at each point using that clock. For an event occurring at any given point P that is fixed in O''s frame, the time O' will assign to that event is the time t at which a light ray originating at P at the time of the event reaches O''s origin, less d/c, where d is the distance from the point P to O''s origin. Let us compare the time shown on O's clock when O passes through the point P with the time O' assigns to that event. In other words, we want to compare O's clock at the instant when O passes through the point P with the clock at P synchronized with the one at O''s origin. The two clocks will not show the same time when O passes through P. To see why, suppose O passes through P at the time t_0 in O's frame. Suppressing the unnecessary y and z coordinates, O assigns to this event the coordinates $(t_0, 0)$. To O', that event has coordinates $(t'_0, x'_0) = (\alpha t_0, -\alpha u t_0)$, and so O' assigns time αt_0 to that event. Since $\alpha > 1$, we conclude that O *records a smaller time interval between events occurring at his own origin than O' records.*

Now let us consider the arrival time of information from that event at O''s origin. That time will be

$$\alpha t_0 + \alpha |u t_0/c| = \begin{cases} t_0\sqrt{\frac{c+u}{c-u}} & \text{if } t_0 > 0\,, \\ t_0\sqrt{\frac{c-u}{c+u}} & \text{if } t_0 < 0\,. \end{cases}$$

We emphasize that Observer O' assigns this time to the *arrival of information* about the original event, but assigns to the original event at O's origin the time $\alpha t_0 = ct_0/\sqrt{c^2 - u^2}$. The "Newtonian correction" for the speed of information is the difference in those two times, which is $u|t_0|/\sqrt{c^2 - u^2} = \alpha u|t_0|/c$, precisely the time required for the light to travel from the location of the first event $(-\alpha u t_0)$ to the location of the second event (0).

5.2. The relativistic Doppler shift. To pursue this line of thought one step farther, we can imagine O broadcasting a signal in the form of a sine wave whose peaks are at times nt_0, $n = 0, \pm 1, \pm 2, \ldots$ and hence have frequency $\nu = 1/|t_0|$. Assuming $t_0 > 0$, we see that those peaks will reach O' at times $nt_0\sqrt{(c+u)/(c-u)}$ and hence the received signals will have frequency $\nu\sqrt{(c-u)/(c+u)}$. This is a "red shift" since the frequency of the received signals is smaller than the frequency of the transmitted signal. That of course is because the two observers are moving apart for positive values of time. The reader can verify that for negative values of t_0, when the two observers were approaching each other, the shift is a "blue shift" by the factor $\sqrt{(c+u)/(c-u)}$.[14]

In summary, we have two time intervals being measured by two different observers. There is the time between the transmission of the two peaks at O's origin, which O measures as $|t_0|$ and O' as $\alpha|t_0|$. And there is the time interval between the reception of the two peaks at O''s origin, which, if $t_0 > 0$, O' measures as $t_0\sqrt{(c+u)/(c-u)}$ and O as $\alpha t_0\sqrt{(c+u)/(c-u)} = ct_0/(c-u)$. (These two events both occur at the same point of space in O''s frame, and hence the time interval that O measures between them is α times the interval measured by O', that follows from the symmetry of the principle enunciated above.) From O's point of view, the reception times differ by more than the transmission times because O' is moving away, and each successive peak has farther to travel in order to reach O'. In fact, O perceives the Doppler shift in the frequency of the signal O' is receiving to be by a factor $1 - u/c$ when the two observers are moving apart and by a factor $1 + u/c$ when they are moving toward each other. This is precisely the classical Doppler shift. To compare these quantities, we have the following inequalities for a signal

---

[14]This factor characterizes the *relativistic Doppler shift*, which differs from the acoustic Doppler shift that applies for signals transmitted as waves in a stationary medium. In the latter case, the shift in frequency is by the factor $1 \pm u/c$ for observer speeds u smaller than the speed c of the signal. The Doppler shift is named after the Austrian physicist Christian Doppler (1803–1853), who identified it in 1842.

broadcast by O and received by O' when the two are moving apart:

$$\nu \; = \; \frac{1}{t_0} = \text{measurement by } O \text{ of the broadcast frequency}$$

$$> \; \nu\sqrt{1 - \frac{u^2}{c^2}} = \text{measurement by } O' \text{ of the broadcast frequency}$$

$$> \; \nu\sqrt{\frac{c-u}{c+u}} = \text{measurement by } O' \text{ of the received frequency}$$

$$> \; \nu\left(1 - \frac{u}{c}\right) = \text{measurement by } O \text{ of the received frequency}.$$

The analysis just given may prove helpful in working Problem 1.2 below.

5.3. Spatial contraction. If we imagine O' reading the coordinates that O assigns to the points on the x-axis at a given instant of time, say $t' = 0$ as judged by O', the second equation of the inverse Lorentz transformation says $x = \alpha x'$, and since $\alpha > 1$, it follows that $|x'| < |x|$. Thus O' perceives O's description of the distance between two points on the line of common motion as too large. Effectively, O' thinks O's "yardstick" has shrunk and is giving readings that are too large by the same factor by which he perceives O's clock to be assigning time intervals that are too small between events at O's origin. Notice that the conflict is over the time intervals between events transpiring at a fixed location in O's system and spatial intervals between points that are fixed in O's system. To phrase this fact another way, what O is calling one meter of length (x) is, according to O', actually less than a meter (x'): In particular, O', undertaking a journey between two fixed points in O's system, will consider the journey to be shorter than O considers it.

A symmetric phenomenon arises when O measures the time intervals between events at a location that is fixed in O''s coordinates or the spatial intervals between points that are fixed in O''s coordinates. Confusion arises unless one keeps firmly in mind whose coordinates are fixed. This caution is needed especially in connection with the twin paradox, discussed below.

Length contraction along the line of mutual motion has an interesting effect on the measurement of volume:

THEOREM 1.2. *The volumes V_0 and V of a solid body B measured respectively in a system of coordinates at rest with respect to B and one in motion at constant speed u with respect to B are related by*

$$V = V_0\sqrt{1 - \frac{u^2}{c^2}}\,.$$

PROOF. For any cube with sides parallel to the axes, one of the sides shrinks by the factor $\sqrt{1 - u^2/c^2}$ when measured by the moving observer. The other two stay the same, so that the cube as measured by the moving observer is a parallelepiped and the volumes of the cube and parallelepiped are related as stated in the theorem. Since any volume can be approximated with arbitrary precision by a union of such cubes, this relation must hold for all volumes.   □

6. Composition of Parallel Velocities

Suppose O' describes the passage of a particle from a point a on the x'-axis at time t'_0 to point b on the same axis at time $t'_1 > t'_0$. Then O' will say the journey required time $t_1' - t_0'$ and covered the distance $b - a$. In contrast, O will say that the particle passed from the point $\alpha(a + ut_0')$ at time $t_0 = \alpha(t_0' + ua/c^2)$ to the point $\alpha(b + ut_1')$ at time $t_1 = \alpha(t_1' + ub/c^2)$, that the passage required time $t_1 - t_0 = \alpha(t_1' - t_0' + (b-a)u/c^2)$ and covered distance $\alpha\big(b - a + u(t_1' - t_0')\big)$. In terms of the speed the two observers ascribe to the particle, O' thinks it is $v = (b-a)/(t_1' - t_0')$, while O thinks it is $w = \big(u(t_1' - t_0') + (b-a)\big)/\big((t_1' - t_0') + u(b-a)/c^2\big)$. Replacing $b - a$ by $v(t_1' - t_0')$ and cancelling $t_1' - t_0'$, the reader can easily verify the fundamental equation

$$w = (u + v)/(1 + uv/c^2).$$

This formula gives the velocity of the particle relative to O, given that its velocity relative to O' is v and O''s velocity relative to O is u. It works only in the particular case when the two velocities are parallel.

REMARK 1.4. The alert reader will have observed the similarity of the formula

$$w = \frac{u + v}{1 + \frac{uv}{c^2}}$$

to the addition formula for the hyperbolic tangent function:

$$\tanh(x + y) = \frac{\tanh(x) + \tanh(y)}{1 + \tanh(x)\tanh(y)}.$$

This similarity suggests a useful mapping from the space of relativistic velocities along a line, which can be thought of as the interval $(-c, +c)$ (negative values meaning motion to the left and positive values to the right, as seen by a fixed observer), to the real line $(-\infty, +\infty)$, namely the mapping $u \mapsto \operatorname{arctanh}(u/c)$. If we introduce a unit of length k, the function $U = k\operatorname{arctanh}(u/c)$, whose inverse is $u = c\tanh(U/k)$, associates velocities with lengths, and the speed c would correspond to an infinite length. Before we develop this idea further, we wish to make a few comments about the use of numbers to represent measurable quantities that are continuous, such as distance, time, mass, velocity, acceleration, force, pressure, and many others.

Mathematical functions do not accept what we call concrete numbers or variables—those with a geometric or physical dimension attached to them—as input. When we develop a function as a power series, for example,

$$f(x) = a_0 + a_1 x + a_2 x^2 + \cdots,$$

all the terms must have the same geometric/physical dimension. Otherwise, they cannot be added. If the coefficients a_n are all to be pure numbers, then x must also be a pure number. Thus, when we give the value of a continuous physical quantity, what we are really giving is its ratio to a *standard unit* of that category. In the present context, we are talking about speed, and in the MKS system, the standard unit of speed is one meter per second. In relativity, however, c, the speed of light serves as a much more natural standard of speed, since it is the same for all observers. When we substitute the speed u of an object into a formula, what we are really inserting, usually without thinking about it, is the ratio u/c, which is a dimensionless real number between -1 and 1.

With that dimensional consideration in mind, we are now going to make an association that will turn out to have interesting consequences later in this chapter. Since we like to draw pictures representing velocities as lengths, we shall associate with each speed u a length U, referred to a standard length k, making the following correspondence:

$$u = c \tanh\left(\frac{U}{k}\right),$$

$$U = k \operatorname{arctanh}\left(\frac{u}{c}\right) = k \ln \sqrt{\frac{c+u}{c-u}}.$$

(See Problem 1.8 below.)

Under this correspondence we find that

$$\frac{u+v}{1+\frac{uv}{c^2}} \leftrightarrow c \tanh\left(\frac{U}{k} + \frac{V}{k}\right).$$

A further advantage of this convenient substitution is that it gives an elegant expression for the Lorentz magnification factor $\alpha = c/\sqrt{c^2-u^2}$, namely $\alpha = \cosh(U/k)$. That relation could be used to simplify the study of relativistic velocities in two dimensions. For other reasons, however, we shall work laboriously through our transformations using only the language of velocities and only later (mostly in Appendix 1) reveal its astonishing connection with hyperbolic geometry.

Thus, under the pairing $u \leftrightarrow c \tanh(U/k)$ relativistic velocities (positive, negative, and zero) are in one-to-one correspondence with lengths, and the usual geometric addition of collinear lengths corresponds to the relativistic addition of the corresponding velocities. Moreover, the speeds $\pm c$ correspond to the points $\pm\infty$ on the extended real line.

7. The Twin Paradox

One of the interesting oddities of special relativity is popularly known as the *twin paradox*. To state it colorfully, we shall ask our twins Mary and John to reprise their roles in the 21st century, assuming the roles of our two observers O and O'. We shall let John represent O, the stay-at-home twin, and Mary the traveling astronaut O'. This being no longer the eighteenth century, they no longer communicate by letters sent across an ocean. Instead, they keep in touch by radio, making direct conversation practical for a little while. As the distance between them increases, however, messages take even longer to go from one to the other than they did in the eighteenth century, when Mary was on the surface of the Earth. As this paradox is stated, Mary leaves the Earth in a rocket ship bound for a planet orbiting a nearby star. Finding the planet unsuitable even for a vacation, much less as a place to settle down, she reverses direction immediately upon arrival and returns at the same uniform speed. Since the ship travels very fast, her brother John back home, observing the clock on the ship, sees it losing time in comparison with his own clock, both on the outward journey and on the return journey. (To visualize this effect, imagine that John has set a clock synchronized with his own at every point that Mary passes through. He can carry out this synchronization, since he and the clocks remain at rest relative to one another. If he watches Mary's journey through a telescope, he can directly compare the time on her clock with the time on the fixed clock as she passes it, and Mary's clock will indeed keep losing time.) Moreover, what is true of the clock on the ship is true of all physical and chemical processes, including the aging process in Mary. Thus, when she returns to the Earth, she really is younger than her twin brother.

If that were all there is to the paradox, we might be surprised, but not astonished. But, if we believe in the relativity of motion, we should be flabbergasted, or so it seems at first sight. After all, in the frame of reference used by Mary, it was her brother John who went traveling, along with the rest of the universe, and reversed directions just as the star arrived at her rocket ship. It was his clock that was running slow. Why isn't John the younger of the two?[15] This paradox has been discussed in many popular accounts of relativity. Quite often, the discussion invokes the general theory of relativity, according to which clocks do run slower in a gravitational field. However, this phenomenon was discussed by Einstein before he developed the general theory of relativity; and, as we shall show below, the fact that Mary winds up younger in this scenario follows from the Lorentz transformation alone and does not require the general theory or any invocation of inertial forces to explain it.

At the heart of the matter lies the FitzGerald–Lorentz contraction. The two siblings agree as to their relative speed, which we have called u and will assume constant here. But they do *not* agree on the distance between the point where they passed each other and the star that later came by Mary's ship. If John measures that distance as d, Mary measures it as d/α, where $\alpha = c/\sqrt{c^2 - u^2}$. This is the case on both the outward and return journeys, so that to Mary the journey covered a total distance $2d/\alpha$ and since she was traveling at speed u relative to the Earth-bound frame, the total elapsed time for the round-trip journey was $t = 2d/(\alpha u)$. That is the amount of proper time elapsed on the rocket ship during the round-trip journey, whether measured by a mechanical clock on board the ship or by the aging process in her body. But from John's point of view, Mary traveled a distance d and back at speed u, so that his clock records an elapsed time of $2d/u$. Since $\alpha > 1$, the twins will agree that Mary's clock shows less time elapsed than John's.

This scenario does not have the symmetry that is sometimes erroneously said to follow from the relativity of motion. To be sure, John also observes that the distance between two *fixed* points in Mary's frame of reference is shorter than Mary herself measures them to be. *But the star and Mary are not relatively fixed. The distance between them is constantly changing.* Taking the primary axis for both observers to be their line of mutual motion, so that the star lies on that axis, we see that when the siblings pass each other the first time, Mary is not setting out to reach the point on her own axis that John regards as being at distance d from Mary's origin at that instant. In fact, to make the story an accurate reflection of relativity, Mary can't reach *any point* other than the origin in her own space, since we imagine that is where she stays located. Her spatial coordinates move along with her. If the twins agree to synchronize their clocks at time 0 when they pass each other, the two clocks they are keeping at Mary's destination will not agree as to the time when Mary's journey began. Since the star is fixed as far as John is concerned, he will say it is also time 0 on the star, which is at distance d. In other words, if he has placed a clock at that star synchronized with his own and looks at it through a telescope, he will always see that clock showing time $t - d/c$, when his own Earth-bound clock shows time t.

---

[15] If you look on the Internet, you will not have any difficulty finding people, some even claiming to have advanced degrees in physics, who deny the paradox for exactly that reason. Such people ought to know better. The reality of the twin paradox has been confirmed by experiment.

Now, what John regards as the event $(t; x, y, z)$ is seen by Mary as the event $\big(\alpha(t - ux/c^2); \alpha(x - ut/c), y, z\big)$. John, as just remarked, has a clock at the distant star synchronized with his own. At the instant $(t = 0)$ when Mary passes him, he records the event at the distant star as $(0, d, 0, 0)$, the event being that John's local clock at the star reads zero. Mary, however, will record this same event as $(-\alpha u d/c^2; \alpha d, 0, 0)$, so that in her coordinates John's clock at her destination read zero when the star was at the distance αd from her origin, and that was at the *earlier* time $-\alpha u d/c^2$. Between that earlier time and the time when she passed her brother, which both agree was time 0, the distance between Mary and the star shrank from αd to $\alpha d - \alpha u^2 d/c^2 = \alpha d(1 - u^2/c^2) = d/\alpha$; that is the distance she plans to travel.

True, Mary can regard herself as remaining in one place, at rest, while her brother moves at speed $-u$, but she does not measure the distance that John moved as d. When the twins meet for the second time, Mary will say that John traveled only the distance $2d/\alpha$, and her clock will show elapsed time $2d/(\alpha u)$, just as we found when analyzing the situation from John's point of view. The difference is that the star *did not move* in John's frame of reference, whereas it did move in Mary's frame of reference.

To summarize: Mary, whose distance to the turnaround point is changing over time, is the one who stays younger, because her biological clock is slow. As some writers express the matter, the traveling twin moves out of the frame of the Earth-bound twin, then moves back in again.[16] It is the bookkeeping involved in making those transitions that causes the loss of time for the traveling twin.

And yet... one still has the feeling of something-wrong-here. After all, according to Mary, the clock her Earth-bound brother uses has been running slow for the entire time of the journey. Why then does it show a *later* time when the twins meet for the second time? How can the tortoise, who always runs slower than the hare, nevertheless win the race? How can someone go into a revolving door behind you and emerge ahead of you? It seems to be a sleight-of-hand trick, like the conjuror who asks you to draw a card from a deck and hold onto it, then suddenly pulls it out of the deck that you drew it from. Somehow, when you weren't looking, the card got switched, but how?

To make the matter as plain as possible, "the card was switched" before the journey even started. In fact, Mary cannot synchronize her clock with *any* accurate clock on the star because the two are in relative motion. As long as we focus our attention on the Earth-centered frame of reference, we do not notice the switch, and then we are surprised when we look at Mary's clock and discover that it is running slow as she arrives at the star.

The algebraically simplified form in which we derived the Lorentz transformation requires a common event as origin, an event that can be regarded as having both space and time coordinates equal to zero in both systems. For this hypothetical journey, let O, using time-space coordinates (t, x), be John, and let O', using coordinates (t', x') be Mary. There are three events that we use as anchors here. Event 0, when Mary passes Earth heading to the star; Event 1, when she arrives at

---

[16]This statement assumes that Mary began and ended the journey standing on the Earth; it ignores the (perhaps tiny) portion of the journey during which acceleration was needed. We prefer to picture Mary as constantly whizzing around and just happening to pass her brother going in opposite directions at two different times, while hurtling past him. That way, we avoid any need to discuss acceleration.

the star and immediately heads back toward Earth at the same speed; and Event 2, when she passes Earth for the second time. We have to change origins at Event 1 in order to use Lorentz coordinate transformations on the return journey. For any event E occurring at John's origin at time t, we have O-coordinates $(t, 0)$ and O'-coordinates (t', x'), where $t' = \alpha t$ and $x' = -\alpha u t$. Thus, during the outbound portion of the journey Mary does record later times for all events at John's origin than John records for them. In that sense, Mary can say that John's clock is slow. But Event 1 *does not occur at John's origin*. We have focused our attention on the wrong place and by so doing missed the "card switch." To John, Event 1 has coordinates, say (t_1, d), where $d = u t_1$. According to the Lorentz transformation, Mary assigns coordinates (t_1', d') to that event, where $d' = \alpha(d - u t_1) = 0$—that is, it occurs at Mary's origin, as we already knew—and $t_1' = \alpha(t_1 - ud/c^2) = \alpha t_1(1 - u^2/c^2) = t_1/\alpha$, which is *smaller* than t_1. It is the ambiguity in the meaning of the term *simultaneous* that causes the surprise. By John's measurements, Event 1 occurred at time $t_1 = d/u$; by Mary's, it occurred at the earlier time $t_1' = t_1/\alpha$. It is the same event, of course, namely Mary's arrival at her destination.

Starting with Event 1, John and Mary both need to "zero out" their coordinate systems for Mary's return journey, and the velocity u needs to become $-u$. The Lorentz transformation that is in effect for this part of the journey is as follows:

$$t' - t_1' = \alpha\left((t - t_1) + \frac{u(x - d)}{c^2}\right),$$
$$x' = \alpha\left(x - d + u(t - t_1)\right).$$

Then, when Event 2 occurs (Mary arrives back at Earth), John's coordinates for this event will be $t = 2d/u = 2t_1$, $x = 0$, so that Mary's will be $t' = t_1' + \alpha(t_1 - ud/c^2) = t_1/\alpha + \alpha t_1 - u^2 t_1 \alpha/c^2 = t_1 \alpha(1/\alpha^2 + 1 - u^2/c^2) = 2t_1/\alpha$ and $x' = \alpha(-d + u t_1) = 0$. Thus, we can rigorously compute that the time on Mary's clock will be $1/\alpha$ times the time on John's clock: Mary will be younger by this factor.

Switching to our hare-and-tortoise analogy, we recall that the tortoise won the race because the hare took a nap. We are the ones who were caught napping in this race, focusing our attention on the Earth-bound clock, since the episode began at Event 0 and ended at Event 2, both of which took place on the Earth. We should have looked at the clocks at Mary's destination.

To phrase the matter in terms of the conjuring-trick analogy, the assertion that O's clock is running slow is the kind of distraction the conjuror uses to get you to focus on the card you thought was in your hand (John's clock on the Earth) but which was in fact *always* in the deck (John's clock on the star), not in your hand. The switching is revealed when your attention finally focuses instead on the deck of cards. If you would like to see the trick performed in slow-motion, work Problem 1.2 below.

8. Relativistic Triangles

With the advent of special relativity, the clean, simple Newtonian system, with its absolute, universal space and time, became untenable. It became impossible to determine the angle made by two lines at a given time, since observers in relative motion to each other may agree on the location of one point at a given time, while disagreeing as to the location of a second point at that same instant of time.

As a result, two observers will probably disagree as to what points constitute the vertices of a given triangle at a given time. Although each individual observer is using his own proper space, which is Euclidean, when that observer's findings are communicated to a second observer, the lines that the first observer regards as fixed are moving when viewed by the second observer, and that motion can cause perpendicular directions to go askew. The only kind of perpendicularity agreed upon by two observers moving at constant speed and in a constant direction relative to each other, is one particular system of coordinate axes consisting of the line of their mutual motion (that is, the line from each to the other), and the planes perpendicular to it. Even in that case, they do not agree about the unit of length along the first axis.

The absence of absolute simultaneity makes it difficult to discuss the "triangle whose vertices are P, Q, and R." While those points may be fixed in the spatial coordinates of Observer O, if O and O' are moving relative to one another, Observer O' will see a triangle of very different shape. Thus, triangles of position are slippery objects, impossible to define in an observer-independent way. What we are going to call *relativistic velocity triangles* are much better behaved in this regard.

We shall almost always find it easier to assume that suitable rotations of coordinates have already been performed by O and O', that is, that O and O' are already using the "privileged" coordinate systems in which they share the first spatial axis. A vector equation used by O to express a relationship between vectors measured only by O (for example, Maxwell's laws) is independent of the orthonormal basis used by O, since rotations preserve the vector operations of addition and scalar multiplication, as well as the dot and cross product on $\mathbb{R}^3$. But O *cannot* rotate axes and reinterpret correctly data received from O'. Transmitting the rotated data back to O' will not produce a rotation of the original data. We shall have occasion to write the laws of mechanics and electromagnetism in vector form and exhibit the transformation of those laws from one observer to another. Such equalities are not truly vector equalities, since they hold only when the vectors are expressed in the privileged bases of the two observers' systems, that is, those that share a common spatial axis along the direction of mutual motion.

As an example of what we have been talking about, notice that O regards the lines whose equations are $y = 2x$, $z = 0$ and $x = -2y$, $z = 0$ as mutually perpendicular. But at any given instant t', O' considers these to be the lines with equations $y' = 2\alpha x' + 2\alpha u t'$, $z' = 0$ and $\alpha x' + \alpha u t' = -2y'$, $z' = 0$, which are not perpendicular unless $u = 0$. They have slopes $m_1 = 2\alpha$ and $m_2 = -\alpha/2$, and the usual condition for perpendicularity ($m_1 m_2 = -1$) implies $\alpha = 1$, which is true only when $u = 0$.

8.1. Composition of nonparallel velocities. If O and O' are using the coordinate systems for which the Lorentz transformation equations were given, so that the velocity of O' relative to O is $\boldsymbol{u} = (u, 0, 0)$ and the velocity of O relative to O' is $-\boldsymbol{u} = (-u, 0, 0)$, and if an observer O'' whose origin also coincided with the origins of O and O' at time 0 is moving relative to O' with velocity $\boldsymbol{v}$ making an angle[17] η with $-\boldsymbol{u}$ according to O', so that O'''s origin at time t' is given by[18] O' as $(-vt' \cos \eta, vt' \sin \eta, 0)$, then the position of O'''s origin at time t' is interpreted by O as the event

$$
\begin{aligned}
t &= \alpha t'(1 - uv \cos \eta / c^2), \\
x &= \alpha t'(u - v \cos \eta), \\
y &= t'v \sin \eta, \\
z &= 0.
\end{aligned}
$$

We find the velocity $\boldsymbol{w}$ of O'' relative to O by dividing each of the spatial coordinates by t:

$$
\begin{aligned}
w_1 &= \frac{u - v \cos \eta}{1 - uv \cos \eta / c^2}, \\
w_2 &= \frac{v \sin \eta}{\alpha(1 - uv \cos \eta / c^2)}, \\
w_3 &= 0.
\end{aligned}
$$

Hence, from the point of view of O, the speed of O'' is

$$
\begin{aligned}
w &= \sqrt{w_1^2 + w_2^2} \\
&= \frac{\sqrt{(u - v \cos \eta)^2 + v^2 \sin^2 \eta / \alpha^2}}{1 - uv \cos \eta / c^2} \\
&= \frac{\sqrt{u^2 - 2uv \cos \eta + v^2 - u^2 v^2 \sin^2 \eta / c^2}}{1 - uv \cos \eta / c^2}.
\end{aligned}
$$

It is easy to verify that w is less than c if u and v are each less than c. In fact (see Problem 1.9), there is an important relation that is satisfied by the three velocities:

$$
\left(1 - \frac{w^2}{c^2}\right)\left(1 - \frac{uv \cos \eta}{c^2}\right)^2 = \left(1 - \frac{u^2}{c^2}\right)\left(1 - \frac{v^2}{c^2}\right).
$$

---

[17]We remark here that this angle is measured by O' using *Euclidean* measuring instruments, even though the resulting triangle will not have sides and angles that satisfy Euclidean relationships. There is no paradox here. An angle is a physically and geometrically dimensionless quantity representing an amount of rotation, and rotation is absolute in all three geometries: elliptic/spherical, parabolic (Euclidean), and hyperbolic. A right angle is one-fourth of a complete rotation. It forms a natural unit of angular measure, and is universally assigned the numerical value $\pi/2$. The reason for using that seemingly arbitrary value comes from analysis; Taylor series become very cumbersome if any other measure is assigned to a right angle. We shall call the resulting values of all angles simply the *numerical values* of those angles. We do not like to use the term *radian measure*, since it suggests what is true only in Euclidean geometry—that the numerical value of the central angle subtended by a circular arc is the ratio of the length of the arc to the length of the radius *and is independent of the radius*. In hyperbolic and elliptic geometry, the numerical value of the angle is not normally equal to that ratio, and the ratio itself varies with the radius of the circle.

[18]Since O is moving along O''s negative first spatial axis, the angle η must be measured *clockwise* from that axis to the line of motion that O' ascribes to O''. That accounts for the negative sign in the first coordinate of the location of O'''s origin, as seen by O'.

REMARK 1.5. This last equality relates the three magnification factors in the Lorentz transformations corresponding to the three velocities. That is, if $\alpha = c/\sqrt{c^2 - u^2}$, $\beta = c/\sqrt{c^2 - v^2}$, and $\gamma = c/\sqrt{c^2 - w^2}$, then

$$\gamma = \alpha\beta\left(1 - \frac{uv\cos\eta}{c^2}\right).$$

We have now established the simple but fundamental fact that if O''''s velocity relative to O' is constant in direction and constant (less than c) in magnitude, as judged by O', while O''s velocity relative to O is similarly constant in direction and constant (less than c) in magnitude relative to O, then O''''s velocity relative to O is also constant in direction and constant (less than c) in magnitude. Therefore the composition of two relativistic velocities is a relativistic velocity and hence a suitable (privileged) pair of coordinate systems that can be chosen by O and O'' should be related to each other by equations of the simple form derived above for the Lorentz transformation. Making that fact computable and verifying that the implied composition is associative if a fourth observer O''' comes along will occupy the last few sections of the present chapter. In the meantime, we wish to develop the trigonometry of these relativistic velocity triangles.

Recalling our remark above that if $u = c\tanh(U/k)$, then $\alpha = \cosh(U/k)$, we have, upon replacing β and γ by $\cosh(V/k)$ and $\cosh(W/k)$, the fundamental relation

$$\cosh(W/k) = \cosh(U/k)\cosh(V/k)(1 - \tanh(U/k)\tanh(V/k)\cos\eta),$$

which, when the parentheses are removed, yields

$$\cosh(W/k) = \cosh(U/k)\cosh(V/k) - \sinh(U/k)\sinh(V/k)\cos\eta.$$

This last equality is precisely the law of cosines in a hyperbolic plane whose radius of curvature is $k\sqrt{-1}$ (see Appendix 1). Putting the matter another way, under the correspondence $u \leftrightarrow U$, where $u = c\tanh(U/k)$ and $U = k\ln\sqrt{(c+u)/(c-u)}$, the relativistic velocity triangle with sides u, v, w corresponds to a triangle with sides U, V, W, and if the latter is regarded as being in the hyperbolic, plane, then the two triangles have the same angles. Thus, at least as far as the law of cosines goes, the trigonometry of a relativistic velocity triangle is identical to the trigonometry of a triangle in a hyperbolic plane. Who could have guessed, two centuries ago, when this geometry was invented, that it would turn out to describe a world of physical laws that had not yet been imagined?

The line from O to O'' lies in the plane of O, O', and O'', and, as will be shown in the optional material that now follows, it makes an angle ξ with the positive x-axis, where

$$\cos\xi = \frac{w_1}{w} = \frac{u - v\cos\eta}{w(1 - uv\cos\eta/c^2)} = \frac{u - v\cos\eta}{\sqrt{u^2 - 2uv\cos\eta + v^2 - u^2v^2\sin^2\eta/c^2}},$$

$$\sin\xi = \frac{w_2}{w} = \frac{v\sin\eta}{(w\alpha(1 - uv\cos\eta/c^2))}$$

$$= \frac{v\sin\eta}{\alpha\sqrt{u^2 - 2uv\cos\eta + v^2 - u^2v^2\sin^2\eta/c^2}}.$$

At this point, we have introduced the Lorentz transformation and discussed some of its interesting properties. That is all the background necessary to a discussion of the special theory of relativity, which begins in Chapter 2. The reader looking for a compact account of special relativity can therefore ignore the remaining sections of the present chapter, which have been written to explore certain matters of interest mostly to mathematicians.

9. Composition of Relativistic Velocities as a Binary Operation*

Consider once again our three observers O, O', and O'', any two of whom have a constant relative velocity. Assuming their origins all three coincide at some time, which each of them can take as the epoch of a calendar (time zero), any observer will see that they form a triangle at any fixed instant of time, and all observers will agree that the length of each side of the triangle is increasing after the initial instant when the three are at the same location. The trouble is, there is no instant except the time when all three origins coincided for which even two of the three observers O, O', and O'' can agree about the locations of all three of their origins, and different observers will observe different rates of increase for each of the sides. This difficulty raises a purely mathematical problem, which we may formulate as follows.

Each observer has a Euclidean space and a separate time line in which to record events $(t; x, y, z)$. In those coordinates O can define the velocity of O' as a vector $\boldsymbol{u}$, and similarly O' can define the velocity of O'' as a vector $\boldsymbol{v}$. Now O'' has some velocity, say $\boldsymbol{z}$ relative to O. How is $\boldsymbol{z}$ related to $\boldsymbol{u}$ and $\boldsymbol{v}$? Schematically, we can write this relation as a binary composition $\boldsymbol{z} = \boldsymbol{u} * \boldsymbol{v}$. In Newtonian mechanics, this binary operation is just the familiar vector addition in $\mathbb{R}^3$, and it is a commutative and associative operation. The symmetry of the formula for composition of relativistic velocities along a line shows that it is also commutative in that context. But the formulas we just gave for composition of nonparallel velocities show that in relativity it is generally not commutative. If we ignore certain difficulties, we can say that this binary operation is associative. That is, if we write $\boldsymbol{u}$ as the pair (O, O') and $\boldsymbol{v}$ as the pair (O', O''), then agree that, in general, if the second element of the first pair is the same as the first element of the second pair, we have the binary operation $(a, b) * (b, c) = (a, c)$. It is then obvious that $\big((a, b) * (b, c)\big) * (c, d) = (a, c) * (c, d) = (a, d) = (a, b) * \big((b, c) * (c, d)\big)$. But that simple computation already points up a seemingly insuperable difficulty. We just said that this composition is commutative when the velocities are collinear. But what can that mean in this notation? How can we compose (b, c) with (a, b)? We would need $c = a$ in order to do so. In terms of computation, what would it mean to assign velocity $\boldsymbol{v}$ to O' relative to O and $\boldsymbol{u}$ to O'' relative to O'?

Vectors have to be written as triples of real numbers, representing length and direction in someone's rectangular coordinate system. Whose system is this? We presume that $\boldsymbol{u}$ represents a vector in O's system and $\boldsymbol{v}$ a vector in O''s system. What does $\boldsymbol{u}$ mean in O''s system? We avoided this difficulty in our derivation of the Lorentz transformation by assuming that the two observers chose their axes so that the x-axis was along their common direction of motion. If we cut them free from this anchor, then we are computationally "adrift." What this means is that we have not yet pinned down the proper algebraic/geometric representation of relativistic velocities. We need some conventions that all of our observers can agree on as to the magnitude and direction of a velocity. Only after we get those conventions, which we shall do using hyperbolic plane trigonometry, will we be in a position to write the concatenation of two relativistic velocities in a form that computers can accept and work with.

There is a second difficulty as well, again connected with the special choice of axes used in deriving the Lorentz transformation. Because of the special choice of coordinates that we made, we got a simple "standard-form" matrix to represent the transformation. It is symmetric, and ten of its twelve nondiagonal entries are equal to zero. But this matrix takes account of only the relative speed of the motion; it assumes that the axes have already been adjusted to take into account the direction of that motion. If two velocities that we are composing are not parallel, multiplying the corresponding matrices does not yield the matrix of the composite velocity, since the simplified matrices relating O and O' to O'' have to be computed in axes that are rotated relative to those used for the matrix relating O to O'. After we solve the first problem, thereby making it possible to say in a computational sense that composition of relativistic velocities is an associative operation, we will be in a position compose the matrices of two standard-form Lorentz transformations and get the standard-form matrix of the composition of the two velocities that they correspond to. (As the reader may already have guessed, the procedure is to sandwich each matrix between two rotations of $\mathbb{R}^4$ representing the alignment of the first spatial axis of each observer with the lines of relative motion of the other two.)

9.1. Relativistic velocity triangles. With so much disagreement between different observers, we may well flee for refuge into the "safe" areas where there is agreement between different observers. What are these areas? Since the Lorentz transformation is linear, it does preserve lines and planes. Thus, each observer, "watching" the other two from the origin of his spatial coordinate system, will see them along two rays from his own origin. Any two of these observers will agree on their mutual speed. These lines do not change over time, and so the plane determined by the three observers is constant for each of them and may be regarded as common to all of them. We can then draw a triangle and label each of its vertices with a symbol representing one of the observers, assigning to each of its sides a "length" (velocity) agreed upon by the two observers at the endpoints. The key to the mutual communication we need for our observers is the fact that angles belong to absolute geometry: A right angle is one-fourth of a complete rotation, whether in elliptic, Euclidean, or hyperbolic geometry and is assigned the "radian" measure $\pi/2$ in all three versions of trigonometry, even though it has the interpretation as the ratio of the subtended arc of a circle to its radius only in Euclidean geometry.[19]

We have seen that two observers in relative motion do not generally agree about the angle between two lines. Nevertheless, each observer is "standing guard" at his own vertex of this triangle and can measure the two sides (velocities) adjacent to

---

[19]This is because circles have an absolute meaning in all three geometries, and arc length on a circle is rotation-invariant, just like the central angles subtended by arcs on the circle. On a given circle in any of these geometries, central angles and the arcs they subtend are proportional. The value $\pi/2$ for a right angle is based on the Euclidean case, where the measure of an angle really is the ratio of the arc it subtends to the radius. While this value is only a convention, it is extremely practical and will be called the *measure* of a right angle (not the *radian measure*, since this ratio is not the same for circles of different radii in elliptic and hyperbolic geometry).

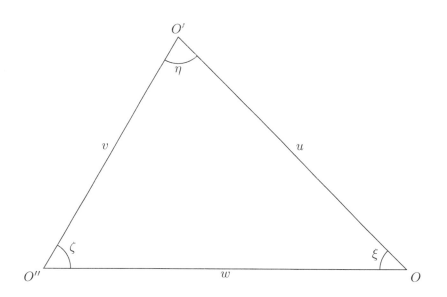

FIGURE 1.3. The velocity triangle

it and the angle between them. The angle measured by the observer located at each vertex is the angle between his lines of sight to the other two observers. Since the velocities are assumed constant, that angle does not change over time, and it is therefore taken as the definition of the angle of the velocity triangle at that vertex.

Each observer knows the two sides and included angle where he is located, and each pair of observers agrees about the magnitude of the side that each measures as the speed of the other. The set of three sides and three angles that results will be called a *relativistic velocity triangle*. If one of the observers computes the magnitude of the side opposite to his vertex using the Euclidean law of cosines, the result will *not* be the value that the other two observers agree on. That computed velocity might well be larger than c when computed in this way. That means only that two objects in an observer's space can move faster than light *relative to each other, as judged by the observer*, but not relative to the fixed frame of axes used by the observer. Since the sides of these triangles do not represent lengths, we need not think of them as situated in the Euclidean space familiar from geometry. The assignments of sides and angles that we have made are what three actual observers would measure in reality, however; and we can transfer them into a geometric space where the sides are lengths via the correspondence $u \leftrightarrow c \tanh(U/k)$, $U = k \operatorname{arctanh}(u/c)$ mentioned earlier, retaining the assignment of angles already made. They now form ordinary triangles in a three-dimensional space, but the geometry of that space is hyperbolic rather than Euclidean.

In Newtonian mechanics, an observer looking at the origins of the three systems at any given time would see them at the vertices of a physical triangle having the angles shown in the velocity triangle and sides proportional to those shown. The relations between the parts of the observed triangle will satisfy the Euclidean law of cosines:

$$w^2 = u^2 + v^2 - 2uv \cos \eta.$$

In relativistic mechanics, however, there is no instant at which any given observer would see a triangle having these angles and also having sides proportional to these sides, and so the triangle shown in Fig. 1.3, while it makes perfectly good sense in the space of velocities, is not similar to any triangle in the space of positions that would be observed by anybody. Its sides and angles have definite values according to the laws of special relativity, but these parts do not satisfy the Euclidean relations. In fact, the law of cosines in the relativistic velocity triangle is, as we have already calculated:

$$(1.5) \qquad w^2 = \frac{u^2 + v^2 - 2uv\cos\eta - u^2v^2\sin^2\eta/c^2}{\left(1 - uv\cos\eta/c^2\right)^2}.$$

This law of cosines was derived in Einstein's 1905 paper on special relativity. Assuming that it holds for each angle in the triangle, it allows us to express the side opposite that angle in terms of the angle and the two sides adjacent to it.

The law of cosines is the basic rule for all trigonometry. Since a triangle is to be determined by any two sides and the included angle (Euclid's fundamental hypothesis about congruence of triangles), trigonometry faces the task of expressing all six parts of a triangle in terms of a given angle and its two adjacent sides. Obviously, it suffices to find an expression for the other two angles, since the third side is already determined by the law of cosines itself, and the two adjacent sides have to be given anyway. That task is achieved by the following formulas.

THEOREM 1.3. *The cosines and sines of angles ξ and ζ in Fig. 1.3 are given in terms of angle η and sides u and v by the following formulas:*

$$(1.6) \qquad \cos\xi = \frac{u - v\cos\eta}{\sqrt{u^2 + v^2 - 2uv\cos\eta - u^2v^2\sin^2\eta/c^2}} = \frac{u - v\cos\eta}{w(1 - uv\cos\eta/c^2)},$$

$$(1.7) \qquad \sin\xi = \frac{v\sin\eta}{\alpha\sqrt{u^2 + v^2 - 2uv\cos\eta - u^2v^2\sin^2\eta/c^2}} = \frac{v\sin\eta}{\alpha w(1 - uv\cos\eta/c^2)},$$

$$(1.8) \qquad \cos\zeta = \frac{v - u\cos\eta}{\sqrt{u^2 + v^2 - 2uv\cos\eta - u^2v^2\sin^2\eta/c^2}} = \frac{v - u\cos\eta}{w(1 - uv\cos\eta/c^2)},$$

$$(1.9) \qquad \sin\zeta = \frac{u\sin\eta}{\beta\sqrt{u^2 + v^2 - 2uv\cos\eta - u^2v^2\sin^2\eta/c^2}} = \frac{u\sin\eta}{\beta w(1 - uv\cos\eta/c^2)}.$$

Here $\beta = c/\sqrt{c^2 - v^2}$. Notice that, when $\eta = \pi/2$ (the case of a right triangle) the expressions for $\cos\xi$ and $\cos\zeta$ agree with the Euclidean definitions, being u/w and v/w respectively. The sines are not the same as for a Euclidean triangle, being off by a factor of α for $\sin\xi$ and β for $\sin\zeta$.

PROOF. We take as our starting point the assumption that the relativistic law of cosines holds for each of the angles of a triangle, and we begin by showing how to derive Eq. (1.6) for the cosine of the angle ξ from this assumption. Applying the assumption to angles η and ξ yields

$$(1.10) \qquad w^2 = \frac{u^2 + v^2 - 2uv \cos \eta - (u^2 v^2 \sin^2 \eta)/c^2}{\left(1 - \frac{uv \cos \eta}{c^2}\right)^2},$$

$$(1.11) \qquad v^2 = \frac{u^2 + w^2 - 2uw \cos \xi - (u^2 w^2 \sin^2 \xi)/c^2}{\left(1 - \frac{uw \cos \xi}{c^2}\right)^2}.$$

Using these equations to express $\cos \xi$ explicitly as a function of u, w, and η by eliminating v between the two equations leads to horrendously complicated algebraic expressions that even *Mathematica* cannot simplify. A simpler procedure is as follows: Let $x = \cos \xi$ and $y = \cos \eta$. Rewrite Eqs. (1.10) and (1.11) as

$$w^2 = \frac{u^2 + v^2 - 2uvy - u^2 v^2 (1 - y^2)/c^2}{\left(1 - \frac{uvy}{c^2}\right)^2},$$

$$v^2 = \frac{u^2 + w^2 - 2uwx - u^2 w^2 (1 - x^2)/c^2}{\left(1 - \frac{uwx}{c^2}\right)^2}.$$

This last equation is a quadratic equation in x:

$$(1.12) \qquad v^2 \left(1 - \frac{uwx}{c^2}\right)^2 = u^2 + w^2 - 2uwx - u^2 w^2 (1 - x^2)/c^2.$$

This is a genuine quadratic equation, not degenerate, since the coefficient of x^2 is $-(u^2 w^2)/(\beta^2 c^2)$, where $\beta = c/\sqrt{c^2 - v^2}$. This coefficient is not zero except in the trivial case when u or w is zero.

Next, eliminate all occurrences of w^2 by using the first equation. The first power w, however, should remain. The result is a new quadratic equation in x that can be solved to express x in terms of u, v, w, and y. This equation, however, is rather complicated, and a computer algebra system such as *Mathematica* is recommended to avoid algebraic mistakes here. One can then simply insert the value of x provided by Eq. (1.6), namely $x = (u - vy)/(w(1 - uvy/c^2))$, and verify that the result is zero. Again, *Mathematica* makes this easier, though not trivially so; it must be coaxed by expanding and putting back together and simplifying before it yields the information that the equation is satisfied. It follows that the two applications of the relativistic law of cosines lead to an equation for $x = \cos \xi$ that is satisfied by two potential values of x, one of which is the value of $\cos \xi$ given by Eq. (1.6). To show that the other root could not be the cosine of any angle, one has only to observe that the constant term of the quadratic equation satisfied by x divided by the coefficient of x^2 is $c^4/(u^2 w^2)$, which is larger than 1. Since this quotient is the product of the roots of the equation, the two roots cannot both lie in the interval $[-1, 1]$, and therefore the other root is not the cosine of any angle. These computations are implemented in *Mathematica* Notebook 1 of Volume 3.   □

To sum up, we shall say that a *relativistic velocity triangle* with sides u, v, and w and opposite angles ζ, ξ, and η respectively, is one for which all three relativistic laws of cosines hold.

$$w^2 = \frac{u^2 + v^2 - 2uv\cos\eta - u^2v^2\sin^2\eta/c^2}{\left(1 - \frac{uv\cos\eta}{c^2}\right)^2},$$

$$v^2 = \frac{u^2 + w^2 - 2uw\cos\xi - u^2w^2(\sin^2\xi)/c^2}{\left(1 - \frac{uw\cos\xi}{c^2}\right)^2)},$$

$$u^2 = \frac{v^2 + w^2 - 2vw\cos\zeta - v^2w^2(\sin^2\zeta)/c^2}{\left(1 - \frac{vw\cos\zeta}{c^2}\right)^2}.$$

These laws imply the following relation between the cosines of the angles ξ and η opposite sides v and w respectively.

$$(1.13) \qquad \cos\xi = \frac{u - v\cos\eta}{w\left(1 - \frac{uv\cos\eta}{c^2}\right)}.$$

All the relations among the velocities and angles measured by our three observers can be deduced from any three of them—which may even be the three angles ξ, η, and ζ—using these trigonometric formulas. What makes these triangles relativistic rather than Euclidean is the fact that each vertex is associated with an observer, whose *Euclidean/Newtonian* measurement of the two sides and angles at that vertex are used to assign measures to these parts of the triangle. Each pair of observers agrees about the velocity (length) of the side joining their vertices, but there is no such agreement about any of the angles. They are arbitrarily defined as the angles measured by their corresponding observers. In general none of the six parts of the triangle is agreed to by all three of the observers. Each observer, we emphasize, is using Euclidean geometry to measure the two sides and the angle at his vertex. That observer will consequently use the Euclidean law of cosines and Euclidean trigonometry to determine the opposite side and the other two angles, and each of the other two observers will generally disagree with him about all three of them. These values inferred from Euclidean geometry are therefore to be discounted when the observers reconcile their observations. By making use of the relativistic trigonometry just discussed, each observer can deduce the measurements the other two are making, and they can then agree on all their relative velocities.

REMARK 1.6. As a final comment on the connection of relativistic velocity triangles with hyperbolic triangles, we shall now show that the formula for $\sin\xi$, when reinterpreted in the hyperbolic plane, is precisely the hyperbolic law of sines. It suffices to show that the formula for $\sin\xi$ translates under the given substitutions to become

$$\sin\xi = \frac{\sin\eta\sinh(V/k)}{\sinh(W/k)}.$$

By Equation (1.7), $u = c\tanh{(U/k)}$, $\alpha = \cosh(U/k)$, $v = c\tanh(V/k)$, $w = c\tanh{(W/k)}$, we have, upon taking account of the relativistic law of cosines,

$$
\begin{aligned}
\sin\xi &= \frac{\tanh{(V/k)}\sin\eta}{\tanh{(W/k)}\cosh(U/k)(1 - \tanh{(U/k)}\tanh{(V/k)}\cos\eta)} \\
&= \frac{\sinh(V/k)\sin\eta\cosh(W/k)}{\sinh(W/k)\cosh(V/k)\cosh(U/k)(1 - \tanh{(U/k)}\tanh{(V/k)}\cos\eta)} \\
&= \frac{\sinh(V/k)\sin\eta\cosh(W/k)}{\sinh(W/k)(\cosh(U/k)\cosh(V/k) - \sinh(U/k)\sinh(V/k)\cos\eta)} \\
&= \frac{\sinh(V/k)\sin\eta}{\sinh(W/k)},
\end{aligned}
$$

as required.

10. Plane Trigonometry*

Having established the values of ξ and ζ, we have now completed our earlier argument showing that the trigonometry of relativistic velocity triangles is simply hyperbolic trigonometry, that is, trigonometry on a sphere of imaginary radius $k\sqrt{-1}$ filtered through the mapping $u = c\tanh(U/k)$. Nevertheless, instead of constantly translating everything we want to say about velocities into statements in the hyperbolic plane, we prefer to deal directly with the formulas derived from the Lorentz transformation. We are not going to derive any facts about relativistic velocity triangles using hyperbolic trigonometry. We do not assume that the reader even knows hyperbolic trigonometry. Everything we prove geometrically is based on ordinary algebra and the Lorentz transformation. In fact, we do the exact opposite in Appendix 1: We use the Lorentz transformation as the key to a very simple treatment of the trigonometry of the hyperbolic plane, a task that is normally requires considerably more machinery. That is all we are going to say about this fascinating connection. For more details about non-Euclidean geometries, see Appendix 1.

In a simpler form, the expression for ξ is

$$(1.14) \qquad \xi = \operatorname{arccot}\left(\frac{\alpha(u - v\cos\eta)}{v\sin\eta}\right) - \operatorname{arccot}\left(\alpha\left(\frac{u}{v}\csc\eta - \cot\eta\right)\right).$$

The angle ξ here is not a signed angle, since we are not at this point concerned with rotating O's coordinate system. It is simply an angle whose universal measure lies between 0 and π. (Again, we are deliberately not calling this universal measure *radian measure*.) The angle ζ that O'' observes between the lines joining his origin to the origins of O and O' can be obtained by interchanging u and v (and hence replacing α with β), and satisfies Eqs. (1.8) and (1.9). Again, expressed more simply

$$(1.15) \qquad \zeta = \operatorname{arccot}\left(\frac{\beta(v - u\cos\eta)}{u\sin\eta}\right) = \operatorname{arccot}\left(\beta\left(\frac{v}{u}\csc\eta - \cot\eta\right)\right).$$

The law of cosines enables us to find the third side of a triangle given two sides and the included angle (see Fig. 1.3). The triangle has six parts—three sides and three angles—but only three of them are needed to determine the triangle. Let us briefly sketch how to infer the other three parts from three that are given.

(1) To solve a triangle given two angles and a side, we use the law of sines given in Problem 1.10 below. This law for relativistic velocity triangles is equivalent to the hyperbolic law of sines stated above.

(2) To find the angles given all three sides, see Problem 1.9 below.

(3) To solve a triangle given only its three angles ξ, η, and ζ opposite sides v, w, and u respectively, we combine the algebraic equations for the cosines of the angles (*Mathematica* is highly recommended for this exercise), and get the following formulas, which hold provided the expressions under the radicals are positive, as they must be if the sum of the three angles is less than two right angles.

$$u = \frac{c\sqrt{\cos^2 \xi + \cos^2 \eta + \cos^2 \zeta + 2\cos\xi \cos\eta \cos\zeta - 1}}{\cos\zeta + \cos\xi \cos\eta},$$

$$v = \frac{c\sqrt{\cos^2 \xi + \cos^2 \eta + \cos^2 \zeta + 2\cos\xi \cos\eta \cos\zeta - 1}}{\cos\xi + \cos\eta \cos\zeta},$$

$$w = \frac{c\sqrt{\cos^2 \xi + \cos^2 \eta + \cos^2 \zeta + 2\cos\xi \cos\eta \cos\zeta - 1}}{\cos\eta + \cos\zeta \cos\xi}.$$

It is easy to verify that u, v, and w, as determined from these equations, are all smaller than c, provided all the denominators are positive, as must be the case if ξ, η, and ζ are to be the angles of a relativistic velocity triangle. In fact, any three positive angles ζ, ξ, and η, the sum of whose measures is less than π form the angles of a unique relativistic velocity triangle whose respective opposite sides are given by the three formulas above. The condition $\zeta + \xi + \eta < \pi$ guarantees that the expression under the radical and all three denominators are positive. (See Problem 1.17.)

10.1. Right triangles. When O and O'' are moving along mutually perpendicular lines as judged by O', these formulas naturally simplify. Taking $\eta = \pi/2$, we get the following equalities:

$$(1.16) \qquad w = \sqrt{u^2 + v^2 - u^2 v^2/c^2} = \frac{c}{\alpha\beta}\sqrt{\alpha^2 \beta^2 - 1},$$

$$(1.17) \qquad \cos\xi = \frac{u}{\sqrt{u^2 + v^2 - u^2 v^2/c^2}} = \frac{u\alpha\beta}{c\sqrt{\alpha^2 \beta^2 - 1}},$$

$$(1.18) \qquad \sin\xi = \frac{v}{\alpha\sqrt{u^2 + v^2 - u^2 v^2/c^2}} = \frac{v\beta}{c\sqrt{\alpha^2 \beta^2 - 1}},$$

$$(1.19) \qquad \cos\zeta = \frac{v}{\sqrt{u^2 + v^2 - u^2 v^2/c^2}} = \frac{v\alpha\beta}{c\sqrt{\alpha^2 \beta^2 - 1}},$$

$$(1.20) \qquad \sin\zeta = \frac{u}{\beta\sqrt{u^2 + v^2 - u^2 v^2/c^2}} = \frac{u\alpha}{c\sqrt{\alpha^2 \beta^2 - 1}}.$$

EXAMPLE 1.1. The vector addition of velocities and displacements was discussed by Galileo in the seventeenth century. To use a modification of his example, suppose a ship is moving at a rate of 12 km per hour relative to the Earth. If you walk from the port side to the starboard side at 5 km per hour, then your actual speed is $\sqrt{12^2 + 5^2} = 13$ km per hour, in a direction that makes an angle $\arctan(5/12) \approx 22.6°$ with the direction the ship is moving. Relativity changes this relation. Since the speed of light is about $300,000$ km per second, if Observer O sees

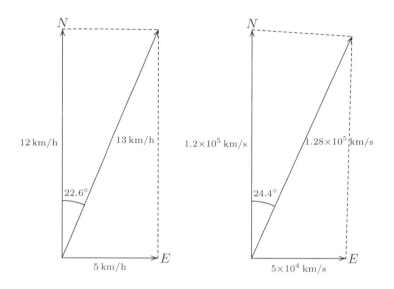

FIGURE 1.4. Galilean addition of velocities (left): A land-based observer sees a ship moving through the water heading due north at 12 km per hour. The pilot of the ship sees a crew member walking from port to starboard (due east) at 5 km per hour. The land-based observer sees that crew member moving at 13 km per hour in a direction 22.6° east of north. Relativistic addition of velocities (right): Three "observers" O, O', and O'' all move through the same origin simultaneously, then diverge. Observer O sees O' moving due north at 120,000 km per second, and O' sees O'' moving due east at 50,000 km per second. Then O sees O'' moving at 128,452 km per second in a direction 24.4° east of north.

observer O' moving at a speed of 120,000 km per second, and O' sees O'' moving at a speed of 50,000 km per second in a direction perpendicular to the direction in which he sees O moving, then O will see O'' moving at a speed of 128,452 km per second in a direction making an angle of 24.4° with with the line of sight to O'.

What can we make of a geometry in which u, v, and w are the sides of a right triangle, with u opposite angle ζ, v opposite angle ξ, and w opposite the right angle? In terms of the famous Pythagorean theorem, these relations would imply $u^2 + v^2 > w^2$. That is, the hypotenuse of a right triangle is shorter than the hypotenuse of a plane right triangle in Euclidean space having legs of the given lengths. That inequality is true of both elliptic non-Euclidean geometry, where the Pythagorean relation is $\cos(C/R) = \cos(A/R)\cos(B/R)$, and hyperbolic non-Euclidean geometry, where the corresponding Pythagorean relation is $\cosh(C/k) = \cosh(A/k)\cosh(B/k)$. (Here R and k are respectively the constant curvature of the elliptic plane and the hyperbolic plane. In elliptic geometry, the sum of the angles of a triangle is larger than two right angles, and in hyperbolic geometry it is less. We shall now show that the angle sum of a right relativistic velocity triangle is less than two right angles, that is, $\xi + \zeta < \pi/2$. We of course already know this fact, since it is a fact of hyperbolic geometry. Nevertheless, we shall give two direct proofs of it, one here for right triangles, and one later for general triangles.

THEOREM 1.4. *Let ξ and ζ be the two acute angles of a relativistic velocity triangle containing a right angle. Then $\xi + \zeta < \pi/2$.*

PROOF. By formulas (1.18) and (1.19), we have $\alpha \sin \xi = \cos \zeta = \sin(\pi/2 - \zeta)$. Since $\alpha = c/\sqrt{c^2 - u^2} > 1$, it follows that $\sin \xi < \sin(\pi/2 - \zeta)$, and therefore $\xi < \pi/2 - \zeta$, that is, $\xi + \zeta < \pi/2$, as asserted.   □

It is not difficult to show (see Problem 1.13) that the sum of the three angles ξ, η, and ζ in any relativistic velocity triangle is less than two right angles.

In the special case when $u = v$, we get what O' regards as an isosceles right triangle, with acute angles ξ given by

$$\xi = \arccos \left(\frac{\alpha}{\sqrt{\alpha^2 + 1}} \right) = \arccos \left(\frac{1}{\sqrt{1 + 1/\alpha^2}} \right).$$

Obviously, ξ is a decreasing function of α. The limiting cases are $u \uparrow c$, in which $\xi \downarrow 0$, and $u \downarrow 0$, in which $\xi \uparrow \pi/4$, the latter being the case when all observers are at rest relative to one another, so that the geometry becomes Euclidean, and simultaneity becomes observer-independent.

REMARK 1.7. Although in general three observers do not agree about the velocities in the triangle whose vertices they form, there is a range of vertex angles η in an *isosceles* triangle for which there is a speed u (depending on η) on each of the two sides of the angle η such that the observer O' at the vertex of the angle will agree with O and O'' as to their mutual speed. The range of angles η for which such a speed u exists is

$$60° = \arccos(1/2) < \eta < \arccos(1/3) < 70.528779365509308630755° \,.$$

Over this range, the corresponding speed u decreases from c to 0. (See Problem 1.11 below.) Observers O and O'' do not agree with O' about any of the angles, and neither of them thinks this is an isosceles triangle. (For example, O does not agree with O' and O'' that their relative speed is $v = u$.)

As an example, take $u = (3\sqrt{2}/5)c$ and $\cos(\eta) = 5/12$. With those values, you will find that

$$\eta \approx 67.056553501352011261° \,.$$

For this case, all three observers agree as to the relative speed of O and O'', namely $0.84c$.

11. The Lorentz Group*

Two problems may be of interest to the mathematically-inclined reader—perhaps less interesting to nonmathematicians. The first is the question of closure: Is the composition of two Lorentz transformations truly a Lorentz transformation? Certainly, if C is moving with constant velocity relative to B and B is moving with constant velocity relative to A, then C is also moving with constant velocity relative to A. Since our relativistic velocity triangles tell us how A and C can rotate coordinates so as to get the matrix of this transformation in the standard form, this question has, in a sense, a trivial answer. But do the computations really work out? We take up that question in Section 12 below and exhibit its computational implementation with *Mathematica* Notebook 5 in Volume 3.

The second question is the associativity of the law of relativistic composition of velocities. Suppose we have four observers, say A, B, C, and D, and that the velocity of B relative to A is $\boldsymbol{u}$, that of C relative to B is $\boldsymbol{v}$, and that of D relative to C is $\boldsymbol{w}$. Then the velocity of C relative to A can be represented as $\boldsymbol{u} +_L \boldsymbol{v}$, where this notation indicates the composition of velocities given by the relativistic velocity triangle determined by A, B, and C, as above. And then, by our computations, the velocity of D relative to A must be $(\boldsymbol{u} +_L \boldsymbol{v}) +_L \boldsymbol{w}$. On the other hand, the velocity of D relative to B is $\boldsymbol{v} +_L \boldsymbol{w}$, and consequently—if we believe the computations that have been performed above, the velocity of D relative to A must also be $\boldsymbol{u} +_L (\boldsymbol{v} +_L \boldsymbol{w})$. Since these two velocities are obviously the same, and we don't think we have made any mistakes in logic during our derivations, we have to conclude that the binary operation $+_L$ is an associative operation:

$$(\boldsymbol{u} +_L \boldsymbol{v}) +_L \boldsymbol{w} = \boldsymbol{u} +_L (\boldsymbol{v} +_L \boldsymbol{w}).$$

This relation would appear to settle the question, but there are subtleties involved when we try to implement this rule computationally. The velocities here are given in three distinct coordinate systems, used by A, B, and C. In what sense can they be added at all? How can you add coordinates in one set of axes to coordinates in a completely different set? We can tame the problem a bit by passing to a three-dimensional hyperbolic space, in which case the associative law becomes a geometric theorem:

THEOREM 1.5. *If the sides and angles of three faces of a tetrahedron satisfy the relations of relativistic velocity triangles, then* (1) *the sides and angles of the fourth face are determined and* (2) *the fourth face is also a relativistic velocity triangle.*

The proof of this theorem forms Section 5 of Appendix 1 in Volume 2. The fact that Lorentz transformations are closed under composition and that their composition is an associative operation makes them into a group, called the *Lorentz group*. It is a six-dimensional Lie group.

11.1. Associativity*. The fact that the fourth face of the velocity tetrahedron is a relativistic velocity triangle when the other three are amounts to the *associative law* for the relativistic composition of velocities. It is quite obvious that velocity $\boldsymbol{0}$ is an identity for this composition, and *it seems* that the inverse of velocity $\boldsymbol{u}$ should be $-\boldsymbol{u}$. And indeed it is, when $-\boldsymbol{u}$ is *defined* as the velocity of O relative to O' given that $\boldsymbol{u}$ is the velocity of O' relative to O. As we have already pointed out above, the associativity of this *verbally described* operation is completely obvious, even without any computations. Thus we have turned the set of physically possible relative velocities in three-dimensional space into a group. We emphasize, however, that both the elements of this group and the group operation have so far only verbal descriptions. Even though we think of these velocities as vectors, they do not transform between observers the way vectors do.

Most importantly, although it is trivial that composition of mappings is an associative operation, there is a subtlety involved in the present case: Before two observers apply the Lorentz transformation as we defined it (using a 4×4 matrix) to convert their space-time coordinates, they must both, in general, perform a rotation

of their spatial axes. The fact that the actual transformation is a "sandwich"
consisting of two rotation matrices with a standard-form Lorentz matrix between
them makes it far from obvious that the composition is associative. Our reasons for
believing that it is, at this point, are physical and geometric. It would be desirable
to have an algebraic proof of the fact. That is difficult to do on the basis of our
definition of a Lorentz transformation; we give an "empirical" verification of it in
Theorem 1.6 below.

To get a *computable* associative operation out of the relativistic addition of
velocities, the triples we need are not the components of velocity in some particular
frame of reference, but rather the relative speed of two observers and the polar
angles along which the two observe each other relative to fixed frames they are
using. Using what we have proved by means of the velocity tetrahedron, we are
now in a position to state the associative law formally and verify that it "computes"
as it should. We still will not quite have made the velocities into a group, even
when we do that, due to the singularity of the polar coordinate system at the origin.
But at least we can allow our three observers to use whatever coordinates they like,
and as long as none of the velocities is zero, we can say exactly how each needs to
rotate its axes in order to communicate with the other two using a vector formula.
We are not going to give a formal proof of the procedure, however, but rather rely
on an "empirical" proof, using *Mathematica* to generate random data and verify
the associativity of the composition. (The formal verification takes too long, even
for *Mathematica*.)

Since we wish to discuss the composition without invoking a privileged coordinate system, we shall make a "sandwich" out of the relative velocity of two observers
O and O', writing it as (θ, u, φ), where θ is the angle measured counterclockwise
from O's first coordinate axis to the line of motion O observes O' to be traversing,
u is the speed of that motion, and φ is the angle measured counterclockwise from
O''s first coordinate axis to the line of motion O' observes O to be traversing. Similarly, let (χ, v, ψ) represent the relative velocity of O' and O''. The speeds u and
v are positive numbers between 0 and c and the four angles are any real numbers,
equality being taken modulo 2π. We wish to define a binary operation "$+_L$" that
we shall call *Lorentz addition* for these two triples of real numbers.

REMARK 1.8. Before we discuss how to implement our velocity triangle as
a binary operation and show that it is associative, we note that we appear to be
defining the set of relativistic velocities in a plane as a *three-dimensional* object. As
we have described it, the set of relativistic velocities in a plane could be represented
geometrically as the product of two circles (represented by the two angles θ and φ)
and the interval $(0, c)$ (representing the speed). You could picture it as a solid torus
(anchor ring) in Euclidean space with its outer surface stripped off and the circle
through the middle of its interior, representing speed 0, also removed. Topologically,
the tricky part of making this object into a computable group is attaching that all-
important identity corresponding to speed 0. When two of these triples happen to
be opposites of each other, that is, $O'' = O$, $\varphi = \chi$, $u = v$, and $\theta = \psi$, the operation
we are going to define does give 0 as the speed, but it also attaches two angles to
that speed. That makes no sense, given that the origin of a polar coordinate system
has no definable polar angle.

These velocities, one would think, are really two-dimensional objects. To make
them two-dimensional, we need some transformation of pairs of angles (θ, φ) that

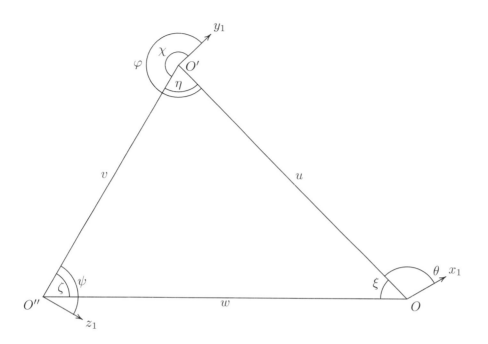

FIGURE 1.5. Lorentz addition of velocities

leaves the Lorentz sum of two velocities invariant. One transformation having this property is the mapping $(\theta, \varphi) \mapsto (d + \theta, d + \varphi)$ for any fixed angle d (see Problem 1.18). Once we have the appropriate transformation, we can take a "quotient space" modulo it and get the two-dimensional object that we need, except for that troublesome problem with the identity.

Defining the addition that we need is not complicated, given that we know how to solve the relativistic velocity triangle in Fig. 1.5

DEFINITION 1.1. The *Lorentz composition* $+_L$ of the speeds u and v in the directions indicated in Fig. (1.5) is the speed w in the direction shown, where we "sandwich" each speed between the two angles from the observer's first spatial axis in the directions of the other two observers. As a formula, replacing the vector velocity $\boldsymbol{u}$ by the triple (θ, u, φ), and similarly for $\boldsymbol{v}$ and $\boldsymbol{w}$, we get

$$(\theta, u, \varphi) \; +_L \; (\chi, v, \psi) = (\theta + \xi, w, \psi - \zeta).$$

Since we have formulas for ξ, w, and ζ, the associativity of this operation ought to be straightforward.

We state it as a formal proposition:

THEOREM 1.6. *The composition* $(\theta, u, \varphi) \; +_L \; (\chi, v, \psi)$ *is an associative operation.*

PROOF. There is a complication resulting from the orientation of the triangle. If the angle $\eta = \varphi - \chi$ is larger than a straight angle, that is, $|\varphi - \chi| > \pi$, then the roles of O and O'' will interchange, and instead of $(\theta + \xi, w, \psi - \zeta)$, we would need $(\theta - \xi, w, \psi + \zeta)$. That will also happen if $|\varphi - \chi|$ is less than π but the y_1-axis is inside the angle η. (Since we represent angles as numbers between 0 and 2π, the angle χ will be larger than φ in this case.) To handle this complication, we need to multiply ξ and ζ by sgn $(\pi - |\varphi - \chi|)$. Once that is done, although the computations are very tedious—so tedious that *Mathematica* will probably run out of memory before it can actually compute the Lorentz sum if the data are given as infinitely precise real numbers—it is possible to demonstrate convincingly through numerical examples that this operation is indeed associative. *Mathematica* Notebook 3 of Volume 3 will provide that convincing proof by generating as many random inputs as one likes. In that notebook, if you input two angles and a speed (the latter in the form ac, where a is a real number in the range $[0, 1)$) for each of two triples, the addition "$+_L$" (called `ladd` in *Mathematica* Notebook 3), can compute the Lorentz sum of the two relative velocities these triples represent, with the caveat that infinitely precise real numbers as data may well lead to a long computation requiring an inordinate amount of computer memory. If you input the data as finite-precision floating-point numbers (again, the speeds must bear the letter c as a suffix, even if they are zero), *Mathematica* will perform the computation in short order. The last command in the notebook checks the associative property by generating five triples of triples representing relative velocities and showing that the composition of three such velocities is the same, no matter how they are grouped.

In dozens of trials, this program always wrote

"Out[4] = {{0,0,0},{0,0,0},{0,0,0},{0,0,0},{0,0,0}}'' .

Thus, by empirical verification, the operation `ladd` is associative.   □

Although the theoretical basis of this program does not include the case of speed 0, the program will accept an input of $0.0c$ for either speed, and the output will be the other term in the sum, except that the two angles in it will be reduced modulo 2π.

12. Closure of Lorentz Transformations under Composition*

In the present section, we shall be considering a number of linear operators on space-time. To keep the algebra simple, it is desirable that the entries in these matrices be physically dimensionless. For that reason, we are going to make use of the technique mentioned earlier, "spatializing" the time coordinate through multiplication by the speed of light. Thus we shall replace time t by the variable $\tau = ct$. Also, to avoid getting too many primes, we shall rename our observers O, O' and O'', referring to them henceforth as X, Y, and Z. Finally, to streamline the notation still further, we assume these three observers are using the space-time coordinates $(\rho; \boldsymbol{\xi}) = (\rho; x^1, x^2, x^3)$, $(\sigma; \boldsymbol{\eta}) = (\sigma; y^1, y^2, y^3)$, and $(\tau; \boldsymbol{\zeta}) = (\tau; z^1, z^2, z^3)$ respectively. As always, we assume that the origins of all three four-dimensional coordinate systems coincide.

We have not actually given a *mathematical* definition of what a Lorentz transformation of $\mathbb{R}^4$ is. Our approach has been through kinematics: It is the transformation of space-time coordinates between two observers in uniform relative motion. What we have derived through simple algebra and Euclidean trigonometry is that, if both observers are using (Euclidean) orthonormal coordinates in $\mathbb{R}^4$ for which (1) the origins coincide, (2) the first axis represents time for both observers, and (3) the second axis is along the line of mutual motion for both observers, this transformation has a matrix of the form

$$
L \sim \begin{pmatrix} \alpha & -\frac{\alpha u}{c} & 0 & 0 \\ -\frac{\alpha u}{c} & \alpha & 0 & 0 \\ 0 & 0 & 1 & 0 \\ 0 & 0 & 0 & 1 \end{pmatrix},
$$

where u is the speed of the second observer relative to the first and $\alpha = (1 - u^2/c^2)^{-1/2}$. More generally, since each observer can perform an orthonormal transformation on the spatial part of $\mathbb{R}^4$, a Lorentz transformation is one whose matrix M is such that there exist two rotation matrices R_1 and R_2 on the spatial portion of $\mathbb{R}^4$ (leaving the time axis fixed) such that $L = R_1 M R_2$ has this form. Or, more practically, we can define a Lorentz transformation to be the *set* of all matrices of this form, where M is a fixed matrix in standard form and R_1 and R_2 vary over all rotations of the spatial portion of space-time. When we use this description as a definition, it is by no means obvious that the Lorentz transformations are even closed under the operation of composition.

Mathematicians generally define a Lorentz transformation on $\mathbb{R}^4$ to be a linear transformation that preserves the quadratic form $\rho^2 - (x^1)^2 - (x^2)^2 - (x^3)^2$. In that form, it *is* obvious that the composition of two Lorentz transformations is a Lorentz transformation. Those concerned with logical rigor are then faced with two choices: (1) Demonstrating that the transformations fitting the description just given are precisely the ones that preserve the given quadratic form; (2) showing directly that if $L_1 = R_{11} M_1 R_{12}$ and $L_2 = R_{21} M_2 R_{22}$ are both Lorentz transformations, then there are rotations S_1 and S_2 of the given form such that $S_1 L_2 L_1 S_2$ has the given form. That is the route we shall follow, using *Mathematica* to avoid having to do messy computations by hand.

We now take up the computational problem just posed: getting the standard-form matrix of the composition of two velocities knowing relative speeds and lines of sight between the pairs of observers. The "sandwich" made by putting the two transformations between two rotations of a special form, which we created to compose relativistic velocities, makes this process computable, though a bit messy.

In our derivation of the Lorentz transformation between X and Y (O and O', as we called them at the time), we assumed that that the x^1- and y^1-axes coincide at all times, that Y is moving in the positive direction with speed u along this axis, as observed by X, without any rotation. This last means that $x^2 = y^2$ and $x^3 = y^3$ for any event E having coordinates $(\rho, x^1, x^2, x^3) = (\rho, \boldsymbol{\xi})$ and $(\sigma; y^1, y^2, y^3) = (\sigma; \boldsymbol{\eta})$, as measured by X and Y respectively. The FitzGerald–Lorentz contraction factor for this transformation is $1/\alpha$, where $\alpha = c/\sqrt{c^2 - u^2}$. In the three-dimensional velocity space used in common by X and Y, the velocity is $\boldsymbol{u} = (u, 0, 0)$, when both observers use the axes just described.

In these two coordinate systems, the Lorentz transformation corresponding to the velocity $\boldsymbol{u}$ can be written as the matrix equation

$$
\begin{pmatrix} \sigma \\ y^1 \\ y^2 \\ y^3 \end{pmatrix} = \begin{pmatrix} \alpha & -\alpha u/c & 0 & 0 \\ -\alpha u/c & \alpha & 0 & 0 \\ 0 & 0 & 1 & 0 \\ 0 & 0 & 0 & 1 \end{pmatrix} \begin{pmatrix} \rho \\ x^1 \\ x^2 \\ x^3 \end{pmatrix},
$$

or equivalently, the set of four equations

$$
\begin{aligned}
\sigma &= \alpha(\rho - ux^1/c)\,, \\
y^1 &= \alpha(-u\rho/c + x^1)\,, \\
y^2 &= x^2\,, \\
y^3 &= x^3\,.
\end{aligned}
$$

REMARK 1.9. The matrix of the Lorentz transformation in these coordinates is obviously symmetric and hence can be diagonalized in $\mathbb{R}^4$ by a simple rotation of the ρx^1-plane (which is also the σy^1-plane) through the angle $\pi/4$ (half of a right angle). Its eigenvalues are 1, 1, $\alpha(1 + u/c) = \sqrt{\frac{c+u}{c-u}}$, and $\alpha(1 - u/c) = \sqrt{\frac{c-u}{c+u}}$. We remark that since all the eigenvalues are positive, this matrix is positive-definite, that is, it can be used to define a positive-definite quadratic form on $\mathbb{R}^4$. What is more important, however, is that this matrix preserves the space-time interval $\rho^2 - (x^1)^2 - (x^2)^2 - (x^3)^2$. It is easy to verify that this form is preserved, that is,

$$
\rho^2 - (x^1)^2 - (x^2)^2 - (x^3)^2 = \sigma^2 - (y^1)^2 - (y^2)^2 - (y^3)^2\,.
$$

The matrix of the inverse of the matrix of the Lorentz transformation in these coordinates is obtained by replacing u with $-u$, as we would expect.

Lorentz transformations are not the only linear transformations that preserve the space-time interval. An obvious group of linear transformations with this property corresponds to the set of matrices of the form

$$
\begin{pmatrix}
1 & 0 & 0 & 0 \\
0 & r_{11} & r_{12} & r_{13} \\
0 & r_{21} & r_{22} & r_{23} \\
0 & r_{31} & r_{32} & r_{33}
\end{pmatrix},
$$

where the r's are the entries in an orthogonal matrix.

12.1. The standard-form matrix of the composite velocity. Suppose now that there is a third observer Z moving relative to Y with velocity $\boldsymbol{v}$, again without any rotation, but that $\boldsymbol{v}$ is not (necessarily) along Y's positive first spatial axis. Suppose it makes angle η with that axis. Then $\boldsymbol{v}$ makes angle $\theta = \pi - \eta$ with $-\boldsymbol{u}$ (the velocity that X has relative to Y). The angle θ is the angle between the lines of sight from Y to X and Z.

As we have seen, the vector representing the velocity of X relative to Z is generally not the negative of the vector representing the velocity of Z relative to X. The discrepancy is due to the changes of coordinates that X and Y must make when shifting from communicating with each other to communicating with Z. Each of the three observers needs to use two sets of coordinates: X will use coordinates $(\rho; \boldsymbol{\xi}') = (\rho; (x')^1, (x')^2, (x')^3)$ when communicating with Z and $(\rho, \boldsymbol{\xi})$ when communicating with Y; Y will use $(\sigma; \boldsymbol{\eta})$ when communicating with X and $(\sigma; \boldsymbol{\eta}') = (\sigma; (y')^1, (y')^2, (y')^3)$, when communicating with Z; and Z will use $(\tau; \boldsymbol{\zeta})$ when communicating with Y and $(\tau, \boldsymbol{\zeta}') = (\tau; (z')^1, (z')^2, (z')^3)$ when communicating with X. We plan to show that if X and Z communicate using the coordinates $(\rho; \boldsymbol{\xi}')$ and $(\tau; \boldsymbol{\zeta}')$, respectively, then the mapping $(\rho; \boldsymbol{\xi}') \mapsto (\tau; \boldsymbol{\zeta}')$ is given in matrix form as

$$
\begin{pmatrix} \tau \\ (z')^1 \\ (z')^2 \\ (z')^3 \end{pmatrix} = \begin{pmatrix} \gamma & -\gamma w/c & 0 & 0 \\ -\gamma w/c & \gamma & 0 & 0 \\ 0 & 0 & 1 & 0 \\ 0 & 0 & 0 & 1 \end{pmatrix} \begin{pmatrix} \rho \\ (x')^1 \\ (x')^2 \\ (x')^3 \end{pmatrix},
$$

where $\gamma = c/\sqrt{c^2 - w^2}$, and the transformation is equivalently described by the set of four equations

$$
\begin{aligned}
\tau &= \gamma(\rho - w(x')^1/c), \\
(z')^1 &= \gamma(-w\rho/c + (x')^1), \\
(z')^2 &= (x')^2, \\
(z')^3 &= (x')^3.
\end{aligned}
$$

These relations are the result of composing the two Lorentz transformations with the rotations in the correct sequence. In particular, the 4×4 matrix of the composite transformation in the proper coordinate systems for O and O'' is the product of five 4×4 matrices, whose entries are rational functions of the entries in the two Lorentz matrices and the sines and cosines of the angles of rotation. The full computation is intimidatingly complicated, but a judicious use of *Mathematica* will verify its correctness.

THEOREM 1.7. *If L_α is the standard-form matrix of the Lorentz transformation corresponding to velocity $\boldsymbol{u}$, and L_β the standard-form matrix of the Lorentz transformation corresponding to velocity $\boldsymbol{v}$, which is to say $\alpha = (1 - u^2/c^2)^{-1/2}$, $\beta = (1 - v^2/c^2)^{-1/2}$, the directions being as in Fig. 1.5, and*

$$
L_\alpha = \begin{pmatrix} \alpha & -\frac{\alpha u}{c} & 0 & 0 \\ -\frac{\alpha u}{c} & \alpha & 0 & 0 \\ 0 & 0 & 1 & 0 \\ 0 & 0 & 0 & 1 \end{pmatrix} \quad \text{and} \quad L_\beta = \begin{pmatrix} \beta & -\frac{\beta v}{c} & 0 & 0 \\ -\frac{\beta v}{c} & \beta & 0 & 0 \\ 0 & 0 & 1 & 0 \\ 0 & 0 & 0 & 1 \end{pmatrix},
$$

then the composition $\boldsymbol{u} +_L \boldsymbol{v}$ has the standard-form matrix

$$
L_\gamma = R_{-\zeta} L_\beta R_0 L_u R_{-\xi},,
$$

where R_ψ represents a rotation of $\mathbb{R}^4$ that leaves the time axis and the third spatial axis fixed, and on the plane of the other two axes is a rotation about the third spatial axis through angle ψ. Moreover,

$$
\gamma = \alpha\beta\left(1 - uv\cos\eta/c^2\right) = \alpha\beta\left(1 + uv\cos\theta/c^2\right),
$$

where $\theta = \pi - \eta$ is the angle between the lines of sight from Y to X and Z.

PROOF. Since we are going to be computing the transformation between X and Z, we shall assume at the outset that both of these observers have, if necessary, rotated their first spatial axes in the direction of each other, X being in the negative direction from Z and Z in the positive direction from X. Consider now the X-coordinates $(\rho; x^1, x^2, x^3)$ of an event. If these are to be transmitted to Y using the standard-form matrix representation of L_α, they must first be multiplied by a rotation matrix of the $x^1 x^2$-plane through angle $-\xi$, where ξ is the angle between the lines of sight from X to Y and Z, that is,[20] by the matrix

$$\begin{pmatrix} 1 & 0 & 0 & 0 \\ 0 & \cos\xi & \sin\xi & 0 \\ 0 & -\sin\xi & \cos\xi & 0 \\ 0 & 0 & 0 & 1 \end{pmatrix}.$$

In order to receive this transmission through the standard-form matrix L_α, Y must use a coordinate system such that X is moving in the negative direction on Y's first spatial axis. Again, to simplify things, we assume that such is the case. The result of this transmission is the 4×1 matrix whose entries are the components of the vector (σ, y^1, y^2, y^3). Then, so that Y and Z can translate their coordinates with the simple 4×4 matrix corresponding to velocity $\boldsymbol{v}$, this 4×1 matrix needs to transformed via a rotation in the y^1y^2-plane through angle $\eta = \pi - \theta$, that is, it needs to be multiplied by the matrix

$$\begin{pmatrix} 1 & 0 & 0 & 0 \\ 0 & \cos\theta & -\sin\theta & 0 \\ 0 & \sin\theta & \cos\theta & 0 \\ 0 & 0 & 0 & 1 \end{pmatrix}.$$

The result of this multiplication is the matrix corresponding to $(\sigma; (y')^1, (y')^2, (y')^3)$, through which Y can communicate with Z via the Lorentz transformation

$$\begin{pmatrix} \tau \\ z^1 \\ z^2 \\ z^3 \end{pmatrix} = \begin{pmatrix} \beta & -\beta v/c & 0 & 0 \\ -\beta v/c & \beta & 0 & 0 \\ 0 & 0 & 1 & 0 \\ 0 & 0 & 0 & 1 \end{pmatrix} \begin{pmatrix} \sigma \\ (y')^1 \\ (y')^2 \\ (y')^3 \end{pmatrix}.$$

Finally, Z must rotate the z^1z^2-plane through the angle $-\zeta$ in order to get the coordinates $(\tau; (z')^1, (z')^2, (z')^3)$ needed to communicate with O. That is, the 4×1 matrix corresponding to $(\tau; \boldsymbol{\zeta})$ is to be multiplied by

$$\begin{pmatrix} 1 & 0 & 0 & 0 \\ 0 & \cos\zeta & \sin\zeta & 0 \\ 0 & -\sin\zeta & \cos\zeta & 0 \\ 0 & 0 & 0 & 1 \end{pmatrix}.$$

When all five of these matrices have been multiplied in the proper order, we do indeed find that the matrix of the mapping $(\rho; \boldsymbol{\xi}') \mapsto (\tau; \boldsymbol{\zeta}')$ is

$$\begin{pmatrix} \gamma & -\gamma w/c & 0 & 0 \\ -\gamma w/c & \gamma & 0 & 0 \\ 0 & 0 & 1 & 0 \\ 0 & 0 & 0 & 1 \end{pmatrix}.$$

The algebraic operations involved in verifying this fact are intimidatingly complex, and *Mathematica* should probably be invoked to shorten the labor. Before verifying that the computation is correct, however, we shall look at a few numerical examples. Again, a *Mathematica* program makes the work much easier, and we shall first present a simple one-cell *Mathematica* notebook (Notebook 4 in Volume 3) that shows how to get the five matrices we need and produces their product instantly. In principle, this program ought to be able to verify that the matrix

---

[20]We previously said X needed to rotate through angle ξ. But that was assuming X started with his principal axis "aimed" at Y. In the present discussion, X begins with his principal axis "aimed" at Z.

product we are discussing is the correct one. The algebra, however, is complicated, and *Mathematica* doesn't seem to be able to simplify it. As a result, we need to go interactive with *Mathematica*, using a new approach. *Mathematica* Notebook 5 in Volume 3 achieves this end.   □

13. Rotational Motion and a Non-Euclidean Geometry*

The ratio of the earth to the heavens is that of a point...

There are some who disagree [with the claim that the Earth does not move] and propose to the contrary, thinking that there is no evidence to contradict them, that the heavens are fixed at a certain distance and that the Earth rotates about the same axis from west to east, making approximately one rotation every day, or that the two rotations [of the Earth and the heavens] can be combined in arbitrary amounts, provided only that they are about the same axis, as we said, and meet each other symmetrically.

It has escaped the attention of these people that, while the simpler mechanics of the stars are equally well explained under this assumption, when considered in relation to the properties that would follow from it as regards the air in which we ourselves live, it is seen to be utterly ridiculous.

Claudius Ptolemy (fl. ca. 130 CE), *Syntaxis (The Almagest)*, Book 1, §§ 6, 7. My translation.

Comment: Several centuries earlier, Archimedes, in his *Sand-reckoner*, in which he calculated the number of grains of sand required to fill up the universe, pointed out that the first statement—obviously made long before Ptolemy lived—was absurd. Archimedes was applying the Euclidean definition of proportion, in which two objects can have a ratio only if they are of the same nature and some multiple of each is greater than the other. No multiple of a point can be larger than a line, but Ptolemy was a practical man, and the image of the ratio of a point to a line is useful in conveying his ideas.

Ptolemy's fear that the earth would disintegrate [if it moved], or that any natural motion—which acts very differently from an artificial or human-caused motion—could cause it to disintegrate, was therefore groundless. He should rather have worried about the universe, which must be moving much faster, being so far away from the earth...

If this ratio [of a point to a line, which Ptolemy gave as the ratio of the Earth to the heavens] held in fact, then the heavens would be infinitely far away... The farther away they are, the faster they must move in order to make a complete circuit in 24 hours. The increasing distance and velocity would each cause the other to become infinite, and it is a principle of Natural Philosophy that what is infinite cannot undergo any change or motion. Therefore, the heavens are at rest.

Nicolaus Copernicus (1473–1543), *On the Revolutions of the Heavenly Spheres*, Book 1, § 8. My translation.

Comment: The reasoning seems to be that, since the Earth is *not* a point, anything having such a ratio to it would have to be infinitely large.

> *In Ptolemy's system, the motions of the heavenly bodies cannot be explained by the action of central forces; celestial mechanics is impossible. The intimate relations that celestial mechanics reveals to us between all the celestial phenomena are true relations; to affirm the immobility of the earth would be to deny these relations; that would be to fool ourselves.*

Henri Poincaré ([**65**], Chapter XI, § VII, p. 352).

> *An observer at the common origin [of two planes mutually revolving about the same axis through that point] who is able to observe clocks located on a given circle with center at the origin would thus see that a rotating clock runs slower than a fixed clock adjacent to it. Since he will not thereby be led to admit that the speed of light along the given radius depends on time, he will interpret the observation by saying that the rotating clock "really does" run slower. He will thus not be able to avoid defining time in such a way that the rate at which a clock runs depends on its location.*

Einstein ([**21**], p. 775). My translation.

> *Even on the most modern views, the question of absolute rotation presents difficulties. If all motion is relative, the difference between the hypothesis that the Earth rotates and the hypothesis that the heavens revolve is purely verbal; it is no more than the difference between "John is the father of James" and "James is the son of John." But if the heavens revolve, the stars move faster than light, which is considered impossible. It cannot be said that the modern answers to this difficulty are completely satisfying...*

Bertrand Russell ([**70**], p. 540).

Many puzzles arise as a result of the relativity of space and time and the FitzGerald–Lorentz contraction. Among them are the car wash puzzle, explored in Problem 1.19 below, and the twin paradox already discussed (see also Problem 1.2 below).

A third puzzle, to which the present section is devoted, involves astronomy: If it truly makes no difference which of two objects is moving, in what sense was Copernicus right and Ptolemy wrong about the solar system? Whether the Earth rotates on an axis or the whole universe revolves around the Earth should make no difference. On the other hand, if the stars revolve around the Earth once a day, then each of them is traversing a gigantic circle many light-years in radius, yet doing so in the space of a single day. How can this seeming contradiction be explained? That is the puzzle we now attempt to solve. The example we are going to give was described qualitatively by Einstein in his 1916 paper on general relativity (see the quotation above). What we intend to do is give the mathematical details *from the standpoint of special relativity*. A full explanation requires the general theory, and one of the purposes of including this example is to make that point. We emphasize that it is the special theory that we are using for this analysis, and that it should not be taken seriously as astronomy. It represents what Eddington ([16], p. 113) called "a crude application of the FitzGerald formula." Using the general theory of relativity, Eddington showed that the actual contraction of the circumference was only one quarter of that obtained from the analysis we are about to give. Lorentz [56] had shown earlier that a rotating solid disk could not contract to the full amount, since elastic forces within it would resist the contraction. Since we are talking about a completely empty astronomical orbit, however, elastic forces would not appear to be involved; and since we are only imagining the scene anyway, it does no harm to use the principles of special relativity only.

In Newtonian terms, if we take Ptolemy's point of view, then the stars are not moving in straight lines at constant speed and therefore some force must be moving them. In the relativistic version of astronomy presented in Chapter 4 below, they are not moving along geodesics (see Appendix 2) even in the geometry we are going to create for Ptolemy's benefit. Thus, even if we manage to prevent Ptolemy's framework from becoming totally absurd, we still leave him "on the hook" for an explanation of the fact that material bodies are moving in circles without any forces in evidence to make them do so.[21] To say this is not to criticize Ptolemy. Neither he nor Copernicus nor Kepler, who refined the heliocentric model by introducing elliptical orbits, aimed at creating a dynamical model of the universe. Their models were purely kinematic, meant to "save the appearances," that is, to hypothesize motions for the heavenly bodies that explain what we see from the Earth. The idea of linking astronomy with the mechanics of forces had to await the recognition of the importance of acceleration. This concept slowly came into focus over a period of time from the thirteenth century on, beginning with the Merton rule for the distance covered by uniformly accelerated bodies, followed by Galileo's application of it to bodies falling near the Earth's surface. Soon after (1644), Descartes stated what we now recognize as the law of inertia and the law of conservation of momentum, all of which Newton wrapped up neatly in his 1687 masterpiece *The Mathematical Principles of Natural Philosophy*. This fundamental change in the way mechanics was treated began in the decade after the death of Kepler, a full century after the death of Copernicus and a millennium and a half after the time of Ptolemy.

We are going to show that the old Ptolemaic (geocentric) system of astronomy can be used to explain celestial *kinematics* (not its *dynamics*) just as completely as the Copernican (heliocentric) system. As for the relativistic dynamics of a rotating universe we shall discuss that subject very briefly in the context of general relativity (without giving any details), in Chapter 7. At that point, we shall have the Einstein tensor at our disposal, making it possible at least to state comprehensibly the fascinating exact solution of the Einstein field equations given by Gödel [34], in which time travel is possible!

In Ptolemy's system, the stars revolve around the Earth once a day. By Euclidean reckoning, light cannot traverse a circle more than 5 billion kilometers in radius in the course of a single day, and the nearest star is more than 30 trillion kilometers away. Therefore some non-Euclidean geometry will be required. To understand the problem, we briefly recap what was done in the case of translational motion, with the Lorentz transformation. The secret of reconciling the points of view of two observers in relative motion along a straight line is that, while each

---

[21] Ptolemy himself, of course, was not bothered by this fact. In his cosmology, circular motion was the natural motion of all bodies in the heavens, and no further explanation of it was needed.

is using separate time and spatial coordinates and the proper space of each is Euclidean, from the point of view of the other observer, those time and space coordinates are intertwined, and the space is, as a result, not Euclidean. The Lorentz transformation was derived for the case of rectilinear motion at constant speed, which classically requires no forces.

The length of the radius from the center of rotation to each point should be the same for both observers, since the motion of that radius is perpendicular to itself at every point, and therefore it does not undergo any FitzGerald–Lorentz contraction.[22] We can assume that our two observers, whom we shall call Ptolemy and Copernicus, share the same clock at the origin. We do not need to worry about the synchronization of clocks that each has situated at what he regards as fixed points on a circle about the origin. Each of them thinks the clocks belonging to the other are whizzing by his own clocks, and, as Einstein remarked, that means each of them thinks the other's clocks are running slow. The two must disagree as to the length of the orbit of a star. Copernicus says the stars are fixed, and therefore the circumference of a circle centered at the Earth and passing through a star at distance R is $2\pi R$. His geometry is, by his measurements, Euclidean. If we are going to tailor our theory to fit the facts in this case, we shall have to allow Ptolemy to shorten that circumference, making it less than the distance that light can travel in a single day. We must picture Ptolemy observing a star careering around its orbit and seeing the path it is going to travel as being shorter than Copernicus measures it.

That is the theoretical basis of what we are about to do. The rest is merely a matter of computational details, which we give below. As it turns out, Ptolemy can say that what Copernicus is regarding as a three-dimensional space partitioned into parallel flat planes is in fact a stack of surfaces of revolution resulting from curling up each of those planes in the same way, the entire stack forming the interior of an infinite cylinder whose axis is the axis of mutual rotation and whose radius is $C/2\pi$, where C is the distance that light can travel during the time of one complete revolution.

13.1. The computational details. We analyze this problem by imagining two observers at the center of the Earth, one (Ptolemy) treating the Earth as fixed, while the other (Copernicus) regards the stars as fixed. For purposes of discussion we focus attention on a single star, which Ptolemy thinks is revolving around him once a day, while Copernicus takes the view that it is Ptolemy who is rotating, the star remaining fixed. In this situation, Ptolemy and Copernicus agree about a number of things, including the distance R from the Earth to the star (since the revolution of each point is perpendicular to the radial coordinate to that point, hence produces no FitzGerald–Lorentz contraction of the radius). They also agree about the local time, since they both have the same coordinate origin at all times. Thus, they agree about the period T of revolution while disagreeing as to who is doing the revolving, and they agree as to the angular velocity of revolution $\theta' = 2\pi/T$.

---

[22]Due to the ambiguity in the concept of "now," however, a radial path through the stars observed at a given instant by one of the two observers would be a spiral as observed by the other. We don't worry about this fact, since we are confining our attention to a single circle with center at the origin.

Where they disagree is in the length C of the orbit of the star. For Copernicus, for whom the star is fixed in a Euclidean space, that length is simply $2\pi R$. Let us temporarily assign length C to the orbit in Ptolemy's scheme of things. The speed u of the star, according to Ptolemy, is $C/T = C\theta'/(2\pi)$. In his 1916 paper on general relativity, Einstein brought up the subject of a rotating frame of reference, and noted that the circumference of a circular orbit would undergo FitzGerald–Lorentz contraction. Developing that idea, we find

$$C = 2\pi R\sqrt{1 - C^2(\theta')^2/(4\pi^2 c^2)}\,.$$

Solving this equation for C, we find

$$C = \frac{2\pi R}{\sqrt{1 + \frac{R^2(\theta')^2}{c^2}}} = cT\frac{2\pi R}{\sqrt{(2\pi R)^2 + (cT)^2}}\,,$$

$$u = \frac{C}{T} = \frac{2\pi Rc}{\sqrt{(2\pi R)^2 + (cT)^2}}\,.$$

These last expressions show that the relativistic length of the orbit C is smaller than cT (the distance light travels over a period of revolution), and that u is smaller than c, no matter how large R becomes.

As $c \to \infty$, or $R \to 0$, both of these expressions approximate the classical expressions for circumference $(2\pi R)$ and speed $(2\pi R/T)$. That is, for small R or large c, we have $cT/\sqrt{(2\pi R)^2 + (cT)^2} \approx 1$.

As $R \to \infty$, we find $C/cT = \frac{2\pi R}{\sqrt{(2\pi R)^2 + (cT)^2}} \to 1$ and $u/c \to 1$.

We can represent the geometry of this relativistically rotating plane as the ordinary Euclidean geometry of a curved surface in three-dimensional space and at the same time make C the circumference of an actual circle in $\mathbb{R}^3$. Introducing the variable $r = C/2\pi$, we claim that such a surface has the following equation in cylindrical coordinates (r, θ, z):

$$z = z(r) = \frac{cT}{4\pi}\int_0^{(2\pi r)^2/(cT)^2}\sqrt{\frac{s^2 - 3s + 3}{(1-s)^3}}\,ds\,,$$

where the domain is $r < \frac{cT}{2\pi}$. This surface is shown in Fig. 1.5. The radius of the orbit described by a point is R, and it is the length of a certain curve from the origin to the orbit. You can compute this length as

$$\int_0^r\sqrt{1 + \left(\frac{dz}{dt}\right)^2}\,dt\,.$$

This integral is elementary, since its integrand is $(1 - (2\pi t/cT)^2)^{-3/2}$, and the integral works out to be

$$\frac{r}{\sqrt{1 - (2\pi r/cT)^2}} = R\,.$$

The orbit of the star is a circular horizontal section of the surface having circumference $2\pi r$.

This example shows how a curved representation of physical space can be useful in physics. After we define curvature in Chapter 5, we will be able to verify that the curvature of this surface at radius r is

$$\kappa = 3\left(\frac{2\pi}{cT}\right)^2\left(1 - \left(\frac{2\pi r}{cT}\right)^2\right)^2\,.$$

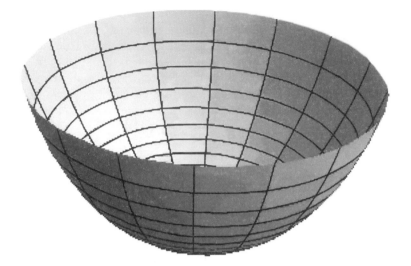

FIGURE 1.6. The relativistic geometry of a rotating plane

This expression shows that the curvature is small (the surface is nearly flat) if c is very large. After we discuss curvature in Chapter 5, we will recognize that this surface has positive curvature, just from its convexity. If the speed of light were infinite, this surface would be a plane. Likewise, if T is very large (the rotation is very slow), then the curvature is also small. When $T = \infty$, the space is not rotating at all, and the equation of the surface is simply $z = 0$.

From Ptolemy's point of view, each star traverses a circle with period T equal to one sidereal day (approximately 86,164 sec). Considering a near star, traversing a circle of radius 4 light years, which gives $R = 4 \times 366.2422cT$, or $R = 3.76158 \times 10^{16}$ m, one finds the circumference of its orbit contracted to length

$$C = 2\pi r \approx cT\Big(1 - \frac{1}{128\pi^2(366.2422)^2}\Big) \approx 0.99999999409861cT \,.$$

The speed u of the star is less than the speed of light by just $5.90138 \times 10^{-9}\, c$, that is, 1.769 m/sec.

As we shall see in the next chapter, the relativistic mass of this star is about 9200 times its rest mass.

REMARK 1.10. We can express the function z as an elliptic integral of second kind. The substitution $s = (1/2)(3 - \sqrt{3}\tan\theta)$ converts the integral

$$\int_0^u \sqrt{\frac{s^2 - 3s + 3}{(1-s)^3}}\, ds$$

into

$$\frac{3}{\sqrt{2}} \int_{\theta_0}^{\frac{\pi}{3}} \sqrt{\frac{\sec^3\theta}{(\sqrt{3}\sin\theta - \cos\theta)^3}}\, d\theta \,,$$

where $\theta_0 = \arctan\big(3 - 2u\big)/\sqrt{3}$. For $0 \le u \le 1$ we have $\pi/6 \le \theta_0 \le \pi/3$.

Using the trigonometric identity

$$\sqrt{3}\sin\theta - \cos\theta = 2\left(\frac{\sqrt{3}}{2}\sin\theta - \frac{1}{2}\cos\theta\right)$$
$$= 2\left(\cos\frac{\pi}{6}\sin\theta - \sin\frac{\pi}{6}\cos\theta\right)$$
$$= 2\sin\left(\theta - \frac{\pi}{6}\right),$$

we now change this integral into

$$\frac{3}{4}\int_{\theta_0}^{\frac{\pi}{3}}\frac{1}{\sqrt{\left(\cos\theta\sin(\theta - \pi/6)\right)^3}}\,d\theta.$$

Now we can use the trigonometric identity

$$\sin a\cos b = \frac{1}{2}\left(\sin(a+b) + \sin(a-b)\right)$$

to write the integral as

$$\frac{3}{\sqrt{2}}\int_{\theta_0}^{\frac{\pi}{3}}\frac{1}{\sqrt{(\sin\left(2\theta - \pi/6\right) - \sin(\pi/6))^3}}\,d\theta.$$

The substitution $\psi = 2\theta - \pi/6$, together with the fact that $\sin\frac{\pi}{6} = 1/2$, then converts the integral to

$$\frac{3}{2\sqrt{2}}\int_{\psi_0}^{\frac{\pi}{2}}\frac{1}{\sqrt{(\sin\psi - 1/2)^3}}\,d\psi =,$$

where $\psi_0 = 2\theta_0 - \pi/6$.

It is more convenient to integrate in the opposite direction, which we can do by replacing ψ by $\varphi = \pi/2 - \psi$. Letting $\varphi_0 = \pi/2 - \psi_0$, we then have

$$\frac{3}{2\sqrt{2}}\int_0^{\varphi_0}\frac{1}{\sqrt{(\cos\varphi - 1/2)^3}}\,d\varphi.$$

Next, we use the identity $\cos\varphi = 1 - 2\sin^2(\varphi/2) = 1 - 2\sin^2\eta$ to write this as

$$\frac{3}{\sqrt{2}}\int_0^{\eta_0}\frac{1}{\sqrt{(1/2 - 2\sin^2\eta)^3}}\,d\eta.$$

Finally, factoring out $2^{-3/2}$ from the denominator, we get

$$6\int_0^{\eta_0}\frac{1}{\sqrt{(1 - 4\sin^2\eta)^3}}\,d\eta.$$

If you ask *Mathematica* to evaluate this last integral, you will be told that it is

$$\text{ConditionalExpression}\left[6\left(-\frac{1}{3}\text{EllipticE}\left[\eta_0, 4\right] + \frac{2\sin\left[2\eta_0\right]}{3\sqrt{-1 + 2\cos\left[2\eta_0\right]}}\right),\right.$$

$$\left.\cos\left[2\eta_0\right] \geq \frac{1}{2}\right].$$

Here, EllipticE $[x,m]$ is the notation *Mathematica* uses for the *elliptic integral of second kind*

$$\int_0^x \sqrt{1 - m \sin^2 t}\, dt\,.$$

The condition imposed on the angle η_0 simply says that $0 \le \eta_0 \le \pi/6$, and this is indeed the case.

This is the first of several times when we shall encounter elliptic functions. They deserve to be better appreciated than they generally are by most physicists and mathematicians. They give exact expressions for solutions to some common differential equations of mathematical physics, solutions that otherwise have to be described qualitatively by inserting corrective terms into expressions involving elementary functions.

14. Problems

PROBLEM 1.1. Solve the equations of the Lorentz transformation for t, x, y, and z in terms of t', x', y', and z', and show that the solution is the same transformation with u replaced by $-u$ (which makes no change in α).

PROBLEM 1.2. Revisit the problem of the twin paradox by imagining that Mary has a telescope trained on the Earth, so that she can constantly observe John's clock. What would she see? Why is it that this clock shows a later time than Mary's own clock when the twins meet at the end of the journey?

PROBLEM 1.3. Consider the vector formulation of the Lorentz transformation given by the mutually inverse relations

$$(t'; \boldsymbol{x}') = \left(\alpha\left(t - \frac{\boldsymbol{u} \cdot \boldsymbol{x}}{c^2}\right); \boldsymbol{x} + \left((\alpha - 1)\frac{\boldsymbol{x} \cdot \boldsymbol{u}}{\boldsymbol{u} \cdot \boldsymbol{u}} - \alpha t\right)\boldsymbol{u} \right)$$

and

$$(t, \boldsymbol{x}) = \left(\alpha\left(t' + \frac{\boldsymbol{u} \cdot \boldsymbol{x}'}{c^2}\right); \boldsymbol{x}' + \left((\alpha - 1)\frac{\boldsymbol{x}' \cdot \boldsymbol{u}}{\boldsymbol{u} \cdot \boldsymbol{u}} + \alpha t'\right)\boldsymbol{u} \right).$$

Verify that these relations really are inverses of each other by inserting the values of $\boldsymbol{x}$ and t from the second relation into the right-hand side of the first relation.

PROBLEM 1.4. Show that the observers O and O' agree about the "space-time metric," that is, show that $(ct^2) - x^2 - y^2 - z^2 = (ct')^2 - x'^2 - y'^2 - z'^2$.

PROBLEM 1.5. Let a and b be dimensionless positive constants. Show that the mutually perpendicular lines $bx = ay$, $z = 0$ and $by = -ax$, $z = 0$ observed by O make the nonobtuse angle $\arccos\left(ab(\alpha^2 - 1)/\sqrt{(a^2\alpha^2 + b^2)(b^2\alpha^2 + a^2)}\right)$ when observed by O' at any given instant s. Show that this is a right angle only if $u = 0$, and that it tends to $0°$ as $u \uparrow c$.

PROBLEM 1.6. Translate the equation of the unit circle $x^2 + y^2 = R^2$, as seen by O, into O''s coordinate system. What kind of curve does this equation represent? How does the shape depend on time?

PROBLEM 1.7. Translate the equation of a general conic section $Ax^2 + 2Bxy + Cy^2 + Dx + Ey + F = 0$, observed by O, into the coordinate system used by O', getting an equation $A'x'^2 + 2B'x'y' + C'y'^2 + D'x' + E'y' + F' = 0$. Show that the discriminant $\Delta = B^2 - AC$ becomes $\Delta' = (B')^2 - A'C' = \alpha^2(B^2 - AC) = \alpha^2\Delta$. In particular, an ellipse ($\Delta < 0$) remains an ellipse (although, as Problem 1.6 shows, a circle may become a more general ellipse), a parabola ($\Delta = 0$) remains a parabola, and a hyperbola ($\Delta > 0$) remains a hyperbola.

PROBLEM 1.8. Consider the mapping $f(u)$ from $(-c, c)$ onto $(-\infty, \infty)$ given by
$$x = f(u) = \log\left(\frac{c+u}{c-u}\right).$$
(The base of the logarithm is not important here. It may be any positive number except 1.)

Find the inverse of the mapping $f(u)$. Also prove that for the Lorentz addition of velocities $u * v = (u + v)/(1 + uv/c^2)$,
$$f(u * v) = f(u) + f(v).$$

Thus, the group of relativistic velocities in one dimension is isomorphic to the additive group of real numbers. (That is no surprise, since, up to isomorphism, this is the only noncompact connected one-dimensional real Lie group that exists.)

PROBLEM 1.9. Suppose that O' observes O and O'' moving away at constant speeds u and v along lines making angle η. Let w be the speed with which O and O'' are moving apart relative to each other, and let
$$\alpha = \frac{c}{\sqrt{c^2 - u^2}}; \quad \beta = \frac{c}{\sqrt{c^2 - v^2}}; \quad \gamma = \frac{c}{\sqrt{c^2 - w^2}}.$$
Show that
$$\gamma = \alpha\beta\left(1 - \frac{uv\cos\eta}{c^2}\right).$$
Thus, the angles of a relativistic velocity triangle are determined by its sides u, v, and w via the formula
$$\cos\eta = \frac{c^2}{uv}\left(1 - \frac{\gamma}{\alpha\beta}\right),$$
for the angle opposite side of length w and the analogous formulas for the other two angles.

PROBLEM 1.10. Show that if ξ and η and v and w are interchanged, then Eqs. (1.5), (1.6), and (1.7) remain valid. Likewise, if ζ and η and u and w are interchanged, then Eqs. (1.5), (1.8), and (1.9) remain valid. It follows from Eqs. (1.7) and (1.9) that
$$\frac{\beta v}{\alpha u} = \frac{v\sqrt{1 - u^2/c^2}}{u\sqrt{1 - v^2/c^2}} = \frac{\sin\xi}{\sin\zeta}.$$
This relation is the law of sines for a relativistic velocity triangle. Formally, it is the ordinary law of sines applied to a triangle in which the sides opposite the two angles are shrunk, each by the FitzGerald–Lorentz contraction factor that would be measured by the observer corresponding to the opposite vertex.

PROBLEM 1.11. If $\eta = 0$ and $u = v$, then $w = |\boldsymbol{u} - \boldsymbol{v}| = 0$. The expression given by Eq. (1.5) for w shows that it is very unlikely that w^2 can ever equal $|\boldsymbol{u} - \boldsymbol{v}|^2 = u^2 - 2uv\cos\eta + v^2$ in any other case. Because of the denominator $\left(1 - uv\cos\eta/c^2\right)^2$, this certainly cannot happen unless η is an acute angle. Show that this can nevertheless occur at any speed in the case of an "isosceles" triangle corresponding to $u = v$, given a suitable vertex angle. To do so, assume $u = v$ and show that the equation $w = |\boldsymbol{u}-\boldsymbol{v}|$ leads to the quadratic equation $2a^2x^2-3x+1 = 0$ for the unknown $x = \cos\eta$, where a ($0 \leq a < 1$) is a dimensionless constant, namely $a = u/c = v/c$. The solutions of this equation are

$$x = \frac{3 \pm \sqrt{9 - 8a^2}}{4a^2} = \frac{2}{3 \mp \sqrt{9 - 8a^2}}.$$

(You will actually get a cubic equation from which the trivial factor $x - 1$ can be divided out.) Show that the positive sign in the first expression for x (corresponding to the negative sign in the second one) is consistent with the relation $x \in [0, 1]$ only when $a = 1$, which is the case when $u = v = c$. In this case, $x = 1$, that is, this is the case $\eta = 0$, which, as we have already remarked, is trivial. Then, for the negative sign on the square root in the numerator, show that x lies in the range $[1/3, 1/2]$ for all values of $a \in [0, 1]$.

Verify the example mentioned in the text, in which $a = 3\sqrt{2}/5$ and $\cos(\eta) = 5/12$, showing that

$$\eta \approx 67.056553501352011261°$$

and that the relative speed of O and O'' is $0.84c$

PROBLEM 1.12. It is well-known that any three lengths u, v, w with $u \leq v \leq w$ are the sides of a Euclidean triangle provided $u+v > w$. (The philosopher Immanuel Kant cited this fact as an example of what he called *synthetic a priori knowledge*.) Is this true for relativistic velocity triangles? If not, what additional conditions are needed?

PROBLEM 1.13. Show that the sum of the angles of a relativistic velocity triangle is smaller than two right angles.

PROBLEM 1.14. This problem has four parts. We define the *angle defect* of a relativistic velocity triangle whose angles are ξ, η, and ζ to be the positive number $\pi - (\xi+\eta+\zeta)$. Consider a triangle with these angles and divide it into two smaller triangles by drawing a line from the vertex a angle η to a point on the opposite side, thereby dividing the triangle into two smaller triangles, one having angles η_1 (part of angle η), ξ, and φ_1 (at the vertex on the side opposite the angle η), and the other having angles η_2, ζ, and $\varphi_2 = \pi - \varphi_1$.

Part 1: Show that the defect of the original triangle is the sum of the defects of the two triangles into which it is divided. (It is not difficult to prove—although you are not being asked to do so—that when a triangle is partitioned into any number of other triangles, its defect is the sum of the defects of the triangles that partition it.) Thus the defect of a triangle is proportional to what we think of as the area of a triangle, and so we shall define the area of a triangle to be c^2 times its defect. We then define the area of a polygon to be the sum of the areas of any set of triangles into which it can be partitioned. It is not difficult to show that this definition is independent of the way in which the polygon is triangulated.

Part 2: Consider an isosceles relativistic velocity triangle having two equal sides of length u with angle η between them, and let the other two angles both be equal to

ξ and the third side equal to w. Show that

$$\cos \xi = \frac{\sin \frac{\eta}{2}}{\sqrt{1 - \frac{u^2}{c^2} \cos^2 \frac{\eta}{2}}},$$

$$\sin \xi = \frac{\cos \frac{\eta}{2}}{\alpha \sqrt{1 - \frac{u^2}{c^2} \cos^2 \frac{\eta}{2}}},$$

$$w = \frac{2u \sqrt{1 - \frac{u^2}{c^2} \cos \frac{\eta}{2}}}{1 - \frac{u^2}{c^2} \cos \eta} \cdot \sin \frac{\eta}{2}.$$

Part 3: Consider a regular polygon P_n consisting of n isosceles triangles having vertex angle $2\pi/n$ glued together along their equal sides, which all have length u. Show that its perimeter $\pi(P_n)$ and its area $A(p_n)$, which is c^2 times its angle defect, are given by

$$\pi(P_n) = \frac{2u \sqrt{1 - \frac{u^2}{c^2} \cos^2 \frac{\pi}{n}}}{1 - \frac{u^2}{c^2} \cos \frac{2\pi}{n}} \cdot \left(n \sin \frac{\pi}{n} \right),$$

$$A(P_n) = \left((n-2)\pi - 2n \arccos \left(\frac{\sin \frac{\pi}{n}}{\sqrt{1 - \frac{u^2}{c^2} \cos^2 \frac{\pi}{n}}} \right) \right) c^2.$$

Part 4: Using the relationship

$$\lim_{\eta \to 0} \frac{\sin \eta}{\eta} = 1,$$

show that

$$\lim_{n \to \infty} \pi(P_n) = 2\pi \alpha u = 2\pi u + \frac{\pi u^3}{c^2} + \frac{3\pi u^5}{8c^4} + \cdots,$$

$$\lim_{n \to \infty} A(P_n) = 2\pi c^2 (\alpha - 1) = \pi \left(u^2 + \frac{3u^4}{8c^2} + \cdots \right).$$

Thus, for small velocities u the circumference of a circle of radius u is asymptotic to the value it would have it u were a length in Euclidean space, and the same is true of the area. Equality holds if $c = \infty$.

PROBLEM 1.15. Show that the sides u, v, and w of a relativistic velocity triangle given in terms of its angles are all less than c.

PROBLEM 1.16. Compute the relative speed w of O and O'', given that O' has speeds $u = 4c/5$ and $v = 3c/5$ relative to them and measures the angle between their trajectories as $3\pi/4$.

PROBLEM 1.17. Let ξ, η, and ζ be three positive angles the sum of whose measures is less than π. Show that the expressions

$$\cos^2 \xi + \cos^2 \eta + \cos^2 \zeta + 2 \cos \xi \cos \eta \cos \zeta - 1$$

and

$$\cos \xi + \cos \eta \cos \zeta$$

are both positive, and hence that the formulas given in the text for the sides of a relativistic velocity triangle having these angles are valid.

PROBLEM 1.18. Suppose $(\theta, u, \varphi) \ +_L \ (\chi, v, \psi) = (\mu, w, \nu)$. Prove that for any angles d, e, and f:

$$
\begin{aligned}
(d + \theta, u, \varphi) \ +_L \ (\chi, v, \psi) &= (d + \mu, w, \nu), \\
(\theta, u, e + \varphi) \ +_L \ (e + \chi, v, \psi) &= (\mu, w, \nu), \\
(\theta, u, \varphi) \ +_L \ (\chi, v, \psi + f) &= (\mu, w, \nu + f).
\end{aligned}
$$

Different operations are being applied to the two addends in each of these cases. Show that if we take $d = e = f$ and combine these results, we obtain a mapping $T_d(\theta, u, \varphi) = (d + \theta, u, \varphi + d)$ that satisfies the equality

$$
T_d(\theta, u, \varphi) \ +_L \ T_d(\chi, v, \psi) = T_d\big((\theta, u, \varphi) \ +_L \ (\chi, v, \psi)\big).
$$

Thus Lorentz addition is invariant under each operation T_d. (*Caution:* This result does not enable us to reduce the dimension of the three-dimensional space we have invented to describe the addition. We cannot, for example, replace (θ, u, φ) by $(0, u, \varphi - \theta)$ and (χ, v, ψ) by $(0, v, \psi - \chi)$, even though $(\theta, u, \varphi) = T_\theta(0, u, \varphi - \theta)$ and $T_\chi(0, v, \psi - \chi) = (\chi, v, \psi)$. The difficulty is that T_θ is not the same operator as T_χ.

PROBLEM 1.19. Suppressing the second and third spatial dimensions, we focus attention on just the time and first spatial axes. The Lorentz transformation is

$$
\begin{aligned}
\tau' &= \alpha\Big(\tau - \frac{u}{c}x\Big), \\
x' &= \alpha\Big(-\frac{u}{c}\tau + x\Big).
\end{aligned}
$$

For a reason that will become clear in a moment, let the angle θ be

$$
\theta = \arccos\left(\sqrt{1 - \frac{u^2}{c^2}}\right) = \arccos\left(\frac{1}{\alpha}\right).
$$

Thus, $\alpha = \sec\theta$.

Solve the second equation of the Lorentz equations for x in terms of x' and τ (and θ), then substitute the result in the first equation, so that τ' and x are expressed in terms of τ and x', yielding

$$
\begin{aligned}
\tau &= (\cos\theta)\tau' + (\sin\theta)x, \\
x' &= -(\sin\theta)\tau' + (\cos\theta)x.
\end{aligned}
$$

Use these equations to solve the car wash puzzle.

PROBLEM 1.20. Here is a variation on the car wash puzzle. Suppose that as the limousine moved through the car wash with speed u, two car wash attendants simultaneously, *as measured by a clock in the car wash*, put scratches in it, one in the front fender, the other in the rear fender. Suppose that they were standing 3 meters apart, *as measured by the car wash attendants themselves*, when they made the scratches. If the limousine is then stopped and measured by the car wash attendants, how far apart will the scratches be?

PROBLEM 1.21. Verify that the space-time interval ds^2 between the two events—the rear of the limousine entering the car wash and the front of it leaving the car wash—is negative (spacelike).

PROBLEM 1.22. Consider the special case of a relativistic velocity triangle when u and v lie along perpendicular directions. For this case, we have $\gamma = \alpha\beta$. Recall, as noted above, that α corresponds to the cosine of the angle of rotation that describes the Lorentz transformation between O and O' when they interchange time coordinates, and likewise β is the cosine of the corresponding angle for the transformation between O' and O'' and γ the angle corresponding to the transformation between O and O''. Show that, when they are regarded as arcs on a sphere, the three angles that provide these geometric representations are the sides of a spherical right triangle in this case.

CHAPTER 2

Relativistic Mechanics

Once we abandon the intuitive absolute space and time of Newtonian mechanics, everything in classical physics becomes "negotiable." A thorough revamping of the entire subject is required, each of its fundamental concepts receiving a new definition. If we assume Lorentz transformations between two observers in motion at constant speed, for example, what becomes of acceleration and force, two fundamental concepts in mechanics? In order to answer that question, we first ask how two observers in uniform relative motion—for notational convenience we shall call them X and Y—reconcile their measurements of the velocity of a particle. Each of them will assign it a velocity vector at each particular "time," but, as we now realize, we need both the time and the particle's location at that time before we can reconcile the two observers' bookkeeping. Our discussion of these topics at first adheres as closely as possible to the Newtonian quantities that would be measured by each individual observer. The proper definition of each of these concepts in more formal special relativity differs in that it treats them as points in four-dimensional space-time and uses proper time rather than observer time to define the basic quantities involved in mechanics. We shall use the familiar spatial three-vectors as much as possible and discuss these interesting four-vectors only in the final section of this chapter.

By studying the motion of a particle, we can get a specific mapping between the two observers' time axes $r \leftrightarrow s$. The two times assigned by the observers to any event involving the particle, such as its collision with another particle, for example, are placed in correspondence with each other. The definitions of the relativistic kinematics of a particle—its position, velocity, and acceleration, are straightforward, except for the complication in reconciling the measurements of these quantities between two observers in relative motion; the details of that operation are consequences of the Lorentz transformation introduced in the previous chapter.

1. The Kinematics of a Particle

Of the three independent physical dimensions that make up classical mechanics—distance, time, and mass—we are concerned only with the first two in this section. Those are the two that get entangled with each other via the Lorentz transformation when we pass from the classical case to special relativity. A particle in motion is idealized as a point $\boldsymbol{x} = x^1(t)\boldsymbol{i} + x^2(t)\boldsymbol{j} + x^3(t)\boldsymbol{k}$ in $\mathbb{R}^3$ whose coordinates are functions of time. So far as each individual observer is concerned, this motion is in no way different from the motion studied in Newtonian mechanics: Velocity and acceleration remain the first and second derivatives of position with respect to time. The differences with the Newtonian principles arise in two situations: (1) when two observers in uniform relative motion compare their observations of velocity and acceleration; (2) when the mass of the particle is taken into consideration. We begin with the first of these.

Suppose that X is using $(r; \boldsymbol{x})$ as space-time coordinates, and from that perspective the motion of the particle is given by a vector-valued function $\boldsymbol{x}(r) = x^1(r)\boldsymbol{i} + x^2(r)\boldsymbol{j} + x^3(r)\boldsymbol{k}$. We shall say that X records a *world-line* for the particle, namely the set of points $(r; \boldsymbol{x})$ in $\mathbb{R}^4$ satisfying

$$\boldsymbol{x} = \boldsymbol{x}(r).$$

Similarly, let us suppose Y is using $(s; \boldsymbol{y})$ as space-time coordinates and that Y records the world-line of the particle as

$$\boldsymbol{y} = \boldsymbol{y}(s),$$

where $\boldsymbol{y}(s) = y^1(s)\boldsymbol{i} + y^2(s)\boldsymbol{j} + y^3(s)\boldsymbol{k}$. We first see how the two observers' measurements of velocity and acceleration are to be reconciled.

1.1. Relativistic velocity of a particle. According to the Lorentz transformation, if Y's velocity relative to X is $\boldsymbol{u} = u\boldsymbol{i}$ and X's velocity relative to Y is $-\boldsymbol{u} = -u\boldsymbol{i}$, we have

$$
\begin{aligned}
s &= \alpha\left(r - \frac{ux^1(r)}{c^2}\right), \\
y^1(s) &= \alpha\left(-ur + x^1(r)\right), \\
y^2(s) &= x^2(r), \\
y^3(s) &= x^3(r).
\end{aligned}
$$

It follows from these equations that, on this world-line,

$$\frac{ds}{dr} = \alpha\left(1 - u(x^1)'(r)/c^2\right) = \alpha\delta,$$

where $\delta = 1 - u(x^1)'/c^2$. We are using primes to denote differentiation of the coordinates in each system *with respect to the time coordinate of that system*, in other words, $(x^1)' = dx^1/dr$ and $(y^1)' = dy^1/ds$.

Similarly,

$$\frac{1}{\alpha\delta} = \frac{dr}{ds} = \alpha\left(1 + \frac{u(y^1)'(s)}{c^2}\right) = \alpha\eta,$$

where $\eta = 1 + \frac{u(y^1)'}{c^2}$. These equations imply that $\alpha^2\delta\eta = 1$, since $\frac{dr}{ds}\frac{ds}{dr} = 1$. The equation $\alpha^2\delta\eta = 1$ can also be verified directly.

If we differentiate both sides of the three spatial coordinate equations with respect to s, we find the following relation between the velocity observed by X and the velocity observed by Y.

THEOREM 2.1. *If X and Y are using a common first spatial axis and the other two pairs of corresponding axes are parallel at all times, then the components of velocity observed by Y are given in terms of those observed by X as follows:*

$$(2.1) \qquad (y^1)'(s) = \frac{1}{\alpha\delta}\alpha\left(-u + (x^1)'(r)\right) = \frac{1}{\delta}\left(-u + (x^1)'(r)\right),$$

$$(2.2) \qquad (y^2)'(s) = \frac{1}{\alpha\delta}(x^2)'(r),$$

$$(2.3) \qquad (y^3)'(s) = \frac{1}{\alpha\delta}(x^3)'(r).$$

The inverse set of equations is

$$
\begin{aligned}
(x^1)'(r) &= \frac{1}{\eta}\big(u + (y^1)'(s)\big), \\[2mm]
(x^2)'(r) &= \frac{1}{\alpha\eta}(y^2)'(s), \\[2mm]
(x^3)'(r) &= \frac{1}{\alpha\eta}(y^3)'(s).
\end{aligned}
$$

PROOF. This is an immediate and trivial computation.   □

It is worth noting that the term $(-u + (x^1)')/(1 - u(x^1)'/c^2)$, which is the part of the right-hand side of Eq. (2.1) that remains after canceling α, is just the relativistic addition of the two collinear velocities $-u$ and $(x^1)'$. Also, the velocity observed by Y is determined by the velocity observed by X and is *independent of the location of the particle.*

THEOREM 2.2. *Under the same assumptions as in Theorem 2.1, the components of acceleration observed by Y are given in terms of those observed by X via the equations*

$$
\begin{aligned}
(y^1)'' &= \frac{(x^1)''}{\alpha^3\delta^3}, \\[2mm]
(y^2)'' &= \frac{(x^2)'' - \frac{u}{c^2}((x^1)'(x^2)'' - (x^1)''(x^2)')}{\alpha^2\delta^3}, \\[2mm]
(y^3)'' &= \frac{(x^3)'' - \frac{u}{c^2}((x^1)'(x^3)'' - (x^1)''(x^3)')}{\alpha^2\delta^3},
\end{aligned}
$$

where the left-hand side is to be evaluated at s and the right-hand side at r.

The inverse equations are

$$
\begin{aligned}
(x^1)'' &= \frac{(y^1)''}{\alpha^3\eta^3}, \\[2mm]
(x^2)'' &= \frac{(y^2)'' + \frac{u}{c^2}((y^1)'(y^2)'' - (y^1)''(y^2)')}{\alpha^2\eta^3}, \\[2mm]
(x^3)'' &= \frac{(y^3)'' + \frac{u}{c^2}((y^1)'(y^3)'' - (y^1)''(y^3)')}{\alpha^2\eta^3},
\end{aligned}
$$

where the left-hand side is evaluated at r and the right-hand side at s.

PROOF. We prove only the first set of equations. This proof is in fact a straightforward computation. All we need to do is differentiate Eqs. (2.1)–(2.3) with respect to Y's time coordinate s, keeping in mind the formula for dr/ds.   □

Here we see a strong departure of special-relativity kinematics from Galilean kinematics. In the latter, the two observers agree about the acceleration of any particle. But, as the equations show, in the case when one observer sees acceleration only along the line of mutual motion $((x^2)'' = 0$, the second observer will find a component of acceleration perpendicular to that line, and the resulting acceleration will be oblique to the line of mutual motion, as is noticeable in Fig. 2.1.

The differences between the classical transformation of velocity and the relativistic transformation are illustrated in Fig. 2.1. This figure shows three views of the trajectory of a particle: (a) by an observer X at rest relative to the Earth; (b)

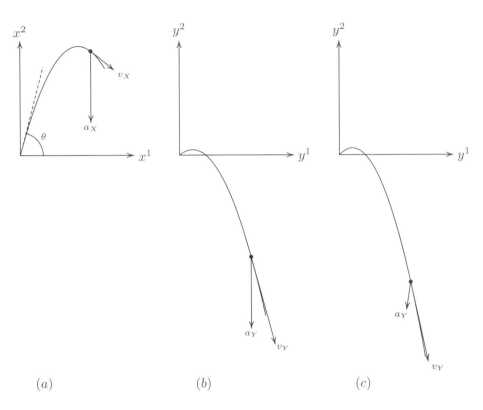

(a) (b) (c)

FIGURE 2.1. A classical projectile observed: (a) from rest; (b) by
an observer moving vertically, according to the Galilean conversion of coordinates; (c) by an observer moving at the same speed
vertically, with a relativistic conversion of coordinates

by a second observer Y moving vertically upwards at four-fifths the speed of light
but using the Galilean coordinate transformation; (c) by the same second observer
Y, using the relativistic transformation. Notice that the Galilean transformation
preserves the magnitude of the acceleration, while the relativistic transformation
shortens it noticeably.[1]

Knowing how to translate X's measurements into Y's measurements, we are
now ready to introduce the other basic concepts of mechanics in terms of X's
measurements alone. But we shall need Y even so, since we are going to be guided
by the principle that both observers need momentum to be conserved.

---

[1]This figure is not realistic, as the projectile is imagined to have been fired at an elevation of
75° with muzzle velocity three-fifths that of light. In order for this projectile to fall as shown in
the figure, the Earth's gravitational field would have to be nearly two million times stronger than
it actually is. We also scaled the velocities by a factor of 5 and the accelerations by a factor of
50. It is legitimate to do so, since these vectors "inhabit" a space that is different from the space
traversed by the projectile. In the spaces these vectors inhabit, the coordinates have physical
dimensions of length/time and length/time-squared respectively.

2. From Kinematics to Dynamics: Mass and Momentum

> *...though motion may be only a form in inert matter, that matter nevertheless has a definite quantity of it, which never increases or decreases, even though there may be more or less of it in various parts. This is why, when one piece of matter moves twice as fast as another and the other is twice as large as the first, we must consider that there is just as much motion in the smaller as in the larger; and whenever the motion of one piece decreases, that of some other piece increases in proportion... if a body has once begun to move, we must conclude that it continues to move, and that it will never stop of its own accord... each piece of matter tends to continue its motion in straight lines, never in curves...*

René Descartes ([**12**], p. 151), my translation.

> *The quantity of matter is the measure of the same, arising from its density and bulk conjointly... The quantity of motion is the measure of the same, arising from the velocity and quantity of matter conjointly.*

Isaac Newton ([**63**], p. 5).

The preceding section contained straightforward consequences of the Lorentz transformation. They were purely geometric, involving only time and space, and the physical concept of mass played no role. We now come to the point of transition from kinematics to dynamics, from *describing* motion (kinematics) to *explaining* it in terms of force. That transition point is the concept that we call *momentum* and what Descartes and Newton in the quotations above called "quantity of motion." Both of these authors defined the momentum of a body to be the product of its mass and velocity. (As will be noted in the next section, this definition is very close to being Aristotle's definition of *force*.) It is interesting that Descartes explicitly stated what we now call Newton's first law of motion (the law of inertia) exactly as we now understand it as well as the fundamental principle we know as conservation of momentum, and he did this when Newton was an infant.

We are presuming as background for this book that the reader has some experience working with momentum in simple problems of Newtonian mechanics. The language of vectors is ideal for this purpose, and we shall use it to get a three-vector version of momentum that applies in special relativity for two observers who are both using their line of mutual motion as a coordinate axis. In the discussion of kinematics given above, the complications caused by the entanglement of space and time (the Lorentz transformation) have required only a slight adjustment of our basic, Newtonian/Euclidean intuition. In the present section, these complications begin to multiply, as the entanglement spreads. Specifically, we shall find that mass is now entangled with speed. That complication forces a larger adjustment in our intuition; and when we introduce the notion of force, we shall find that our intuition has been stretched still farther and that we have a very strange-looking formula.

Since a very successful structure has been built on Newtonian principles, we naturally want to keep as many of the basic concepts as possible close to the Newtonian ones. These basic quantities are distance (length), time, and mass. More complex entities such as momentum, force, and energy are defined in terms of them. The important conservation laws (conservation of momentum, conservation of angular momentum, and conservation of energy) are pillars of mechanics, and our redefinition of mass will be aimed at keeping at least the first two of these laws.

We start with momentum. Imagine that Y sees two particles of identical mass m approaching from opposite directions, each being the same distance away and moving with the same speed v, so that Y ascribes to them velocities $\boldsymbol{v}$ and $-\boldsymbol{v}$. These particles are coated with "superglue," so that when they collide, directly in front of Y, they both come to a dead stop. As far as Y is concerned, both classically and relativistically, this is all fine. The momentum of the pair added up to zero algebraically both before and after the collision. The kinetic energy lost by the particles has to be explained, but let that pass for the moment.

The same reconciliation would be possible for X, whom we imagine as moving along with one of the particles, say the one whose speed is $-v$, provided we took the classical addition of velocities, whereby the other particle is moving with speed $2v$ relative to X. In classical terms the total momentum was $m(2v) + m(0) = 2mv$ before the collision. Afterwards it is $mv + mv = 2mv$, since the two particles remain at Y's origin, while X's origin continues to move with velocity $-v\boldsymbol{i}$.

Now let us look at the situation relativistically. Relative to X the speed of the other particle before the collision is $w = \frac{2v}{1+\frac{v^2}{c^2}}$. Hence the momentum before the collision would seem to be $\frac{2mv}{1+\frac{v^2}{c^2}}$, and afterward it would appear to be $2mv$, which is obviously a larger number than before.

One way out of this difficulty is to assume that a particle is more massive when it is moving than when it is at rest. Let us assign the value m_v to the particle when moving at speed v. We shall assume that both mass and momentum are conserved in this thought experiment. Then from X's point of view the total mass of the system before the collision is $m_0 + m_w$, and afterward it is $2m_v$, since X continues to move in the negative direction at speed v. These two expressions must be numerically equal. (Although mass is not constant in relativity, the mass of a system changes only when some external work is done on it to increase its total energy. No such work occurred in the collision just described.) The momentum before the collision was $m_w w$, and afterward it was $(m_0 + m_w)v$. Setting these expressions equal to each other, we get the equation

$$m_0 v = m_w(w - v).$$

Now, if we solve the equation between w and v for v, assuming $v < c$, we find

$$v = \frac{c^2}{w}\left(1 - \sqrt{1 - \frac{w^2}{c^2}}\right),$$

which—we note for reference below—is equivalent to the relation

$$\sqrt{1 - \frac{w^2}{c^2}} = 1 - \frac{vw}{c^2} = 1 - \frac{2v^2}{c^2 + v^2} = \frac{c^2 - v^2}{c^2 + v^2}.$$

Since $w = \frac{c^2}{w}\frac{w^2}{c^2}$, it follows that

$$w - v = \frac{c^2}{w}\left(\sqrt{1 - \frac{w^2}{c^2}} - \left(1 - \frac{w^2}{c^2}\right)\right) = v\sqrt{1 - \frac{w^2}{c^2}}.$$

This relation then yields immediately the following principle:

THEOREM 2.3. *If momentum is to be conserved for all observers in uniform relative motion, then the mass m_w of a particle moving at speed w must be*

$$(2.4) \qquad m_w = \frac{m_0}{\sqrt{1 - \frac{w^2}{c^2}}}.$$

This equation was derived as the only possible way of retaining conservation of momentum in relativity. The derivation does not prove that momentum will be conserved in all situations, however. Our redefinition of mass was necessary, but has not been proved sufficient, for that purpose. We are going to omit the proof of sufficiency, however, and content ourselves with a necessary condition that determines the equations we need to use.

There is one obvious danger of inconsistency here. The total mass before the collision, according to X, was $m_0 + m_w$. According to Y, the total mass before the collision was $2m_v$. Since Y is at rest relative to the system after the collision and mass is conserved when no external work is done on the system, its new rest mass will be $2m_v$, and X will ascribe to it the mass $2m_v/\sqrt{1 - v^2/c^2}$. But that number is precisely $m_0 + m_w$. For we have $m_0 + m_w = m_0\left(1 + 1/\sqrt{1 - w^2/c^2}\right)$. By the relation derived above, we thus get

$$m_0 + m_w = m_0\left(1 + \frac{c^2 + v^2}{c^2 - v^2}\right) = \frac{2m_0}{1 - v^2/c^2} = \frac{2m_v}{\sqrt{1 - v^2/c^2}}.$$

As far as we have explored, there are no obvious inconsistencies between the two observers if we adopt the formula in Eq. (2.4) as the definition of the mass of a particle of rest mass m_0 when it is moving with speed w. We can define the relativistic (three-)momentum of a particle as $m\boldsymbol{v}$, where $\boldsymbol{v}$ is the observed velocity of the particle and m its observed relativistic mass. One consequence of this fact is that the momentum and velocity of a particle are no longer directly proportional, as they are in Newtonian mechanics; for m is no longer a constant independent of $\boldsymbol{v}$. That difference will have further consequences for the concepts of force, work, and energy in relativity.

If we consider instead of a particle of mass m a mass density ρ, Eq. (2.4) has the following consequence:

THEOREM 2.4. *A solid body having density ρ_0 in a coordinate system with respect to which it is at rest has density*

$$(2.5) \qquad \rho = \alpha^2 \rho_0 = \frac{\rho_0 c^2}{c^2 - u^2}$$

when measured in a system moving with speed u relative to the given system.

PROOF. Consider a portion of the body having mass m_0 in a volume V_0 and thus average density m_0/V_0. When measured in the moving coordinate system m_0 will become $m = \alpha m_0$, and by Theorem 1.2 of Chapter 1, the volume will become $V = V_0/\alpha$. □

We demonstrated above that the mass of a particle measured by an observer depends on its speed relative to that observer. In that connection, we introduce the quantities

$$\beta = \frac{1}{\sqrt{1 - \frac{((y^1)')^2 + ((y^2)')^2 + ((y^3)')^2}{c^2}}} \; ; \; \gamma = \frac{1}{\sqrt{1 - \frac{((x^1)')^2 + ((x^2)')^2 + ((x^3)')^2}{c^2}}} .$$

The equations $\gamma = \alpha\beta\eta$ and $\beta = \alpha\gamma\delta$ can be computed directly, as was done in the preceding chapter when we computed the composition of two Lorentz transformations. (Imagine a third observer Z "riding" the particle. The roles of α, β, and γ are then exactly as in Chapter 1.)

Notice that, since we are not dealing with a constant particle velocity here, the quantities β and γ are not constant. In fact,

$$\beta' = \frac{d\beta}{ds} = \frac{\beta^3}{c^2}(\boldsymbol{a}_y \cdot \boldsymbol{v}_y)\,,$$

where $\boldsymbol{v}_y = \big((y^1)'(s),(y^2)'(s),(y^3)'(s)\big)$ is the velocity of the particle measured by Y and $\boldsymbol{a}_y = d\boldsymbol{v}_y/ds$ the corresponding acceleration, the dot product being taken in Y's coordinates. We shall similarly write $\boldsymbol{v}_x$ and $\boldsymbol{a}_x$ for the velocity and acceleration measured by X, in which case we have

$$\gamma' = \frac{d\gamma}{dr} = \frac{\gamma^3}{c^2}(\boldsymbol{a}_x \cdot \boldsymbol{v}_x)\,.$$

THEOREM 2.5. *Two observers in motion at relative speed u along their common first spatial axis assign components to the momentum $\boldsymbol{p}$ of the particle that have the following relations to each other:*

$$p_{x1} = m_0\gamma(x^1)' = m_0\gamma\frac{(y^1)' + u}{\eta} = \alpha m_0\beta((y^1)' + u) = \alpha p_{y1} + \alpha m_0\beta u\,,$$

$$p_{x2} = m_0\gamma(x^2)' = m_0\beta(y^2)' = p_{y2}\,,$$

$$p_{x3} = m_0\gamma(x^3)' = m_0\beta(y^3)' = p_{y3}\,;$$

$$p_{y1} = m_0\beta(y^1)' = m_0\beta\frac{(x^1)' - u}{\delta} = \alpha m_0\beta((x^1)' - u) = \alpha p_{x1} - \alpha m_0\gamma u\,,$$

$$p_{y2} = m_0\beta(y^2)' = m_0\gamma(x^2)' = p_{x2}\,,$$

$$p_{y3} = m_0\beta(y^3)' = m_0\gamma(x^3)' = p_{x3}\,.$$

These relations are shown in Fig. 2.2. In that figure Observer X sees a particle of unit mass moving around a circle of radius one light-second at a linear speed of $\pi/4$ light-seconds per second (one revolution every eight seconds). Observer Y, moving as usual along the x^1/y^1-axis at constant linear speed equal to three-fifths of light speed, observes the particle doing a loop-the-loop. After 3.7 seconds have elapsed at the particle on a clock synchronized with X's clock (5.35428 seconds on a clock at the particle that is synchronized with Y's clock), the particle is at the point marked by a bullet in the figure. Observer Y perceives this event as occurring later (thinks X's clock is slow). The momenta observed by the two observers are shown as solid arrows (the longer one in the case of the double arrow on the left). For comparison, the trajectory Y would observe using Newtonian physics is shown as a dashed line and the corresponding momentum as a dashed arrow on the right.

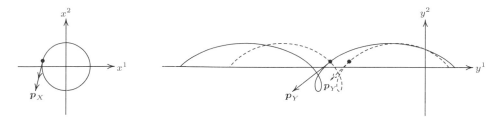

FIGURE 2.2. Left: a particle in uniform circular motion observed by X. The double arrow shows the classical (smaller) and relativistic (larger) momenta. Right (solid curve): the trajectory of the particle as observed by Y, moving at three-fifths the speed of light relative to X; the solid arrow shows its relativistic momentum at the same event where this momentum is shown on the left from X's point of view. Right (dashed curve): the trajectory Y would observe in Newtonian mechanics; the dashed arrow shows its classical momentum at the same point in time as shown on the left.

3. Relativistic Force

If A is the mover and B the thing moved, which has been moved a certain distance Γ in a certain time Δ, then in the same amount of time Δ a force (δύναμις, dynamis) equal to A will move half of B twice the distance Γ or move it the same distance Γ in half of the time Δ. Thus we have a direct proportion.

Aristotle, *Physics*, Book VII. My translation.

Although it is generally risky to "modernize" the words of scientists[2] who lived long ago, we might think of the "thing moved" as either mass or weight. Aristotle would not have had the concept of mass; but what we think of as weight, that is, the force required to lift an object against gravity, would have made sense to him. With that interpretation, Aristotle is saying that the ratio that a force applied to an object bears to the weight of that object is directly proportional to the distance the object moves and inversely proportional to the time required to move that distance. In our terms the "constant of proportionality" (another anachronism in the world of ancient Greek science) must be the reciprocal of a velocity. If, on the other hand, we forge ahead with our modernized interpretation of "the thing moved" as a mass, it appears that Aristotle thought of force as the cause of what we call *momentum*. The quotation shows that he thought that forces were measured by numbers and could be added and subtracted. He does not seem to take account of any *direction* that a force may have. Having no precise definition of instantaneous velocity, he really could handle only mechanical problems where direct proportion could be applied. A quantitative understanding of even the simplest case of accelerated motion—where the acceleration is constant—had to await the scholars of Merton College, Oxford, in the thirteenth century. Even so, Aristotle's use of the term *force* is closer to the everyday *nonscientific* meaning we still give it today. Like

---

[2]Indeed, the very concept of a "scientist" is anachronistic when applied to ancient times.

Aristotle, we think it is necessary to apply a continuous force in order to keep a body moving at a constant rate of speed. Every bicyclist feels this way, even when bicycling on the level, since it is necessary to exert force on the pedals to overcome mechanical friction and wind resistance. Similarly, dragging anything along the ground requires a continuous force to keep overcoming friction.

In mechanics, since the time of Descartes and Newton, it has been found more useful to define force as whatever causes a change in momentum. An Aristotelian push or pull will do that, either speeding up a moving body or slowing it down, or changing the direction in which it moves. The skill of driving an automobile, for example, involves knowing when to do these three things by pressing down on the accelerator or brake or turning the steering wheel. These three actions activate motors (the engine, the power brakes, or the power steering) that cause a suitable change in the momentum of the vehicle.

The idealized situations in Newtonian mechanics, where things are frictionless and the law of inertia really applies, are remote from everyday experience. Classical mechanics relies on a list of standard forces, such as gravity, the elastic force of a spring, electromagnetism, and others. All of these forces are intuitive, and the reason they are is that we picture a force as being a push or pull. Our intuition is based on this experience of our muscles. And that is where special relativity leaves us behind, as we are about to see. Relativistic force is highly *non*-intuitive, and we must start over from the beginning to compile a list of relativistic forces to replace the Newtonian ones. Right now, having defined only momentum, we can guess that complications lie ahead, since relativistic momentum is not directly proportional to velocity, as it is in the classical case. The fact that mass increases with velocity means that a force, in addition to causing a change in speed or direction, can also cause a change in mass. That is a feat it cannot achieve in classical mechanics.

We take the relativistic definition of the momentum of a body of mass m moving with velocity $\boldsymbol{v}$ to be $\boldsymbol{p} = m\boldsymbol{v}$, it being understood that this m is really the relativistic mass that we called m_v above, that is,

$$\boldsymbol{p} = \frac{m_0}{\sqrt{1 - \frac{\boldsymbol{v} \cdot \boldsymbol{v}}{c^2}}} \boldsymbol{v} \, .$$

This is a genuine, coordinate-free vector equation, since both sides are vectors used by the same observer. Such will be the case with all the vectors that arise in the present section, except when we compare the expressions for these vectors used by two observers, at which point we shall once again revert to coordinate-wise notation.

In Newtonian mechanics, since momentum is directly proportional to velocity, it makes no difference whether we write Newton's second law as $\boldsymbol{F} = m\boldsymbol{a} = m\boldsymbol{r}''$ or $\boldsymbol{F} = \boldsymbol{p}'$. But these two definitions are not equivalent in special relativity, and we must choose between them. We choose the equation $\boldsymbol{F} = \boldsymbol{p}'$ as the definition of force. Although the equation is formally the same as the classical definition of force, there are two important differences: (1) the mass depends on the speed, and (2) consequently there is an extra term in the expression for the force. In particular, *the acceleration produced by a force is not generally parallel to the force.*

$$\boldsymbol{F} = m\frac{d\boldsymbol{v}}{dt} + \frac{dm}{dt}\boldsymbol{v} = m\boldsymbol{a} + m'\boldsymbol{v}.$$

For our purposes, we need to rearrange this last equation slightly, and to that end we compute m' explicitly, using the chain rule:

$$m' = m_0\Big(-\frac{1}{2}\Big)\Big(1 - \frac{\boldsymbol{v}\cdot\boldsymbol{v}}{c^2}\Big)^{-\frac{3}{2}}\Big(\frac{-2\boldsymbol{v}\cdot\boldsymbol{a}}{c^2}\Big) = \frac{m(\boldsymbol{v}\cdot\boldsymbol{a})}{c^2 - \boldsymbol{v}\cdot\boldsymbol{v}}.$$

DEFINITION 2.1. The *relativistic force* on a moving particle is given by

(2.6)
$$\boldsymbol{F} = m\Big(\boldsymbol{a} + \frac{\boldsymbol{v}\cdot\boldsymbol{a}}{c^2 - \boldsymbol{v}\cdot\boldsymbol{v}}\boldsymbol{v}\Big).$$

Since the projection of $\boldsymbol{a}$ perpendicular to $\boldsymbol{v}$ is

$$\boldsymbol{a} - \frac{\boldsymbol{v}\cdot\boldsymbol{a}}{\boldsymbol{v}\cdot\boldsymbol{v}}\boldsymbol{v},$$

we can resolve $\boldsymbol{F}$ into components parallel and perpendicular to $\boldsymbol{v}$:

$$\boldsymbol{F} = m\Big(\boldsymbol{a} - \frac{\boldsymbol{v}\cdot\boldsymbol{a}}{\boldsymbol{v}\cdot\boldsymbol{v}}\boldsymbol{v}\Big) + m(\boldsymbol{v}\cdot\boldsymbol{a})\Big(\frac{1}{c^2 - \boldsymbol{v}\cdot\boldsymbol{v}} + \frac{1}{\boldsymbol{v}\cdot\boldsymbol{v}}\Big)\boldsymbol{v}$$
$$= m\Big(\boldsymbol{a} - \frac{\boldsymbol{v}\cdot\boldsymbol{a}}{\boldsymbol{v}\cdot\boldsymbol{v}}\boldsymbol{v}\Big) + m\frac{c^2(\boldsymbol{v}\cdot\boldsymbol{a})}{(\boldsymbol{v}\cdot\boldsymbol{v})(c^2 - \boldsymbol{v}\cdot\boldsymbol{v})}\boldsymbol{v}.$$

We can now express the acceleration in terms of the force and the velocity. The key to doing so is the relation

$$\boldsymbol{F}\cdot\boldsymbol{v} = m\frac{c^2(\boldsymbol{v}\cdot\boldsymbol{a})}{c^2 - \boldsymbol{v}\cdot\boldsymbol{v}} = m_0\alpha^3(\boldsymbol{v}\cdot\boldsymbol{a}) = c^2\frac{dm}{dt},$$

which allows us to express $\boldsymbol{v}\cdot\boldsymbol{a}$ as

$$\boldsymbol{v}\cdot\boldsymbol{a} = \frac{\boldsymbol{F}\cdot\boldsymbol{v}}{mc^2}(c^2 - \boldsymbol{v}\cdot\boldsymbol{v}) = \frac{\boldsymbol{F}\cdot\boldsymbol{v}}{m_0\alpha^3}.$$

We can then solve Eq. (2.6) for the acceleration:

(2.7)
$$\boldsymbol{a} = \frac{\boldsymbol{F}}{m} - \frac{\boldsymbol{v}\cdot\boldsymbol{a}}{c^2 - \boldsymbol{v}\cdot\boldsymbol{v}}\boldsymbol{v} = \frac{\boldsymbol{F}}{m} - \frac{\boldsymbol{F}\cdot\boldsymbol{v}}{mc^2}\boldsymbol{v}.$$

Equation (2.7) shows that the relativistic expression for the acceleration due to a force is the same as the classical expression when the force is perpendicular to the velocity. Such a situation arises, for example, in the case of a charge moving in a magnetic field. But in general, to get a desired acceleration $\boldsymbol{a}$, one must apply a force that has a component perpendicular to $\boldsymbol{a}$. Since one component of the relativistic force is parallel to the velocity rather than the acceleration, we might revert to our driving analogy to say that steering is more complicated than we instinctively think if relativistic forces are involved. A turn of the steering wheel must generally be accompanied by pressure on the brake or the accelerator in order to get the car accelerating in the desired direction. It is fortunate that our automobiles do not move at speeds comparable to c. If they did, steering them would become a very counter-intuitive process and would require a lot of practice before one could be safely licensed to drive.

THEOREM 2.6. *Two observers X and Y using coordinates such that Y has velocity $\boldsymbol{u} = u\boldsymbol{i}$ relative to X and X has velocity $-\boldsymbol{u} = -u\boldsymbol{i}$ relative to Y will reconcile the components of force by the following formulas.*

$$f_{x1} = p'_{x1} = \frac{1}{\alpha\eta}\left(\alpha p'_{y1} + \frac{m_0\beta^3}{c^2}(\boldsymbol{v}_y \cdot \boldsymbol{a}_y)u\right) = \frac{1}{\eta}f_{y1} + \frac{m_0\beta^3}{\alpha\eta c^2}(\boldsymbol{v}_y \cdot \boldsymbol{a}_y)u,$$

$$f_{x2} = p'_{x2} = \frac{1}{\alpha\eta}p'_{y2} = \frac{1}{\alpha\eta}f_{y2},$$

$$f_{x3} = p'_{x3} = \frac{1}{\alpha\eta}p'_{y3} = \frac{1}{\alpha\eta}f_{y3}.$$

PROOF. This is a computation entirely contained in the previous discussion. $\square$

COROLLARY 2.1. *In the special case when the velocity $\boldsymbol{v}_y$ is perpendicular to $\boldsymbol{u}$ (that is, $(y^1)' = 0$), these equations simplify to*

$$(2.8) \quad f_{x1} = p'_{x1} = \frac{1}{\alpha}\left(\alpha p'_{y1} + \frac{m_0\beta^3}{c^2}(\boldsymbol{v}_y \cdot \boldsymbol{a}_y)u\right) = f_{y1} + \frac{m_0\beta^3}{\alpha c^2}(\boldsymbol{v}_y \cdot \boldsymbol{a}_y)u,$$

$$(2.9) \quad f_{x2} = p'_{x2} = \frac{1}{\alpha}p'_{y2} = \frac{1}{\alpha}f_{y2},$$

$$(2.10) \quad f_{x3} = p'_{x3} = \frac{1}{\alpha}p'_{y3} = \frac{1}{\alpha}f_{y3}.$$

PROOF. In this case, we have $\eta = 1$. $\square$

The relativistic forces on the projectile depicted in Fig. 2.1 are shown in Fig. 2.3, both being scaled by a factor of 25. Qualitatively speaking, the departure from the classical downward force is not really noticeable, even with this much scaling. (The force actually points downward and very slightly to the right.) The main departure from Newtonian mechanics is that the two observers, even though in uniform relative motion along a straight line, *do not agree* about the force.

In the case of the uniform circular motion shown in Fig. 2.3, where $\boldsymbol{a}_X$ is perpendicular to $\boldsymbol{v}_X$, the relativistic acceleration and force align with the Newtonian acceleration and force, only the relativistic force is larger because the relativistic mass of the particle is larger. From Y's point of view, $\boldsymbol{a}_Y \cdot \boldsymbol{v}_Y = -0.0465046$ light-seconds-squared per second-cubed, so that the relativistic force does not point in the same direction as the relativistic acceleration (shown magnified by a factor of 5 on the right-hand side of Fig. 2.4). A computation made using *Mathematica* reveals that the two sides of Eq. (2.6) are both equal to $0.872973\boldsymbol{i} - 0.167666\boldsymbol{j}$.

3.1. How are relativistic forces determined? At this point, we have built up the basic machinery of mechanics in the context of special relativity. We have done this by assuming (1) that the speed of light c in empty space is the same for all observers and (2) that any two observers in motion at constant relative velocity must use the same kinematic laws when they both observe a moving particle. These assumptions led to the Lorentz transformation for space-time coordinates and to the relativistic expressions for the transformation of velocity and acceleration of a moving particle. When we adjoined the assumption (3) that the two observers must both agree on the dynamic law known as conservation of momentum, we were led to the relativistic phenomenon—absent from classical mechanics—that the mass of a particle measured by an observer depends on its speed relative to that observer.

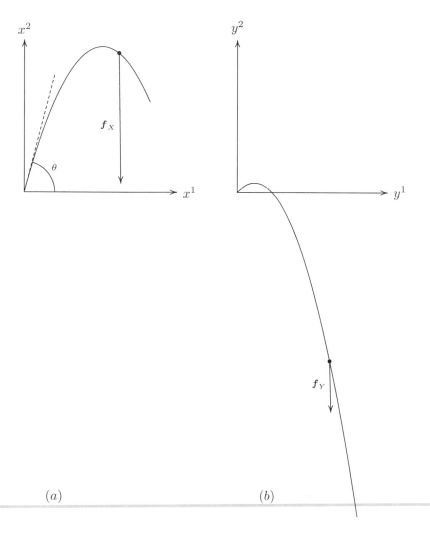

(a) (b)

FIGURE 2.3. Relativistic forces on the projectile of Fig. 2.1. Left:
As observed by X. Right: As observed by Y. The classical force
is vertical, and in this case, the relativistic departure from that
position is barely noticeable.

That fact in turn led us to the relativistic version of the fundamental second law
of motion $\boldsymbol{F} = \boldsymbol{p}'$.

Newtonian mechanics is useful only when we have a complete list of the relevant
forces acting on a body. The net force acting on a specific moving particle cannot
be *defined* to be simply $m\boldsymbol{a}$. That definition leads to the highly uninteresting,
though indisputable, equation $m\boldsymbol{a} = m\boldsymbol{a}$. To get any practical value out of the
theory for explaining the motion of a pendulum or a vibrating string or membrane
or an elastic spring or a planet, we need to bring in a list of specific forces such
as tension, friction, air resistance, and gravity so that we get a second expression
for the force. By setting that second expression equal to $m\boldsymbol{a}$, we get a system of

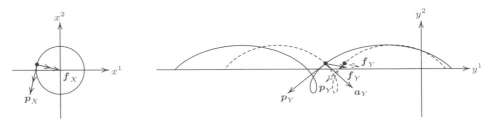

FIGURE 2.4. Forces on a particle of unit mass observed by X to be traversing a circle of radius 1 light-second at a uniform rate of one revolution every 8 seconds. Momentum and force are shown at time 3.7 seconds after the particle has crossed the horizontal axis traveling upward. As in the classical case, the force is centripetal and perpendicular to the momentum. However, the relativistic momentum and force are larger than the corresponding classical quantities, as shown by the two arrow lengths on the left. Right: the same momentum and force as observed by Y using the classical conversion (dashed curve and arrows) and relativistic conversion (solid curve and arrows). In the latter case, the force does not point in the same direction as the acceleration.

second-order differential equations whose solution will describe the motion of the body. All that is well known, and brings us to a significant question: *How do we get that second expression for the force so that we can obtain a system of equations of motion for a particle?*

In some cases—for example, the Lorentz equation $\boldsymbol{F} = q(\boldsymbol{E} + (1/c)\boldsymbol{v} \times \boldsymbol{B})$ for a charged particle moving in an electromagnetic field—we simply take the classical definition, setting this force equal to the relativistic force $\boldsymbol{p}'$. Since the force is now the relativistic force and in general vectors are not invariant under the Lorentz transformation, the assumption of this law in relativity necessitates a revision in the equations of transformation for the electric and magnetic fields if two observers in uniform relative motion are to agree on the basic physical law. Einstein carried out that revision in his first paper on special relativity [**17**], and one consequence of it was a stunning simplification in Maxwell's laws, essentially a reduction from four equations, two of which were vector equations, to a single pair of scalar equations. This reduction will be discussed in Chapter 3.

Assuming we have solved the problem of postulating the relativistic force on a particle—and we have emphatically not done this yet for the important case of gravitational forces—the description of the motion of a particle under the influence of a relativistic force is more complicated than in the classical case, since the relativistic expression for force introduces terms into the differential equations involving not only $\boldsymbol{r}''$ but also $\boldsymbol{r}'$, which is frequently absent from the equations describing mechanical processes.

The force on a charged particle given by the Lorentz equation in electromagnetic theory is independent of the mass of the particle. It depends on the charge of the particle rather than its mass. The Lorentz law, and Coulomb's law of electrostatics differ from the inverse-square law of gravity given by Newton, where the force is directly proportional to the mass of the moving particle:

$$\boldsymbol{F} = -\frac{GMm}{|\boldsymbol{r}|^3}\boldsymbol{r}\,.$$

Here G is the universal gravitational constant, given in MKS units as $G = 6.67384 \times 10^{-11}$ meters-cubed per second-squared per kilogram,[3] M is the mass of a central particle, regarded as fixed at the origin of the coordinate system, and m is the mass of the orbiting particle, which is at location $\boldsymbol{r}(t)$ at time t. In this case, both sides of the equation $\boldsymbol{F} = m\boldsymbol{a}$ are directly proportional to the mass m of the orbiting particle, which therefore cancels out of the classical differential equation. As a result, the predicted Newtonian orbit is independent of the mass of that particle. Unlike mass, charge does not vary with velocity, and this crucial difference between mass and charge poses a challenge for those seeking a Unified Field Theory, in which the laws of gravity and electromagnetism would both be special cases of some more general principle (along with certain subatomic forces—the weak and strong forces—that do not concern us at all in this book).

For us, the challenge will be to find a law of gravity that will reduce to Newton's law in the non-relativistic limit (as $c \to \infty$). We want to go beyond even privileged inertial systems, any two of which are in uniform motion relative to each other. We want a law that anybody can use, no matter what coordinates are being used to describe space-time. We shall take up that challenge beginning in Chapter 4. In the meantime, we are going to take the development of relativistic mechanics a step further and exhibit an application of the relativistic expression for force that will justify the labor involved in defining it.

4. Work, Energy, and the Famous $E = mc^2$

We continue our extension of the basic concepts of mechanics into special relativity. The concepts of work and energy, which were introduced over the century and a half following the publication of Newton's *Principia*, play a very large role in the transition to modern physics, and relativistic principles naturally force some modifications in them. We shall lay out the Newtonian and relativistic principles side by side for easy comparison.

4.1. Newtonian mechanics. In Newtonian mechanics, the work $W(\gamma)$ done by the net force $\boldsymbol{F}(\boldsymbol{r})$ moving a particle over a path γ parameterized as $\boldsymbol{r}(t)$, $a \le t \le b$, is

$$W(\gamma) = \int_\gamma \boldsymbol{F} \cdot d\boldsymbol{r} = \int_a^b \boldsymbol{F}\big(\boldsymbol{r}(t)\big) \cdot \boldsymbol{r}'(t)\, dt\,.$$

The work $W(\gamma)$ is the sum of infinitesimal terms, each of which is of the form "force $\times$ distance." The result of doing such work is a change ΔK in the kinetic energy $(1/2)mv^2 = m|\boldsymbol{r}'(t)|^2/2$. That is, since $\boldsymbol{F} = m\boldsymbol{r}''$, the work can be computed using time t as a parameter if the particle is at location $\boldsymbol{r}(t)$ at time t. Then

$$W(\gamma) = \int_a^b m\boldsymbol{r}''(t)\cdot\boldsymbol{r}'(t)\, dt = \int_a^b \frac{d}{dt}\Big(\frac{1}{2}m\boldsymbol{r}'(t)\cdot\boldsymbol{r}'(t)\Big)\, dt = \frac{1}{2}m|\boldsymbol{r}'(b)|^2 - \frac{1}{2}m|\boldsymbol{r}'(a)|^2\,.$$

---

[3]These physical dimensions are practically impossible to remember. In Chapter 4, we shall see that it is more convenient to think of them as length times velocity-squared divided by mass.

Thus, the work is equal to the change in kinetic energy of the particle. We observe that it is only the portion of the force tangential to the path that does any work on the particle. The normal component is there only to keep the particle *on* the path. If there is no normal force, then the particle moves along a straight line.

Caution: The final expression here is determined by the values of the velocity at just the endpoints of the path. But the value of $r'(b)$ may very well (in general, does) depend on the particular path joining $r(a)$ to $r(b)$. We cannot in general fix a and then define the work done as a function of a point r. There is, however, one case in which we can do that, and it is an important one that we shall discuss after we describe work and kinetic energy in relativity.

4.2. Special relativity. The relativistic definition of the work done by a force $F(t)$ acting over a path parameterized by a function $x(t)$, $t_0 \le t \le t_1$, follows the Newtonian pattern, using the line integral of the force over the path:

$$W = \int_{t_0}^{t_1} F(t) \cdot x'(t)\, dt\,.$$

It will be convenient to think of the parameter t as time in our discussion, although it makes no difference mathematically, since the formulas for a change of variable in an integral remain valid independently of any physical interpretation. Consider a particle of rest mass m_0 being moved along a path by a force F so that its position at time t is $x(t)$. Since $x'(t)$ is the velocity of the particle at time t, and

$$F(t) = \frac{d p}{dt} = \frac{d}{dt}(m v)\,,$$

we have

$$W = \int_{t_0}^{t_1} \frac{d}{dt}(m v) \cdot v\, dt = \int_{t_0}^{t_1} \frac{d}{dt}(m v \cdot v) - (m v \cdot a)\, dt\,.$$

By the fundamental theorem of calculus, the first of the integrals on the right-hand side of this equation is simply the difference

$$m_{v_1} v_1^2 - m_{v_0} v_0^2\,,$$

where we have written v_1 for $|x'(t_1)|$ and v_0 for $|x'(t_0)|$. It remains to consider the last integral:

$$
\begin{aligned}
\int_{t_0}^{t_1} m v \cdot a\, dt &= \int_{t_0}^{t_1} \frac{m_0 (v \cdot v')}{\sqrt{1 - \frac{v \cdot v}{c^2}}}\, dt \\
&= -m_0 c^2 \int_{t_0}^{t_1} \frac{d}{dt}\sqrt{1 - \frac{v \cdot v}{c^2}}\, dt \\
&= -m_0 c^2 \left(\sqrt{1 - \frac{v_1^2}{c^2}} - \sqrt{1 - \frac{v_0^2}{c^2}} \right) \\
&= -c^2 \left(m_{v_1}\left(1 - \frac{v_1^2}{c^2}\right) - m_{v_0}\left(1 - \frac{v_0^2}{c^2}\right) \right) \\
&= m_{v_0}(c^2 - v_0^2) - m_{v_1}(c^2 - v_1^2)\,.
\end{aligned}
$$

Subtracting this result from the integral obtained previously, we find the following result:

THEOREM 2.7. *The work done by a relativistic force moving a particle over a path is determined by the speed of the particle at the initial and terminal points. If those speeds are v_0 and v_1 respectively, the work is*

$$(2.11) \qquad W = (m_{v_1} - m_{v_0})c^2 \,.$$

In other words, the work done in accelerating the particle from speed v_0 to speed v_1 is the relativistic increase in the mass of the particle multiplied by the square of the speed of light. We *define* this expression to be the change in kinetic energy of the particle. Thus, if we assign to the particle a "rest energy" of $m_0 c^2$, then its kinetic energy at speed v is $(m_v - m_0)c^2$, and its total energy is given by the most famous equation of twentieth-century physics:

$$(2.12) \qquad E = mc^2 \,.$$

Equation (2.12) expresses the equivalence of mass and energy, undoubtedly the best-known of all the consequences of the special theory of relativity. As a consequence of Eq. (2.12), the relativistic kinetic energy T is slightly larger than the Newtonian, which is $m_0 v^2/2$. It is

$$
\begin{aligned}
T = mc^2 - m_0 c^2 &= m_0 c^2 (\alpha - 1) \\
&= m_0 c^2 \left(-1 + (1 - v^2/c^2)^{-1/2} \right) \\
&= m_0 c^2 \left(\frac{1}{2} \frac{v^2}{c^2} + \frac{3}{8} \frac{v^4}{c^4} + \frac{5}{16} \frac{v^6}{c^6} + \cdots \right) \\
&= \frac{1}{2} m_0 v^2 + \frac{3 m_0 v^4}{8 c^2} + \frac{5 m_0 v^6}{16 c^4} + \cdots \,.
\end{aligned}
$$

(See Problem 2.4 below.)

In the limit as c tends to infinity, the relativistic kinetic energy approaches the Newtonian. The difference is hardly measurable at ordinary speeds, where $v < 10^{-7}c$. That fact gives us confidence that we have adopted the correct relativistic definition of kinetic energy.

5. Newtonian Potential Energy

We have separated the relativistic energy of a particle into two components, its rest energy $m_0 c^2$ and its kinetic energy $(m - m_0)c^2$. In Newtonian mechanics there is no rest energy, but there often is a second kind of energy, the energy of position in a force field, called *potential energy*. The potential energy of a particle at a location r is defined as the work done against the field in moving the particle from a given base point to the point r. That definition makes sense only if the work is the same over all paths from the base point to r. We define a force field $\boldsymbol{F}$ to be *conservative* if the work $W(\gamma)$ done by the field in moving a particle over the path γ depends only on the endpoints of the path γ and not on the particular route chosen to get from one to the other. In that case, we can fix a base point $\boldsymbol{r}_0$ and define a *potential energy* function $V(\boldsymbol{r})$ for the particle by the equation

$$V(\boldsymbol{r}) = - \int_{\boldsymbol{r}_0}^{\boldsymbol{r}} \boldsymbol{F} \cdot d\boldsymbol{r} \,.$$

The negative sign appears in this definition because the potential energy $V(\boldsymbol{r})$ is the work done *against* the field $\boldsymbol{F}$ in moving a particle from the base point $\boldsymbol{r}_0$ to the variable point $\boldsymbol{r}$.

Since the integral is independent of the path and its parameterization, we can let the path consist of the concatenation of the straight-line segments from $\boldsymbol{r}_0 = (x_0, y_0, z_0)$ to (x_1, y_0, z_0), then from (x_1, y_0, z_0) to (x_1, y_1, z_0), and finally from (x_1, y_1, z_0) to $\boldsymbol{r} = (x_1, y_1, z_1)$. The result is that for a force $\boldsymbol{F}(x, y, z) = P(x, y, z)\,\boldsymbol{i} + Q(x, y, z)\,\boldsymbol{j} + R(x, y, z)\,\boldsymbol{k}$,

$$V(x_1, y_1, z_1) = -\int_{x_0}^{x_1} P(s, y_0, z_0)\, ds - \int_{y_0}^{y_1} Q(x_1, s, z_0)\, ds - \int_{z_0}^{z_1} R(x_1, y_1, s)\, ds\,.$$

It follows from this equation that

$$\frac{\partial V}{\partial z} = -R\,.$$

By changing the order in which we traverse the line segments from (x_0, y_0, x_0) to (x_1, y_1, z_1), we can show likewise that

$$\frac{\partial V}{\partial y} = -Q$$

and

$$\frac{\partial V}{\partial x} = -P\,.$$

To summarize, we have

$$\operatorname{grad} V(x, y, z) = \nabla V(x, y, z) = -\boldsymbol{F}(x, y, z)\,,$$

where the nabla-symbol (∇) denotes the *gradient operator*. This operator will be used often in what follows, especially in connection with the Euler equations in the calculus of variations and harmonic functions in mechanics. William Rowan Hamilton introduced the symbol in 1837 in the form ▷. After vector analysis was distilled from Hamilton's quaternions, the symbol was found useful by Peter Guthrie Tait (1831–1901), an enthusiastic champion of vector analysis. In an address to the Edinburgh Physical Society in 1889, published the following year ([**79**], p. 92), he said that Maxwell had an aversion to the name, but preferred it to the more descriptive, but vulgar-sounding names *sloper* and *grader*. Tait goes on to say that *nabla* is the Hebrew word for an ancient Assyrian harp—that word is nowadays transliterated as *nevel*—and that the name was suggested by the orientalist William Robertson Smith (1846–1894).

The resemblance of the symbol ∇ to an upside-down version of the Greek letter Δ, led Maxwell to have some fun with Tait. On 7 November 1870, he wrote to Tait, "What do you call this? Atled"? And then on 23 January 1871, he asked Tait, "Still harping on that Nabla?" Those details and a fuller history can be found at the following website.

www.mat.univie.ac.at/~neum/contrib/nabla.txt

The multiplicity of potential functions corresponding to a given force (a different one for every base point) is not a problem, since they all have the same gradient, namely the negative of the force itself.

If the kinetic energy of a particle in motion whose position at time t is $\boldsymbol{r}(t)$ is defined as $T(t) = m|\boldsymbol{r}'(t)|^2/2$, we have $T'(t) = m\boldsymbol{r}''(t) \cdot \boldsymbol{r}'(t) = \boldsymbol{F}(\boldsymbol{r}(t)) \cdot \boldsymbol{r}'(t)$. As a result, we have proved the following fundamental fact:

THEOREM 2.8. *In Newtonian mechanics, if the force $\boldsymbol{F}(\boldsymbol{r})$ depends only on the position and is the negative of the gradient of a potential function $V(\boldsymbol{r})$, then the total energy $V(t) + T(t)$ of a moving particle at any time t is constant:*

$$V'(t) = \nabla V\big(\boldsymbol{r}(t)\big) \cdot \boldsymbol{r}'(t) = -\boldsymbol{F}\big(\boldsymbol{r}(t)\big) \cdot \boldsymbol{r}'(t) = -T'(t) \,.$$

This theorem is the law of *conservation of energy* and is the reason for using the term *conservative* to describe a force field that is the negative of the gradient of a potential. Since, as we shall prove below, the Newtonian force of gravitation is conservative, it follows that the gravitational field does no net work when a particle makes one revolution in a periodic orbit.

5.1. The criterion for a field to be conservative. *Locally,* a force field is conservative if it is *irrotational.* That means its curl is $\boldsymbol{0}$: If $\boldsymbol{F} = P\,\boldsymbol{i} + Q\,\boldsymbol{j} + R\,vk$, then

$$\boldsymbol{0} = \operatorname{curl} \boldsymbol{F} = \nabla \times \boldsymbol{F} = \Big(\frac{\partial R}{\partial y} - \frac{\partial Q}{\partial z}\Big)\boldsymbol{i} + \Big(\frac{\partial P}{\partial z} - \frac{\partial R}{\partial x}\Big)\boldsymbol{j} + \Big(\frac{\partial Q}{\partial x} - \frac{\partial P}{\partial y}\Big)\boldsymbol{k} \,.$$

Any field that is the negative of the gradient of a potential function is conservative, that is $\operatorname{curl} \operatorname{grad} V = \nabla \times (\nabla V) = \boldsymbol{0}$. That fact is a consequence of the equality of mixed partial derivatives, as one can see by replacing P, Q, and R in the last equation by their values as partial derivatives of V.

This will not do globally, as we can see by considering the field $\boldsymbol{F}(x, y, z) = \big(-y/(x^2 + y^2)\big)\boldsymbol{i} + \big(x/(x^2 + y^2)\big)\boldsymbol{j}$. The curl is given by

$$\nabla \times \boldsymbol{F} = -\frac{\partial\big(x/(x^2 + y^2)\big)}{\partial z}\boldsymbol{i} - \frac{\partial\big(y/(x^2 + y^2)\big)}{\partial z}\boldsymbol{j} +$$

$$+ \Big(\frac{\partial\big(x/(x^2 + y^2)\big)}{\partial x} - \Big(\frac{\partial\big(-y/(x^2 + y^2)\big)}{\partial y}\Big)\Big)\boldsymbol{k}$$

$$= \Big(\frac{1}{x^2 + y^2} - \frac{2x^2}{(x^2 + y^2)^2} + \frac{1}{x^2 + y^2} - \frac{2y^2}{(x^2 + y^2)^2}\Big)\boldsymbol{k} = \boldsymbol{0} \,.$$

It is therefore an irrotational field. But if we parameterize a closed path γ_1 that begins at $(1, 0, 0)$ and ends at $(-1, 0, 0)$ as $\boldsymbol{r}(t) = \cos t\,\boldsymbol{i} + \sin t\,\boldsymbol{j}$, $0 \le t \le \pi$, we have

$$\int_{\gamma_1} \boldsymbol{F}\cdot d\boldsymbol{r} = \int_0^\pi \sin^2 t + \cos^2 t\,dt - \pi \,,$$

while over the path γ_2 given by $\boldsymbol{r}(t) = \cos t\,\boldsymbol{i} - \sin t\,\boldsymbol{j}$, $0 \le t \le \pi$, we get

$$\int_{\gamma_2} \boldsymbol{F}\cdot d\boldsymbol{r} = \int_0^\pi -\sin^2 t - \cos^2 t\,dt = -\pi \,.$$

The discrepancy is a consequence of the topology of $\mathbb{R}^3$ with the z-axis removed, since the vector field $\boldsymbol{F}$ is not defined on that axis. On a *simply connected* domain, in which every loop can be continuously shrunk to a point, the condition $\nabla \times \boldsymbol{F} = 0$ is necessary and sufficient for the force $\boldsymbol{F}$ to be conservative. This applies in particular on $\mathbb{R}^3 \setminus \{\boldsymbol{0}\}$, which is simply connected.

Newton's equation of motion for a particle of mass m can now be written in terms of the potential function as

$$\boldsymbol{r}'' + \frac{1}{m}\nabla V = \boldsymbol{0} \,.$$

Componentwise, setting $\psi = V/m$, so that ψ represents the potential energy per unit mass, we have

(2.13) $$(x^i)'' + \frac{\partial \psi}{\partial x^i} = 0\,.$$

5.2. Potential energy as a measure of distance. In the previous chapter, we mentioned that the true measure of distance is often the time required to traverse a path or the energy required to do so. A good illustration of this principle can be found in a famous play from the late twentieth century.

> BOATMAN: *From Richmond to Chelsea, it's a quiet float down-stream, from Chelsea to Richmond, it's a hard pull upstream. And it's a penny halfpenny either way. Whoever makes the regulations doesn't row a boat.*

> Robert Bolt, *A Man for All Seasons*, Act 1.

The geometric concept of Euclidean distance, while useful in laying out build-ings and level racetracks, is not always the best measure of the interval between two places. More often, as we have already noted, we need to know how much time is required to go from place to place. In commercial enterprises such as taxis and airlines we need to know how much fuel is required. A flight from Boston to Seattle, for example, generally takes longer and requires more fuel than a flight from Seattle to Boston. That is because the prevailing winds across North America tend to be from West to East, reducing the load on the airplane's engines in one direction and increasing it in the other. Similarly, a 1500-meter race would not be fair if one of the contestants ran around a level track and the other had to go over a hill. The time spent running downhill would not compensate for the longer time spent running uphill, just as the blue Doppler shift on an approaching object does not compensate for the redshift when the same object moves an equal distance in the opposite direction. (See Problem 1.2 of Chapter 1.) Thus, the poor boatman in the quotation above understands through hard experience that the distance up-stream is not the same as the distance downstream, whatever an odometer might say. He knows that a simplistic application of geometry is not the proper measure of distance in this situation. His customer might well measure the distance by the fare and say it is the same both ways. The boatman disagrees.

The concept of potential energy is what the boatman intuitively understands and the regulators who set the rates do not. In our geometrization of physics, we shall need to use this concept in defining the space-time metric that will allow us to replace the model of motion subject to force with that of motion along a geodesic. We begin with the Newtonian potential energy due to the gravitational field of a massive particle.

5.3. Gravitational force. Newton's inverse-square law for a gravitating par-ticle, that is,

$$\boldsymbol{F} = -\frac{GMm}{r^3}\boldsymbol{r}\,,$$

provides us with a gravitational potential function $V(\boldsymbol{r}) = -GMm/|\boldsymbol{r}|$ for the force on a particle of mass m. One can easily verify that (1) $V(\boldsymbol{r})$ is the negative of the time integral of the dot product of force and velocity, assuming that the body is infinitely distant at time $t = 0$ and at $\boldsymbol{r}(t)$ at time t, and (2) the gradient of $V(\boldsymbol{r})$

is the negative of the force given by this equality. The potential and its gradient are undefined at $\mathbf{0}$, but otherwise perfectly well-defined at every other point of $\mathbb{R}^3$. As mentioned above, that makes the condition $\nabla \times \mathbf{F} = \mathbf{0}$ the necessary and sufficient condition for the potential $V(\mathbf{r})$ to be defined. (We do not need the general theorem, however, since we already have an explicit expression for $V(\mathbf{r})$.)

This particular force has the interesting property that not only its curl $\nabla \times \mathbf{F}$ but also its divergence $\nabla \cdot \mathbf{F}$ vanishes:

$$\operatorname{div} \mathbf{F} = \nabla \cdot \mathbf{F} = \frac{\partial F^1}{\partial x^1} + \frac{\partial F^2}{\partial x^2} + \frac{\partial F^3}{\partial x^3} = 0\,.$$

This means that the potential function $V(\mathbf{r})$ is a *harmonic* function. It satisfies *Laplace's equation*, named after Pierre-Simon Laplace (1749–1827):

$$\nabla^2 V = \nabla \cdot \nabla \times V = -\operatorname{div} \operatorname{grad} V = \operatorname{div} \mathbf{F} \equiv 0\,.$$

It is well-known that there is only one function harmonic in the interior of a closed surface (such as a sphere) that has a given set of values on the boundary of the sphere. If we didn't already know the Newtonian potential, we could get Newton's equations of motion by looking for a potential function that is a spherically symmetric harmonic function of $r = \sqrt{x^2 + y^2 + z^2}$, say $V(r)$. For such a function Laplace's equation says that

$$V''(r) + \frac{2V'(r)}{r} = 0\,,$$

and the general solution of this equation is $V(r) = a + b/r$. Since additive constants make no difference to a potential function, we see that $V(r)$ could be taken to be of the form b/r, as in fact it is.

The value of potential functions is not really computational. In the end, they send us back to the Newtonian forces when we take their gradients. But they have great value when combined with the kinetic energy function in what is called *Hamilton's principle*, which we shall discuss below.

To gain further insight, we replace the particle at the origin by a continuous density[4] $\rho(\mathbf{r})$. In that case, there is still a potential function, since we can compute the gravitational field due to this density by passing to the limit of sums of point masses $\rho(\mathbf{r})\,dV$; for the resultant (vector sum) of conservative forces is conservative. When the gravitating mass is given by a density, the potential energy per unit mass $\varphi(\mathbf{r}) = V(\mathbf{r})/m$ satisfies *Poisson's equation*, named after Siméon-Denis Poisson (1781–1840):

(2.14) $$\nabla^2 \varphi(\mathbf{r}) = 4\pi G \rho(\mathbf{r})\,.$$

This last relation is directly computable if the density $\rho(\mathbf{r})$ is spherically symmetric, that is, can be written as $\rho(r)$ where $r = |\mathbf{r}|$. For that case, Newton showed that the gravitational attraction exerted on a particle of mass m located

---

[4]It is most unfortunate that, as remarked in the preceding chapter, the symbol ρ is used to denote both the radial coordinate when spherical coordinates are used and the density of mass or charge. The usage is so well established, however, that our formulas would only look peculiar if we chose a different symbol. The reader is hereby warned once again to be alert for the ambiguity. Also, having ousted the symbol from its role as the radial variable in spherical coordinates on $\mathbb{R}^3$, we are forced to replace it with r, which is normally the radial coordinate in $\mathbb{R}^2$.

at the point x by a spherical shell of radius s with center at the origin and having constant two-dimensional density $\tilde{\rho}$ per unit area satisfies

$$F(x) = \begin{cases} -\frac{4\pi\tilde{\rho}Gms^2}{r^3}x, & \text{if } r = |x| > s, \\ 0, & \text{if } r = |x| < s. \end{cases}$$

That is, the force on a particle of mass m outside the sphere is the same as would be exerted by a particle having the same mass $M = 4\pi\tilde{\rho}s^2$ as the whole spherical shell but situated at the origin, while there is no net force at any point inside the shell. We can define a volume density ρ on an infinitely thin shell of thickness ds by taking $\rho(s)\,ds = \tilde{\rho}(s)$ and consider the infinitesimal increment of mass $dM = \rho(s)\,dV = 4\pi s^2\rho(s)\,ds$ and the infinitesimal increment of force given by

$$dF(x) = -\left(\frac{Gm}{r^3}\,dM\right)x = -\frac{Gm}{r^3}(4\pi s^2\rho(s)\,ds)\,x = -\frac{4\pi Gm}{r^2}(s^2\rho(s)\,ds)\,\xi,$$

where $\xi = x/|x| = x/r$ is a unit vector having the direction of x.

We then integrate from 0 to x along any radial line consisting of the points $s\xi$, $0 \le s \le r$, in order to get the resultant attraction on a particle at x due to all the shells. When we do, we find that the potential function $V(r)$ is also spherically symmetric, and the potential per unit mass is

$$\varphi(r) = 4\pi G \int_0^r \rho(s)\left(s - \frac{s^2}{r}\right)ds.$$

(See Problem 2.7.) The potential per unit mass φ satisfies the Poisson equation (2.14), which is directly computable in this case.

In the special case we are going to consider, where $\rho(s) = \rho$ is constant, we have

$$\varphi(r) = \frac{2\pi G\rho r^2}{3}.$$

The force on a particle of mass m at a point x on the sphere of radius r is therefore

$$F(x) = -\frac{4}{3}\pi G\rho m x.$$

Throughout this section, we have considered potential energy functions that depend only on distance, not velocity. They correspond to forces that depend only on the position of a particle and not on its velocity. As a result, our discussion does not apply, for example, to a charged particle in a magnetic field, on which the force is given by the Lorentz law $F = (v/c) \times B$, where v is the velocity of the particle and B the magnetic field.

6. Hamilton's Principle

Up to now, we have adopted the simplest progression to develop the mathematics of special relativity, following in the strict Newtonian sequence of concepts: time (t), position $(r = (x, y, z))$, velocity (v), acceleration (a), mass (m), momentum (p), and force (F). Then the velocity is $v(t) = r'(t)$, the acceleration is $a(t) = r''(t)$, the momentum is $p(t) = mv(t) = mr'(t)$, and whatever force F is hypothesized must satisfy Newton's second law $F(t) = p'(t) = mr''(t) = ma(t)$. In order to analyze any mechanical phenomenon and determine the position of the particle $r(t)$, it is necessary (1) to know what the net force $F(t)$ is, and (2) to solve the system of differential equations $mr''(t) = F(t)$. In the case that we are most interested in,

$\boldsymbol{F}$ depends only indirectly on time, being a function of the position $\boldsymbol{r}$. The force is Newton's law of gravity, introduced above.

The formulation $\boldsymbol{F} = m\boldsymbol{a}$ has a nice intuitive feel to it, but does not lend itself readily to the kind of geometrization required for the general theory of relativity. As a first step on the way to that geometrization, we have adjoined three concepts introduced over the eighteenth and nineteenth centuries, mostly by Continental mathematicians, namely the concepts of work (W), kinetic energy (T), and potential energy (V). We then applied these concepts relativistically, noting that the work done on a particle by a force, which in Newtonian mechanics equals the change in its kinetic energy, could be interpreted as a change in its mass through the famous equation $E - mc^2$, where $m = m_0\alpha = mc/\sqrt{c^2 - v^2}$, and m_0 is the rest mass.

Newton's formulation of physics does not involve the notion of energy, which was foreshadowed as *vis viva* (living force $= mv^2 =$ the double of what we now call kinetic energy), introduced by Newton's contemporary Leibniz. Their successors in Britain and on the Continent introduced the concept of energy and showed that Newton's second law could be formulated by saying that a particle subject to a conservative force moves so that the integral of the difference between the kinetic energy and potential energy per unit mass is minimized or maximized. As we are primarily interested in the case of gravitation, we shall describe the computations for that case.

7. The Newtonian Lagrangian

Besides the total energy $T + V$, we are also interested in the difference $T - V$, which is called the *Lagrangian* $L(x, y, z, x', y', z')$. We have

$$T - V = L(x, y, z, x', y', z') = \frac{1}{2}m\big((x')^2 + (y')^2 + (z')^2\big) + \frac{GMm}{\sqrt{x^2 + y^2 + z^2}}.$$

The Lagrangian depends only on (1) the masses M and m of the two bodies, (2) the universal gravitational constant G, and (3) the position and speed of the particle. In the case of gravitational force, the potential energy V depends only on the position and the kinetic energy T only on the speed. It turns out that when x, y, and z vary with time (the particle moves), Newton's equations of motion are equivalent to the statement that the integral of the Lagrangian with respect to time is "stationary" (an extremal). That fact is known as *Hamilton's principle*, named after William Rowan Hamilton (1805–1865). We state it as a formal theorem, although we keep the context limited to potential functions that are independent of velocity:

THEOREM 2.9. *If the force acting on a particle is conservative, then over a time interval $t_0 \leq t \leq t_1$, the particle will move over a path $\boldsymbol{r}(t)$ for which the integral*

$$\int_{t_0}^{t_1} T\big(\boldsymbol{r}'(t)\big) - V\big(\boldsymbol{r}(t)\big)\, dt$$

is stationary (usually a minimum).

PROOF. By Euler's equations in the calculus of variations (see Appendix 2), the assertion of the theorem is equivalent to the following system of equations of motion:

$$\frac{d}{dt}\left(\frac{\partial L}{\partial x'}\right) = \frac{\partial L}{\partial x},$$
$$\frac{d}{dt}\left(\frac{\partial L}{\partial y'}\right) = \frac{\partial L}{\partial y},$$
$$\frac{d}{dt}\left(\frac{\partial L}{\partial z'}\right) = \frac{\partial L}{\partial z},$$

and these equations say that

$$mx'' = -\frac{\partial V}{\partial x},$$
$$my'' = -\frac{\partial V}{\partial y},$$
$$mz'' = -\frac{\partial V}{\partial z},$$

which are equivalent to the vector equation

$$m\boldsymbol{r}'' = -\operatorname{grad} V = \boldsymbol{F}.$$

This is exactly Newton's second law. $\qquad\qquad\square$

To formalize Hamilton's principle still further, we introduce what we are calling the *velocity-gradient* of a function $L(\boldsymbol{r}, \boldsymbol{r}')$, which we denote $\nabla_{\boldsymbol{r}'} L$ and define as

$$\nabla_{\boldsymbol{r}'} L = \frac{\partial L}{\partial x'}\boldsymbol{i} + \frac{\partial L}{\partial y'}\boldsymbol{j} + \frac{\partial L}{\partial z'}\boldsymbol{k}.$$

Here, as in the calculus of variations, the symbols x', y', and z' are to be regarded as independent variables.[5] The fact that in applications they will be taken as the time derivatives of the variables x, y, and z is to be ignored. By our definition of Newtonian kinetic energy T, we have

$$\nabla_{\boldsymbol{r}'} T = m\boldsymbol{r}' = \boldsymbol{p}.$$

Thus, in this notation, momentum is the velocity-gradient of kinetic energy.

The standard gradient, which is normally denoted by an unadorned nabla-symbol (∇), will be called the *position-gradient* and denoted[6] $\nabla_{\boldsymbol{r}}$. It is defined as

$$\nabla_{\boldsymbol{r}} L = \frac{\partial L}{\partial x}\boldsymbol{i} + \frac{\partial L}{\partial y}\boldsymbol{j} + \frac{\partial L}{\partial z}\boldsymbol{k}.$$

By definition of the potential function V, the equation $\nabla_{\boldsymbol{r}} V = -\boldsymbol{F}$ holds, where F is the (conservative) force. Euler's equations can now be written as

$$\frac{d}{dt}(\nabla_{\boldsymbol{r}'} L) = \nabla_{\boldsymbol{r}} L,$$

---

[5]They are usually denoted using dots rather than primes, a notation that the British had nearly abandoned when it was resurrected by Lagrange for the calculus of variations. It is still much commoner than our primed notation.

[6]The subscripted nabla-notation here is too neat to resist. It does, however, conflict with the use of subscripted nablas to denote covariant derivatives, a notation found in all textbooks on relativity and one that we shall introduce in Chapter 5. Our use of it in the present context is restricted to this chapter and should not be allowed to interfere with its standard use when it reappears later on.

and, since $L = T - V$, where T depends only on $\boldsymbol{r}'$ and V only on $\boldsymbol{r}$, we can rewrite this as

$$\frac{d}{dt}(\nabla_{\boldsymbol{r}'} T) = -\nabla_{\boldsymbol{r}} V\,.$$

This equation in turn gives us

$$\frac{d\boldsymbol{p}}{dt} = \boldsymbol{F}\,,$$

which is (again) Newton's second law.

7.1. Connection with differential geometry. The value of Hamilton's principle is not computational, since we are still dealing with the same system of differential equations that we got directly from Newton's law. On the theoretical level, however, it pleases the physicists because of its resemblance to Fermat's principle, which asserts that a ray of light moving through a variably refractive medium follows a path of least time (see Chapter 8). It pleases the mathematicians because it brings in the calculus of variations, which has intimate connections with the geometry of surfaces. In particular, the problem of finding the "straightest possible" path from one point to another on a curved surface is a variational problem, to be solved by applying the Euler equations to the integral that gives the length of a path on the surface. The general problem is framed as follows for a three-dimensional manifold.

A curved 3-dimensional manifold, a map of which is somehow (miraculously) given to us in terms of coordinates (x^1, x^2, x^3), has a metric tensor giving the squared infinitesimal interval between two of its points ds^2 in the form

$$ds^2 = \sum_{i,j=1}^{3} g_{ij}(x^1, x^2, x^3)\, dx^i\, dx^j\,.$$

We think of the interval s obtained by evaluating the integral $\int_\gamma ds$ as a measure of the distance along the curve γ from its initial point to its terminal point. It may be length in the ordinary geometric sense. It may also be the time required for the trip, or the energy needed to make it, or any other suitable measure of the separation of the two points. The generalized "length" $\ell(\gamma)$ of a parameterized path $\gamma(t) = \big(r^1(t), x^2(t), x^3(t)\big)$, $t_0 \le t \le t_1$, is given by the integral

$$\ell(\gamma) = \int_0^{\ell(\gamma)} ds = \int_{t_0}^{t_1} \frac{ds}{dt}\, dt = \int_{t_0}^{t_1} \sqrt{\sum_{i,j=1}^{3} g_{ij}\big(x^1(t), x^2(t), x^3(t)\big) \frac{dx^i}{dt}\frac{dx^j}{dt}}\, dt\,.$$

To get what Gauss called the "shortest path" and what we now call a geodesic, we use the Euler equations to write down the differential equations of a stationary value of this integral. Here we have the same variational problem that we just stated as Hamilton's principle in the context of mechanics, only now it is in the context of pure geometry. Thus, we can set Hamilton's principle alongside the problem of finding a geodesic and treat them as being exactly the same kind of mathematical problem. The "geometrization" of mechanics is now in sight. In general, geodesics are not simple to compute, even on a surface as simple as the hyperbolic paraboloid whose equation is $z = xy$. The reader will find a number of examples in Chapters 5 and 6. It is easy to verify that in Euclidean space and in four-dimensional space-time, they are given by linear relations among the variables.

Normally we simplify things by taking s as the parameter along the path, so that $dt = ds$ and the integrand has the constant value 1. When $t = s$, the Euler equation on x^k is

$$\frac{d}{dt}\left(\sum_{j=1}^{3} g_{kj}\left(x^1(t), x^2(t), x^3(t)\right)\frac{dx^j}{dt}\right) = \sum_{i,j=1}^{3} \frac{\partial g_{ij}}{\partial x^k}\frac{dx^i}{dt}\frac{dx^j}{dt}\,.$$

In the frequently encountered case where $g_{ij} = 0$ if $i \neq j$, this expression simplifies to

$$\frac{d}{dt}\left(g_{kk}\left(x^1(t), x^2(t), x^3(t)\right)\frac{dx^k}{dt}\right) = \sum_{i=1}^{3} \frac{\partial g_{ii}}{\partial x^k}\left(\frac{dx^i}{dt}\right)^2\,.$$

This set of equations has an obvious resemblance to those that arise in the physical problem where the integrand is the Lagrangian $L = T - V$. In the latter case, the Euler equation is

$$\frac{d}{dt}\left(m\frac{dx^k}{dt}\right) = \frac{\partial V}{\partial x^k}\,.$$

The resemblance between the right-hand sides of these two equations strongly suggests that the metric coefficients g_{ij} are generalized potential functions. The right-hand side further suggests that kinetic energy might be expressed by a bilinear operator on pairs of velocities, a step we shall take in Section 8 below, but only in the context of four-dimensional space-time. The left-hand sides of the two equations give a hint that the metric coefficients on the diagonal of the matrix $\left(g_{ij}\right)$ are essentially masses, which are forms of energy in relativistic mechanics.

To summarize, we now have two applications of the Euler equations, one in mechanics and one in geometry. Is it possible to "map" the mechanical problem into the geometric problem and thereby unify the two subjects? Geometric optics had already invoked a least-time principle in the study of light rays. How would that be done? In the mechanical problem, the integrand of the time integral being minimized is $T - V$, where T and V are the kinetic and potential energies of (say) a moving particle. Can we think of these energies as the metric coefficients describing the shape of a three-dimensional manifold (or, more generally, an n-dimensional manifold)? We are especially interested in gravitation and we want to study it in the context of a four-dimensional space-time. The two sets of Euler equations have enough similarities to make this project tantalizing, but also enough differences to make it challenging. In Chapter 4, we shall make an attempt to carry out this program and see that it comes very close to succeeding. It will finally succeed when we pass to general relativity and work in the context of space-time. Right now we carry away from this analysis the insight that the metric coefficients we need to geometrize mechanics must be expressed in terms of potential and kinetic energy and mass, which is relativistically equivalent to energy.

After this sojourn in the simpler world of Newtonian mechanics we now return to the mechanics of special relativity, in which the force on a particle does depend on its velocity, since the Newtonian equation $\boldsymbol{F} = m\boldsymbol{r}''$ has been reinterpreted to take account of the variation of mass with speed.

8. The Relativistic Lagrangian

When we last left the topic of energy in special relativity, introducing its equivalence with mass under the famous $E = mc^2$, we seem to have caused the concept of potential energy to be swallowed up and lost. Potential energy appears to be more complicated in special relativity, since kinetic energy is $(m - m_0)c^2$, and m is determined by the observed speed of the particle, independently of its position. The rest energy $m_0 c^2$ is also independent of position, whereas we like to think of gravitational potential energy as being determined by position alone. But all the energy seems to be accounted for in Eq. (2.12). Is there yet some *other* energy that we could call the potential energy? Would it add still more energy to the particle, beyond what we have already assigned to it?

In Newtonian mechanics, the potential energy represents the work done against a conservative force field in bringing a particle from a base point to any other point. We can already see a difficulty in adapting this definition to relativity. As the particle accelerates, its mass changes, since mass depends on the velocity. It would appear that the force must also depend on velocity, and therefore could not be a function of position alone, as it is in the Newtonian case. That seeming obstacle, however, is only apparent. We are certainly free to imagine a force field that depends only on position. The difficulty comes in imagining how it can be conservative, since it changes the mass of a particle moving through it, and there are infinitely many possible masses. Without trying to solve that difficulty, let us simply hypothesize that we have a function of position $W(\boldsymbol{r})$ whose gradient $\nabla_{\boldsymbol{r}} W$ equals the negative of the *relativistic* force at each point. We can make up such a scenario *ad hoc*. Take *any* function of position $W(\boldsymbol{r})$, and *define* the relativistic force to be $\boldsymbol{F} = -\nabla_{\boldsymbol{r}} W$. Whether the model will fit any actual physical situation or not, we can at least study the mathematics of it. But the function W will probably *not* be what we wish to call the potential in this case, for the following reason.

As we showed above, Newton's second law is equivalent to the Euler equation $(d/dt)(\nabla_{\boldsymbol{r}'} T) = -\nabla_{\boldsymbol{r}} V$, where V is the potential energy and T the kinetic energy. We already have a relativistic expression for the kinetic energy, namely $T = m_0 c^2 (\alpha - 1)$, where $\alpha = (1 - \boldsymbol{r}' \cdot \boldsymbol{r}'/c^2)^{-1/2}$. What we are lacking is a relativistic expression for potential energy. In order to formulate it, we observe that the velocity-gradient of the relativistic kinetic energy is *not* the relativistic momentum $\boldsymbol{p} = \alpha m_0 \boldsymbol{r}'$. We actually have

$$\nabla_{\boldsymbol{r}'} T = \alpha^2 \boldsymbol{p} = \frac{c^2}{c^2 - |\boldsymbol{r}'|^2} \boldsymbol{p} \,.$$

If we intend to express the basic law of relativistic mechanics using a Lagrangian, something has to be modified, and we don't want it to be our definition of kinetic energy, since that definition reduces very neatly to the Newtonian version as c tends to infinity.

We can get a potential function V that will provide the proper Lagrangian, but it must be a function of both position and velocity. We'll get it by modifying the *arbitrary* function of position $W(\boldsymbol{r})$ postulated above.

DEFINITION 2.2. If $W(\boldsymbol{r})$ is the Newtonian potential for a given particle acted on by a force $\boldsymbol{F}$, the *relativistic potential* for the same particle is

$$V(\boldsymbol{r}, \boldsymbol{r}') = W(\boldsymbol{r}) + \left(\alpha - 1 + \frac{1}{\alpha}\right) m_0 c^2 \,,$$

where, as usual, $\alpha = (1 - \boldsymbol{r}' \cdot \boldsymbol{r}'/c^2)^{-1/2}$.

DEFINITION 2.3. The *relativistic Lagrangian* is $L(\boldsymbol{r}, \boldsymbol{r}')$, given by

$$L(\boldsymbol{r}, \boldsymbol{r}') = T(\boldsymbol{r}') - V(\boldsymbol{r}, \boldsymbol{r}') = -W(\boldsymbol{r}) - \frac{m_0 c^2}{\alpha} \,,$$

With this definition, $\nabla_{\boldsymbol{r}} L = -\nabla_{\boldsymbol{r}} W = \boldsymbol{F}$, while $\nabla_{\boldsymbol{r}'} L = -m_0 c^2 \nabla_{\boldsymbol{r}'}(1/\alpha) = \alpha m_0 \boldsymbol{r}' = \boldsymbol{p}$. Thus, Euler's equation $(d/dt)\nabla_{\boldsymbol{r}'} L = \nabla_{\boldsymbol{r}} L$ says $d\boldsymbol{p}/dt = \boldsymbol{F}$, which is the relativistic definition of force. It is therefore mathematically feasible to set up a relativistic Lagrangian, although it seems rather artificial. We have introduced it only for the sake of completeness, and will not mention it again.

9. Angular Momentum and Torque

Angular momentum is to rotational motion what linear momentum is to translational motion. The modification required to establish this analogy comes ultimately from Archimedes' law of the lever, which we nowadays express by saying that two forces $\boldsymbol{F}_1$ applied at a point $\boldsymbol{r}_1$ and $\boldsymbol{F}_2$ applied at a point $\boldsymbol{r}_2$ will have the same tendency to produce rotation about the center $\boldsymbol{0}$ if they produce the same *torque*[7] about the point. Torque, in turn, is neatly expressed in the language of vector analysis as the cross product of the radius vector from the center to the point where the force is applied and the force itself. The equality of torques just mentioned is expressed as the vector equation

$$\boldsymbol{r}_1 \times \boldsymbol{F}_1 = \boldsymbol{r}_2 \times \boldsymbol{F}_2 \,.$$

For more details on this subject and its application to the rotation of a rigid body, see Appendix 5.

9.1. Newtonian angular momentum. Since the force $\boldsymbol{F}$ is the time derivative of momentum $\boldsymbol{p}$, the preceding equation can be written as

$$\boldsymbol{r}_1 \times (\boldsymbol{p}_1)' = \boldsymbol{r}_2 \times (\boldsymbol{p}_2)' \,.$$

In Newtonian mechanics, a particle of mass m moving in a plane in such a way that its location at time t has polar coordinates $\big(r(t), \theta(t)\big)$—that is, $\boldsymbol{r}(t) = r(t) \cos\big(\theta(t)\big) \boldsymbol{i} + r(t) \sin\big(\theta(t)\big) \boldsymbol{j}$—will have an instantaneous *angular momentum about the origin*, denoted $\boldsymbol{l}$, given by

$$\boldsymbol{l} = \boldsymbol{r}(t) \times \big(m\boldsymbol{r}'(t)\big) = mr^2(t)\theta'(t) \, \boldsymbol{k} \,.$$

The angular momentum is perpendicular to the plane of motion, and its magnitude $mr^2(t)\theta'(t)$ is equal to the mass m times *twice the rate at which area is swept out*. We establish this claim informally as follows.

$$dA = dx\,dy = (\cos\theta\,dr - r\sin\theta\,d\theta)(\sin\theta\,dr + r\cos\theta\,d\theta)$$
$$= r(\cos^2\theta + \sin^2\theta)\,dr\,d\theta = r\,dr\,d\theta \,.$$

This is an oriented area in the sense that $dx\,dy = -dy\,dx$ and $dr\,d\theta = -d\theta\,dr$. As a result $dr\,dr = d\theta\,d\theta = 0$. The geometric justification of these last relations is that each of the expressions $dr\,dr$ and $d\theta\,d\theta$ represents an infinitesimal parallelogram whose adjacent sides are also parallel, hence has area 0. If we want the area swept

---

[7]From the Latin word *torquor* meaning *I twist*. It is the source of our word *torture*.

out by the radius vector from the origin to a point moving along a curve whose polar equation is $r = r(\theta)$, starting at a fixed point where $\theta = \theta_0$, we simply integrate:

$$A(\theta) = \int_{\theta_0}^{\theta} \int_0^{r(\varphi)} s\, ds\, d\varphi = \int_{\theta_0}^{\theta} \frac{1}{2}\left(r^2(\varphi)\right) d\varphi\,.$$

As a result, we find

$$\frac{dA}{d\theta} = \frac{1}{2}r^2(\theta).$$

If θ in turn is a function of time t, then

$$\frac{dA}{dt} = \frac{1}{2}r^2\big(\theta(t)\big)\frac{d\theta}{dt}\,.$$

The angular momentum per unit mass therefore has a geometric interpretation in Newtonian mechanics as twice the rate at which area is swept out by the radius vector from the origin to the moving particle. Kepler's second law of planetary motion asserts that this rate is constant for each of the planets, and so we can interpret that law as *conservation of angular momentum*.

9.2. Relativistic angular momentum. We define the relativistic angular momentum of a particle about the origin in the natural way, as $l = r \times p$, and the relativistic torque as $N = r \times F$. It follows that

$$\frac{dl}{dt} = r' \times p + r \times \frac{dp}{dt} = r \times F = N\,.$$

Just as in Newtonian mechanics, torque is the time derivative of angular momentum, in the same way that force is the time derivative of linear momentum. The term $r' \times p$ vanishes since, even though p is no longer a *constant* scalar multiple of r', it remains a scalar multiple of it. Thus, angular momentum is conserved when the torque is zero.

We shall show below that angular momentum is conserved in the motion of one particle orbiting another under the influence of gravity. As just mentioned, in the classical case, conservation of angular momentum per unit mass is precisely Kepler's second law, which we should not expect to observe in relativity. If it held, there would be no relativistic precession of the perihelion of a planet, since Kepler's second law in Newtonian mechanics implies a constant elliptical orbit. In the relativistic case, the angular momentum per unit *rest* mass will be conserved under the influence of gravity.

For a particle of mass m moving in a plane (the only case we need to consider) having polar coordinates $\big(r(t), \theta(t)\big)$ at time t, conservation of angular momentum in Newtonian mechanics, where mass remains constant, amounts to the fact that $r^2\theta'$ is constant. That is, its time derivative $2rr'\theta' + r^2\theta''$ vanishes.

In special relativity, the magnitude of the angular momentum is $m_v r^2\theta'$, where the derivative is taken with respect to laboratory time t. Now let us replace laboratory time t with proper time s. To reiterate what we have said above, the infinitesimal increment ds in proper time is given when rectangular and spherical coordinates are used on the spatial portion of space-time by the expression

$$ds = \sqrt{dt^2 - \frac{1}{c^2}(dx^2 + dy^2 + dz^2)} = \sqrt{dt^2 - \frac{1}{c^2}(d\rho^2 + \rho^2\, d\varphi^2 + \rho^2 \sin^2\varphi\, d\theta^2)},$$

and this interval is the same for all observers whose space-time coordinates are related by a Lorentz transformation. Again, as already pointed out, this equation implies that

$$\frac{ds}{dt} = \sqrt{1 - \frac{\boldsymbol{v} \cdot \boldsymbol{v}}{c^2}} = \frac{1}{\alpha},$$

where $\boldsymbol{v} = \boldsymbol{r}'$ is the Newtonian velocity, obtained by differentiating $\boldsymbol{r}$ with respect to laboratory time t. Thus our differentiation operator d/dt is $\sqrt{1 - \boldsymbol{v} \cdot \boldsymbol{v}/c^2}\, d/ds$, and conversely, $d/ds = \left(1/\sqrt{1 - \boldsymbol{v} \cdot \boldsymbol{v}/c^2}\right) d/dt = \alpha\, d/dt$. We also recall from Chapter 1 that proper time on a moving particle is the time that would be recorded by a clock moving with the particle. It is smaller than laboratory time.

We now have a very elegant expression for the mass m_v of a particle moving at speed v whose rest mass is m_0:

$$m_v = m_0 \frac{dt}{ds}.$$

Angular momentum is useful in the solution of the differential equations of mechanics, since its conservation provides an *integral* of those equations. An integral of a set of differential equations is a function of the variables in the equation that assumes a constant value when the time and position satisfy the set of differential equations. See Appendix 5 for further explanation of integrals.

The fact that relativistic angular momentum is conserved—that is, $m_v r^2 d\theta/dt$ is constant—can also be expressed by saying that $m_0 r^2 d\theta/ds$ is constant.

9.3. Conservation of angular momentum under a central force. Angular momentum is conserved under a wide class of forces, including any *central* (centripetal or centrifugal) force of the form $\boldsymbol{F} = \varphi(\boldsymbol{r})\boldsymbol{r}$, where $\boldsymbol{r}$ is the location. This fact will follow from pure geometry—well, actually, algebra—when we replace force by curvature in Chapter 4. Newton himself (see [**63**], Proposition 1, Theorem 1, pp. 32–33) gave an infinitesimal proof of it that is very geometric, based on Fig. 2.5.

Now in relativity, we know that the acceleration of a particle is not normally parallel to the force, and therefore we can't rely on the geometric infinitesimal argument given by Newton. Conservation of angular momentum nevertheless holds under a central force. Let us state this fact as a formal theorem.

THEOREM 2.10. *Angular momentum is conserved when a particle moves under the influence of a central Newtonian or relativistic force* $\boldsymbol{F}$.

PROOF. We shall show that the vector $\boldsymbol{l} = \boldsymbol{r} \times \boldsymbol{p}$ is constant. If the force (whether Newtonian or relativistic does not matter) is $\boldsymbol{F} = \varphi(\boldsymbol{r})\,\boldsymbol{r}$, where $\varphi(\boldsymbol{r})$ is a scalar-valued function of position, then

$$\boldsymbol{l}' = \boldsymbol{r}' \times \boldsymbol{p} + \boldsymbol{r} \times \boldsymbol{p}' = \boldsymbol{0} + \boldsymbol{r} \times \boldsymbol{F} = \boldsymbol{r} \times \big(\varphi(\boldsymbol{r})\boldsymbol{r}\big) = \boldsymbol{0}.$$

Here again, the term $\boldsymbol{r}' \times \boldsymbol{p}$ vanishes because $\boldsymbol{p} = m\boldsymbol{r}'$, and m is a scalar, whether it represents Newtonian or relativistic mass. □

COROLLARY 2.2. *Under a central Newtonian or relativistic force field, the orbit of a particle lies in a plane.*

PROOF. The radius vector $\boldsymbol{r}$ from O to the particle is always perpendicular to the constant angular momentum $\boldsymbol{l}$. Therefore the particle stays in the plane through O perpendicular to $\boldsymbol{l}$. □

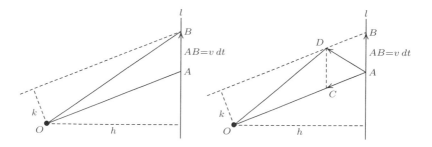

FIGURE 2.5. Newton's proof of the conservation of angular momentum. Left: When no force acts on a particle, it moves according to Newton's first law, along the line l at constant speed v. About any fixed center O at distance h from the line l, it sweeps out area at the rate $hv/2$. In time dt, it sweeps out OAB, which has the area $(1/2)hv\,dt = (k/2)\overline{OA}$. Right: If a force directed toward O would carry the particle from A to C in time dt, then on this infinitesimal level, it actually moves to D, the corner of the parallelogram $ABDC$ diagonally opposite to A. Hence it sweeps out the triangle OAD, whose area is also $(k/2)\overline{OA}$. Thus, a central force does not change the rate at which area is swept out, and that rate is therefore constant.

10. Four-Vectors and Tensors*

In an effort to adhere as closely as possible to the Newtonian picture of the physical world while at the same time taking advantage of vector analysis, we have used a three-dimensional space of locations as the foundation for six other three-component quantities: velocity, acceleration, momentum, force, angular momentum, and torque. All of these quantities have been expressed in terms of mass and the time derivatives of the spatial coordinates. We have taken this approach, assuming the reader knows something about classical mechanics and the application of vector methods.

Where Newtonian mechanics is concerned, the familiar three-dimensional vector analysis is a very convenient tool to use, since space and time are independent of each other when two observers reconcile their measurements. In relativity, where the time coordinates of two observers are entangled with with the spatial coordinates along the line of mutual motion, however, it leads to some awkward asymmetries in the expressions that we have derived, the most glaring of which is the expression for relativistic force. It imposes a frequent need to resort to component-by-component computations using specially selected coordinates. The complication arises because the vector dot and cross products are not invariant under the Lorentz transformation. Although we shall continue the somewhat cumbersome process of analyzing our equations component by component in the next two chapters, it will eventually be necessary to find a more streamlined way of talking about relativistic mechanics. We shall bring up this topic again in Chapter 4, and Chapters 5 and 6 will be devoted to the development of the techniques that are required. At that point, the concepts of the present section will be useful. The reader may wish to postpone reading this section until the need for it arises. We place this material here because it is logically most closely related to the material of this chapter and would be a distraction to explain on the fly when it is needed.

10.1. The four-velocity. We can best explain the revised language by bringing back the two hypothetical observers X and Y from Section 1. To get the kind of unity we want, we will assign them systems of space-time coordinates (x^0, x^1, x^2, x^3) and (y^0, y^1, y^2, y^3) respectively, where the coordinates x^0 and y^0 represent time and the other three in each system are spatial coordinates in an orthonormal basis of $\mathbb{R}^3$. As usual, we assume that Y is moving along the common x^1-axis (y^1-axis) at speed u relative to X, so that the coordinates are related by the Lorentz transformation

$$
\begin{aligned}
y^0 &= \alpha\left(x^0 - \frac{u}{c^2}x^1\right), \\
y^1 &= \alpha(-ux^0 + x^1), \\
y^2 &= x^2, \\
y^3 &= x^3,
\end{aligned}
$$

where $\alpha = 1/\sqrt{1 - u^2/c^2}$. This system of equations can be written in matrix form as

$$
Y = LX,
$$

where

$$
Y = \begin{pmatrix} y^0 \\ y^1 \\ y^2 \\ y^3 \end{pmatrix} \quad L = \begin{pmatrix} \alpha & -\frac{u\alpha}{c^2} & 0 & 0 \\ -\alpha u & \alpha & 0 & 0 \\ 0 & 0 & 1 & 0 \\ 0 & 0 & 0 & 1 \end{pmatrix} \quad \text{and} \quad X = \begin{pmatrix} x^0 \\ x^1 \\ x^2 \\ x^3 \end{pmatrix}.
$$

We make the seemingly trivial remark that the matrix L is actually the *Jacobian*[8] matrix of the change of coordinates $(x^0, x^1, x^2, x^3) \mapsto (y^0, y^1, y^2, y^3)$:

$$
L = \begin{pmatrix} \frac{\partial y^0}{\partial x^0} & \frac{\partial y^0}{\partial x^1} & \frac{\partial y^0}{\partial x^2} & \frac{\partial y^0}{\partial x^3} \\ \frac{\partial y^1}{\partial x^0} & \frac{\partial y^1}{\partial x^1} & \frac{\partial y^1}{\partial x^2} & \frac{\partial y^1}{\partial x^3} \\ \frac{\partial y^2}{\partial x^0} & \frac{\partial y^2}{\partial x^1} & \frac{\partial y^2}{\partial x^2} & \frac{\partial y^2}{\partial x^3} \\ \frac{\partial y^3}{\partial x^0} & \frac{\partial y^3}{\partial x^1} & \frac{\partial y^3}{\partial x^2} & \frac{\partial y^3}{\partial x^3} \end{pmatrix}.
$$

This remark appears trivial, since L is a constant matrix. That is, the change of coordinates $(x^0, x^1, x^2, x^3) \mapsto (y^0, y^1, y^2, y^3)$ is a linear transformation. That linearity is precisely what makes the special theory of relativity "special." The general theory differs from it in allowing more general changes of coordinates. The Jacobian matrix associated with a change of coordinates is one of two matrices that will occur constantly beginning in Chapter 4. The other one is the matrix of the space-time metric, described below.

For two events E_1 and E_2, to which X assigns coordinates $(x_1^0, x_1^1, x_1^2, x_1^3)$ and $(x_2^0, x_2^1, x_2^2, x_2^3)$ respectively while Y assigns them coordinates $(y_1^0, y_1^1, y_1^2, y_1^3)$ and $(y_2^0, y_2^1, y_2^2, y_2^3)$, the proper time interval Δs between the two events is the same for both observers, being the square root of

$$
\begin{aligned}
\Delta s^2 &= (x_2^0 - x_1^0)^2 - \frac{1}{c^2}\left((x_2^1 - x_1^1)^2 + (x_2^2 - x_1^2)^2 + (x_2^3 - x_1^3)^2\right) \\
&= (y_2^0 - y_1^0)^2 - \frac{1}{c^2}\left((y_2^1 - y_1^1)^2 + (y_2^2 - y_1^2)^2 + (y_2^3 - y_1^3)^2\right).
\end{aligned}
$$

---

[8]Named after Carl Gustav Jacobi (1804–1851)

We now call in a third observer Z. We do not need *all* of the coordinates of this observer, since we are going to take Event E_i, $i = 1, 2$, to be the appearance of a time s_i on a clock at the origin of the coordinate system used by Z. Thus, we can think of Z as a person "riding on" a particle whose motion is being observed by X and Y. As of right now, Z must be moving at constant speed along a straight line, but we shall soon remove that restriction. Both observers X and Y keep a record of the time and place where the particle is when Z's clock reads s and record these measurements in the form of world-lines $\left(x^0(s), x^1(s), x^2(s), x^3(s)\right)$ and $\left(y^0(s), y^1(s), y^2(s), y^3(s)\right)$.

The spatial coordinates of both events in the coordinate system used by Z are all 0, so that for Z, the proper time interval is just $s_2 - s_1$. We therefore get the equations

$$(s_2 - s_1)^2 = (x_2^0 - x_1^0)^2 - \frac{1}{c^2}\left((x_2^1 - x_1^1)^2 + (x_2^2 - x_1^2)^2 + (x_2^3 - x_1^3)^2\right)$$
$$= (y_2^0 - y_1^0)^2 - \frac{1}{c^2}\left((y_2^1 - y_1^1)^2 + (y_2^2 - y_1^2)^2 + (y_2^3 - y_1^3)^2\right).$$

If we divide these equations by $(s_2 - s_1)^2$ and let s_2 approach s_1, which we relabel as s, we get the interesting equations

$$1 = \left(\frac{dx^0}{ds}\right)^2 - \frac{1}{c^2}\left(\left(\frac{dx^1}{ds}\right)^2 + \left(\frac{dx^2}{ds}\right)^2 + \left(\frac{dx^3}{ds}\right)^2\right)$$
$$= \left(\frac{dy^0}{ds}\right)^2 - \frac{1}{c^2}\left(\left(\frac{dy^1}{ds}\right)^2 + \left(\frac{dy^2}{ds}\right)^2 + \left(\frac{dy^3}{ds}\right)^2\right).$$

At this point, we no longer need the restriction that Z is moving a constant speed and direction. Taking derivatives automatically "linearizes" equations, in the sense that the best linear approximation to any function has exactly the same derivative as the function itself. In this way, we have found one observer-independent quantity that can stand in for the velocity vector of a moving particle, even one whose observed velocity is not constant. Although X and Y measure the velocity differently, both can compute the *velocity four-vector* of Z, which we define to be the quadruple that X computes as

$$\boldsymbol{u}_4 = (u^0, u^1, u^2, u^3) = \left(\frac{dx^0}{ds}, \frac{dx^1}{ds}, \frac{dx^2}{ds}, \frac{dx^3}{ds}\right)$$

and Y computes as

$$\boldsymbol{v}_4 = (v^0, v^1, v^2, v^3) = \left(\frac{dy^0}{ds}, \frac{dy^1}{ds}, \frac{dy^2}{ds}, \frac{dy^3}{ds}\right).$$

We use the subscript 4 here to distinguish a four-velocity from an ordinary velocity. In particular $\boldsymbol{u}_4$ represents a velocity that X records for Z, whereas previously $\boldsymbol{u}$ was the velocity X assigned to Y.

The components of these two vectors are not numerically the same. What is the same in both is the meaning of the symbol s, and one other important quantity, which we state as a theorem (already proved):

THEOREM 2.11. *The "squared space-time length" of the velocity four-vector* $\boldsymbol{u}_4$ *is the same as that of* $\boldsymbol{v}_4$, *namely* 1. *In symbols*

$$
\begin{aligned}
1 = |\boldsymbol{u}_4|^2 &= (u^0)^2 - \frac{1}{c^2}\left((u^1)^2 + (u^2)^2 + (u^3)^2\right) \\
&= \left(\frac{dx^0}{ds}\right)^2 - \frac{1}{c^2}\left(\left(\frac{dx^1}{ds}\right)^2 + \left(\frac{dx^2}{ds}\right)^2 + \left(\frac{dx^3}{ds}\right)^2\right) \\
&= \left(\frac{dy^0}{ds}\right)^2 - \frac{1}{c^2}\left(\left(\frac{dy^1}{ds}\right)^2 + \left(\frac{dy^2}{ds}\right)^2 + \left(\frac{dy^3}{ds}\right)^2\right) \\
&= (v^0)^2 - \frac{1}{c^2}\left((v^1)^2 + (v^2)^2 + (v^3)^2\right) = |\boldsymbol{v}_4|^2 \,.
\end{aligned}
$$

The "space-time length" is a dimensionless quantity, namely the absolute constant 1. This is because the component u^0 is dimensionless, while u^2, u^3, and u^4 are velocities, which become dimensionless when divided by c. To keep notation to a minimum, we are using the symbol $|\boldsymbol{u}_4|^2$ for this quantity, but the reader must bear in mind that it can be negative; it must be distinguished from the Euclidean norm of a vector in $\mathbb{R}^4$, for which the same notation might be used; that square-norm (for which we have no use) would be $(u^0)^2 + (u^1)^2 + (u^2)^2 + (u^3)^2$. Throughout this section, the square norm of any four-vector $\boldsymbol{v}_4 = (v^0, v^1, v^2, v^3)$ will always have this pseudo-Euclidean value $|\boldsymbol{v}_4|^2 = (v^0)^2 - (1/c^2)\left((v_1)^2 + (v_2)^2 + (v_3)^2\right)$. This will make sense only for four-vectors in which all the ratios v^i/v^0, $i = 1, 2, 3$, have the dimension of velocity.

The four-velocity offers a number of advantages for the further development of relativity. Within special relativity it provides a common language for two observers in uniform relative motion. Each of them can use ordinary time and ordinary Euclidean space just as in Newtonian mechanics to measure the velocity of a particle. While the relation between those two velocities is rather messy, each of them can compute the four-velocity, and the two four-velocities are related quite simply. We introduce column matrices U and V corresponding to the two four-velocities $\boldsymbol{u}_4$ and $\boldsymbol{v}_4$. Since the Jacobian matrix L of the Lorentz transformation has constant entries,

$$
V = \frac{d}{ds}Y = \frac{d}{ds}(LX) = L\frac{d}{ds}X = LU \,.
$$

Since the entry in row $i + 1$, column $j + 1$ of L is $\partial y^i / \partial x^j$, $i, j = 0, 1, 2, 3$, we can write this relation componentwise as

$$
(2.15) \qquad\qquad v^i = \sum_{j=0}^{3} \frac{\partial y^i}{\partial x^j} u^j \,.
$$

A quantity whose components transform in this way under a change of coordinates is called a *contravariant vector* and is also known as a *tensor of type* $(1, 0)$. Since the duality between covariant and contravariant tensors can become confusing, we suggest the following way of telling them apart. Notice that Eq. (2.15) amounts to the chain rule for partial derivatives, which asserts that

$$
\frac{\partial}{\partial x^j} = \sum_i \frac{\partial y^i}{\partial x^j}\frac{\partial}{\partial y^i} \,.
$$

In this last equation, the left-hand side is to be expressed in X's coordinate system. On the right-hand side, the partial derivative is a quantity defined in Y's coordinate system. The coefficients linking the two are $\partial y^i/\partial x^j$, that is, they are derivatives of Y's coordinates with respect to X's and therefore *normally expressed as functions of X's coordinates*. Thus, the right-hand side mixes up the two sets of coordinates, and that mixing accounts for the prefix *contra-* that occurs in the name.

In contrast, the differential dx^j satisfies

$$dx^j = \sum_i \frac{\partial x^j}{\partial y^i}\, dy^i,$$

and here the right-hand side contains both coefficients and differentials expressed in Y's coordinates. The differential is therefore a *covariant* vector, often called a *covector*.

The danger of confusion arises because, looking at Eq. (2.15), one can see that the left-hand side is expressed in Y's coordinates and both the components u^j and the coefficients $\partial y^i/\partial x^j$ are expressed in X's coordinates. It appears then that this equation defines a covariant vector. The explanation is that it truly does, but the covariant vector is not the directional derivative operator $\sum v^i \partial/\partial y^i = \sum u^j \partial/\partial x^j$. It is rather the *coefficient* v^i, which can be thought of as a *linear functional* operating on directional derivative operators. That is where the duality arises: a linear operator on contravariant vectors is a covariant vector, and vice versa.

In practice, it is easy to tell contravariant tensors from covariant ones, since by a now-universal convention,[9] the coefficients in the expansion of a contravariant tensor in any basis are denoted with superscripts and those of a covariant tensor are subscripts. There are also mixed tensors, whose coefficients have both subscripts and superscripts. Finally, a scalar is regarded as a tensor of type $(0,0)$ (rank 0) and can be treated as either covariant or contravariant. When visual clues are lacking—say a tensor is described as an operator rather than being given as a set of coefficients—just keep in mind that a contravariant vector on an abstract manifold is a directional derivative operator that transforms between coordinate systems via the rule

$$\boldsymbol{u} \leftrightarrow \sum_j u^j \frac{\partial}{\partial x^j} = \sum_j u^j \sum_i \frac{\partial y^i}{\partial x^j}\frac{\partial}{\partial y^i} = \sum_i \left(\sum_j \frac{\partial y^i}{\partial x^j} u^j\right)\frac{\partial}{\partial y^i} = \sum_i v^i \frac{\partial}{\partial y^i},$$

As this equation shows, the two four-velocity vectors regarded as equivalent correspond to the same directional derivative operator. That is a good reason for insisting on this transformation law when coordinates are changed. If two quantities transform in this way, they can be regarded as different representations of the same object.

---

[9]One must be cautious here, however. It is not difficult to find expositions on-line in which the differentials dx^i are called contravariant and the partial derivatives $\partial/\partial x^i$ are called covariant. More important than the name attached to each is the fact that they are interacting with each other in dual ways. The *components* (coordinates) of a contravariant vector transform covariantly, but a *basis* of the space of contravariant vectors transforms contravariantly. Another possible source of confusion, especially pronounced among earlier writers such as Eddington ([**16**], p. 43) is the failure to distinguish between an operation and the result of applying that operation—that is, between the operator $\partial/\partial x^i$ and the corresponding partial derivative that results when this operator is applied to a function: $\partial\varphi/\partial x^i$. In the place just cited, Eddington refers ambiguously to "a set of four quantities," and one can't be sure whether he means the quantities that we now write as (dx^1, dx^2, dx^2, dx^4) or the coefficients in a linear combination of these four quantities.

The real reason for the term *contravariant* comes from the theory of manifolds, explained in Appendix 4. Given a C^∞-mapping $\varphi : \mathfrak{M} \to \mathfrak{N}$ from one manifold to another, associated with it is a mapping $d\varphi$ of their tangent spaces. If $\varphi(P) = Q$, then tangent vectors at P and Q are linear mappings, say X and Y from the space of locally C^∞-functions at P and Q respectively having the product-rule property: $X(fg) = f\,X(g) + g\,X(f)$ and $Y(hk) = h\,Y(k) + k\,Y(h)$ for locally C^∞-functions f and g at P and h and k at Q. (It is shown in Appendix 4 that any such mapping is a linear combination of partial derivative operators.) The mapping $d\varphi$ allows X to operate on C^∞-functions near Q by the rule $d\varphi(X)(h) = X(h \circ \varphi)$. Thus we associate with the mapping

$$\mathfrak{M} \xrightarrow{\ \varphi\ } \mathfrak{N}\,,$$

the *covariant* functional $d\varphi$:

$$T_P \xrightarrow{\ d\varphi\ } T_{\varphi(P)}\,,$$

which maps in the *same direction* as φ, that is, it is "covariant" with φ. Since $d\varphi$ is a covariant linear operator—in local coordinate systems at P and Q, its matrix is the Jacobian matrix of the mapping φ—the vectors (linear combinations of partial derivative operators) on which it operates are contravariant.

REMARK 2.1. This appears to be a suitable point to introduce the *Einstein summation convention*, which we shall re-introduce in Chapter 4 for the benefit of readers who postpone the reading of this section until the end. Whenever the same index appears both "up" and "down" in a term, summation is understood to be performed over that index (and hence it is a "dummy index" which can be replaced by any convenient symbol without changing the meaning of the term). In this case, the repeated indices are i and j which appear as superscripts ("up") and in the denominators of the partial derivative operators ("down"). That convention saves huge amounts of writing. (Actually, in Einstein's day, the usage of subscripts and superscripts was less rigid, and he simply said that one should sum over a repeated index, unless there was an explicit instruction not to do so.)

Tensors turn out to be the best possible language for general relativity. A quantity that transforms contravariantly (or covariantly, as we shall shortly note) can form the basis for a common discussion by two different observers. Tensors are a sort of *lingua franca* of physics. Each observer has a "native" language consisting of observed time and space, but is separated from other observers by the specificity of that language. One can picture two observers attempting to talk about the motion of a particle. If one is speaking Italian and the other Romanian, there will be some difficulty in communicating. But if they agree to converse in French, they can talk, confident that they are both discussing the same thing.[10] Einstein believed that physical laws should be stated as tensor relations, since that language made it possible for two observers to reconcile their measurements. At present, contravariance is restricted to linear changes of coordinates, but in Chapters 5 and 6, we shall remove that restriction and allow any smooth coordinate changes and require (nearly) all physically meaningful quantities to be tensors.

---

[10]In the eighteenth century, Prussians, Russians, Britons, Scandinavians, Italians, and others often spoke and wrote to one another in French; that is the origin of the term *lingua franca*. The "*lingua franca*" of Europe had previously been Latin (which had displaced Greek in the West), and nowadays it is English.

The four-velocity vector has an additional virtue besides its simple contravariant behavior under changes of variable. It is easily translated into the "native" language of the observer. Dividing by its zeroth component, which is the (dimensionless) reciprocal of the FitzGerald–Lorentz contraction factor, we get a four-vector whose "time" component is the absolute, dimensionless constant 1 and whose three spatial components constitute the velocity three-vector of classical mechanics:

$$\left(1, \frac{\frac{dx^1}{ds}}{\frac{dx^0}{ds}}, \frac{\frac{dx^2}{ds}}{\frac{dx^0}{ds}}, \frac{\frac{dx^3}{ds}}{\frac{dx^0}{ds}}\right) = \left(1, \frac{dx^1}{dx^0}, \frac{dx^2}{dx^0}, \frac{dx^3}{dx^0}\right) = (1, \boldsymbol{r}'),$$

where $\boldsymbol{r}(t) = \left(x^1(t), x^2(t), x^3(t)\right)$ and $t = x^0$. As a result, we have the equation

$$\boldsymbol{u}_4 = \frac{dx^0}{ds}\left(1, \boldsymbol{r}'(t)\right) = \frac{dt}{ds}\left(1, \boldsymbol{r}'(t)\right).$$

To summarize, the four-velocity carries within it the Newtonian velocity in the form of the ratios of the last three components to the zeroth, and that zeroth component is just the reciprocal of the FitzGerald–Lorentz contraction factor. That is, it is the quantity that appears in the Lorentz transformation equations as α, β, or γ in Chapter 1. We shall denote it by α in the present discussion.

To prove that the zeroth component has this value, observe that the world-line of the particle can be parameterized by observed time $t = x^0$ as well as proper time s. That means that $t = x^0(s)$ is an increasing function of s. It therefore has a local inverse, and $ds/dt = 1/(dt/ds)$. Moreover, as one can easily see from the definition of proper time,

$$\left(\frac{ds}{dt}\right)^2 = \left(\frac{ds}{dx^0}\right)^2 = 1 - \frac{|\boldsymbol{r}'(t)|^2}{c^2},$$

so that we could write

$$\boldsymbol{u}_4 = \alpha\left(1, \boldsymbol{r}'(t)\right),$$

where α is the reciprocal of the FitzGerald–Lorentz contraction factor for a body moving at velocity $\boldsymbol{r}'(t)$ relative to X, that is, $\alpha = c/\sqrt{c^2 - \boldsymbol{r}' \cdot \boldsymbol{r}'}$.

10.2. The metric tensor. At this point, we need to introduce the second matrix mentioned above, the one that defines the metric of space-time. In standard space-time coordinates $(t; x, y, z)$ it is the simple diagonal matrix with constant entries

$$M = \begin{pmatrix} 1 & 0 & 0 & 0 \\ 0 & -\frac{1}{c^2} & 0 & 0 \\ 0 & 0 & -\frac{1}{c^2} & 0 \\ 0 & 0 & 0 & -\frac{1}{c^2} \end{pmatrix}.$$

In contrast to the Jacobian matrix, the matrix M operates *within* a given coordinate system rather than connecting two different ones. It is regarded as a bilinear function on pairs of contravariant four-vectors $\boldsymbol{u}$ and $\boldsymbol{v}$. These need not be velocity four-vectors, but we remind the reader that it is necessary for all the ratios u^i/u^0 and v^i/v^0, $i = 1, 2, 3$, to have the physical dimension of velocity, so that all the terms that are to be added in the following expression have the same physical dimension.

$$M(\boldsymbol{u}, \boldsymbol{v}) = g_{ij}u^i v^j,$$

where the entry in row $i + 1$ and column $j + 1$ of M is g_{ij}. That is, $g_{00} = 1$, $g_{ii} = -1/c^2$ if $i = 1, 2, 3$ and $g_{ij} = 0$ if $i \neq j$. The off-diagonal elements in the first row and first column are thought of as having dimensions of (length $\times$ time)$^{-1}$.

This fact is theoretically important, since the sum that defines $M(\boldsymbol{u}, \boldsymbol{v})$ has to make sense, but it does not affect any computations in the present case since the numbers are all zero anyway. If $\boldsymbol{u} = \boldsymbol{u}_4$, as defined above, then

$$M(\boldsymbol{u}_4, \boldsymbol{u}_4) = 1.$$

The bilinear form defined by this matrix is covariant, and is a *tensor of type* $(0, 2)$, since it operates on a pair of contravariant vectors. If we change the coordinate system from X to Y, the vectors $\boldsymbol{u}_4$ and $\boldsymbol{v}_4$ will transform into $\tilde{\boldsymbol{u}}_4$ and $\tilde{\boldsymbol{v}}_4$, where

$$\tilde{u}_4^i = \frac{\partial y^i}{\partial x^k} u^k \quad \text{and} \quad \tilde{v}_4^j = \frac{\partial y^j}{\partial x^l} v_4^l.$$

If we want the metric to continue measuring things as before, we must replace M by $\widetilde{M}$ such that

$$\tilde{g}_{ij} \tilde{u}_4^i \tilde{v}_4^j = \tilde{g}_{ij} \frac{\partial y^i}{\partial x^k} \frac{\partial y^j}{\partial x^l} u_4^k v_4^l = g_{kl} u_4^k v_4^l.$$

Thus, instead of multiplying the coordinates of $\boldsymbol{u}_4$ and $\boldsymbol{v}_4$ by the Jacobian matrix whose entries are $\partial y^i / \partial x^k$ when we change from X to Y coordinates, we need to multiply the coordinates g_{kl} by the *inverse* of the Jacobian, whose entries, by the chain rule, are $\partial x^k / \partial y^i$. And, since i and k range from 0 to 3, it is the same as $\partial x^l / \partial y^j$ with the same range of indices. We therefore have

$$\tilde{g}_{ij} = \frac{\partial x^k}{\partial y^i} \frac{\partial x^l}{\partial y^j} g_{kl}.$$

The contrast between covariant and contravariant tensors can now be seen. The functions g_{kl} and $\tilde{g}_{ij}$ are the *coordinates* of the tensor M and therefore have variance opposite to that of M itself. In the equation connecting the two sets of coordinates, those on the right are expressed in X's coordinates and the connecting coefficients $\partial x^l / \partial y^j$ are in Y's coordinates. Therefore, the coordinates of this tensor are contravariant, and so the tensor M itself is covariant.

The inverse of the metric tensor M is an example of a contravariant tensor of type $(2, 0)$, and so we shall write its components as m^{ij} to show this contravariant nature. There are also mixed tensors, for example of type $(1, 1)$, an example of which is the matrix product MM^{-1}, which is the identity matrix I. We write its entries as δ_i^j to show the mixed nature of this tensor. Here δ_i^j is known as the *Kronecker delta*:[11]

$$\delta_i^j = m_{jk} m^{ki} \begin{cases} 1, & \text{if } i = j \\ 0, & \text{if } i \neq j. \end{cases}$$

One also occasionally needs a purely covariant or purely contravariant identity matrix, and these have entries that can be denoted δ_{ij} and δ^{ij}. Numerically, all three Kronecker deltas are equal, being 1 if $i = j$ and 0 if $i \neq j$.

---

[11] Named after Leopold Kronecker (1823–1891), who introduced it in a paper bearing the title "Über bilineare Formen" that he read before the Royal Prussian Academy of Sciences on 15 October 1866. This paper was published in the *Monatsberichte* (*Monthly Bulletin*) of the Academy for that year, pp. 597–612, and again in Crelle's *Journal für die reine und angewandte Mathematik*, Bd. 68, pp. 273–285. The relevant passage can be found on p. 150 of the first volume of the 1895 Teubner edition of his collected works, edited by Kurt Hensel.

10.3. Momentum, force, and energy. We have defined relativistic momentum to be a three-component vector $m\boldsymbol{r}'(t)$, where m is the relativistic mass and $\boldsymbol{r}'(t)$ the velocity measured by an observer. In other words, it is $\alpha m_0 \boldsymbol{r}'(t)$, where m_0 is the rest mass of the particle. Since $\alpha = dt/ds$, this relativistic three-momentum is the spatial component of the *four-momentum*

$$\boldsymbol{p}_4 = m_0 \boldsymbol{u}_4.$$

The zeroth component of $\boldsymbol{p}_4$ is the relativistic mass $\alpha m_0 = m$. Also, since $|\boldsymbol{u}_4|^2 = 1$, we have the relation $|\boldsymbol{p}_4|^2 = m_0^2$. (We remind the reader that this norm is a pseudo-Euclidean norm, not the usual one on $\mathbb{R}^4$.)

The reader can guess what comes next. We have replaced three-dimensional Newtonian objects with four-dimensional relativistic objects related to them in the obvious way—by adjoining a zeroth component corresponding to the time coordinate and taking derivatives with respect to proper time instead of coordinate time. We need only take one more step in this direction to get the four-vector $\boldsymbol{F}_4$ corresponding to force:

$$\boldsymbol{F}_4 = \frac{d\boldsymbol{p}_4}{ds} = \alpha \frac{d\boldsymbol{p}_4}{dt} = (F^0, F^1, F^2, F^3).$$

In terms of X's coordinates, the spatial part of the force four-vector is

$$\alpha \frac{d}{dt}(m\boldsymbol{r}'(t)) = \alpha \boldsymbol{F}(t),$$

where $\boldsymbol{F}(t)$ is the relativistic force we defined as a three-vector in Section 3 above.

The time coordinate of $\boldsymbol{F}_4$ is immediately seen to be dm/ds, since the time coordinate of $\boldsymbol{p}_4$ is $m_0 \alpha = m$.

10.4. The equation $E = mc^2$ revisited. As is the case with the four-velocity and four-momentum, the ratios F^i/F^0 here for $i = 1, 2, 3$ have the physical dimension of velocity. We can therefore consider the effect of applying the metric tensor M as a bilinear operator on the pair $(\boldsymbol{u}_4, \boldsymbol{F}_4)$. Since the metric coefficients g_{ij} are constant and symmetric in the indices i and j, we can differentiate the relation $(M(\boldsymbol{u}_4, \boldsymbol{p}_4) = M(\boldsymbol{u}_4, m_0\boldsymbol{u}_4) = m_0$ and get

$$0 = m_0 g_{ij} \frac{d}{ds}(u_4^i u_4^j) = 2m_0 g_{ij} u_4^i \frac{du_4^j}{ds} = 2M(\boldsymbol{u}_4, \boldsymbol{F}_4).$$

Canceling the 2 and taking account of the definition of $\boldsymbol{F}_4$, we get the following relation, after canceling a factor of α on both sides.

$$F^0 = \frac{\alpha}{c^2} \boldsymbol{F} \cdot \boldsymbol{r}',$$

where, again, $\boldsymbol{F}$ is the relativistic three-force defined earlier and the prime denotes differentiation with respect to t.

Using the explicit form of F^0, we find that

$$\boldsymbol{F} \cdot \boldsymbol{r}' = m_0 \alpha^3 \boldsymbol{r}'(t) \cdot \boldsymbol{r}''(t) = c^2 \frac{d}{dt}(m_0 \alpha) = c^2 \frac{dm}{dt}.$$

Now the left-hand side of this equation is just dW/dt, where

$$W = \int_{t_0}^{t} \boldsymbol{F}(s) \cdot \boldsymbol{r}'(s)\, ds$$

represents the work done on the particle between a fixed (laboratory) time t_0 and time t. Comparing the two relations, we find, as we did before, that

$$mc^2 = C + W \,,$$

where C is a constant representing the energy mc^2 at time t_0. In particular, if the particle starts from rest at time t_0, then $C = m_0 c^2$, and the work is once again seen to be the change in energy associated with the moving particle. We thus arrive once again at the equation $E = mc^2$.

10.5. Stress-energy-momentum tensors. We now rewrite our momentum four-vector as an *energy-momentum* four-vector

$$\boldsymbol{p}_4 = \left(\frac{E}{c^2}, \boldsymbol{p} \right),$$

where $E = mc^2$.

By the relation $|\boldsymbol{p}_4|^2 = m_0^2$, we find a sort of "Pythagorean mass-momentum-energy relation" that we can write as $m_0^2 = m^2 - |\boldsymbol{p}|^2/c^2$, or

(2.16) $$E^2 = c^2 |\boldsymbol{p}|^2 + m_0^2 c^4 \,.$$

In the previous chapter, we found that the simple Euclidean formula for distance $ds^2 = dx^2 + dy^2 + dz^2$ was concealing the more general metric tensor (g_{ij}), which appeared only when general coordinates were used. Something similar happens here and leads to a very interesting and useful class of tensors in four variables.

We saw above that the time component of the four-force is

$$F^0 = \frac{m_0 \alpha^4 \boldsymbol{v} \cdot \boldsymbol{a}}{c^2} = \frac{\alpha}{c^2} \boldsymbol{F} \cdot \boldsymbol{v} = \frac{1}{c^2} \boldsymbol{F} \cdot \frac{d\boldsymbol{r}}{ds} \,.$$

Except for the factor $1/c^2$, this is the rate at which the force is doing work, measured in *proper time*, since $d\boldsymbol{r}/ds = \alpha d\boldsymbol{r}/dt = \alpha \boldsymbol{v}$. Thus, the time component of the four-force represents the rate at which the mass m of a moving particle is changing relative to a clock attached to the particle, both particle and clock being observed by someone in a "laboratory" frame. (From the point of view of an observer moving with the particle, the clock is keeping perfect time and the mass never changes.) This suggests a further enlargement of mechanics into space-time. Kinetic energy can be expressed as the quadratic form associated with a bilinear form T that operates on pairs of velocities $(\boldsymbol{u}, \boldsymbol{v})$, where $\boldsymbol{u} = (u^1, u^2, u^3)$ and $\boldsymbol{v} = (v^1, v^2, v^3)$, by the equation

$$T(\boldsymbol{u}, \boldsymbol{v}) = t_{ij} u^i v^j \,.$$

Here, $t_{ij} = (m/2) g_{ij}/2$, where the metric in these coordinates is given as $ds^2 = \sum g_{ij} dx^i \, dx^j$.

If we take $u^0 = dt/ds = \alpha$, we can then consider an analogous representation of energy as a bilinear form T operating on four-velocity vectors, which we shall call a *stress-energy-momentum* tensor. In matrix form, with the usual colloquial expression of the coordinates $(x^0; x^1, x^2, x^3)$ as $(t; x, y, z)$ we can write analogous expressions to represent various forms of energy. The simplest of these is the "rest-energy" tensor $T = (t_{ij}) = m_0 M$ whose entries are $t_{ij} = m_0 g_{ij}$:

$$\left(\frac{dt}{ds} \quad \frac{dx}{ds} \quad \frac{dy}{ds} \quad \frac{dz}{ds} \right) \begin{pmatrix} t_{00} & t_{01} & t_{02} & t_{03} \\ t_{10} & t_{11} & t_{12} & t_{13} \\ t_{20} & t_{21} & t_{22} & t_{23} \\ t_{30} & t_{31} & t_{32} & t_{33} \end{pmatrix} \begin{pmatrix} \frac{dt}{ds} \\ \frac{dx}{ds} \\ \frac{dy}{ds} \\ \frac{dz}{ds} \end{pmatrix} .$$

To see the effect of this definition, apply this tensor to a pair of velocity four-vectors $\boldsymbol{u}_4$ and $\boldsymbol{v}_4$. The result is

(2.17) $$T(\boldsymbol{u}_4, \boldsymbol{v}_4) - m_0 g_{ij} u^i v^j .$$

When $\boldsymbol{u}_4 = \boldsymbol{v}_4$, this becomes

$$T(\boldsymbol{u}_4, \boldsymbol{u}_4) = m_0 |\boldsymbol{u}_4|^2 = m_0 .$$

Thus, the associated quadratic form, when applied to a four-velocity, returns the rest mass as output.

As another example, let ρ_0 be the rest density of a distribution of matter in space, and consider the simple tensor whose matrix in standard coordinates is

$$T = \begin{pmatrix} \rho_0 & 0 & 0 & 0 \\ 0 & 0 & 0 & 0 \\ 0 & 0 & 0 & 0 \\ 0 & 0 & 0 & 0 \end{pmatrix} .$$

See Eddington ([**16**], p. 102). Theorem 2.4 now implies that when the matter in the distribution is moving with four-velocity $\boldsymbol{u}_4$, we have

$$T(\boldsymbol{u}_4, \boldsymbol{u}_4) = \rho_0 \alpha^2 = \rho .$$

Stress-energy-momentum tensors play a role analogous to the various forces that appear in applications of Newton's second law. The analog of this law in general relativity is provided by the *Einstein field equations*

$$G_{\mu\nu} + \Lambda g_{\mu\nu} = \frac{8\pi G}{c^4} T_{\mu\nu} ,$$

Here $G_{\mu\nu}$ is the *Einstein tensor*, which we are going to call "Ein" in Chapter 7. It is formed in turn from the *Ricci tensor*, which we shall call "Ric." (It is generally denoted simply $R_{\mu\nu}$.) The exact relation is Ein = Ric + Rg, where g is the metric tensor $g = (g_{ij})$, and R the contraction of the Ricci tensor using the inverse of the metric tensor, that is, $R = g^{ij} R_{ij}$. The Einstein and Ricci tensors are explained in Chapters 6 and 7. The constant Λ is the *cosmological constant*, and G is Newton's gravitational constant (and is the reason why we are not using the letter G for the Einstein tensor). The left-hand side of this equation is a one-size-fits-all expression analogous to mr'' in Newton's second law. It represents the "response" of the system to the tensor on the right-hand side, which corresponds to the force $\boldsymbol{F}$, which has to be postulated individually for each motion that is to be explained. In the particularly simple case of empty space where no matter is present and "nothing is happening," the stress-energy-momentum tensor can be chosen with ultimate simplicity, as the zero tensor. It is stunning that such a simple assumption explains with incredible precision the precession of a planetary orbit and the deflection of light around a star. All that will be discussed in Chapter 4 below.

What all stress-energy-momentum tensors have in common is a certain "architecture" whereby the time-time component in the upper left corner represents energy, the time-space components along the remainder of the first row and column represent momentum, and the space-space components below and to the right represent stresses. Hence the name "stress-energy-momentum" tensor. This is analogous to the fact that the $\boldsymbol{F}$ in an application of Newton's second law must have the physical dimension of force. We are not going to delve deeply into this topic, but we do need to go far enough to see why Einstein was led to reformulate mechanics as he did. In particular, we shall not go into the problem of internal stresses on a solid body, which greatly complicate the theory of gravitation. We shall confine ourselves to particles orbiting a massive body, also idealized as a particle.

10.6. Incompressible flows. Conservation of mechanical energy in the classical case requires that the body be incompressible, that is, the divergence of its momentum density must vanish. That is the criterion we shall need in Chapter 7, when we justify a relativistic formulation of the law of gravity for a continuously distributed mass density. At that point, we shall introduce the appropriate curved-space version of the divergence and get the standard field equations. In the present chapter, we propose to put forward a heuristic argument that this criterion is reasonable.

Suppose we are given a velocity field $\boldsymbol{v}(t; x, y, z)$ that gives the velocity of the particle in a distribution of flowing matter that is at the point (x, y, z) at time t. Let the density at this point and time be $\rho(t; x, y, z)$. Consider now any closed surface $\Sigma \subset \mathbb{R}^3$ such as a sphere. For convenience, picture its interior as a convex region B. Consider the momentum density $\rho \boldsymbol{v}$ at a point $\boldsymbol{\xi} \in \Sigma$ at time t. This momentum density can be separated into components tangential and normal to Σ. The tangential component represents a flow *over* the surface and does not tend to change the mass $M(t)$ in the interior region B at time t, which is simply

$$M(t) = \iiint\limits_B \rho(t; x, y, z)\, dx\, dy\, dz \,.$$

The component normal to the surface represents an outward flow of mass dm through a small patch of surface $d\Sigma$ over an infinitesimal time interval dt given by

$$dm(t) = -\big(\rho(t; \boldsymbol{\xi}) \boldsymbol{v}(t; \boldsymbol{\xi}) \cdot d\boldsymbol{A}(\boldsymbol{\xi})\big)\, dt \,,$$

where $d\boldsymbol{A} = (dA)\, \boldsymbol{n}(\boldsymbol{\xi})$, dA is the infinitesimal element of surface area (that is, the area of the patch $d\Sigma$), and $\boldsymbol{n}(\boldsymbol{\xi})$ is the outward unit normal vector. The negative sign is taken because an outward flow decreases the mass in the region B. Summing up these flows over all the infinitesimal patches $d\Sigma$, we get the infinitesimal change in the mass in the region B:

$$dM(t) = -\iint\limits_\Sigma \rho(t; \boldsymbol{\xi}) \boldsymbol{v}(t; \boldsymbol{\xi}) \cdot d\boldsymbol{A}\, dt \,,$$

that is,

$$M'(t) = -\iint\limits_\Sigma \rho(t; \boldsymbol{\xi}) \boldsymbol{v}(t; \boldsymbol{\xi}) \cdot d\boldsymbol{A} \,.$$

By the divergence theorem, this means

$$M'(t) = -\iiint_B \nabla \cdot \big(\rho(t; x, y, z)\boldsymbol{v}(t; x, y, z)\big)\, dx\, dy\, dz\,.$$

On the other hand, differentiating inside the triple integral in the expression for $M(t)$, we obviously have

$$M'(t) = \iiint_B \frac{\partial \rho(t; x, y, z)}{\partial t}\, dx\, dy\, dz\,.$$

Since the region B is arbitrary, the integrands in the last two expressions must be equal, and we get the *equation of continuity*

$$\frac{\partial \rho}{\partial t} + \nabla \cdot (\rho \boldsymbol{v}) = 0\,.$$

Because of this equation, if the condition $\nabla \cdot (\rho \boldsymbol{v}) = 0$ holds, then $\rho(t; x, y, z)$ is independent of t. That is, mass does not have any tendency to accumulate or rarify at any point. The density at a given point does not change over time, although it may vary from one point to another.

The equation of continuity suggests a definition of the divergence of the momentum density four-vector $\boldsymbol{\rho}_4 = (\rho, \rho\, dx/dt, \rho\, dy/dt, \rho\, dz/dt)$ as

$$\begin{aligned}
\operatorname{div} \boldsymbol{\rho}_4 &= \frac{\partial \rho}{\partial t} + \frac{\partial\big(\rho \frac{dx}{dt}\big)}{dx} + \frac{\partial\big(\rho \frac{dy}{dt}\big)}{dy} + \frac{\partial\big(\rho \frac{dz}{dt}\big)}{dz} \\
&= \frac{\partial \rho}{\partial t} + \nabla \cdot (\rho \boldsymbol{v})\,.
\end{aligned}$$

If that is done, the equation of continuity merely says $\operatorname{div} \boldsymbol{\rho}_4 = 0$. For more details, see, for example, the book by Eddington, ([**16**], p. 117).

A similar increase in notational efficiency can be introduced into electromagnetic theory using the stress-energy-momentum tensor and reformulating the electric and magnetic fields as tensors. When that is done, Maxwell's four equations (two curl equations and two divergence equations) become just two tensor equations. Since we are aiming at a minimal presentation of the principles of general relativity, with a heavy emphasis on the geometry, we shall not attempt to "tensorize" electromagnetism. Instead, in the following (optional) chapter, we shall return to our three-vectors and show through direct application of the Lorentz transformation how Maxwell's two curl equations can be derived from the two divergence equations.

11. Problems

PROBLEM 2.1 Show that Eqs. (2.1)–(2.3) can be written in vector form as

$$\boldsymbol{v}_x(r) = \frac{1}{\alpha \eta} \boldsymbol{v}_y(s) + \frac{\alpha - 1}{\alpha \eta}\left(\frac{\boldsymbol{v}_y(s) \cdot \boldsymbol{u}}{\boldsymbol{u} \cdot \boldsymbol{u}}\right)\boldsymbol{u} + \frac{1}{\eta}\boldsymbol{u},$$

where $\boldsymbol{v}_x = (x^1)'\boldsymbol{i} + (x^2)'\boldsymbol{j} + (x^3)'\boldsymbol{k}$ and $\boldsymbol{v}_y = (y^1)'\boldsymbol{i} + (y^2)'\boldsymbol{j} + (y^3)'\boldsymbol{k}$ are the velocity vectors the two observers assign to the particle. Here the dot product is taken by Y. The vector $\boldsymbol{u}$ is common to the two, in accordance with our convention that a vector equation of this type is to be used only in the privileged coordinate systems where both observers take $\boldsymbol{i} = \boldsymbol{u}/u$.

PROBLEM 2.2. Show that $\boldsymbol{v}_y$ is constant (in s-time) if and only if $\boldsymbol{v}_x$ is constant (in r-time).

PROBLEM 2.3. Verify that as c approaches infinity, all the equations of relativistic mechanics become the classical equations of Newtonian mechanics. In particular, show that $m_w \to m_0$ as $c \to \infty$.

PROBLEM 2.4. Use the binomial expansion

$$\left(1 - \frac{v^2}{c^2}\right)^{-\frac{1}{2}} = 1 + \frac{1}{2}\frac{v^2}{c^2} + \frac{3}{8}\frac{v^4}{c^4} + \cdots$$

to verify that, as mentioned in connection with Eq. (2.12), the relativistic kinetic energy is

$$(m_v - m_0)c^2 = \frac{1}{2}m_0 v^2 + \frac{3m_0 v^4}{8c^2} + \frac{5v^6}{16c^4} + \cdots.$$

Deduce that, as $c \to \infty$, the relativistic kinetic energy approaches the Newtonian kinetic energy $\frac{1}{2}m_0 v^2$.

PROBLEM 2.5. Show that the increase in rest mass observed by Y in the two-particle collision discussed in the text can be accounted for by saying that each particle converted the kinetic energy of the other into mass. Thus, both mass and energy are conserved in this case.

PROBLEM 2.6. Consider a particle moving along a straight line, so that both $\boldsymbol{v}$ and $\boldsymbol{a}$ also have the direction of this line. With that direction fixed, we can regard velocity, acceleration, and force as scalars. Show that in this case

$$F = \alpha^3 m_0 a,$$

where, as usual, $\alpha = c/\sqrt{c^2 - v^2}$.

PROBLEM 2.7. Prove "Newton's lemma" that the gravitational attraction exerted by a spherical shell of constant density (per unit *area*) is zero inside the sphere, while at points outside the sphere it is equal to that of a particle at the center of the sphere having mass equal to the total mass of the shell. Then show that the force exerted on a body of unit mass by a continuous, spherically symmetric mass density $\rho(r)$ (per unit *volume*) is equal to

$$\boldsymbol{F}(\boldsymbol{x}) = -\frac{4\pi G}{r^3}\left(\int_0^r \rho(s)s^2\,ds\right)\boldsymbol{x},$$

where $r = |\boldsymbol{x}|$. Finally, show that the potential function $\varphi(\boldsymbol{x})$ for the force exerted on a body of unit mass is

$$\varphi(\boldsymbol{x}) = -4\pi G \int_0^r \rho(s)\left(s - \frac{s^2}{r}\right)ds,$$

and that

$$\nabla^2 \varphi(\boldsymbol{x}) = 4\pi G \rho(r).$$

CHAPTER 3

Electromagnetic Theory*

It is clear from our previous considerations that the (special) theory of relativity has grown out of electrodynamics and optics. In these fields it has not appreciably altered the predictions of the theory, but it has considerably simplified the theoretical structure, i.e. the derivation of laws, and—what is incomparably more important— it has considerably reduced the number of independent hypotheses forming the basis of the theory. The special theory of relativity has rendered the Maxwell–Lorentz theory so plausible, that the latter would have been generally accepted by physicists even if experiment had decided less unequivocally in its favour.

Einstein ([**23**], p. 44).

One motive for Einstein's 1905 paper on special relativity was a desire to improve the handling of the Maxwell equations.[1] The Lorentz transformation, besides explaining the peculiarity that electromagnetic radiation propagates with constant velocity relative to all observers, also turns out to make the equations of transformation for electric and magnetic fields more symmetric. In order to show how this simplification comes about, we have to explain how two observers reconcile their measurements of charge, charge density, current density, and electric and magnetic fields when one is moving with constant velocity u relative to the other. Throughout, we assume that they are using spatial coordinates with one axis along the common line of motion. This assumption enables us to use vector language for the transformations, even though the dot and cross product are not generally invariant under Lorentz transformations. As the ultimate goal of the present chapter, we plan to use the classical language of three-component vectors and the special-relativity equations of transformation of electric and magnetic fields to reduce the Maxwell curl equations to the Maxwell divergence equations, thus essentially replacing eight simultaneous equations in the coordinates of the fields with two equations.

The reader is warned that the present chapter is highly algebraic, with just a small admixture of elementary physical theory and almost no geometry (no figures at all!). Those who are impatient with arguments that proceed by concatenating formulas might prefer to accept the main result as given, omit the proofs, and enjoy contemplating the beautiful and fruitful interaction of special relativity with Maxwell's laws.

---

[1]Named for James Clerk Maxwell (1831–1879).

1. Charge and Charge Density

The logical place to begin the study of electrodynamics is with the notion of electric charge. It is possible to produce charged particles by well-defined processes. (Rubbing an inflated balloon on someone's hair is one method familiar to everyone.) Charge manifests itself by exerting a force of attraction or repulsion on other charges. By observation, there are two types of charge, arbitrarily labeled positive and negative. They might equally well have been labeled left and right or northern and southern. Charge is an independent physical dimension, like mass, length, and time. Charges are measured quantitatively based on the empirical relation

$$F = k q_1 q_2 / r^2 ,$$

where F is the magnitude of the force the two particles bearing charges q_1 and q_2 exert on each other, r is the distance between them, and k a constant of proportionality. It is presumably possible to measure the amount of charge q_1 on one of the particles relative to a fixed charge q_0 arbitrarily taken as a unit by using this equation with $q_2 = q_0$. The constant of proportionality will depend on the units of force and can then be determined from the equation by taking both q_1 and q_2 equal to the unit q_0. Throughout this chapter, we shall use Gaussian units. The Maxwell equations have a slightly different form in some of the alternative systems used; we prefer this particular one because it requires no new constants, only letters for charge density, current density, electric field, and magnetic field, plus the symbol c for the speed of light, which we have already used and the very familiar number π. In particular, we have no need to mention the fields $\boldsymbol{D}$ and $\boldsymbol{H}$, or the dielectric permittivity ε and the magnetic permeability μ. We mention only in passing that these last two quantities can be determined by laboratory measurement, and their product turns out to be c^{-2}. That coincidence convinced Maxwell in 1861 that light is an electromagnetic phenomenon. The Gaussian system was used by Einstein in his 1905 paper. It has the additional advantage that it allows the electric and magnetic fields to be treated symmetrically.

Since the force is inversely proportional to the square of the distance, it resembles Newton's law of gravity. Like that law, it can be geometrized by imagining that a charged particle creates a radial field of force called its electric field, which we shall denote $\boldsymbol{E}$. This field is then imagined to act on charged particles just as a gravitational field acts on matter. The only difference is that gravitation is unipolar—it always attracts—while an electric field attracts particles opposite in sign to the particle that produces it and repels those having the same sign. Thus, the electric field created by a point charge q located at the origin is, when suitable units are chosen,

$$(3.1) \qquad\qquad \boldsymbol{E}(\boldsymbol{\xi}) = \frac{q}{(\boldsymbol{\xi} \cdot \boldsymbol{\xi})^{\frac{3}{2}}} \boldsymbol{\xi} .$$

Like the gravitational field due to a point mass, the electric field due to a charged particle requires a second particle of the same type if it is to be detected and measured. If there were only one particle in the universe that had mass or charge, that particle would have difficulty proving that it was generating a gravitational or electric field.

Since the electric field $\boldsymbol{E}$ is normal to a sphere of radius r centered at the location of the particle producing it, the vector integral of the field over that sphere is just area of the sphere $4\pi r^2$ times the normal component of $\boldsymbol{E}$, which is q/r^2.

That amounts to the equation

$$\int_{\mathbb{S}_r} \boldsymbol{E} \cdot d\boldsymbol{A} = 4\pi r^2 \times \frac{q}{r^2} = 4\pi q\,.$$

Note that this result is independent of the radius of the sphere. Even more is true. The sphere can be replaced by any closed surface enclosing the charged particle. The reason is that one can imagine a small sphere centered at the location of the charged particle between the particle and the enclosing surface. Then, by the divergence theorem, the difference between the integrals of $\boldsymbol{E}$ over the two surfaces equals the integral of the divergence $\nabla \cdot \boldsymbol{E}$ over the region between the two surfaces. But by direct and trivial computation, $\nabla \cdot \boldsymbol{E} = 0$ throughout that region.

These results imply the following theorem, known as *Gauss's Theorem*. It applies either to discrete charges inside a closed surface or to a continuous charge density[2] ρ.

THEOREM 3.1. *If a closed surface encloses a number of point charges, or a continuous distribution of charge producing an electric field $\boldsymbol{E}$, then the integral of $\boldsymbol{E}$ over that surface—called the* flux *of $\boldsymbol{E}$ through the surface—equals 4π times the total amount of charge enclosed.*

COROLLARY 3.1. *In a region of space containing a continuous charge distribution with density ρ, the divergence of the electric field is*

$$\nabla \cdot \boldsymbol{E} = 4\pi\rho\,.$$

This corollary is one of Maxwell's equations.

It is a fact well established by experiment that the charge on a particle, unlike the mass of the particle, is independent of the observer. Thus, if Observer X detects a point charge q as an event $(r; x_1, x_2, x_3)$, then Observer Y will also detect that same point charge q at the corresponding event $(s; y_1, y_2, y_3)$ given by the Lorentz transformation.

Suppose that observer O detects a lattice of equal charges of magnitude q that are permanently located at the points (jd, kd, ld), where j, k, and l range over all the integers, and d is a fixed length. For that observer there is a charge density

$$\rho = \frac{q}{d^3}\,.$$

The physical dimension of ρ is charge per unit volume. Now imagine that observer X sees the charges moving with velocity $\boldsymbol{v} = v_1\boldsymbol{i} + v_2\boldsymbol{j} + v_3\boldsymbol{k}$, so that X has velocity $-\boldsymbol{v}$ relative to observer O. The charge density observed by X is easy to compute, since the charge is invariant, but each volume d^3 is decreased by the factor $1/\gamma$, where $\gamma = c/\sqrt{c^2 - |\boldsymbol{v}|^2}$. Thus, X observes a charge density $\rho_x = \gamma\rho$. (This γ is the same quantity introduced in § 2 of the previous chapter, but with $\boldsymbol{x}'$ replaced by $\boldsymbol{v}$.)

Now suppose that Y is moving with velocity $\boldsymbol{u} = u\boldsymbol{i}$ relative to X, and as usual, Y and X are sharing their first axis, which is the line of mutual motion, and assigning the same second and third coordinates to each point. Then, according to

---

[2]Just as a reminder: The symbol ρ is used here for charge density, while in previous chapters it denoted the "spatialized time" rc, or mass density, or the radial coordinate in spherical coordinates.

Y, the charges are moving with velocity $\boldsymbol{w}$, which is the relativistic composition of $\boldsymbol{v}$ and $-\boldsymbol{u}$. By Eqs. (2.1)–(2.3) of the previous chapter, we have

$$w^1 = \frac{-u + v^1}{\delta},$$

$$w^2 = \frac{1}{\alpha\delta}v^2,$$

$$w^3 = \frac{1}{\alpha\delta}v^3,$$

where $\alpha = c/\sqrt{c^2 - |\boldsymbol{u}|^2}$ and $\delta = 1 - uv^1/c^2 = 1 - \boldsymbol{u}\cdot\boldsymbol{v}/c^2$.

Thus, Y will observe a charge density $\rho_y = \beta\rho$, where $\beta = c/\sqrt{c^2 - |\boldsymbol{w}|^2}$. Again, this β is the same quantity introduced in § 2 of the previous chapter, only with $\boldsymbol{y}'$ replaced by $\boldsymbol{w}$.

2. Current and Current Density

Since from X's point of view, the charges are moving with velocity $\boldsymbol{v}$, we say that X is observing a *convection current density* $\boldsymbol{J}_x = \rho_x\boldsymbol{v}$, and we need to see how this current is related to the charge density and convection current density observed by Y. The answer is fairly obvious: From Y's point of view, the particles are moving at the velocity $\boldsymbol{w}$. As we know from Chapter 2, $\beta = \alpha\gamma\delta$. Thus we have

$$(3.2) \qquad \rho_y = \beta\rho = \alpha\gamma\delta\rho = \alpha\rho_x(1 - uv^1/c^2) = \alpha(\rho_x - uJ_{x1}/c^2).$$

When X makes these observations, Y will observe the current density $\boldsymbol{J}_y = \rho_y\boldsymbol{w}$. Componentwise, this means

$$J_{y1} = \rho_y w^1 = \beta\rho\frac{1}{\delta}\big(-u + v^1\big) = \alpha\gamma\big(-u + v^1\big)\rho = \alpha\big(J_{x1} - u\rho_x\big),$$

$$J_{y2} = \rho_y w^2 = \beta\rho\frac{1}{\alpha\delta}v^2 = \gamma\rho v^2 = \rho_x v^2 = J_{x2},$$

$$J_{y3} = \rho_y w^3 = \beta\rho\frac{1}{\alpha\delta}v^3 = \gamma\rho v_3 = \rho_x v^3 = J_{x3}.$$

REMARK 3.1. It is interesting that the scalar-vector pair $(\rho_x; \boldsymbol{J}_x)$ transforms to $(\rho_y; \boldsymbol{J}_y)$ precisely by means of the Lorentz transformation associated with the velocity $\boldsymbol{u}$, as if charge density were like a point in time and current density like a point in space. In retrospect, we should not be surprised by this fact, since the pair $(\rho_x; \boldsymbol{J}_x)$ equals $(\gamma\rho; \gamma\rho\boldsymbol{v}) = \rho\boldsymbol{v}_4$, where $\boldsymbol{v}_4 = (\gamma, \boldsymbol{v})$ is the velocity four-vector discussed in the previous chapter corresponding to the three-velocity $\boldsymbol{v}$. In other words, this four-vector is a direct analogue of the momentum four-vector $\boldsymbol{p}_4 = m_0\boldsymbol{v}_4$ discussed there, with the "rest" charge density ρ replacing the rest mass m_0.

2.1. The divergence of the current density. We are now going to look at current density from the point of view of a single observer. The physical units of current density $\boldsymbol{J}$ are charge per unit area per unit time. The statement that $\boldsymbol{J}$ has magnitude J in a particular direction, means that in an infinitesimal amount of time dt, the amount of charge passing through an infinitesimal element of area dS perpendicular to the direction of $\boldsymbol{J}$ is $J\,dS\,dt$. As a consequence, in a volume R of space enclosed by a surface S, the surface integral

$$\iint_S \boldsymbol{J} \cdot d\boldsymbol{n} = \iint_S J\,dS,$$

where $\boldsymbol{n}$ is the unit outward-pointing normal to the surface, gives the amount of charge passing *out of* the region R per unit of time. If $Q(t)$ is the total amount of charge enclosed at time t, we thus have

$$\iint\limits_{S} \boldsymbol{J} \cdot d\boldsymbol{n} = -Q'(t) \,.$$

We can now repeat the argument in Subsection 10.6 of Chapter 2. By the divergence theorem,

$$\iint\limits_{S} \boldsymbol{J} \cdot d\boldsymbol{n} = \iiint\limits_{R} \nabla \cdot \boldsymbol{J} \, dV \,.$$

Now

$$Q(t) = \iiint\limits_{R} \rho(t) \, dV \,,$$

so that

$$Q'(t) = \iiint\limits_{R} \frac{\partial \rho}{\partial t} \, dV \,.$$

If we divide the first and last of these triple integrals over R by the volume of the region R, and then let that volume shrink to zero, we arrive once again at the *equation of continuity* that we encountered at the end of the previous chapter when discussing a flow of matter:

$$\nabla \cdot \boldsymbol{J} = -\frac{\partial \rho}{\partial t} \,.$$

3. Transformation of Electric and Magnetic Fields

We have given some facts about electric fields here, but we are going to assume that the basic facts of magnetic fields are known. In particular, the reader is assumed to know that magnetism, like electricity, is bipolar. A bar magnet possesses two poles labeled north and south (not positive and negative or right and left, as could have been the case).[3] All that we need to know about magnetic fields (assuming again, a consistent system of units is being used) is the following *Lorentz law* [4] for the force on a particle bearing charge q and moving with velocity $\boldsymbol{v}$:

$$\boldsymbol{F} = q\left(\boldsymbol{E} + \frac{\boldsymbol{v}}{c} \times \boldsymbol{B}\right).$$

That is, a magnetic field $\boldsymbol{B}$ is detected and measured by putting a moving charged particle into it and observing how the particle is deflected. Maxwell's laws provide two connections between the fields $\boldsymbol{E}$ and $\boldsymbol{B}$ (the two curl equations). The other two Maxwell equations (the divergence equations) uncouple the two fields.

---

[3] Obviously, this terminology comes from the magnetic compass, whose needle aligns itself in the north-south direction when free to do so. Since like poles repel and unlike poles attract, the linguistically odd result is that the north geographical pole of the Earth is approximately its south magnetic pole and vice versa.

[4] Named for the same Hendrik Antoon Lorentz (1853–1928) for whom the Lorentz transformation is named. Lorentz's derivation of the law, however, had been anticipated by both James Clerk Maxwell and Oliver Heaviside (1850–1925).

3.1. Newtonian transformation of the fields. The Newtonian equations of transformation for the electric and magnetic fields $\boldsymbol{E}_x$, $\boldsymbol{E}_y$ and $\boldsymbol{B}_x$, $\boldsymbol{B}_y$ observed in two reference frames when Y is moving with velocity $\boldsymbol{u}$ relative to X (and—automatically in Newtonian mechanics—X with velocity $-\boldsymbol{u}$ relative to Y) exhibit the asymmetry remarked on by Einstein:

THEOREM 3.2. *If observer Y is moving with velocity $\boldsymbol{u}$ relative to observer X, then, in Newtonian mechanics, the electric and magnetic fields observed by the two are related as follows:*

$$\boldsymbol{E}_y = \boldsymbol{E}_x + \frac{\boldsymbol{u}}{c} \times \boldsymbol{B}_x\,;$$
$$\boldsymbol{B}_y = \boldsymbol{B}_x\,.$$

PROOF. In a pair consisting of an electric field of intensity $\boldsymbol{E}$ and a magnetic field $\boldsymbol{B}$, the force on a moving charged particle is given by the Lorentz law stated above.

This being Newtonian mechanics, the Lorentz law and the two equations just written are all genuine vector equations. If X observes fields $\boldsymbol{E}_x$ and $\boldsymbol{B}_x$, the force on a charged particle moving along with Y must be $q(\boldsymbol{E}_x + (\boldsymbol{u}/c) \times \boldsymbol{B}_x)$. But for Y, the velocity of the particle is zero, so that a magnetic field will not exert any force on it. Thus for Y, the force on the particle must be entirely due to an electric field $\boldsymbol{E}_y$. Since the two observers, being in uniform motion relative to each other, must agree on all forces (this, we repeat, is Newtonian mechanics), we have $\boldsymbol{E}_y = \boldsymbol{E}_x + (\boldsymbol{u}/c) \times \boldsymbol{B}_x$. The first of the two equations in the theorem is now proved, and by symmetry, $\boldsymbol{E}_x = \boldsymbol{E}_y - (\boldsymbol{u}/c) \times \boldsymbol{B}_y$.

For a particle moving with velocity $\boldsymbol{v}$ relative to Y and hence velocity $\boldsymbol{u} + \boldsymbol{v}$ relative to X, we must have

$$\boldsymbol{E}_y + \frac{\boldsymbol{v}}{c} \times \boldsymbol{B}_y = \boldsymbol{E}_x + \frac{\boldsymbol{v} + \boldsymbol{u}}{c} \times \boldsymbol{B}_x$$
$$= \boldsymbol{E}_y + \frac{\boldsymbol{v}}{c} \times \boldsymbol{B}_x\,.$$

It follows that $\boldsymbol{v} \times (\boldsymbol{B}_y - \boldsymbol{B}_x) = \boldsymbol{0}$ for all velocities $\boldsymbol{v}$, and hence $\boldsymbol{B}_y = \boldsymbol{B}_x$. □

We remark that if c were infinite, the electric fields observed by X and Y would also be equal.

3.2. The relativistic transformations. In his 1905 paper, Einstein derived the relativistic version of these equations, in which the electric and magnetic fields are treated symmetrically. We shall now derive these relativistic equations by assuming the Lorentz law for the force on a charge moving in electric and magnetic fields. Of course, it will be the relativistic force that we assume, but the law will remain verbally the same. We remind the reader, however, that we are assuming Gaussian units here, and in other systems of measurements the equations look slightly different.

THEOREM 3.3. *If the space-time coordinates of observers X and Y are related by the standard Lorentz transformation in the usual privileged coordinates, where Y is moving with velocity $\boldsymbol{u} = v\boldsymbol{i}$ relative to X, then the electric and magnetic fields* *observed by X and Y are related as follows:*

$$(3.3) \qquad \boldsymbol{E}_y = \alpha(\boldsymbol{E}_x + \frac{\boldsymbol{u}}{c} \times \boldsymbol{B}_x) + \frac{(1-\alpha)(\boldsymbol{E}_x \cdot \boldsymbol{u})}{\boldsymbol{u} \cdot \boldsymbol{u}} \boldsymbol{u};$$

$$(3.4) \qquad \boldsymbol{B}_y = \alpha\Big(\boldsymbol{B}_x - \frac{\boldsymbol{u}}{c} \times \boldsymbol{E}_x\Big) + \frac{(1-\alpha)(\boldsymbol{B}_x \cdot \boldsymbol{u})}{\boldsymbol{u} \cdot \boldsymbol{u}} \boldsymbol{u}.$$

PROOF. Let our two observers X and Y both measure the force on the same unit charge $q = 1$ in their electric and magnetic fields $\boldsymbol{E}_x$, $\boldsymbol{E}_y$, $\boldsymbol{B}_x$, $\boldsymbol{B}_y$ using the Lorentz law. First assume that the charge is stationary relative to observer Y, so that $\boldsymbol{v}_y = \boldsymbol{0}$ and $\boldsymbol{v}_x = \boldsymbol{u}$. Comparing the three components of the force, as given by Eqs. (2.8)–(2.10) in the previous chapter, we have, since $E_{x1} = ((\boldsymbol{E}_x \cdot \boldsymbol{u})/(\boldsymbol{u} \cdot \boldsymbol{u}))\boldsymbol{u}$ and $\boldsymbol{v}_y = \boldsymbol{0}$,

$$E_{y1} = F_{y1} = F_{x1} = E_{x1} = \alpha E_{x1} + (1-\alpha)\frac{\boldsymbol{E}_x \cdot \boldsymbol{u}}{\boldsymbol{u} \cdot \boldsymbol{u}} u,$$

$$E_{y2} = F_{y2} = \alpha F_{x2} = \alpha(E_{x2} - \frac{u}{c}B_{x3}),$$

$$E_{y3} = F_{y3} = \alpha F_{x3} = \alpha(E_{x3} + \frac{u}{c}B_{x2}).$$

The first component of $\boldsymbol{u} \times \boldsymbol{B}_x$ is zero, since this cross product is perpendicular to $\boldsymbol{u}$, which points along X's first coordinate axis. Now since the two observers agree that the coordinates of $\boldsymbol{u}$ are $(u,0,0)$, it follows that

$$\boldsymbol{E}_y = \alpha(\boldsymbol{E}_x + \frac{\boldsymbol{u}}{c} \times \boldsymbol{B}_x) + (1-\alpha)\frac{\boldsymbol{E}_x \cdot \boldsymbol{u}}{\boldsymbol{u} \cdot \boldsymbol{u}} \boldsymbol{u}.$$

By symmetry, we have

$$\boldsymbol{E}_x = \alpha(\boldsymbol{E}_y - \frac{\boldsymbol{u}}{c} \times \boldsymbol{B}_y) + (1-\alpha)\frac{\boldsymbol{E}_y \cdot \boldsymbol{u}}{\boldsymbol{u} \cdot \boldsymbol{u}} \boldsymbol{u}.$$

By looking at the last two components of this last equation, we can get an expression for B_{y2} and B_{y3}. We have, since $\boldsymbol{u} \cdot \boldsymbol{u} \times \boldsymbol{B}_x = 0$,

$$\begin{aligned}
\frac{\boldsymbol{u}}{c} \times \boldsymbol{B}_y &= \boldsymbol{F}_y - \frac{1}{\alpha}\boldsymbol{E}_x + \Big(\frac{1}{\alpha} - 1\Big)\frac{\boldsymbol{E}_y \cdot \boldsymbol{u}}{\boldsymbol{u} \cdot \boldsymbol{u}} \boldsymbol{u} \\
&= \Big(\alpha(\boldsymbol{E}_x + \frac{\boldsymbol{u}}{c} \times \boldsymbol{B}_x) + (1-\alpha)\frac{\boldsymbol{E}_x \cdot \boldsymbol{u}}{\boldsymbol{u} \cdot \boldsymbol{u}} \boldsymbol{u}\Big) - \frac{1}{\alpha}\boldsymbol{E}_x \\
&\quad + \Big(\frac{1}{\alpha} - 1\Big)\Big(\alpha\frac{\boldsymbol{E}_x \cdot \boldsymbol{u}}{\boldsymbol{u} \cdot \boldsymbol{u}} \boldsymbol{u} + (1-\alpha)\frac{\boldsymbol{E}_x \cdot \boldsymbol{u}}{\boldsymbol{u} \cdot \boldsymbol{u}} \boldsymbol{u}\Big) \\
&= \alpha\Big(1 - \frac{1}{\alpha^2}\Big)\boldsymbol{E}_x + \alpha\Big(\frac{\boldsymbol{u}}{c} \times \boldsymbol{B}_x\Big) - \alpha\Big(1 - \frac{1}{\alpha^2}\Big)\frac{\boldsymbol{E}_x \cdot \boldsymbol{u}}{\boldsymbol{u} \cdot \boldsymbol{u}} \boldsymbol{u} \\
&= \alpha\Big(\frac{\boldsymbol{u}}{c} \times \boldsymbol{B}_x + \frac{u^2}{c^2}\Big(\boldsymbol{E}_x - \frac{\boldsymbol{E}_x \cdot \boldsymbol{u}}{\boldsymbol{u} \cdot \boldsymbol{u}} \boldsymbol{u}\Big)\Big).
\end{aligned}$$

The last two components of this equation assert that

$$-uB_{y3} = \alpha\Big(-uB_{x3} + \frac{u^2}{c}E_{x2}\Big),$$

$$uB_{y2} = \alpha\Big(uB_{x2} + \frac{u^2}{c}E_{x3}\Big).$$

The result is the set of equations

$$B_{y2} = \alpha\left(B_{x2} + \frac{u}{c}E_{x3}\right),$$
$$B_{y3} = \alpha\left(B_{x3} - \frac{u}{c}E_{x2}\right).$$

It remains now only to obtain an expression for B_{y1}. To do that, we consider a unit charge moving so that $y_1' = y_2' = 0$, but $y_3' = z \neq 0$. In terms of the quantities α, β, γ, δ, and η introduced in the previous chapter in connection with the kinematics of a particle, we have $\eta = 1$ for this motion. As a result, $x_1' = u$, $x_2' = 0$, and $x_3' = y_3'/\alpha = z/\alpha$. The *second* component of the electric and magnetic forces on such a particle measured by Y and X respectively is given by $E_{y2} + (z/c)B_{y1}$ and $E_{x2} - (u/c)B_{x3} + \big(z/(c\alpha)\big)B_{x1}$. Eq. (2.9) now implies that

$$E_{y2} + \frac{z}{c}B_{y1} = F_{y2} = \alpha F_{x2} = \alpha E_{x2} - \alpha\frac{u}{c}B_{x3} + \frac{z}{c}B_{x1}.$$

However, as we saw above, $E_{y2} = \alpha E_{x2} - \alpha(u/c)B_{x3}$, so that

$$B_{y1} = B_{x1} = \alpha B_{x1} + (1-\alpha)\frac{\boldsymbol{B}_x \cdot \boldsymbol{u}}{\boldsymbol{u}\cdot\boldsymbol{u}}u.$$

In other words, our three transformation equations can be written

$$\boldsymbol{B}_y = \alpha\left(\boldsymbol{B}_x - \frac{\boldsymbol{u}}{c}\times\boldsymbol{E}_x\right) + (1-\alpha)\frac{\boldsymbol{B}_x\cdot\boldsymbol{u}}{\boldsymbol{u}\cdot\boldsymbol{u}}\boldsymbol{u}.$$

$\square$

Already we see a reason for preferring the relativistic electrodynamics to classical electrodynamics, since it treats the electric and magnetic fields symmetrically. We are about to give an even more convincing reason for this preference.

4. Derivation of the Curl Equations from the Divergence Equations

We now know how to transform all the principal electromagnetic quantities ρ, $\boldsymbol{J}$, $\boldsymbol{E}$, and $\boldsymbol{B}$ between two observers in relative motion. The Lorentz law for the force on a charge and Maxwell's equations are now assumed to hold for all observers equally, provided they are in uniform motion with respect to a given (arbitrary) observer.

In Gaussian units, the Maxwell equations are

$$(3.5) \qquad \nabla\cdot\boldsymbol{B} = 0,$$
$$(3.6) \qquad \nabla\cdot\boldsymbol{E} = 4\pi\rho,$$
$$(3.7) \qquad \nabla\times\boldsymbol{B} = \frac{1}{c}\left(4\pi\boldsymbol{J} + \frac{\partial\boldsymbol{E}}{\partial t}\right),$$
$$(3.8) \qquad \nabla\times\boldsymbol{E} = -\frac{1}{c}\frac{\partial\boldsymbol{B}}{\partial t}.$$

The velocity c that appears in these equations turns out to be the speed of light. In informal terms, the first of these equations says there are no magnetic charges. The second is the corollary to Gauss's Theorem, proved at the beginning of this chapter (Corollary 3.1). The third is Ampère's law:[5] Currents (and—Maxwell's

---

[5]Named for Adrién-Marie Ampère (1775–1836).

correction of the law—changing electric fields) produce magnetic fields. The fourth is Faraday's law:[6] A changing magnetic field produces an electric field.

Now it is very easy to derive the divergence equations from the curl equations and suitably given initial values for $\boldsymbol{B}$ and $\boldsymbol{E}$ (see Problems 3.2 and 3.4 below). The basic fact involved is that the divergence of the curl is zero. This can be done in either a Newtonian or a relativistic setting.

The more difficult task, which we now undertake, is to show that if we have the relativistic equations of transformation for the electric and magnetic fields, we can make the converse deduction of the curl equations from the divergence equations. That is one of the most fascinating features of special relativity: its implication that there are really only two laws here. If everybody observes the divergence equations, then everybody will also observe the curl equations. Formally, we have the following theorem.

THEOREM 3.4. *Let X be a fixed but arbitrary observer. If every observer Y in uniform motion with respect to X observes that Eq. (3.6) holds, then X will observe that Eq. (3.7) holds. Likewise, if every observer Y in uniform motion with respect to X observes that Eq. (3.5) holds, then X will observe that Eq. (3.8) holds.*

PROOF. Since we are dealing with vector expressions used by two observers here, it is best to resort once again to coordinate-wise reasoning in a privileged coordinate system. We note that the partial derivative operator along the direction of motion for the two observers transforms as follows:

$$\frac{\partial}{\partial y^1} = \frac{\partial x^1}{\partial y^1}\frac{\partial}{\partial x^1} + \frac{\partial r}{\partial y^1}\frac{\partial}{\partial r} = \alpha\left(\frac{\partial}{\partial x^1} + \frac{u}{c^2}\frac{\partial}{\partial r}\right).$$

The other two partial derivative operators are the same for both observers. Here r is the time variable used by X. For aesthetic reasons, we shall replace it with the more traditional t.

We now assume that all observers agree on the equation $\nabla \cdot \boldsymbol{E} = 4\pi\rho$, that is,

$$\frac{\partial E_{y1}}{\partial y^1} + \frac{\partial E_{y2}}{\partial y^2} + \frac{\partial E_{y3}}{\partial y^3} = 4\pi\rho_y,$$

$$\frac{\partial E_{x1}}{\partial x^1} + \frac{\partial E_{x2}}{\partial x^2} + \frac{\partial E_{x3}}{\partial x^3} = 4\pi\rho_x.$$

Writing $\boldsymbol{E}_y$ in terms of $\boldsymbol{E}_x$ and $\boldsymbol{B}_x$ and the partial derivative operators $\frac{\partial}{\partial y_1}$, $\frac{\partial}{\partial y_2}$, and $\frac{\partial}{\partial y_3}$ in terms of $\frac{\partial}{\partial t}$, $\frac{\partial}{\partial x_1}$, $\frac{\partial}{\partial x_2}$, and $\frac{\partial}{\partial x_3}$, we find, since $E_{y1} = E_{x1}$,

$$
\begin{aligned}
4\pi\rho_y = \nabla_y \cdot \boldsymbol{E}_y &= \alpha\left(\frac{\partial E_{x1}}{\partial x^1} + \frac{u}{c^2}\frac{\partial E_{x1}}{\partial t}\right) \\
&\quad + \alpha\left(\frac{\partial E_{x2}}{\partial x^2} - \frac{u}{c}\frac{\partial B_{x3}}{\partial x^2}\right) + \alpha\left(\frac{\partial E_{x3}}{\partial x^3} + \frac{u}{c}\frac{\partial B_{x2}}{\partial x^3}\right) \\
&= \alpha\nabla_x \cdot \boldsymbol{E}_x + \alpha\frac{u}{c}\left(\frac{1}{c}\frac{\partial E_{x1}}{\partial t} - \left(\frac{\partial B_{x3}}{\partial x^2} - \frac{\partial B_{x2}}{\partial x^3}\right)\right) \\
&= \alpha\left(4\pi\rho_x + \frac{u}{c}\cdot\left(\frac{1}{c}\frac{\partial \boldsymbol{E}_x}{\partial t} - \nabla_x \times \boldsymbol{B}_x\right)\right).
\end{aligned}
$$

(3.9)

We now have

$$4\pi\alpha\left(\rho_x - \frac{\boldsymbol{u}\cdot\boldsymbol{J}_x}{c^2}\right) = 4\pi\rho_y = \alpha\left(4\pi\rho_x + \frac{u}{c}\cdot\left(\frac{1}{c}\frac{\partial \boldsymbol{E}_x}{\partial t} - \nabla_x \times \boldsymbol{B}_x\right)\right).$$

---

[6]Named for Michael Faraday (1791–1867).

This equation can be written as

$$0 = \boldsymbol{u} \cdot \left(\frac{4\pi \boldsymbol{J}_x}{c} + \frac{1}{c} \frac{\partial \boldsymbol{E}_x}{\partial t} - \nabla_x \times \boldsymbol{B}_x \right).$$

Since $\boldsymbol{u}$ may be any velocity whatever, it follows that

$$\nabla_x \times \boldsymbol{B}_x = \frac{1}{c} \left(4\pi \boldsymbol{J}_x + \frac{\partial \boldsymbol{E}_x}{\partial t} \right).$$

The proof that Eq. (3.8) follows from Eq. (3.5) is similar and is left as an exercise (Problem 3.3 below). $\square$

5. Problems

PROBLEM 3.1. Prove that the convection current density $\boldsymbol{J}_y = J_{y1}\boldsymbol{i} + J_{y2}\boldsymbol{j} + J_{y3}\boldsymbol{k}$ detected by Y is given by the equations

$$\begin{aligned} J_{y1} &= \alpha \big(J_{x1} - u\rho_x \big), \\ J_{y2} &= J_{x2}, \\ J_{y3} &= J_{x3}. \end{aligned}$$

PROBLEM 3.2. Show that if $\nabla \times \boldsymbol{E} = -(1/c)\partial \boldsymbol{B}/\partial t$, then $\partial(\nabla \cdot \boldsymbol{B})/\partial t \equiv 0$. (This means that $\nabla \cdot \boldsymbol{B}$ is constant over time at each point, and hence identically zero at a given point if it vanishes at that point for even one value of t.)

PROBLEM 3.3. Show that if every "Observer Y" observes that $\nabla_y \cdot \boldsymbol{B}_y = 0$, then X will observe that

$$\nabla_x \times \boldsymbol{E}_x = -\frac{1}{c} \frac{\partial \boldsymbol{B}_x}{\partial t}.$$

PROBLEM 3.4. Derive the divergence equation for the electric field $\boldsymbol{E}$ from the curl equation for the magnetic field $\boldsymbol{B}$, assuming that at each point there is a time when $\nabla \cdot \boldsymbol{E} = 4\pi\rho$ at that point.

PROBLEM 3.5. Assume that the charge density and current density are zero. Show that in this case, all the components of $\boldsymbol{B}$ and $\boldsymbol{E}$ satisfy the *homogeneous three-dimensional wave equation*

$$\begin{aligned} \frac{\partial^2 u}{\partial t^2} &= c^2 \nabla \cdot \nabla u = c^2 \left(\frac{\partial^2 u}{\partial x^2} + \frac{\partial^2 u}{\partial y^2} + \frac{\partial^2 u}{\partial z^2} \right) \\ &= c^2 \big(\nabla^2 u \big), \end{aligned}$$

where $\nabla^2 u$ is the *Laplacian* operator.

REMARK 3.2. If we do not assume that charge and current density are zero, this equation can still be written as

$$\square u = f(t; \boldsymbol{x})$$

for some function $f(t; \boldsymbol{x})$ where $\square u$ is the *d'Alembertian operator*

$$\square u = \frac{\partial^2}{\partial t^2} - c^2 \big(\nabla^2 u \big).$$

Notice the analogy between the d'Alembertian operator and the space-time interval:

$$\frac{\partial^2}{\partial t^2} - c^2 \left(\frac{\partial^2}{\partial x^2} + \frac{\partial^2}{\partial y^2} + \frac{\partial^2}{\partial z^2} \right) \leftrightarrows t^2 - \frac{1}{c^2} \big(x^2 + y^2 + z^2 \big).$$

Part 2

The General Theory

Introduction to Part 2

In Chapters 4–7, we are going to explain the rudiments of general relativity. The amount of geometry involved in this undertaking is considerable, and so we begin in Chapter 4 by exhibiting two of the simplest nontrivial applications, given by Einstein around the year 1916. These are (1) an explanation of the previously unaccounted-for portion of the precession of the perihelion of Mercury and (2) the relativistic deflection of light passing near a star, which is about double what could be predicted on the basis of classical physics (and, by astonishing coincidence, about double what Einstein himself had predicted in 1911!). After that introductory chapter, in which the reader is asked to accept without proof the Einstein law of gravity—namely the vanishing of the Ricci tensor in an intuitively reasonable metric of space-time—we proceed in the two chapters that follow to expound the requisite differential geometry, following a roughly historical order of development.

To give a simple principle that summarizes very roughly what is going on in all this geometrical physics, we invoke the informal equivalence we have already mentioned several times, namely force = curvature. By Newton's first law, an object not at rest and not subject to any force will move in a straight line at constant speed. In practice, however, we don't observe forces; we postulate them when we observe objects that are not moving in a straight line at constant speed. If an object is not at rest and not moving in a straight line, then obviously it is moving in a curved line. Thus force = curvature in this instance. An object may also move in a straight line, but at a nonconstant speed. In that case, the graph of its position against time—its "world-line"—will be curved. The principle is thus established that force occurs where a world-line is curved.

Since it is really the curving that we observe and not generally the force itself, it seems reasonable to dispense with force and formulate the laws of mechanics in terms of curved lines. The question is: Which curved lines correspond to which phenomena? When we see objects moving in curves, how do we formulate a general physical law that explains why each one is following its particular curved world-line? One clue comes from geometric optics. According to Fermat's principle, a ray of light follows a "path of least resistance", which in practice means that it takes minimal time to move from one point to another given the restrictions on its speed imposed by the media it is traveling through. A second clue is the fact that, as we saw in Chapter 2, Newton's law of gravity can be interpreted as saying that an object moves so that the time integral of the difference between its kinetic and potential energies assumes a stationary value. (For our purposes, a stationary value is simply one that satisfies the Euler equations of the calculus of variations.) Both of these principles invoke the calculus of variations, one of whose geometric applications is to find the shortest paths (geodesics) on a curved surface in three-dimensional space. In this way, we are led to look for geodesics instead of forces to give a physical explanation of nonlinear motion.

We are now going to illustrate this principle with a very simple, unrealistic example taken from the theory of objects falling near the surface of the Earth. (Realistic examples are never simple enough to lead off the discussion of a technical problem.) The motion of a falling object is vertical, and we shall assume the x-axis is directed downward. The acceleration of a body (neglecting many complications) is the constant $g = 9.81 \, \mathrm{m/sec^2}$, and its elevation at time t, since the x-axis is directed downward is $x(t) = (g/2)t^2 + v_0 t + x_0$, where v_0 is the initial (downward) speed and x_0 the initial elevation. We have just given a dynamical description of this motion, and the resulting world-line is a parabola, not a straight line, due to the presence of the constant gravitational force g per unit mass. We would now like to impose an infinitesimal metric ds on the two-dimensional space of time (t) and distance fallen (x), in which this world-line is a geodesic. We can't quite do this in general, but we can do it if we consider a restricted class of falling bodies, namely those for which the initial speed v_0 is $\sqrt{2gx_0}$.

In this case, we let the metric be $ds^2 = dt^2 - (1/(4gx)) \, dx^2$ on the quadrant where $x > 0$ and $t > 0$. By Euler's equation, we have

$$\frac{d}{ds}(2t') = 0 \, ,$$

so that t is a linear function of s, say $t = ls$. (We can assume time t and proper time s agree at 0.) Since obviously $dt^2 > ds^2$, we must have $l > 1$.

We now have

$$1 = l^2 - \frac{1}{4gx(s)}\left(x'(s)\right)^2 \, ,$$

which yields

$$\frac{x'(s)}{\sqrt{x(s)}} = \sqrt{4g(l^2 - 1)} \, .$$

As it happens, this differential equation is the integral of the other (second-order) Euler equation for this variational problem. As a result

$$x(s) = \frac{1}{4}\left(\sqrt{4g(l^2 - 1)}s + 2\sqrt{x_0}\right)^2 = \frac{g(l^2 - 1)}{l^2}t^2 + \frac{\sqrt{4g(l^2 - 1)x_0}}{l}t + x_0 \, .$$

By choosing $l = \sqrt{2}$, we get this to be

$$x(t) = \frac{1}{2}gt^2 + \sqrt{2gx_0}t + x_0 \, .$$

Thus the world-lines deduced on the basis of Newtonian principles for this restricted class of motions are in fact geodesics in a suitable metric of two-dimensional space-time.

The reader may be wondering how this metric was chosen. The explanation is that these coefficients determine how "complicated" or "winding" a geodesic must be. In that respect they act as the geometric analog of the potential functions of Newtonian mechanics, which determine the resistance of (say) a gravitational field to a force acting against it. The potential function for Newton's law of gravity is essentially $1/x$ along this line, and so it is not unreasonable to include a coefficient of $1/x$ in the metric. The gravitational acceleration g is needed to keep the physical dimensions of the terms in balance, all being time-squared.

CHAPTER 4

Precession and Deflection

A serious difficulty thus arises from comparing the theories of the Earth and Mercury, which seem to imply different values for the mass of Venus. If one takes the mass given by the observations of Mercury, one must conclude that the secular variation of the obliquity of the ecliptic derived from observation contains errors that are implausible. Either that, or the obliquity must be changing as the result of other causes as yet unknown. If, on the other hand, one views the variation in the obliquity and the causes that produce it as being well established, one will be led to conjecture that the excess precession of the perihelion of Mercury is due to some other effect, as yet unknown: cui theoriæ nundum accesserit [for which no theory has yet been made available].

Urbain Le Verrier ([**50**], pp. 381–382). My translation.

It is in the motion of Mercury that the effect [acceleration of the mean motion] will be most perceptible, because it is the planet that has the highest velocity. Tisserand formerly made a similar calculation, admitting Weber's law... Tisserand found that if the Newtonian attraction took place in conformity with Weber's law, there would result, in the perihelion of Mercury, a secular variation of 14″, *in the same direction as that which has been observed and not explained, but smaller, since the latter is* 38″... *This cannot be regarded as an argument in favour of the new Dynamics, since we still have to seek another explanation of the greater part of the anomaly connected with Mercury...*

Henri Poincaré ([**66**], pp. 180–183).

An unconstrained particle will then move in a straight line in [the metric of special relativity]. But if arbitrary new space-time coordinates $x_1, \ldots, x_4$ *are introduced, the* $g_{\mu\nu}$ *will no longer be constant, but rather functions of space and time. At the same time, the motion of an unconstrained particle in the new coordinates will be curved... We shall thus consider this motion as being due to the influence of a gravitational field.*

Einstein ([**21**], p. 779). My translation.

The present book aims to take the reader from the familiar Newtonian basic law of mechanics $\boldsymbol{F} = m\boldsymbol{a}$ to the Einstein field equation that replaces it, traditionally written as

$$G_{\mu\nu} = \frac{8\pi G}{c^4} T_{\mu\nu}\,.$$

(Our own notation for this relation, in Chapter 7 below, will appear bizarre; it will certainly not be found in any other book. We have been led to this peculiar notation in a desperate—and perhaps futile—effort not to use the symbol G in more than one sense within the covers of a single book, and especially not within a single equation, as is done here.)

The road from Newton's second law of motion to the Einstein field equation is long and arduous. In a sense, the change in point of view arises out of Newton's first two laws. The notion of force is an elusive one, since many forces (gravitational, electric, magnetic, and the like) are invisible. By Newton's two laws, the way we know a force is acting on a particle is to observe it and see if it moves in a straight line at constant speed. If it doesn't, some force is required to explain its motion. If we introduce four-dimensional space-time, we can see that the unforced motion of a particle is graphed as a straight line in that space-time. Since a straight line is a geodesic—for present purposes, the shortest path joining two of its points, we can rephrase Newton's first law by saying that an unforced motion is along a geodesic in the Euclidean metric. We thereby get a rough equivalence: a curved world-line is the indicator of a net force. But, since force itself is invisible, what we observe is the curvature. Here we have the germ of an idea: Get rid of force entirely, and let the motion simply be along a geodesic in a curved space-time. That is, impose a non-Euclidean metric on space-time in which the observed path *is* a geodesic.

That road, as we say, is very long. For that reason, in the present chapter we are going to get ahead of the story and see how the end result (the Einstein field equation) explains the precession of Mercury and the deflection of light around the Sun. We shall reserve the motivation for the field equations to later chapters. Since this departure from the Newtonian view is a radical one, for psychological reasons, we consider whether some less drastic approach might work. There are two such approaches that might come to mind. The first would be to retain Newton's equations of motion, but interpret force as in special relativity.[1]

The second would be to accept the geometric point of view, but try to impose the metric on three-dimensional Newtonian space rather than four-dimensional space time with its unusual metric.[2]

I confess in advance that both of these approaches fail to explain the precession of the perihelion of Mercury. Nevertheless, they do bring out some important concepts that are of use in general relativity. We emphasize that these two explorations are present for psychological and pedagogical purposes only, and the reader can choose to omit them without loss of continuity. We make no claim that they provide a logical transition or connection between special and general relativity. General relativity is not merely a perturbation of special relativity.

---

[1] An anonymous reviewer, to whom I am grateful, pointed out that this idea was anticipated and fully developed by the Dutch physicist Willem de Sitter (1872–1934) in 1911 [10].

[2] As with the previous approach, I am grateful to an anonymous reviewer who pointed out to me that what I have done here is subsumed as a small part of two papers ([6], [7]) by Elie Cartan (1869–1951).

1. Gravitation as Curvature of Space

The investigation of relativity has so far been confined to the dynamics of a particle under an unspecified force. Where we have considered specific forces, those forces have been electrical and magnetic. The most important force in celestial mechanics—gravity—has not yet been considered. It was Einstein's idea to account for gravity in terms of a suitable choice of the metric of space-time, by showing that the observed paths followed by physical bodies and the path of a light beam coincide with the geodesics in the metric.[3] Our aim is to expound this idea—it will be the fourth of our four analyses of the orbit of Mercury—and show how it accounts for the precession of Mercury's perihelion in a way that Newton's law of gravity and two modifications of it that might occur to someone cannot.

We are going to discuss only the computations, leaving the theoretical justification for choosing the particular gravitational law of general relativity for later chapters. The argument, which will not be complete until the end of Chapter 7, consists of four parts:

(1) The basic principle that force and acceleration are to be replaced by a least-action principle asserting that a particle in a gravitational field is simply moving along a path of minimal proper time in a metric of space-time that has been modified from its flat-space value $ds^2 = dt^2 - (dx^2 + dy^2 + dz^2)/c^2$ due to the presence of a second particle that is fixed at the origin of this space-time.

(2) A mathematical hypothesis based on symmetry considerations giving the general form of this metric.

(3) A physical hypothesis (the law of gravity proper) from which the specific form of the metric can be derived.

(4) The use of variational principles to find the orbit as a geodesic in the metric that results from the mathematical and physical hypotheses.

In this chapter, we discuss the first, second, and fourth of these procedures, giving only the statement of the third and postponing the justification of it to the three chapters that follow. We shall derive the equations of motion of a particle in the gravitational field created by a fixed point mass and then apply those equations to explain the anomalous precession of the perihelion of Mercury mentioned in the epigram above. Le Verrier's remark that the required theory was yet to be developed turned out to be prophetic.[4]

---

[3]The germ of this idea had occurred much earlier, to Einstein's Swiss compatriot Leonhard Euler (1707–1783). In his 1744 work on the calculus of variations, written while he was at the Prussian Academy of Sciences, he showed that a particle moving on a surface and not subject to any tangential forces would move along a geodesic. For information on Euler's equation in the calculus of variations and on geodesics, see Appendix 2. We really ought to be using Gauss' term *shortest path* rather than *geodesic* at this point, since a shortest path remains a shortest path when reparameterized, but a geodesic has only one parameterization. Nevertheless, we shall save time and words by just using the term *geodesic*, since the second-order differential equation from which they are constructed is the same in both cases.

[4]Le Verrier had noted the unexplained precession of the perihelion of Mercury by an amount that he calculated to be 38 seconds of arc per century more than the approximately 5557 seconds that could be accounted for through perturbations caused by the other planets. He thought the discrepancy might indicate the presence of an undiscovered planet near the Sun. He tended to think along these lines, having been the co-discoverer of Neptune, to which he was led by calculating its perturbation of the orbit of Uranus.

As our first analysis of planetary orbis, we present the classical Newtonian explanation of the two-body problem using Newton's famous second law of motion, popularly referred to as "$F = ma$." Our presentation saves time by using the vector analysis invented in the late nineteenth century. Once the presentation is complete, we'll attempt to update it, first by giving the force the meaning it has in special relativity—a technique that shed wonderful light on the Maxwell equations in Chapter 3—and then by attempting to construct a perturbation of the standard Euclidean metric on $\mathbb{R}^3$ such that the Euler equations for geodesics will coincide with Newton's differential equations of motion. Neither of these projects will work, as it turns out, and we shall be forced to invoke general relativity to solve the problem.

2. First Analysis: Newtonian Orbits

We interpret the Newtonian computation of the orbit as a two-particle problem. (The gravitational equivalence of a rigid body to a particle of the same mass located at the center of gravity of the body was proved by Newton himself.) In the formulation of the problem that we shall use, the Sun is regarded as a heavy particle of mass M fixed in space at the origin, so that its location as a vector is $\mathbf{0}$ at all times. We treat the planet—which we shall refer to as Mercury for the sake of definiteness, although the argument is of course not specific to that planet—as a particle of mass m, so much smaller that its gravitational effect on the Sun is negligible. This assumption is justified by the fact that $M = 30,000\,m$. By neglecting the gravitational effect on the Sun, we are using what Sternberg ([**78**], p. 298) calls a *passive equation.*[5]

If $\mathbf{r}(t)$ is the location of Mercury at time t, We recall from Chapter 2 that Newton's second law, together with his law of gravity, says

$$(m\mathbf{r}')' = -\frac{GMm}{r^3}\mathbf{r}\,,$$

where m is the mass of the orbiting particle, G is the universal gravitational constant, equal to 6.67384×10^{-11} meters-cubed per second-squared per kilogram, M is the mass of the Sun, equal to 1.98892×10^{30} kg, and $r = |\mathbf{r}| = \sqrt{\mathbf{r} \cdot \mathbf{r}}$. We begin by noting a simple property of orbits under this law, which despite its simplicity is sufficiently noteworthy that we state it as a theorem.

THEOREM 4.1. *The orbit of a particle moving subject to Newton's law of gravity is independent of the (rest) mass of the particle, whether the force is interpreted as Newtonian or relativistic.*

PROOF. Writing $m = \alpha m_0$, where $\alpha = (1 - \mathbf{r}' \cdot \mathbf{r}'/c^2)^{-1/2}$ in the relativistic case and $\alpha = 1$ in the Newtonian case, we can simply cancel m_0 from the equation of motion and get an equation that is independent of m_0. □

REMARK 4.1. Theorem 4.1 states a fundamental principle known as the *principle of equivalence*, that is, the inertial mass m that appears in the equation $\mathbf{F} = m\mathbf{r}''$ is the same as the gravitational mass m that appears in the formula for gravitational

---

[5]Sternberg distinguishes two kinds of physical laws, which he calls *source equations* and *passive equations*. The former are the broad general laws of physics, like the field equations of general relativity, applicable to idealized models and arrived at by inverting what Sternberg calls a Legendre transformation. The latter are particular applications of them intended to apply in practical cases, with the above-mentioned exclusion of small effects.

attraction $-(GMm/r^3)\boldsymbol{r}$. This principle, in relation to falling bodies, was stated by Galileo, and there is a famous legend of his having dropped two weights from the Leaning Tower of Pisa to test it. The principle distinguishes gravitational force from other forces such as, for example, electrostatic forces, which are determined by charge rather than mass. (Correspondingly, charge, unlike mass, is independent of velocity.) Everyone has noticed that when a traffic light turns from red to green, a large truck will accelerate more slowly than a small sports car next to it in the adjacent lane. That is because the motive forces per unit mass in the two are different. But if the two vehicles were driven off a cliff simultaneously, each would experience the same acceleration, and they would remain side-by-side all the way to the bottom.

According to Clifford M. Will ([**86**], Chapter 2), Newton tested the principle, since it implied that the period of a pendulum was determined only by its length and was independent of the weight and material of the pendulum bob. Later, the Hungarian nobleman Baron Roland von Eötvös[6] (1848–1919) made a more sophisticated test, using the fact that the gravitational attraction of the Earth and the centrifugal force due to its rotation, at any point between the poles and the equator, are not lined up. The two components of the acceleration of a body due to these forces are respectively inversely proportional to the gravitational and inertial mass. Eötvös found no difference in the two.

A corollary of this principle is that one can use gravitational acceleration as a unit of force, as we do when we talk about g-force. The apparent increase in weight that we feel when an elevator we are in begins to go upward or an airplane we are aboard takes off is an example of this equivalence. We feel heavier until the elevator or airplane reaches its "cruising" speed, as if the gravitational field of the Earth had suddenly increased and (in the case of the airplane) changed direction.

2.1. Kepler's second law: conservation of angular momentum. Our first observation is that, because the force is central, the angular momentum per unit mass $\boldsymbol{l}$, which is $\boldsymbol{r} \times \boldsymbol{r}'$, is conserved, as we proved in Chapter 2 (Theorem 2.10). As a result (Corollary 2.2), the motion is confined to the plane perpendicular to the constant vector $\boldsymbol{l}$, in which we shall henceforth use polar coordinates (r, θ). The vector equation now becomes a system of two differential equations. Taking $\boldsymbol{r}(t) = r(t)\cos\theta(t)\,\boldsymbol{i} + r(t)\sin\theta(t)\,\boldsymbol{j}$, we find that these two equations are

$$r'' - r(\theta')^2 = -\frac{GM}{r^2},$$
$$2r'\theta' + r\theta'' = 0.$$

The second of these equations merely expresses the conservation of angular momentum per unit mass, as established in Theorem 2.10. This is Kepler's second law.

Because of this law, after we prove that the orbit is an ellipse, we shall be able to express the magnitude of the angular momentum per unit mass l as follows, using only the average distance (a) from the Sun, the eccentricity of the orbit (e), and the period (T)

$$l = \frac{2\pi ab}{T} = \frac{2\pi a^2\sqrt{1-e^2}}{T},$$

---

[6]Pronounced "Utvush." His name in Hungarian was actually Loránd.

where a and b are respectively the maximum and minimum distances from Mercury to the Sun (the distances at aphelion and perihelion respectively), and T is the time elapsed between successive perihelia. When we work the same problem using the theory of relativity, we shall be able to eliminate l and T entirely from the differential equation of the orbit. But we need these constants now in order to determine a certain distance that is required to make the relativistic computation possible.

The average distance from the Sun to the planet is interpreted as the average of the distance r_{aph} at aphelion and its distance r_{peri} at perihelion. That distance is the semi-major axis of the elliptical orbit, denoted a here. It will follow once we have proved that the orbit is in fact an ellipse (Kepler's first law) that the semi-minor axis is $b = a\sqrt{1 - e^2} = 5.66715 \times 10^{10}$ meters. The eccentricity e is $(r_{\mathrm{aph}} - r_{\mathrm{peri}})/(r_{\mathrm{aph}} + r_{\mathrm{peri}})$, so that

$$1 - e^2 = \frac{4r_{\mathrm{aph}}r_{\mathrm{peri}}}{(r_{\mathrm{aph}} + r_{\mathrm{peri}})^2}.$$

We can therefore write the angular momentum per unit mass as

$$l = \frac{\pi(r_{\mathrm{aph}} + r_{\mathrm{peri}})\sqrt{r_{\mathrm{aph}}r_{\mathrm{peri}}}}{T}.$$

The fact that the motion is planar (direction of the angular momentum vector is constant) and that angular momentum is conserved (its magnitude is also constant) both follow from the fact that the force is central. In the relativistic model, the analogous feature of the problem is that the space-time metric is *radial*, that is, the coefficients of the differentials depend only on the radial coordinate ρ.

When we replace θ' by l/r^2 in the first of Newton's equations, we get the fundamental differential equation for r:

$$r'' = \frac{1}{r^3}(l^2 - GMr).$$

To get an estimate of the magnitudes we are dealing with here, let us calculate some numbers specific to the orbit of Mercury. The observational data are as follows:

Distance from the Sun at perihelion: $r_{\mathrm{peri}} = 4.60012 \times 10^{10}$ meters;

Distance from the Sun at aphelion: $r_{\mathrm{aph}} = 6.98169 \times 10^{10}$ meters;

Average distance from the Sun: $a = \frac{1}{2}(r_{\mathrm{peri}} + r_{\mathrm{aph}}) = 5.7909 \times 10^{10}$ meters;

Eccentricity (since, as we are about to prove, the orbit is an ellipse):

$$e = \frac{r_{\mathrm{aph}} - r_{\mathrm{peri}}}{r_{\mathrm{aph}} + r_{\mathrm{peri}}} = 0.20563.$$

Orbital period: $T = 87.969 \times 86,400 = 7.60052 \times 10^6$ seconds.

From these data, we calculate that:

$$\begin{aligned} GMr_{\mathrm{peri}} &= 6.10939 \times 10^{30}\,\mathrm{m^2/s^2}, \\ GMr_{\mathrm{aph}} &= 9.27234 \times 10^{30}\,\mathrm{m^2/s^2}, \\ l^2 &= 7.3603 \times 10^{30}\,\mathrm{m^4/s^2}. \end{aligned}$$

Our computation of l^2 is obtained by dividing twice the area of the ellipse ($2\pi ab$) by the orbital period (T) to get the angular momentum l per unit mass, then squaring the result. Since the angular momentum l per unit mass is twice the rate at which area is swept out and is a constant $r^2\theta'$, it follows that the tangential speed $r\theta'$ is l/r at all times. In particular, at perihelion we have $r\theta' = 58{,}976.4$ meters per second, a little less than one five-thousandth of the speed of light. Since $r' = 0$ at this point, this is the actual speed v with which Mercury is moving. Thus, the derivative of proper time with respect to laboratory time in the Sun-centered frame of reference, which is $\sqrt{1 - (v^2/c^2)}$, is not significantly smaller than 1. In fact, it is about 0.999999981. The same argument applies at aphelion, where the derivative is about 0.999999992. It follows that, without any sensible error, we can take the relativistic angular velocity per unit mass equal to the Newtonian. This way of computing l^2 is not quite satisfactory, since it is the only quantity appearing in our computations that actually requires knowledge of the period T. When we do the same computation using relativistic considerations, we will be able to eliminate this dependence on T, expressing the angular velocity in terms of a certain fixed distance ρ_s and the perihelion and aphelion distances of Mercury from the Sun.

As these numerical data show, the quantity GMr is larger than l^2 at aphelion and smaller at perihelion. Such would have to be the case, naturally, since r has a minimum at perihelion and a maximum at aphelion. At these points, $r' = 0$, and r'' must be negative at aphelion and positive at perihelion.

2.2. Kepler's third law: period vs. mean radius. Because r' is missing from the differential equation, we can execute the usual trick, setting $p = dr/dt$, so that

$$r'' = dp/dt = (dp/dr)(dr/dt) = p\,(dp/dr)\,.$$

We then have the differential equation

$$p\,dp = \left(\frac{l^2}{r^3} - \frac{GM}{r^2}\right)dr\,.$$

If we integrate this from the perihelion point, where $p = 0$, $r = r_{\mathrm{peri}}$, we find

$$\frac{1}{2}p^2 = \frac{GM}{r} - \frac{l^2}{2r^2} - \frac{GM}{r_{\mathrm{peri}}} + \frac{l^2}{2r_{\mathrm{peri}}^2}\,.$$

That is,

$$\left(\frac{dr}{dt}\right)^2 = \frac{2GMr - l^2}{r^2} - \frac{2GMr_{\mathrm{peri}} - l^2}{r_{\mathrm{peri}}^2}\,.$$

This equation has no real solutions unless some value $r(t)$ is such that the right-hand side is positive. If this condition is met, then the solution $r(t)$ will always stay between the perihelion and aphelion values. As it approaches one of these values, its derivative tends to zero, and its second derivative, as shown above, is negative at aphelion and positive at perihelion, meaning that r will return to the region after reaching its extreme value, that is, to the region where the right-hand side is positive, rather than crossing into the region where it is negative.

Taking $r = r_{\mathrm{aph}}$, we find, since $dr/dt = 0$ at this point also, that

$$2GM\left(\frac{1}{r_{\mathrm{peri}}} - \frac{1}{r_{\mathrm{aph}}}\right) = l^2\left(\left(\frac{1}{r_{\mathrm{peri}}}\right)^2 - \left(\frac{1}{r_{\mathrm{aph}}}\right)^2\right),$$

that is,

$$
\begin{aligned}
2GM &= l^2\left(\frac{r_{\text{aph}} + r_{\text{peri}}}{r_{\text{aph}}r_{\text{peri}}}\right) = \frac{2l^2}{a(1 - e^2)} \\
&= \frac{\pi^2 r_{\text{aph}} r_{\text{peri}}(r_{\text{aph}} + r_{\text{peri}})^2}{T^2}\left(\frac{r_{\text{aph}} + r_{\text{peri}}}{r_{\text{aph}}r_{\text{peri}}}\right) \\
&= \frac{\pi^2(r_{\text{aph}} + r_{\text{peri}})^3}{T^2},
\end{aligned}
$$

and this means (since $a = (r_{\text{peri}} + r_{\text{aph}})/2$) that

$$
\frac{a^3}{T^2} = \frac{GM}{4\pi^2},
$$

that is, the cube of the average distance to the Sun divided by the square of the period is a constant determined by the mass of the Sun and the gravitational constant. It is independent of the particular planet whose orbit is being calculated. This is Kepler's third law, historically a very important one, since it suggested the inverse-square law of gravity in the first place. For Mercury, $a^3/T^2 = 3.36165 \times 10^{18}$ cubic meters per second-squared, and $GM/(4\pi^2) = 3.36228 \times 10^{18}$ cubic meters per second-squared. Considering that all four of the quantities in these two equations are obtained from independent measurements, this is remarkably good agreement between theory and observation.

2.3. Kepler's first law: conic-section orbits. Since θ' is positive, we can change variables from t to θ to discuss the shape of the orbit. The conversion comes from the chain rule: Since $d\theta/dt = l/r^2$, it follows that $d/dt = (l/r^2)d/d\theta$. We then get the equation

$$
\frac{l^2}{r^4}\left(\frac{dr}{d\theta}\right)^2 = \left(\frac{dr}{dt}\right)^2 = \left(\frac{2GMr - l^2}{r^2}\right) - \left(\frac{2GMr_{\text{peri}} - l^2}{r_{\text{peri}}^2}\right),
$$

which we rewrite as

$$
\left(\frac{1}{r^2}\frac{dr}{d\theta}\right)^2 = \left(\frac{2GM}{l^2 r} - \frac{1}{r^2}\right) - \left(\frac{2GM}{l^2 r_{\text{peri}}} - \frac{1}{r_{\text{peri}}^2}\right).
$$

Since θ is dimensionless, both sides here have the physical dimension of length to the power -2. We now use a device that is always applied at this point, replacing r by $1/u$. This equation then becomes

$$
\left(\frac{du}{d\theta}\right)^2 = \left(\frac{2GM}{l^2}\right)u - u^2 - \left(\frac{2GM}{l^2}\right)u_{\text{peri}} + u_{\text{peri}}^2.
$$

We can eliminate the constant here by differentiating this equation with respect to θ, then dividing it by $2(du/d\theta)$:

$$
(4.1) \qquad\qquad u'' + u = \frac{GM}{l^2}.
$$

This is a linear equation, and as such, its general solution is obtained by adding to a particular solution—say the constant solution $u = \frac{GM}{l^2}$—the general solution of the homogeneous equation $u'' + u = 0$, which, as is well-known, is $u_0 \cos(\theta + \theta_0)$, for arbitrary constants u_0 and θ_0. Thus,

$$
\frac{1}{r} = u = \frac{GM}{l^2} + u_0 \cos(\theta + \theta_0).
$$

If we measure θ from perihelion, we can set $\theta + \theta_0 = 0$ at that point, and then we have

$$u_0 = \frac{1}{r_{\text{peri}}} - \frac{GM}{l^2}.$$

In that case, multiplying by $l^2/GM = a(1 - e^2)$, we have

$$r(1 + e\cos\theta) = a(1 - e^2).$$

If we measured θ from aphelion instead, the equation would be

$$r(1 - e\cos\theta) = a(1 - e^2).$$

These last two equations, as the reader may know, are polar forms of the equation of an ellipse with one focus at $r = 0$, eccentricity e, and semi-major axis a. Actually, Newtonian theory in general says only that the orbit will be a conic section, which might *a priori* be a parabola or hyperbola. But if we assume that the orbit has an aphelion point, it can only be an ellipse or circle. One can verify that the equation $r(p + q\cos\theta) = n$ is the equation of an ellipse if $p^2 > q^2$. It is the equation of a hyperbola when $p^2 < q^2$ and the equation of a parabola when $p^2 = q^2$. Conservation of energy explains the three forms. If the sum of gravitational potential energy and the initial kinetic energy of the planet is larger than the energy needed to move the planet to infinity against the gravitational field, the orbit will be a hyperbola. If the two energies are exactly equal, it will be a parabola. Elliptic orbits result when the planet does not have enough energy to escape to infinity. Here we invoke the empirical fact that the orbit of a planet is bounded to conclude that $q^2 < p^2$. To see why this is the equation of an ellipse, write it as $pr = n - qr\cos\theta = n - qx$ and convert it to its rectangular form by squaring, replacing r^2 with $x^2 + y^2$, completing the square on x, and dividing by the right-hand side to get

$$\frac{\left(x + \frac{nq}{p^2-q^2}\right)^2}{\left(\frac{np}{p^2-q^2}\right)^2} + \frac{y^2}{\left(\frac{n}{\sqrt{p^2-q^2}}\right)^2} = 1.$$

The semi-major axis is $np/(p^2-q^2)$, and the eccentricity is $\sqrt{1 - (p^2 - q^2)/p^2} = |q/p|$. Since $p = 1$, $q = -e$, and $n = a(1 - e^2)$, this does indeed confirm that the semi-major axis is a and the eccentricity is e. Thus, we have now proved everything we set out to prove in this case.

3. Second Analysis: Newton's Law with Relativistic Force

We are eventually going to be forced to consider general relativity because the problem we have set for ourselves is to explain the *observed fact* that the perihelion of Mercury was precessing at a rate about 1% higher than could be accounted for using Newton's laws together with perturbations of the orbit due to the other planets. We are going to show, as Einstein did, that this effect can be explained using general relativity. But general relativity is a radical step beyond even special relativity. Indeed, general relativity contradicts the basic assumption of special relativity, in that the speed of light in empty space is *not* constant. That assumption needs to be qualified by postulating the absence of any gravitational field. Before bringing in "big guns" to pacify the problem, we should explore what can be done using only special relativity. Retaining Newton's law of gravity, but taking the force to be the relativistic force of Chapter 2, we face the system of differential equations

$$\frac{d\boldsymbol{p}}{dt} = -\frac{GMm}{(\boldsymbol{r} \cdot \boldsymbol{r})^{3/2}}\boldsymbol{r},$$

where now $m = \alpha m_0$, $\boldsymbol{p} = m\boldsymbol{r}'$, and $\alpha = \left(1 - (\boldsymbol{r}' \cdot \boldsymbol{r}')/c^2\right)^{-1/2}$. Because the force is central, the motion will be planar; that is, relativistic angular momentum will be conserved. Although the computations are tedious—*Mathematica* can save a great deal of work here (see Problem 4.21 below)—one does eventually get the interesting second-order differential equation

$$(4.2) \qquad r\left(\frac{d\theta}{dt}\right)^2\left(r^2 + 2\left(\frac{dr}{d\theta}\right)^2 - r\frac{d^2r}{d\theta^2}\right) = GM.$$

Letting $u = 1/r$, and recalling that $(\theta')^2 = l^2/(\alpha^2 r^4)$, where l is the constant relativistic angular momentum, we are led to the equation

$$u'' + u = \frac{GM\alpha^2}{l^2},$$

where the independent variable is the angle θ. We shall assign the Newtonian value $\sqrt{GMa(1 - e^2)}$ to the relativistic angular momentum l (see Problem 4.22), since we are going to be doing approximations. With a little reflection, we find that α^2 can be eliminated from this equation, leading to the final version of it:

$$(4.3) \qquad u'' + u = \frac{GM}{l^2} + \frac{GM}{c^2}\left(u^2 + (u')^2\right).$$

This relation shows that the relativistic equation is essentially a perturbation of the Newtonian equation, the perturbation term being of the order of u^2, since typically $(u')^2$ is very small, unless the eccentricity of the orbit is very large. Perturbations of that magnitude will also appear in the next two modifications of Newton's equation of motion. The use of the word *perturbation* here is not intended to imply that general relativity is in any sense a mere perturbation of special relativity, much less of Newtonian mechanics.

Even *Mathematica* finds a closed-form solution of this differential equation to be beyond its power and simply gives it back when asked to solve it. We are not interested in solving it exactly at the moment and have presented it only to show what happens when we try to investigate planetary orbits using relativistic mechanics—what Poincaré in the quotation above called the "new Dynamics."[7] Nevertheless, there is a way of solving it that one might call the "Catch-22" method. By trying plausible forms for the *inverse* function $\theta(u)$, one finds that this inverse function can be can be expressed in quadratures (see Problem 4.23), leading to an intriguing reduction of this equation to an equivalent form that *Mathematica* can solve.[8] But one has to give the equivalent form to *Mathematica*, which does not discover it on its own. Even then, *Mathematica* only gives back the inverse function that was used to produce the reduction in the first place. In other words, *Mathematica* is of no help in solving this equation. The reason for the bizarre name we have given this method is that it is, to say the least, unusual to require

---

[7]Poincaré himself, inspired by the work of Lorentz, created a considerable portion of special relativity independently of Einstein.

[8]And not only *Mathematica*. The equivalent form does not contain u' and therefore can be solved by the standard technique of letting $u' = p$, $u'' = p\,dp/du$; any sophomore can do it.

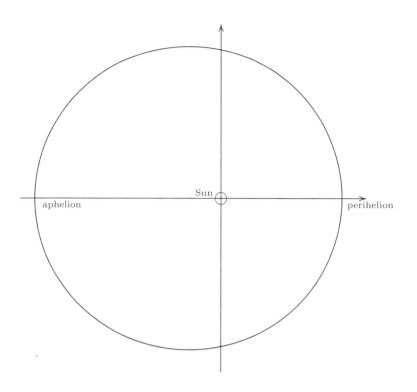

FIGURE 4.1. Orbit of Mercury traversed three times, interpreting
the Newtonian gravitational force as in special relativity.

that the solution to a problem be known in advance in order to create a method of
solving the problem.[9] Despite these oddities, it turns out—again, see Problem 4.23
below—that the converted equation is equivalent to a much simpler equation in
which the exponential function appears as a coefficient.

The coefficient GM/c^2 turns out to be exactly half of the Schwarzschild radius
r_s that will be introduced below. If the Sun were to become a nonrotating black hole
(it won't), its radius would be r_s. Numerical solution of Eq. (4.3) over a range of
values of the angle θ from 0 (at perihelion) to 20 (slightly more than three complete
orbits) using the empirical data for Mercury leads to the graphical representation
of the orbit shown in Fig. 4.1. Notice that, despite having traced the orbit three
times, there is no evidence of any deviation from the Newtonian model. Of course,
we should we not expect any precession to be apparent to the unaided eye over
such a short period.

On the other hand, numerical study of differential equations of the form $u'' + u =
a + b\big(u^2 + (u')^2\big)$ shows that noticeable precession does occur if ab has a value
larger than about 0.01. Figure 4.2 shows the case $a = 2$, $b = 0.02$ with an assumed

---

[9]Another example of the method comes from the Galois group of an equation, from which
one can determine whether or not the equation is solvable through the application of a finite
number of algebraic operations. The trouble is, in most cases (not all), one has to know the roots
already in order to compute the group. Once you know the roots, it is probably not an interesting
question whether they could have been obtained by algebraic operations.

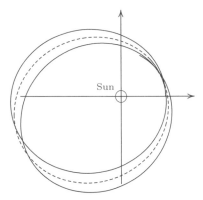

FIGURE 4.2. Orbit of a planet when the reciprocal of its distance from the Sun (denoted $u(\theta)$) satisfies the differential equation $u''(\theta) + u(\theta) = 2 + 0.02\big((u(\theta))^2 + (u'(\theta))^2\big)$ with initial conditions $u(0) = 3$ and $u'(0) = 0$. The orbit is shown for $0 \leq \theta \leq 20$, a little more than three complete cycles of the angular variable θ and exhibits a noticeable precession. The second cycle is shown as a dashed curve.

initial value $u(0) = 3$. In the case shown in Fig. 4.1, with actual data for the orbit of Mercury, we have $a = 1.80309 \times 10^{-11}\,\mathrm{m}^{-1}$ and $b = 1476.9\,\mathrm{m}$, so that $ab = 2.66298 \times 10^{-8}$, which is obviously much less than 0.01.

4. Third Analysis: Newtonian Orbits as Geodesics

The inverse-square law, which is a plausible law of weakening of a supposed absolute force, becomes quite unintelligible (and indeed impossible) when expressed as a restriction on the intrinsic geometry of space-time; we have to substitute some law obeyed by the tensors which describe the world-conditions determining the natural geometry.

Eddington ([**16**], p. 82).

Before we pass to general relativity we wish to introduce geodesics into the classical problem we have just analyzed, so that the relativistic version we are going to analyze will not seem so radically different as it sometimes appears. We return to the Newtonian interpretation of the equations of motion we have just studied. One advantage to be gained from this approach is that the Schwarzschild radius just mentioned can be made to appear in Newtonian mechanics, and it does so when we make the equations of motion into the equations of geodesics in a perturbed metric. What we get is no longer Newton's basic equation of motion, but a perturbation of it by a term that is proportional to this "Newtonian Schwarzschild" radius. There is no deep mystery about all this. It is the connection of the distance with potential energy that causes that particular radius to be distinguished. As a result, in both Newtonian and relativistic physics, the passage from a flat space to a curved one involves perturbing some of the metric coefficients by terms proportional to the potential energy in the Newtonian case, or the metric coefficients—which are analogs of potential energy—in the relativistic case.

4.1. From force to energy. The Newtonian mathematical model of particle motion is based on the concept of force, a physical concept of very ancient lineage that goes all the way back to Aristotle. Archimedes probably would have found that Newton's treatment of mechanics had some resemblance to his own ways of thinking.[10] The foundations of modern mechanics were laid down by Newton out of the philosophical legacy that he inherited from Descartes and the Medieval philosophers and (ultimately) Aristotle. Only the foundations, however. Much of the above-ground part of the edifice, which is still taught to university students today as "classical mechanics," was built in the eighteenth century and mostly by mathematicians on the Continent, such as Euler and Lagrange.

In the previous chapter, we saw that Newton's differential equations of motion can be obtained from Euler's equations for the minimum of the time integral of the difference between kinetic and potential energy. In a sense, this result describes the geometry of a highly abstract space, in which the points represent the *energies* of the mechanical system. Mechanics was developed along these lines by Heinrich Hertz (1857–1894) in his posthumously published treatise [**39**]. The integrand in the variational problem, however, is not given as the element of arc length on a surface, which is the more geometric sense in which we usually think of a geodesic. Hertz geometrized mechanics beautifully, in a treatise that contains remarkably few formulas for a work of mathematical physics. He showed, for example, that the square of the acceleration of a particle can be written as $c^2 v^4 + \dot{v}^2$, where v is the speed of the particle, $\dot{v} = v'(t)$ is its time derivative, which he showed was equal to the tangential component of the acceleration, and c is the curvature[11] of the path along which the particle is moving, expressed in angular units per unit length. Since angles are dimensionless, both cv^2 and $\dot{v}$ have the correct physical dimensions of length divided by time-squared. For a path of minimum acceleration, hence minimal force on the particle, one needs $\dot{v} = 0$, so that v is constant, and then the curvature c needs to be as small as possible; that is, the path requiring least action is a maximally straight path traversed at a uniform rate of speed—a geodesic.

Even with this elegant formulation of mechanics, Hertz found it necessary to talk about force, however, and he did not derive Newton's equations of motion from the element of arc length on a path in physical space traversed by the particle. We shall make an effort in that direction to see where it leads. (It leads to interesting things.)

---

[10]The Greeks, especially Eudoxus and Archimedes, were able to carry out some infinitesimal reasoning, in a rather ponderous way. By considering all possible integer multiples of two quantities of the same type, they were able to say what a ratio was with infinite precision. Actually, that last statement is not quite accurate. They were able to give meaning to the statement that the *vaguely defined, intuitive* quantity they called a ratio was equal to another such ratio, with infinite precision. That is, they could talk meaningfully about exact proportion. Assuming a trichotomy law for any two ratios, they were then able to reason by *reductio ad absurdum* using what we would call rational approximations to the ratios to prove that two ratios were equal to each other in that infinitely precise sense. Newton, one of the pioneers of calculus, used infinitesimal reasoning, saying explicitly that he was doing so to avoid the tedium of having to write everything out according to the method of the ancients.

[11]The choice of c to denote curvature looks natural in English, although it actually is a bit odd, since Hertz wrote in German, where the word is *Krümmung*. We shall repay the linguistic compliment by using the letter κ for curvature, since we are already using the letter c for the speed of light.

By the late nineteenth century, Henri Poincaré was writing that geometry and physics were to some degree interchangeable at their interface. If one wanted space to be Euclidean, that could be arranged by choosing a suitable list of forces to explain motion on the basis of Newton's second law of motion. But Poincaré also realized that the opposite was possible. The notion of force can be eliminated entirely if a suitable non-Euclidean geometry is assumed. The trajectory of a particle can then be described as a geodesic. Such a principle (Fermat's principle) works well in explaining the refraction of light, for example, where the concept of force does not enter. Mathematically, such an equivalence is to be expected, since the Euler equations for extremals in the calculus of variations are second-order partial differential equations, just like the dynamic equations used to formulate the laws of mechanics.

The Euler equations are explained in Appendix 2 of Volume 2. Their application in geometry will be explained piecewise as we proceed. The best way to begin is to note that on the infinitesimal level, the Pythagorean theorem allows us to rectify a curve in $\mathbb{R}^3$ by integrating its element of arc length ds, given by the equation

$$ds^2 = dx^2 + dy^2 + dz^2 \,,$$

so that the length of a curve is

$$\int \sqrt{\left(\frac{dx}{ds}\right)^2 + \left(\frac{dy}{ds}\right)^2 + \left(\frac{dz}{ds}\right)^2}\, ds \,,$$

and volumes are given similarly on the infinitesimal level as

$$dV = dx\, dy\, dz \,.$$

When we change coordinates, for example, to use spherical co-latitude (φ) and longitude (θ) along with the radius (ρ), these formulas become

$$
\begin{aligned}
ds^2 &= d\rho^2 + \rho^2\, d\varphi^2 + \rho^2 \sin^2 \varphi\, d\theta^2 \,, \\
dV &= \rho^2 \sin \varphi\, d\rho\, d\varphi\, d\theta \,.
\end{aligned}
$$

These formulas are typical of a general rule that allows us to define distance and n-dimensional volume on a manifold[12] of dimension n parameterized by coordinates $(x^1, \ldots, x^n)$. To do so, we use the matrix M that we called the *matrix of metric coefficients* in the preceding chapter. (Einstein called it the *fundamental tensor* and denoted it g. This is the last time we can use the letter M for it, since we are henceforth reserving that letter for the mass of the Sun.)

$$M = \begin{pmatrix} g_{11} & \cdots & g_{1n} \\ \vdots & \ddots & \vdots \\ g_{n1} & \cdots & g_{nn} \end{pmatrix} \,.$$

We then define the elements of arc length and volume as

$$
\begin{aligned}
ds^2 &= \sum_{i=1}^{n} \sum_{j=1}^{n} g_{ij} dx^i\, dx^j \,, \\
dV &= \sqrt{\det(M)}\, dx^1 \cdots dx^n \,.
\end{aligned}
$$

---

[12]See Appendix 4 in Volume 2 for an explanation of manifolds in general.

We shall mostly be using coordinates in which the matrix M is diagonal, so that these formulas will not be as messy as their general form suggests. These generalizations, based on work of Gauss that will be explained in the next chapter, were introduced by Bernhard Riemann (1826–1866).

By late nineteenth century the theory of special relativity was being developed by Lorentz with advice from Poincaré. The notion of absolute time and a three-dimensional absolute space was already becoming obsolete. In the years from 1910 to 1915, Einstein was publishing papers on the geometrization of gravity, and his great triumph was based on four-dimensional space-time. We are now about to embark on our final attempt to explain planetary orbits using a curved three-dimensional space, taking "the road less traveled."[13]

4.2. From energy to geometry. The minimization problem for the Lagrangian is not completely geometric, since the two energies are not interpreted as lines or areas or volumes. Since the assigned task is to minimize the time required to traverse an arc of the orbit, we need to invoke our "temporization" of distance using the fixed but arbitrary speed v_0 introduced in Chapter 1, which we called the speed of information. The infinitesimal increment in time dt is then given in Cartesian coordinates by the quadratic form

$$dt^2 = \frac{1}{v_0^2} dx^2 + \frac{1}{v_0^2} dy^2 + \frac{1}{v_0^2} dz^2 \,.$$

We would like to nudge this variational formulation of Newton's laws over the border into pure geometry. Accordingly, we are setting the following challenge:

Impose a non-Euclidean metric on $\mathbb{R}^3$ in which the infinitesimal element of time dt on a particle whose spherical coordinates at time t are $\rho(t)$, $\varphi(t)$, $\theta(t)$ is of the form

$$dt^2 = \frac{1}{v_0^2} \left(g(\rho) \, d\rho^2 + \rho^2 \, d\varphi^2 + \rho^2 \sin^2 \varphi \, d\theta^2 \right),$$

and is such that Euler's geodesic equations

$$\frac{d}{dt}\left(2g(\rho)\rho'\right) = g'(\rho)\left(\rho'\right)^2 + 2\rho\left(\varphi'\right)^2 + 2\rho \sin^2 \varphi (\theta')^2 \,,$$

$$\frac{d}{dt}\left(2\rho^2 \varphi'\right) = 2\rho^2 \sin \varphi \cos \varphi \, (\theta')^2 \,,$$

$$\frac{d}{dt}\left(2\rho^2 \sin^2 \varphi \, \theta'\right) = 0 \,,$$

are Newton's equations of motion for a particle in the gravitational field due to a massive particle M located at the origin.

REMARK 4.2. In setting Newton's law of gravity as our goal and making it the equation of a geodesic, we are assuming the principle of equivalence mentioned in Remark 4.1 above. That is, the geodesic traversed by a particle will be independent of its mass, determined only by the metric in space. That principle will thus be part of our geometrized celestial mechanics. The same will be true, when we reach our goal of giving the relativistic law of gravity. Thus the principle of equivalence will be a testable part of general relativity.

---

[13]Less traveled because Einstein built a superhighway to bypass it. But not untraveled; see, for example, the recent book of Sternberg ([**78**], pp. 292–294).

The reason for modifying only the coefficient of the radial coordinate is that gravitational force is a central force and acts along radial lines. Although our quest will fail, it will nevertheless achieve a considerable amount, stopping just short of the result that we want.

The geodesics in the unperturbed metric are easily computed, being the mappings $t \mapsto \big(x(t), y(t), z(t)\big)$ that minimize the integral

$$\int dt = \frac{1}{v_0} \int \left(\left(\frac{dx}{dt} \right)^2 + \left(\frac{dy}{dt} \right)^2 + \left(\frac{dz}{dt} \right)^2 \right)^{\frac{1}{2}} dt \, .$$

Since t, x, y, and z do not appear in the integrand, Euler's equations say that there are constants c_x, c_y, c_z, x_0, y_0, and z_0 such that

$$\begin{aligned} x(t) &= x_0 + c_x t \, , \\ y(t) &= y_0 + c_y t \, , \\ z(t) &= z_0 + c_z t \, . \end{aligned}$$

In other words, the geodesics are straight lines in the Cartesian sense, traversed at a constant speed, as we already knew.

Let us now switch to spherical coordinates, in which the unperturbed metric has an infinitesimal time increment given by

$$dt^2 = \frac{1}{v_0^2} \big(d\rho^2 + \rho^2 \, d\varphi^2 + \rho^2 \sin^2 \varphi \, d\theta^2 \big) \, .$$

We are going to modify this metric by perturbing the coefficient of ρ only, since gravitational force has no tangential component. How shall this be done? We expect the influence of gravity to wane with distance, so that the perturbation should disappear at infinity. Thus, the function $g(\rho)$ that we assumed above ought to tend to 1 as $\rho \to \infty$. It is simpler to deal with the logarithm of $g(\rho)$, which will tend to 0 as $\rho \to \infty$. Accordingly, we assume that the perturbed metric has the form

(4.4) $$dt^2 = \frac{1}{v_0^2} \big(e^{\lambda(\rho/\rho_s)} \, d\rho^2 + \rho^2 \, d\varphi^2 + \rho^2 \sin^2 \varphi \, d\theta^2 \big) \, ,$$

where $\lambda(\rho/\rho_s)$ is written as a function of the ratio of ρ to a fixed distance ρ_s, since mathematical functions generally accept only pure numbers as input. The fixed distance ρ_s can be any unit. We are going to choose it so as to get the closest reasonable approximation to Newton's equations. The value we choose will be called, for reasons that will become clearer as we proceed, the "Newtonian Schwarzschild radius." (The relativistic version of it, corresponding to $v_0 = c$, is an extremely important quantity.) If v_0, the speed of information, were infinite,[14] we would be forced to set $dt = 0$ and replace the indeterminate form $v_0 \, dt = \infty \cdot 0$ by a finite spatial metric ds. In that case, we would have $\rho_s = 0$, $\rho/\rho_s = \infty$, and $e^{\lambda(\rho/\rho_s)} \equiv 1$ for all r; the square-distance metric would then be the usual Euclidean metric. (As we have already remarked, there is a sense in which $v_0 = \infty$ in Newtonian mechanics.) But we are going to depart from that principle and assume temporarily that v_0 is finite, since we don't want $dt = 0$. Our object is to choose the function λ so as to get Newton's equations of motion out of the Euler equations for a geodesic. Before beginning this project, we note that two

---

[14] Imagine the "subspace transmissions" that science-fiction writers sometimes invoke to solve the problem of communicating over intergalactic distances.

preliminary results from the Newtonian theory *already, independently of the choice of* λ, follow from the equations for a geodesic.

First, the Euler equation on the co-latitude angle φ is

$$\frac{d}{dt}\left(2\rho^2\varphi'\right) = 2\rho^2 \sin\varphi \, \cos\varphi \left(\theta'\right)^2.$$

This equation has a unique solution given initial values of the co-latitude $\varphi(t_0)$ and its derivative $\varphi'(t_0)$. Without stretching our imaginations too much, we can suppose that there is some time t_0 at which $\varphi(t)$ achieves a local minimum or maximum value φ_0.[15] If the axes are rotated suitably, we can arrange that $\varphi_0 = \pi/2$. Since the point is a local extremum, we have $\varphi'(t_0) = 0$. Now the equation just written is certainly satisfied by the constant function $\varphi(t) = \pi/2$, and this solution also satisfies the initial conditions. Therefore it is the only solution. Thus, φ drops out of the discussion, and so we have achieved the reduction to a planar problem. This is the same reduction we got by Newtonian reasoning from the vector relation

$$\frac{d}{dt}(\boldsymbol{r} \times \boldsymbol{r}') = \mathbf{0}.$$

Being now in the plane, our spherical coordinates ρ and θ are the same as polar coordinates r and θ. Accordingly, we change ρ to r and ρ_s to r_s. Our metric then becomes

$$dt^2 = \frac{1}{v_0^2}\left(e^{\lambda(r/r_s)}\,dr^2 + r^2\,d\theta^2\right).$$

Our first preliminary result is now established.

The second preliminary result is Kepler's second law (conservation of angular momentum). It follows from Euler's equation on the variable θ, which says

$$\frac{d}{dt}\left(2r^2\theta'\right) = 0.$$

As a consequence, the angular momentum per unit mass given by $l = r^2\theta'$, is constant. Written out in full, Euler's equation says

$$4r\frac{dr}{dt}\frac{d\theta}{dt} + 2r^2\frac{d^2\theta}{dt^2} = 0.$$

This equation is trivially the same as the second of Newton's two equations of motion, which, we recall from our proof of Kepler's first law, have the following form in polar coordinates:

$$(4.5) \qquad \frac{d^2r}{dt^2} = -\frac{GM}{r^2} + r\left(\frac{d\theta}{dt}\right)^2,$$

$$(4.6) \qquad \frac{d^2\theta}{dt^2} = -\frac{2}{r}\frac{dr}{dt}\frac{d\theta}{dt}.$$

Again, this result is independent of any choice of λ.

It now looks as if we have already come very near to the goal of geometrizing the motion of a planet, and we have not yet introduced any perturbation into the metric. We just need to choose λ so as to get the first of Newton's two equations,

---

[15]If not, by the intermediate-value theorem for derivatives, the co-latitude must be forever increasing or forever decreasing. But such a phenomenon has never been observed in any planet.

and for this purpose we have two tools. One of them is Euler's equation on the variable r, which says

$$\frac{d}{dt}\left(2e^{\lambda(r/r_s)}\frac{dr}{dt}\right) = \frac{1}{r_s}\lambda'\left(\frac{r}{r_s}\right)e^{\lambda(r/r_s)}\left(\frac{dr}{dt}\right)^2 + 2r\left(\frac{d\theta}{dt}\right)^2,$$

$$2e^{\lambda(r/r_s)}\frac{d^2r}{dt^2} = -\frac{1}{r_s}\lambda'\left(\frac{r}{r_s}\right)e^{\lambda(r/r_s)}\left(\frac{dr}{dt}\right)^2 + 2r\left(\frac{d\theta}{dt}\right)^2,$$

$$\frac{d^2r}{dt^2} = -\frac{1}{2r_s}\lambda'\left(\frac{r}{r_s}\right)\left(\frac{dr}{dt}\right)^2 + re^{-\lambda(r/r_s)}\left(\frac{d\theta}{dt}\right)^2.$$

The other tool is the result of dividing the equation for the metric by dt^2. This equation can be solved to show that

$$\left(\frac{dr}{dt}\right)^2 = e^{-\lambda(r/r_s)}\left(v_0^2 - r^2\left(\frac{d\theta}{dt}\right)^2\right).$$

(This equation is really just an integral of the preceding one.)

When we insert this value of $(dr/dt)^2$ into the previous equation, we have the second-order differential equation

$$\frac{d^2r}{dt^2} = -\frac{v_0^2}{2r_s}\lambda'\left(\frac{r}{r_s}\right)e^{-\lambda(r/r_s)} + re^{-\lambda(r/r_s)}\left(1 + \frac{r}{2r_s}\lambda'\left(\frac{r}{r_s}\right)\right)\left(\frac{d\theta}{dt}\right)^2.$$

We need this last equation to be Newton's other equation of motion, that is, Eq. (4.5). Hence the right-hand side must be

$$-\frac{GM}{r^2} + r\left(\frac{d\theta}{dt}\right)^2.$$

The first term in the equation for d^2r/dt^2 is

$$\frac{v_0^2}{2}\frac{d}{dr}e^{-\lambda(r/r_s)},.$$

It appears then that we need to choose λ so that

$$\frac{d}{dr}e^{-\lambda(r/r_s)} = -\frac{2GM}{v_0^2 r^2}.$$

By integrating, we see that we need

$$e^{-\lambda(r/r_s)} = C + \frac{2GM}{rv_0^2} = C + \frac{2GM}{r_s v_0^2}\frac{r_s}{r}.$$

Since the left-hand side of this last equation tends to 1 as $r \to \infty$, we see that $C = 1$. We have not yet chosen the Newtonian Schwarzschild radius r_s. We now do so, setting

$$r_s = \frac{2GM}{v_0^2}.$$

Then the perturbation factor $e^{\lambda(r/r_s)}$ is $r/(r + r_s)$. Thus, we have

$$\lambda(r/r_s) = \ln\left(\frac{r}{r + r_s}\right),$$

and

$$\lambda'\left(\frac{r}{r_s}\right) = \frac{r_s}{r} - \frac{r_s}{r + r_s} = \frac{r_s^2}{r(r + r_s)}.$$

We now have one of the two terms we were seeking, and the function λ is completely determined.

We now have only to calculate the remaining term, which is

$$re^{-\lambda(r/r_s)}\left(1 + \frac{r}{2r_s}\lambda'\left(\frac{r}{r_s}\right)\right)\left(\frac{d\theta}{dt}\right)^2.$$

Unfortunately, computation reveals that this expression is

$$\left(r + \frac{3}{2}r_s\right)\left(\frac{d\theta}{dt}\right)^2.$$

If not for the presence of the second term in parentheses here, this would be exactly Newton's differential equation of motion. Thus, we must reluctantly confess that our project has failed.

4.3. A possible new approach to gravity? Despite this failure, our effort to geometrize the physics of planetary orbits as geodesics in a non-Euclidean metric has yielded one positive result toward such a geometrization, namely conservation of angular momentum. And it comes very close to predicting the actual orbit of a planet. If r_s is very small compared with the average distance from the planet to the Sun (and it assuredly is if v_0 is the speed of light, as we shall see), that term can be neglected, and the geodesic equations will coincide with Newton's laws.

Perhaps we were not sufficiently imaginative when we made it our goal to get the equations Newton left us. This new, geometric point of view has brought to our attention a perturbation of them that may well yield the same predictions within the limits of physical measurement. Dare we hope that it might even do *better* and explain phenomena for which Newton's law of gravity has not proved adequate? Let us explore this possibility. We begin by stating the result we have found as a formal theorem:

THEOREM 4.2. *Let v_0 be an arbitrary fixed speed, used as an "exchange rate" to convert distances into time intervals, and let r_s be the "Newtonian Schwarzschild radius" given by $r_s = 2GM/v_0^2$, where G is the universal gravitational constant and M the mass of a particle fixed at the origin of a Cartesian coordinate system. Suppose further that a particle is moving along a path of minimal length (as measured by the time required to traverse it at speed v_0) in the planar metric in which the infinitesimal element of time dt along any path for which the particle has coordinates $(r(t), \theta(t))$ at time t is given by*

(4.7)
$$dt^2 = \frac{1}{v_0^2}\left(\frac{r}{r + r_s}\,dr^2 + r^2\,d\theta^2\right).$$

Then the functions $r(t)$ and $\theta(t)$ on a geodesic satisfy the system of perturbed Newtonian equations

(4.8)
$$\frac{d^2r}{dt^2} = -\frac{GM}{r^2} + \left(r + \frac{3}{2}r_s\right)\left(\frac{d\theta}{dt}\right)^2,$$

(4.9)
$$\frac{d^2\theta}{dt^2} = -\frac{2}{r}\frac{dr}{dt}\frac{d\theta}{dt}.$$

The non-Euclidean geometry that this metric imposes on the plane forces us to reinterpret the meaning of the radial coordinate r. It is no longer the distance from a point bearing that coordinate to the origin. As measured by the time standard, that distance is now

$$
\begin{aligned}
v_0 t_r &= v_0 \int_0^{t_r} dt = \int_0^r \sqrt{\frac{s}{s + r_s}}\, ds \\
&= \sqrt{r(r + r_s)} + r_s \ln\left(\frac{\sqrt{r_s}}{\sqrt{r} + \sqrt{r + r_s}}\right) \\
&= r\left(\sqrt{1 + \frac{r_s}{r}} + \frac{r_s}{r} \ln\left(\frac{\sqrt{r_s/r}}{1 + \sqrt{1 + r_s/r}}\right)\right).
\end{aligned}
$$

As the integral with respect to s shows, this number is *smaller than* r, so that $v_0 t_r$, the actual radius of the circle whose points have radial coordinate r, is less than r. The circumference, on the other hand, remains Euclidean. If t_c is the time required to traverse it at speed v_0, its length must be given by

$$
v_0 t_c = v_0 \int_0^{t_c} dt = r \int_0^{2\pi} d\theta = 2\pi r.
$$

The ratio of circumference to radius is therefore $t_c/t_r > 2\pi$.[16]

The new coefficient of dr^2 forces a change in the volume element on $\mathbb{R}^3$. In the Euclidean case that element is

$$
dV = \rho^2 \sin\varphi\, d\rho\, d\varphi\, d\theta = \sqrt{(d\rho)^2 (\rho\, d\varphi)^2 (\rho \sin\varphi\, d\theta)^2}.
$$

This volume element gets perturbed and becomes

$$
dV = \sqrt{\frac{\rho}{\rho + \rho_s}}\, \rho^2 \sin\varphi\, d\rho\, d\varphi\, d\theta.
$$

This change would hardly affect a particle, one might think, since a particle has no volume to perturb. But if a solid body were to move through this non-Euclidean space, it would experience internal stresses as its shape changed. Those internal stresses amount to work being done by the metric on the body, and thus need to be considered in the energy balance of any description of the motion.

Another way to look at the coefficient of $d\rho^2$ in the modified metric is to insert the definition of ρ_s into it. When that is done, the metric becomes

$$
dt^2 = \frac{1}{v_0^2}\left(\frac{1}{1 + \frac{GM}{\frac{1}{2}\rho v_0^2}}\, d\rho^2 + \left(\rho^2\, d\varphi^2 + \rho^2 \sin^2\varphi\, d\theta^2\right)\right).
$$

Thus, the perturbed coefficient of $d\rho^2$ is now seen to be a fractional-linear (Möbius) function of the ratio of two energies, which we write as V/T. Here $V = GM/\rho$ is $-V_\rho$, where V_ρ is the Newtonian potential energy per unit mass at distance ρ, while $T = t_{v_0} = v_0^2/2$ is the kinetic energy per unit mass for a particle moving at the speed of information v_0. Both energies per unit mass are constants independent of the particular particle whose orbit is being computed. Thus, we can express the metric in two different ways: "geometrically," using a ratio of two lengths, or "physically," using a ratio of two energies. We now have a connection

---

[16]This fact makes the geometry of the perturbed metric difficult to visualize. If the ratio t_c/t_r were less than 2π, we could imagine the circles being circles of constant latitude on a sphere, with the radius of a circle being the latitude in question. But how does extra length get "crammed into" a circle without increasing its radius?

between geometry and energy, in the form of the following relation, which amounts to a direct proportion between two energies and two distances:

$$(4.10) \qquad \rho_s T + \rho V = 0.$$

Equation (4.10), which says $\rho = -\rho_s T/V$, is the best clue we will have when we geometrize gravity in Chapter 7. The most difficult challenge we will face is that of adjusting our mathematical formulation of physical intuition so as to replace the Newtonian relation $\boldsymbol{F} = m\boldsymbol{r}''$ with the differential equations of a geodesic. The connecting link turns out to be the fact that the Newtonian potential energy is inversely proportional to ρ, which is the variable that occurs in the metric coefficients of the manifold that represents space-time in the presence of an attracting particle. The metric coefficients that we shall encounter will be recognized as analogs of potential energy, and the equations of motion will be expressed in terms of them and their partial derivatives.

4.4. Geometrized astronomy. Let us now explore this new approach and see how well it explains the things we already know. Our fundamental relation is the metric relation, which for a particle moving in a plane, becomes, after being divided by $d\theta^2$,

$$\frac{r}{r+r_s}\left(\frac{dr}{d\theta}\right)^2 = v_0^2\left(\frac{dt}{d\theta}\right)^2 - r^2 = \frac{v_0^2 r^4}{l^2} - r^2.$$

Here l is the constant angular momentum per unit mass, depending on the particular orbit the particle is traversing. We rewrite this relation as

$$\left(\frac{dr}{d\theta}\right)^2 = r(r+r_s)\left(\left(\frac{v_0 r}{l}\right)^2 - 1\right).$$

Since $r > 0$ and $r_s > 0$, we see that the equation for $(dr/d\theta)^2$ imposes a lower limit on r, namely $r \geq l/v_0$. Circular orbits where r has this lower limit as a constant value do satisfy the geodesic equation. But this same expression gives us the very bad news that *elliptic orbits are impossible!* In an elliptic orbit, we must have $dr/d\theta = 0$ at two *different* positive values of r, namely perihelion and aphelion. That is impossible, given that v_0 and l are constant for a given orbit. That is a disastrous failure of our project,[17] since it means we cannot get Kepler's first law.

Despite this setback, there is insight of importance to our project that can be gained by pursuing the analysis a little longer. As in the analysis we gave based on Newton's laws, we find that this equation looks better if we replace r by $u = 1/r$, then differentiate with respect to θ and cancel $2\,du/d\theta$. The result of doing so is the equation

$$u'' + u = \frac{v_0^2 r_s}{2l^2} - \frac{3}{2}r_s u^2 = \frac{GM}{l^2} - \frac{3}{2}r_s u^2.$$

This equation is Newton's equation derived above, only with the perturbation term $-(3/2)r_s u^2$. Of course, when we do the usual trick of replacing u'' by $p\,dp/du$, where $p = du/d\theta$ and integrating from a value of u where $p = 0$ (in this case the value $u = u_1 = v_0/l$), we get the previous equation back, and so we can apparently

---

[17]In case the reader has forgotten, our new project was begun after the failure of our attempt to get Newton's equations of motion out of geometry, and represented an attempt to study astronomy using the perturbed Newtonian equations that we were able to derive from geometry.

not get much comfort from the fact that this equation is a small perturbation of the Newtonian equation.

Nevertheless, it will be useful to consider this equation in the abstract, since it will appear again in the relativistic discussion. We state the main results as a formal theorem.

THEOREM 4.3. *Consider the second-order ordinary differential equation*

$$u''(\theta) + u(\theta) = \alpha + \beta \big(u(\theta)\big)^2,$$

where $\alpha > 0$, β is an arbitrary real number (positive, negative, or zero), and $u(\theta)$ is required to assume only positive values. Let $u_1 = u(\theta_1) > 0$ be a local maximum value of a solution $u(\theta)$, Then, either $u(\theta)$ decreases for all $\theta > \theta_1$, or it assumes a positive minimum value $u_2 = u(\theta_2)$ and thereafter remains in the range $u_2 \leq u(\theta) \leq u_1$ for all $\theta > \theta_1$.

If u_1 is a local minimum value of u and $\beta < 0$, then $u(\theta)$ increases toward a finite value $u_2 > u_1$ that it cannot exceed. If it reaches that value, it will thereafter remain in the interval $u_1 \leq u(\theta) \leq u_2$. If $\beta > 0$, there may or may not be such a value u_2. If there is not, $u(\theta)$ will increase without bound.

PROOF. Since we are going to be taking u to be the reciprocal of the radius vector from the Sun to an orbiting object, only positive values of u can be considered. From basic geometry, the orbit necessarily has a perihelion, where the reciprocal of the radius vector is maximized at value u_1. We then look for places where $du/d\theta = 0$ and $u < u_1$. (It would be of no use to find such values with $u > u_1$, since those values are inaccessible from the local maximum u_1.)

We now set $u'' = p(dp/du)$, where $p = du/d\theta$, as we have done twice before. Since $p = 0$ when $u = u_1$, we have

$$
\begin{aligned}
\frac{1}{2}p^2 &= \int_{u_1}^{u} (\beta u^2 - u + \alpha)\, du = \frac{\beta}{3}(u^3 - u_1^3) - \frac{1}{2}(u^2 - u_1^2) + \alpha(u - u_1) \\
&= (u - u_1)\Big(\frac{\beta}{3}(u^2 + u_1 u + u_1^2) - \frac{1}{2}u - \frac{1}{2}u_1 + \alpha\Big) \\
&= (u - u_1)\Big(\frac{\beta}{3}u^2 + \Big(\frac{\beta}{3}u_1 - \frac{1}{2}\Big)u + \Big(\frac{\beta}{3}u_1^2 - \frac{1}{2}u_1 + \alpha\Big)\Big).
\end{aligned}
$$

We need to know if the quadratic factor here has a positive root u_2 that is less than u_1. To that end, let $u = w + u_1$. We rewrite the quadratic factor as

$$\frac{\beta}{3}(w + u_1)^2 + \Big(\frac{\beta}{3}u_1 - \frac{1}{2}\Big)(w + u_1) + \Big(\frac{\beta}{3}u_1^2 - \frac{1}{2}u_1 + \alpha\Big) = \frac{\beta}{3}w^2 + \Big(\beta u_1 - \frac{1}{2}\Big)w + D,$$

where $D = \beta u_1^2 - u_1 + \alpha = u''(\theta_1)$. Since $u(\theta_1)$ is a local maximum of u, we see that $D < 0$. The product of the roots of this quadratic expression in w is $3D/\beta$.

If $\beta > 0$, that product is negative, and so there is exactly one negative root $w_2 = u_2 - u_1$, where $u_2 < u_1$, and one positive root $u_3 - u_1$, where $u_3 > u_1$. In this case, since the leading coefficient $(\beta/3)$ of this quadratic polynomial is positive, the polynomial is negative between the roots, which means it is negative at least for $w_2 \leq w \leq 0$, that is, $u_2 \leq u \leq u_1$. Multiplying it by $u - u_1$ makes it positive again in this range, and hence makes it a possible value of $p^2/2$.

Since u must begin to decrease (as a function of θ) after passing through the local maximum, it will remain in a range where p can have real values (that is, $p^2 > 0$) unless it reaches the local minimum value $u_2 < u_1$. The only possible difficulty in that case would occur if u_2 were negative. That might well happen, and in that case, we would be describing a nonclosed trajectory, since u cannot cross 0. It might not get arbitrarily close to zero; at least conceivably, it might approach a positive lower limit but never reach it.[18]

If $\beta < 0$, the product of the roots of the quadratic function in w is positive, and so we have either no roots with $w < 0$ or two such roots. In the former case, u_1 is the smallest (and perhaps only) real root of the original cubic polynomial. Since its leading coefficient is negative, this polynomial is negative for $u_1 < u < u_2$, where u_2 is the next root of the cubic, if any, and equal to infinity if there are no other roots. Either way, the polynomial cannot yield possible values of $p^2/2$ in this range. Thus $u(\theta) \le u_1$ for all θ. Since the solution has attained the value u_1 at θ_1, it must be always less than or equal to u_1. Unless it is constant, it cannot reach any minimum value, since p would vanish at such a value. It must therefore be monotonically decreasing for $\theta > \theta_1$.

If there are two such roots with $w < 0$, then u_1 is the largest of the three roots of the cubic, say $u_1 > u_2 > u_3$. Since the leading coefficient of the cubic polynomial is negative, the polynomial is positive for $u_2 < u < u_1$. It follows that if u is in this range, the polynomial provides possible real values for p, and u_1 can be a local maximum. As in the case $\beta > 0$, the function will decrease for $\theta > \theta_1$ until it either reaches the value u_2 or approaches 0. Again, the latter case corresponds to a nonclosed trajectory.

The situation when u_1 is a local minimum of $u(\theta)$ is handled similarly and is left to the reader. (See Problem 4.19.)   □

In the particular model we have been developing, we have been unlucky. We have $u_1 = v_0/l$, $\alpha = v_0^2 r_s/(2l^2)$, and $\beta = -3r_s/2$, so that $\beta u_1^2/3 - u_1/2 + \alpha = -v_0/(2l)$, which is negative. It follows that there is no zero u_2 of $du/d\theta$ between 0 and u_1, as we already knew from the explicit expression we had for $du/d\theta$. The project was sound, but nature did not cooperate. Facts are stubborn things, and we must reconcile ourselves to this outcome. This is easier to do when we realize that relativity enables us to carry out the idea to a most satisfying conclusion that is an improvement on Newton.

4.5. Noncircular orbits. Before leaving this topic, we shall wring the last tiny details out of the project we just aborted, details that will also find an echo in the relativistic solution. Let us introduce a new angle φ—*not* the co-latitude angle we used previously—via the relation

$$r = \frac{l}{v_0} \sec \varphi,$$

We then have

$$\frac{l^2}{v_0^2} \sec^2 \varphi \, \tan^2 \varphi \left(\frac{d\varphi}{d\theta}\right)^2 = \left(\frac{dr}{d\theta}\right)^2 = \frac{l^2}{v_0^2} \sec^2 \varphi \, \tan^2 \varphi (1 + (r_s v_0/l) \cos \varphi),$$

---

[18]A little more can be said in this case. The solution cannot be a *uniformly almost-periodic function* of time unless it is constant (that is, the corresponding orbit is a circle), since it can have at most one local extremum. As Sternberg ([**77**], pp. 1–14) has pointed out, Ptolemy's program of explaining planetary motion by means of epicycles can be neatly interpreted—in a language that would have been incomprehensible to Ptolemy—by saying that planetary orbits are uniformly almost-periodic functions of time. (See Problem 4.18.)

so that

$$d\theta = \frac{d\varphi}{\sqrt{1 + \mu \cos \varphi}},$$

and

$$\theta = \int_0^\theta dt = \int_0^\varphi \frac{dt}{\sqrt{1 + \mu \cos t}},$$

where $\mu = r_s v_0 / l$ and we choose the positive sign on the square root, since we can orient the angles θ and φ to suit ourselves. Typically, as we shall see, μ is very small. In fact, for a circular orbit of radius r, we see easily that $\mu = r_s/r$. The relation between θ and φ just written is well known in the theory of elliptic functions, expressed in the language of *Mathematica* by the mutually inverse relations[19]

$$\theta = \frac{2}{\sqrt{1 + \mu}} \, \mathrm{EllipticF} \left(\frac{\varphi}{2}, \frac{2\mu}{1 + \mu} \right),$$

$$\varphi = 2 \, \mathrm{JacobiAmplitude} \left(\frac{\theta \sqrt{1 + \mu}}{2}, \frac{2\mu}{1 + \mu} \right).$$

The function $\mathrm{EllipticF}\,(\varphi, m)$ is defined by the relation

$$\mathrm{EllipticF}\,(\varphi, m) = \int_0^\varphi \frac{1}{\sqrt{1 - m \sin^2 t}} \, dt.$$

In our case, $m = 2\mu/(1+\mu)$, so that $\mu = m/(2-m)$. The Maclaurin expansions of these functions are

$$\mathrm{EllipticF}\,(\varphi, m) = \varphi + \frac{m}{3!} \varphi^3 +$$
$$+ \frac{9m^2 - 4m}{5!} \varphi^5 + \frac{225m^3 - 180m^2 + 16m}{7!} \varphi^7 + \cdots,$$

$$\mathrm{JacobiAmplitude}\,(\theta, m) = \theta - \frac{m}{3!} \theta^3 +$$
$$+ \frac{m^2 + 4m}{5!} \theta^5 - \frac{m^3 + 44m^2 + 16m}{7!} \theta^7 + \cdots.$$

The integrand that defines the function $\mathrm{EllipticF}[x, \, m]$ is of period π, which means that the difference $\mathrm{EllipticF}\,[x + \pi, m] - \mathrm{EllipticF}\,[x, m]$ is a positive constant. Thus, this function increases by a constant amount $2K[m]$ over each interval of length π, where $K[m] = \mathrm{EllipticF}\,[\pi/2, m]$ is a constant known as the *complete elliptic integral of first kind*. The graph of the function $\mathrm{EllipticF}\,[x, m] - x$ is shown in Fig. 4.3 for $m = 0.6$. This is a rather large parameter value compared to what we will actually have in our astronomical application, but for very small values of m, the function increases too slowly to be noticeable over a short portion of the graph. In the astronomical examples we shall be considering, m will be of the order 10^{-7} for a typical planetary orbit when $v_0 = c$. Hence the equation $\theta = \varphi$ is very nearly exact. The function $\mathrm{EllipticF}\,[x, m] - 2K[m]x/\pi$ has period π.

If we imagine that (r, φ) are polar coordinates that closely approximate the standard coordinates (r, θ), we see that the new family of geodesics we have introduced are simply the vertical lines $x = r \cos \varphi = l/v_0$. This result is not very impressive, although, when we replace φ by θ, these lines do bend, thereby forming very shallow, roughly hyperbolic shapes, but with "wiggles." Unfortunately

---

[19]In the older language introduced by Jacobi, $\mathrm{EllipticF}\,(\varphi, m)$ was called the *elliptic integral of first kind* with modulus $k = \sqrt{m}$. The function $\mathrm{JacobiAmplitude}\,(\theta, m)$ was its inverse and written simply $\mathrm{am}\,(\theta)$.

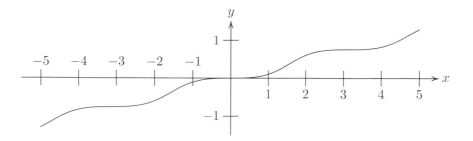

FIGURE 4.3. Graph of the equation $y = \text{EllipticF}\,[x, 0.6] - x$.

also, they bend *in the wrong direction*, as if the attracting particle was actually a repelling particle.

Let us summarize what we have achieved.

- A perturbed version of Newton's equations of motion of a particle in the gravitational field due to a particle of mass M located at the origin can be obtained from the Euler equations for a geodesic, provided the element of time dt in polar coordinates (r, θ) is required to satisfy

$$dt^2 = \frac{1}{v_0^2}\left(\frac{r}{r + r_s}\,dr^2 + r^2\,d\theta^2\right).$$

- The distance r_s equals $2GM/v_0^2$ and is called the *Newtonian Schwarzschild radius*. The ratio r_s/r at a typical point in the orbit is the ratio of the negative of the potential energy of the moving particle to its kinetic energy at speed v_0. (We cannot simply incorporate the potential energy into our equation, since a nonhomogeneous mathematical function can accept as input only dimensionless variables. Thus we need a fixed energy as a unit, one that can be specified in advance of any particular orbit. The kinetic energy per unit mass at speed v_0 seems like the only reasonable candidate.)

- Besides the circular geodesic orbits predicted by both the standard Newtonian theory and the approach via geodesics, there are other geodesics that are asymptotic to straight lines and can be expressed exactly in terms of a new polar angle φ such that the standard polar angle θ can be expressed in terms of φ as an elliptic integral of first kind:

$$\theta = (2/\sqrt{1+\mu})\,\text{EllipticF}\,(\varphi/2, 2\mu/(1+\mu)).$$

- The geodesic approach unfortunately rules out elliptical orbits and hence must be abandoned as a serious attempt at mathematical astronomy. It can be retained, however, as a useful "toy" geometry exhibiting some features that will prove to be important in the relativistic model we are now going to study.

REMARK 4.3. The Newtonian Schwarzschild radius ρ_s can also be described as the radius such that the escape velocity at distance ρ_s equals the standard unit of speed v_0. In relativity, where $v_0 = c$, it is the radius of a (nonrotating) black hole of mass M, and it will reappear in exactly this form when we discuss relativistic gravitation. Right now, we note that a material particle moving at speed v_0 has equal kinetic and potential energies at the distance ρ_s from the fixed particle of mass M. Without any force being applied to it, it could escape to infinity if located outside that radius, but not if it were inside.

The Schwarzschild radius of the Sun (with $v_0 = c$) is about 3 km, as will be seen below. In general, this radius tends to be very small compared with other quantities involved in astronomy. If the Sun were replaced by a body of this radius, when seen from Earth it would subtend an angle of less than $1/120^{\text{th}}$ of an arc-second, far too small to be observed, even by the Hubble telescope, whose resolving power is at best $1/20^{\text{th}}$ of an arc-second.

On several occasions, we have indulged in mathematical whimsy, investigating a relativistic version of Ptolemaic astronomy in Chapter 1 and two unfruitful ways of analyzing planetary orbits in the present section and the one preceding. We shall now indulge in an even more fanciful misapplication of physical laws, extrapolating them ludicrously beyond the range of measurements for which they are valid. We do so to make the simple point that *the Schwarzschild radius is a very tiny dimension!* The difference in the order of magnitude of the Schwarzschild radius of an object and the dimensions of that object is seen in stark relief on the subatomic level.

Obviously, we do get information about the structure of atomic nuclei; an atomic nucleus is not a black hole. It follows that the radius of an atomic nucleus— tiny though it be—has to be much larger than its Schwarzschild radius. In fact, the formula for a nuclear radius is $r_n = 1.25 \times 10^{-15} \sqrt[3]{n}$ meters, where n is the number of protons and neutrons in the nucleus. Since the mass of a proton or neutron is approximately 1.674×10^{-27} kg, and $G = 6.674 \times 10^{-11}$ m³/kg-s², while $c = 3 \times 10^8$ m/s, you can see that the Schwarzschild radius of a nucleus could exceed the actual radius only if the nucleus contained roughly 1.1×10^{58} protons and neutrons. (Of course, as we warned, this number is ludicrously outside the range of atomic weights for which the nuclear radius formula is known to be approximately true, and there is not the slightest reason to take this bit of fantasy as a serious scientific argument.) For comparison, the number of protons in the observable universe (the *Eddington number*[20]) is 1.58×10^{79}. Since there are about one-fifth as many neutrons as protons, the total number of protons and neutrons is about 1.9×10^{79}.

REMARK 4.4. Looking ahead to the concepts of Chapter 5, we note that the Gaussian curvature of the punctured plane with this metric is, if expressed in units of length rather than time,

$$\kappa(r) = \frac{r_s}{2r^3}.$$

Looking still farther ahead, to Chapter 6, we note that this plane is *Riemannian* since its matrix of metric coefficients is positive-definite. Its Laplace–Beltrami operator is

$$\nabla^2 f = \frac{2GM}{r_s r^2}\left(r(r + r_s)\frac{\partial^2 f}{\partial r^2} + (r + r_s/2)\frac{\partial f}{\partial r} + \frac{\partial^2 f}{\partial \theta^2}\right).$$

---

[20]Named after Sir Arthur Stanley Eddington (1882–1944).

5. Fourth Analysis: General Relativity

We now come to one of the crucial retrodictions[21] of the general theory of relativity. We wish to compute the orbit of the planet Mercury and show that its perihelion precesses in the direction of its revolution about the Sun, and by 43 seconds of arc per century as a result of relativistic effects. Its total precession of 5600 seconds per century is more than 99% explainable by ordinary Newtonian mechanics and the perturbations of other planets. The relativistic solution to this problem amounts to computing the geodesics in a space-time metric that is modified from the flat metric of special relativity. Some of the constants can be determined by both observation and computation, and of course we expect the two outcomes to be the same. By comparing the relativistic computation with the Newtonian computation that we have just completed, we will be able to see how the relativistic theory is an adjustment of the Newtonian theory, with a small perturbation term added to the differential equation that has to be solved.

5.1. Space-time in general relativity. When one uses spherical coordinates (ρ, φ, θ) on $\mathbb{R}^3$, the infinitesimal increment of proper time ds on a moving particle is given by

$$ds^2 = dt^2 - \frac{1}{c^2}d\rho^2 - \frac{\rho^2}{c^2}\, d\varphi^2 - \frac{\rho^2 \sin^2 \varphi}{c^2}\, d\theta^2 \, .$$

If we assume that a particle of mass M is located at the origin of the spatial coordinates at all times, our goal is to explain gravity as a perturbation in this metric. We have just done that for the classical case when time and space are independent of each other. With that experience as a guide, we assume a relativistic perturbation that changes the coefficients of both dt^2 and $d\rho^2$.

We replace the metric of special relativity by a metric of the general form

$$ds^2 = f(\rho)\, dt^2 - \frac{1}{c^2}g(\rho)\, d\rho^2 - \frac{\rho^2}{c^2}\, d\varphi^2 - \frac{\rho^2 \sin^2 \varphi}{c^2}\, d\theta^2 \, .$$

As with our Newtonian analysis, the reason for assuming a metric of this form is provided by symmetry considerations. It seems reasonable that the perturbation would be the same in all directions from the origin and at all times, hence should depend only on the radial coordinate ρ and not on time or on the longitude or co-latitude. It also seems reasonable that it would not change the infinitesimal length on a path at right angles to a line from the origin near the point where the path and the line intersect. Thus we apply the perturbation factors $f(\rho)$ and $g(\rho)$ only in the two coordinates expressing the time t and the radial distance ρ. A space-time metric of this form will henceforth be called *radial*. It corresponds to a central attractive force in classical mechanics.

Again, since we would expect this metric to tend to coincide with the flat special relativity metric at large distances, we assume that $f(\rho)$ and $g(\rho)$ tend to 1 as $\rho \to \infty$. For that reason, as in the Newtonian case, we shall write them in a form that will turn out to be simpler for computation, namely $f(\rho) = e^{\lambda(\rho)}$ and

---

[21] A retrodiction is a proposition deduced from a theory that explains a previously unexplained phenomenon, in this case part of the precession of the perihelion of Mercury.

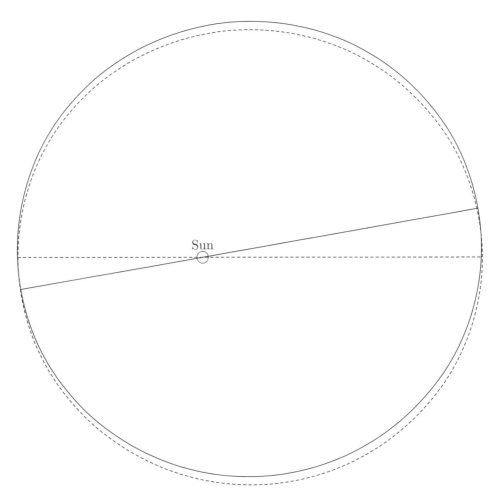

FIGURE 4.4. Precession of the perihelion of Mercury. Starting from the dashed ellipse, relativistic effects cause the major axis to rotate about the Sun as a center, at a rate of about 43 seconds of arc per century of Earth time, becoming at some point the solid ellipse. The effect is extremely small, and the amount of precession shown here (10 degrees) would require nearly 84,000 years if due to relativistic effects alone. Actually, because of perturbations due to the other planets, this much precession occurs in about 650 years.

$g(\rho) = e^{\nu(\rho)}$, where $\lambda(\rho)$ and $\nu(\rho)$ tend to zero as $\rho \to \infty$. Thus we assume that[22]

$$ds^2 = e^{\lambda(\rho)}dt^2 - \frac{1}{c^2}e^{\nu(\rho)}d\rho^2 - \frac{\rho^2}{c^2}\,d\varphi^2 - \frac{\rho^2\sin^2\varphi}{c^2}\,d\theta^2\,.$$

---

[22] As the formulas are already sufficiently complicated, we are going to ignore the pedantic point that the exponential function really needs dimensionless variables as its input. We could de-dimensionalize by replacing ρ with ρ/ρ_s, as we did above, but the final result would be the same. The reader is hereby alerted to this caveat, and can now ignore it.

In order to proceed we need to find the explicit forms of $\lambda(\rho)$ and $\nu(\rho)$. For that, we need a system of differential equations that they must satisfy. Where can we find them?

In the Newtonian case, we were guided by the principle that we wanted the Euler equations for a geodesic to produce Newton's equations of motion for the gravitational field. Since it is precisely that system of equations of motion that we are now planning to replace, we cannot use them to determine λ and ν. Without the intuitive "push and pull" of Aristotelian forces, it is not obvious how gravitation can be explained in purely geometric terms. Our effort to do so on the basis of absolute time and space shows that even a tiny deviation from the exact Newtonian system of differential equations causes great disruption in the predictions of the theory, excluding elliptic orbits, for example. We shall take that sobering lesson as evidence of the value of relativistic refinements when we find (as we shall) that the same approach works extremely well when the relativistic refinements are taken into account. One thing is clear, however: We'd certainly be lost if we looked for any explanations that did not involve second-order differential equations, preferably linear ones in first approximation. That makes the geodesic approach seem more plausible, since Euler's equations for geodesics are exactly of that type.

Very well, we can agree that we are looking for second-order differential equations. But on what *intuitive* basis can we derive them? Einstein argued that (1) the laws of physics must have the same form for all observers, and hence must be stated as tensor equations and (2) apart from the Riemann curvature tensor, whose vanishing implies that the space is flat (see below), there is only one tensor of the same type $(0, 2)$ as the tensor of metric coefficients and involving at most second-order partial derivatives of the metric coefficients, namely the Ricci tensor[23] we are about to define.

Again, we can say, "Very well, we need the Ricci tensor to derive the differential equations of motion." But that admission still leaves us essentially clueless: What *intuitive physical quantity* are we supposed to set equal to the Ricci tensor in order to get the equations? Since for a gravitational field in empty space-time that physical quantity turns out to be identically zero, we can postpone answering the harder question until we have done the computations. At the moment, we merely accept that we are going to require the Ricci tensor to vanish for some reasonable perturbation of the flat-space metric and then determine the perturbation from that principle. In the back of our minds we hold the thought that force corresponds to curvature, as in Newton's first law (where the absence of force means an absence of curvature), and hence curvature may be used to replace force in physical theory. As Einstein mentioned, there is another important tensor involving only second-order derivatives of the metric coefficients. That is the metric of type $(3, 1)$, known as the *Riemann curvature tensor*. It will be defined in the next chapter; but requiring it to vanish would be tantamount to requiring the space to be flat. The Ricci tensor, which is obtained from the Riemann curvature tensor by the operation of contraction, can vanish on a curved space, and so it represents a reasonable compromise, which very fortunately achieves the desired explanation of the precession of the perihelion of Mercury. We now turn to the actual computations.

This geometric approach to the problem involves some quantities from differential geometry that we cannot explain at the moment. We must take it on faith

---

[23]Named after Gregorio Ricci-Curbastro (1853–1925).

that these quantities, the *Christoffel symbols*[24] Γ^i_{jk}, determine the curvature of any manifold for which they are computed. That is the arcane part that we are temporarily skipping. The actual computation of these symbols is not difficult.

5.2. The Christoffel symbols. The infinitesimal proper time interval ds is a quadratic form on $\mathbb{R}^4$:

$$ds^2 = g_{il}\, dx^i\, dx^l,$$

where[25] $x^1 = t$, $x^2 = \rho$, $x^3 = \varphi$, and $x^4 = \theta$, and by the Einstein summation convention the terms on the right-hand side are summed for each index over the range from 1 to 4.

We then have

$$g_{11} = e^{\lambda(\rho)}, \quad g_{22} = -\frac{1}{c^2}e^{\nu(\rho)}, \quad g_{33} = -\frac{\rho^2}{c^2}, \quad g_{44} = -\frac{\rho^2 \sin^2\varphi}{c^2},$$

and $g_{il} = 0$ if $i \neq l$. It is therefore very easy to find the inverse of the matrix (g_{ij}), namely

$$g^{11} = e^{-\lambda(\rho)}, \quad g^{22} = -c^2 e^{-\nu(\rho)}, \quad g^{33} = -\frac{c^2}{\rho^2}, \quad g^{44} = -\frac{c^2}{\rho^2 \sin^2\varphi},$$

and $g^{il} = 0$ if $i \neq l$.

From this information we can compute the sixty-four Christoffel symbols Γ^i_{jk}, which are defined as follows:

$$(4.11) \qquad \Gamma^i_{jk} = \frac{1}{2}g^{il}\left(\frac{\partial g_{jl}}{\partial x^k} + \frac{\partial g_{lk}}{\partial x^j} - \frac{\partial g_{jk}}{\partial x^l}\right).$$

By the Einstein summation convention, the terms on the right are summed on the index l from 1 to 4.

REMARK 4.5. The coordinates used in physics usually come with some physical dimension attached to them, meaning that their numerical value corresponds to the measurement of a physical quantity: length, mass, time, force, energy, and the like. For rectangular space-time coordinates $(t; x, y, z)$, the dimension of t is time, and the dimension of x, y, and z is length. In spherical coordinates $(t; \rho, \varphi, \theta)$, the dimension of ρ is length, but φ and θ are dimensionless, being angles measured as the ratio of an arc to a radius, both of which are lengths. It is a commonplace that the arithmetic operation of addition can be performed only on terms that have the same physical dimension. Thus one can see the need for the compensating coefficients $1/c^2$, ρ^2, and $\rho^2 \sin^2\varphi$ in the expression for ds^2.

Suppose now that we use coordinates (x^1, x^2, x^3, x^4) on space-time. Let the dimension of x^i be denoted $[i]$. That is also the dimension of its infinitesimal increment dx^i. If we assign a dimension $[d]$ to the space-time interval ds—which may be time, as we have chosen, or length, or energy, or any other suitable measure of the interval between two events—then the dimension of g_{ij} needs to be $[d]^2/([i][j])$ in order for the equation defining ds^2 to make sense. It is then easy to compute that the dimension of g^{ij} must be $[i][j]/[d]^2$. This dimensionality has the rather odd consequence that, while the entries in the matrix products $g_{ij}g^{jk}$ and $g^{ij}g_{jk}$ are both *numerically* equal to the Kronecker delta δ^j_i and δ^i_j, the dimension of this Kronecker delta should theoretically be taken into account: $[\delta^j_i] = [j]/[i]$ and $\delta^i_j = [i]/[j]$.

---

[24]Named after Elwin Bruno Christoffel (1829–1900).

[25]We shall henceforth let time be the first coordinate, rather than the zeroth, as it was in Chapter 2.

This is dimensionless in the only important case, when $i = j$. Otherwise, $\delta_i^j = 0$ numerically, and the dimension of this zero is irrelevant.

With each partial derivative operator $\partial/\partial x^i$ we associate the dimension $1/[i]$, so that the dimension of $\partial g_{ij}/\partial x^k$ will be $[d]^2/([i][j][k])$. It is therefore permissible to add the three terms in parentheses in Eq. (4.11). As a result, the dimension of the Christoffel symbol Γ_{jk}^i will be $[i]/([j][k])$. This dimension is independent of the dimension chosen to measure the space-time interval ds. In fact, as one can easily see, the Christoffel symbols are unaffected if all of the metric coefficients g_{ij} are multiplied by the same constant. When we develop this topic more fully in the next two chapters, we shall regard the tensors dx^i and $\partial/\partial x^j$ as operating on each other to produce pure numbers, 0 if $i \neq j$ and 1 if $i = j$. When the vector $u^i(\partial/\partial x^i)$ is operated on by the covector $v_j\, dx^j$, the result is the scalar $u^i v_i$, which has dimension equal to the product of the dimension of u^i and the dimension of v_i. (Usually, both of these are dimensionless. Whether or not that is the case, their product will have the same dimension for all i, being the dimension of $u^i\, \partial/\partial x^i$, which must be the same for all i if these terms are to be added, times the dimension of $v_i dx^i$, which must be the same for all i.)

Nature is kind to us in this regard, in that all the tensors we need to consider can be subjected to certain tensor operations, such as raising and lowering an index and contracting on two indices, without any violation of the principles of dimensionality. That is, the operations to be performed and the equations that result are dimensionally consistent. For example, the coordinate R_{jkl}^i of the Riemann curvature tensor has dimension $[i]/([j][k][l])$. We can then use the metric tensor g_{ij} to "lower" the index i and obtain thereby the *covariant* Riemann curvature tensor Rie where

$$\text{Rie}_{ijkl} = g_{im} R_{jkl}^m\,.$$

The dimension of the coefficient Rie_{ijkl} is

$$\left([d]^2/([i][m])\right)\left([m]/([j][k][l])\right) = [d]^2/([i][j][k][l])\,.$$

When we contract the Riemann curvature tensor R_{jkl}^i on the indices i, and k, we get what will be called the *Ricci tensor* and denoted Ric_{jl}:

$$\text{Ric}_{jl} = R_{jil}^i\,.$$

The dimension of Ric_{jl} is $1/([j][l])$.

We can use the *inverse* g^{ij} of the metric tensor "raise" an index on the Ricci tensor, getting a mixed Ricci tensor of type $(1,1)$:

$$\text{Ric}_i^j = g^{jk}\text{Ric}_{ik}\,,$$

where the dimension of the coefficient Ric_i^j is $[j]/([d]^2[i])$. If we then contract by setting $i = j$ and summing, we get a scalar whose dimension $1/[d]^2$ is the dimension that we shall ascribe to curvature. We shall denote this scalar simply as R. It is called the *scalar curvature* (see Chapter 6):

$$R = \text{Ric}_i^i\,.$$

Because the physical dimension of R is $1/[d]^2$, the tensor $R\,g_{ij}$ has the same dimension, namely $1/([i][j])$, as the Ricci tensor Ric_{ij}, and hence we can consider linear combinations of the two. This result will turn out to be important in Chapter 7.

REMARK 4.6. Physicists like to streamline their equations by assuming units are chosen so that certain constants have the value 1 and hence can be omitted. This is always done, for example, in the case of Newton's second law, which says $F = kma$ for a constant k. We use the MKS system of units and define the unit of force (the newton) so as to get $k = 1$. In relativity, it is usually assumed that the units result in $c = 1$ and $G = 1$, since these quantities appear so often in equations. Since we are exploring this area for the first time, we are not going to adopt that convention. Very often, it results in equations that appear dimensionally absurd, and one is then forced to work out where factors of c and/or G need to be inserted to make them sensible.

REMARK 4.7. In an earlier language, the Christoffel symbols Γ_{ij}^k were called *Christoffel's three-index symbols*. The original convention, followed by Einstein [21], Russell [72], and others used the following notation for them:

$$\Gamma_{ij}^k = \left\{ \begin{matrix} i \ j \\ k \end{matrix} \right\} .$$

Eddington [16] used the notation

$$\Gamma_{ij}^k = \{ij, k\} ,$$

and also considered what we might denote by a triple subscript as

$$\Gamma_{ijk} = [ij, k] = g_{kl}\{ij, l\} .$$

He also remarked (p. 59) that the process of moving back and forth between $\{ij, k\}$ and $[ij, k]$ was equivalent to raising or lowering an index in a tensor and said that the notation Γ_{ij}^k "might be convenient," but chose to adhere to what was at the time the standard notation.

REMARK 4.8. Einstein set great store by equations between "tensors" because of their invariance when translated from any set of coordinates to any other. Generally, we expect to find a tensor wherever we find a multi-indexed symbol, such as g_{ij} or Γ_{jk}^i. For that reason, we need to point out that the Christoffel symbols are *not* tensors, because they do not transform via the Jacobian matrix when coordinates are changed. (See Appendix 6.) Even so, some authors, for example Wald ([83], p. 34), refer to them as tensors. Wald remarks immediately, however, that these symbols in different coordinate systems are not related in the usual way, "since we change tensors as well as coordinates."

Fortunately, because the matrix (g_{il}) is symmetric, the Christoffel symbols are also symmetric in the two lower indices, so that only 40 of them are potentially different. Because the matrices (g_{il}) and (g^{il}) are diagonal, only one term in the sum that defines Γ_{jk}^i is nonzero. In the end, only nine of the Christoffel symbols are different and only 13 of them are nonzero. We shall compute just two of them and then simply list the values of the others. Let us do just Γ_{11}^1 and Γ_{12}^1. (The others are computed in exactly the same way.) We have

$$\Gamma_{11}^1 = \frac{1}{2}g^{11}\left(\frac{\partial g_{11}}{\partial x^1} + \frac{\partial g_{11}}{\partial x^1} - \frac{\partial g_{11}}{\partial x^1}\right) = 0 ,$$

since $x^1 = t$ and all of the matrix entries g_{il} are independent of t.

Then

$$\Gamma_{21}^1 = \frac{1}{2}g^{11}\left(\frac{\partial g_{12}}{\partial x^1} + \frac{\partial g_{11}}{\partial x^2} - \frac{\partial g_{12}}{\partial x^1}\right) = \frac{1}{2}c^{-2}e^{-\lambda(\rho)}\left(c^2e^{\lambda(\rho)}\lambda'(\rho)\right) = \frac{1}{2}\lambda'(\rho) .$$

By patiently working through all 40 of these computations, we obtain the following table of Christoffel symbols:

$$\Gamma^1_{12} = \frac{1}{2}\lambda'(\rho) = \Gamma^1_{21}, \quad \Gamma^2_{11} = \frac{1}{2}c^2 e^{\lambda(\rho)-\nu(\rho)}\lambda'(\rho), \quad \Gamma^2_{22} = \frac{1}{2}\nu'(\rho),$$

(4.12)
$$\Gamma^2_{33} = -\rho e^{-\nu(\rho)}, \quad \Gamma^2_{44} = -\rho e^{-\nu(\rho)}\sin^2\varphi, \quad \Gamma^3_{32} = \frac{1}{\rho} = \Gamma^3_{23},$$

$$\Gamma^3_{44} = -\sin\varphi\cos\varphi, \quad \Gamma^4_{24} = \frac{1}{\rho} = \Gamma^4_{42}, \quad \Gamma^4_{34} = \cot\varphi = \Gamma^4_{43}.$$

All the other Christoffel symbols are zero.

Mathematica Notebook 6 in Volume 3 does these computations in a flash.

6. Einstein's Law of Gravity

The desired equations of a matter-free gravitational field are therefore satisfied when all $B^\rho_{\mu\sigma\tau}$ [coordinates of the Riemann curvature tensor] vanish. But this condition is too broad. Indeed, it is clear, for example, that the gravitational field due to a material point cannot be "modified" through any choice of coordinates to the case when the $g_{\mu\nu}$ are constant in a neighborhood of the particle... Thus the idea suggests itself that one should require the vanishing of the symmetric [Ricci] tensor $B_{\mu\nu}$ derived from [$B^\rho_{\mu\sigma\tau}$ through contraction]... It should be noted that [this choice] leads to a minimum of arbitrariness. For $B_{\mu\nu}$ is the only tensor of second rank formed from the $g_{\mu\nu}$ and their derivatives of order no higher than 2 that is linear in the second-order derivatives... The fact that [this law] provides the Newtonian law in first approximation and explains the motion of the perihelion of Mercury discovered by Leverrier should, in my view, be regarded as convincing evidence of the physical correctness of the theory.

Einstein ([**21**], pp. 803–804). My translation.

To make all this work useful we need a gravitational law that will impose conditions on these symbols in the form of differential equations that we can solve to get explicit functions in place of $\lambda(\rho)$ and $\nu(\rho)$. Einstein decided that this law of gravity should be expressed by the vanishing of the Ricci tensor, which is obtained by contracting the Riemann curvature tensor on two indices. The reason Einstein gave for choosing that particular law seems rather vague: It was the simplest tensor that could be formed from metric coefficients and their first and second partial derivatives whose vanishing would not require space-time to be flat. As we shall see in Chapters 6 and 7, once you have the program of "geometrizing" forces by incorporating them into the metric, the metric coefficients that determine the geometry of space-time play a role analogous to the potential functions whose gradients give the forces in Newtonian mechanics. The Newtonian potentials are characterized by being solutions of Laplace's equation. As it turns out (see Corollary 6.4), when we use a preferred "normal" coordinate system, the curved-space analog of Laplace's equation for these generalized potentials amounts to the vanishing of the Ricci tensor.

That is all we intend to say at the moment. The geometric significance of these tensors will be discussed in Chapters 5 and 6. In both of those chapters and in Appendix 4, we are going to use the word *tensor* rather freely, with minimal explanation. As to what actually constitutes a tensor, the details are given in Section 1 of Appendix 6. Right now, we are merely carrying out the computations. Leaving aside the questions of what the Riemann curvature tensor is and why this is an appropriate law to assume, we merely state the fact that the Ricci tensor on a four-dimensional manifold is given as a bilinear functional acting on two vectors $\boldsymbol{u} = (u^1, u^2, u^3, u^4)$ and $\boldsymbol{v} = (v^1, v^2, v^3, v^4)$. Without the Einstein summation convention, this tensor would be written as[26]

$$\mathrm{Ric}\,(\boldsymbol{u}, \boldsymbol{v}) = \sum_{j=1}^{4} \sum_{l=1}^{4} \sum_{i=1}^{4} \left(\frac{\partial \Gamma_{lj}^i}{\partial x^i} - \frac{\partial \Gamma_{ij}^i}{\partial x^l} + \sum_{m=1}^{4} \left(\Gamma_{im}^i \Gamma_{lj}^m - \Gamma_{lm}^i \Gamma_{ij}^m \right) \right) u^j v^l \,.$$

This tensor is determined by the 4×4 tableau $\{\mathrm{Ric}_{jl}\}_{j,l=1}^4$, (where again, and for the *last* time, the Einstein summation convention is not invoked)

$$(4.13) \qquad \mathrm{Ric}_{jl} = \sum_{i=1}^{4} \left(\frac{\partial \Gamma_{lj}^i}{\partial x^i} - \frac{\partial \Gamma_{ij}^i}{\partial x^l} + \sum_{m=1}^{4} \left(\Gamma_{im}^i \Gamma_{lj}^m - \Gamma_{lm}^i \Gamma_{ij}^m \right) \right),$$

and j and l range from 1 to 4.

The physical dimension of Ric_{jl} in coordinates (x^i) having dimensions $[i]$ is $1/([j][l])$. As noted above, we *cannot* add the Ricci tensor Ric_{jl} to the metric tensor g_{jl}, even though they are both tensors represented by 4×4 matrices of real numbers, since the latter has dimension $[d]^2/([j][l])$. It does make sense, however, to consider $\mathrm{Ric}_{jl} + \lambda g_{jl}$ if λ has the physical dimension $1/[d]^2$, which is to say, the dimension of curvature. It will be important to do so in Chapter 7. At the moment, we need not worry about these considerations, since we are going to assume that the Ricci tensor vanishes.

From now on, the Einstein convention is in effect: When an index appears as both superscript and subscript in a single term, summation is to be performed on that index over an appropriate range, the range being 1 to 4 in this case on the indices i and m. The indices j and l in Eq. (4.13) also range from 1 to 4, but no summation occurs over them, since each of them occurs only as a subscript.

The analysis of a planetary orbit in general relativity consists of two components: First, modifying the flat metric of space-time so as to get a curved space-time for which the Ricci tensor vanishes. Having found the Ricci tensor through that requirement, we then postulate that the world-line of the planet (idealized as a particle) is a geodesic in the curved metric. The present section is devoted to the first of these tasks, and the one following to the second.

Setting the sixteen components of the Ricci tensor equal to zero in order to find the explicit form of only the two functions $\lambda(\rho)$ and $\nu(\rho)$ seems like overkill. Only two such equations would actually be needed to determine $\lambda(\rho)$ and $\nu(\rho)$. From another point of view, in fact, the very possibility would appear to be remote. We introduced only two arbitrary functions. Why would we think it is possible

---

[26]Einstein and others denoted the Ricci tensor by $B_{\mu\nu}$ and later $G_{\mu\nu}$. There is here a regrettable collision of symbols, since we have already used the symbol G for the universal gravitational constant. Modern physicists often denote the Ricci tensor by $R_{\mu\nu}$, which is better. However, they also denote the Riemann curvature tensor by $R_{\rho\sigma\nu}^{\mu}$. We prefer for the sake of clarity to invent a special notation just for the present book. Henceforth the *covariant* Riemann tensor will be denoted Rie and the Ricci tensor Ric. The tensor commonly denoted $R_{\rho\sigma\nu}^{\mu}$ will be treated as a trilinear mapping of vectors into a vector and denoted $R(\boldsymbol{u}, \boldsymbol{v}, \boldsymbol{w})$ (see Chapter 6).

to choose them in such a way as to annihilate sixteen quantities? In general, trying to get two variables to satisfy sixteen equations looks like a very dubious enterprise. In fact, however—perhaps we are seeing here an example of a pre-established harmony between mathematics and the world—these sixteen equations yield *only* two independent equations among the whole lot. First of all, because $\mathrm{Ric}\,(\boldsymbol{u}, \boldsymbol{v}) = \mathrm{Ric}\,(\boldsymbol{v}, \boldsymbol{u})$, only ten of the sixteen coordinates can be independent.[27] But, as you can verify, in fact the equation $\mathrm{Ric}_{j,l} = 0$ is a trivial identity when $j \neq l$, so that six of the ten possible equations yield no information at all. That leaves only the four equations $\mathrm{Ric}_{jj} = 0$, $j = 1, 2, 3, 4$. The equations $\mathrm{Ric}_{33} = 0$ and $\mathrm{Ric}_{44} = 0$ are essentially identical, however, and both are consequences of the equations $\mathrm{Ric}_{11} = 0$ and $\mathrm{Ric}_{22} = 0$. There is therefore no redundancy at all in the metric we have assumed.

The output from the first extension of *Mathematica* Notebook 6 (Volume 3) is

$$\left\{ \left\{ c^2 \frac{e^{\lambda(\rho)-\nu(\rho)} \big(\lambda'(\rho)(4 + \rho\lambda'(\rho) - \rho\nu'(\rho)) + 2\rho\lambda''(\rho) \big)}{4\rho}, 0, 0, 0 \right\}, \right.$$

$$\left\{ 0, \frac{1}{4}\left(\frac{4\nu'(\rho)}{\rho} + \lambda'(\rho)\big(-\lambda'(\rho) + \nu'(\rho) \big) - 2\lambda''(\rho) \right), 0, 0 \right\},$$

$$\left\{ 0, 0, 1 - \frac{1}{2}e^{-\nu(\rho)}\big(2 + \rho\lambda'(\rho) - \rho\nu'(\rho) \big), 0 \right\},$$

$$\left. \left\{ 0, 0, 0, \frac{1}{2}\sin^2(\varphi)\big(2 + e^{-\nu(\rho)}(-2 - \rho\lambda'(\rho) + \rho\nu'(\rho)) \big) \right\} \right\}.$$

Setting the first two diagonal entries equal to zero, we get two differential equations that are very easy to solve. These are

(4.14) $$\lambda''(\rho) + \frac{1}{2}\lambda'(\rho)\big(\lambda'(\rho) - \nu'(\rho) \big) + \frac{2\lambda'(\rho)}{\rho} = 0$$

and

(4.15) $$\lambda''(\rho) + \frac{1}{2}\lambda'(\rho)\big(\lambda'(\rho) - \nu'(\rho) \big) - \frac{2\nu'(\rho)}{\rho} = 0.$$

6.1. Determination of $\lambda(\rho)$ and $\nu(\rho)$. Equations (4.14) and (4.15) are independent, and they determine λ and ν. By subtracting one from the other and canceling $\frac{2}{\rho}$, we find that

$$\lambda'(\rho) + \nu'(\rho) = 0,$$

from which it follows that $\lambda(\rho) + \nu(\rho)$ is constant.

Since this function tends to zero as ρ tends to infinity, the constant must be zero. In other words, $\nu(\rho) = -\lambda(\rho)$. The effect of that relation is that the two perturbing functions f and g are reciprocals of each other. As we shall see in Chapter 6 when we discuss the Ricci tensor in detail, this effect should be expected when the Ricci tensor vanishes. It means that the determinant of the 4×4 matrix of metric coefficients is unchanged by the perturbation. That is, the four-dimensional volume element

$$dV = \sqrt{-\rho^4 \sin^2 \varphi \, dt^2 \, d\rho^2 \, d\varphi^2 \, d\theta^2}$$

---

[27]The reader can verify this visually in the case of the space-time metric we are using. The general fact is not trivial. It is proved below as Theorem 6.12.

is not changed.[28] The Ricci tensor is actually is a measure of the amount by which this volume element departs from its flat-space value, as we shall see in Chapter 6. When it vanishes, volumes—space-time volumes, in this case—are unchanged from their flat-space values.

Replacing $\nu(\rho)$ by $-\lambda(\rho)$ in Eq. (4.14), we find

$$\lambda''(\rho) + \left(\lambda'(\rho)\right)^2 + \frac{2\lambda'(\rho)}{\rho} = 0\,.$$

Letting $\mu(\rho) = \lambda'(\rho)$, we find the first-order equation

$$\mu' + \mu^2 + \frac{2\mu}{\rho} = 0\,,$$

which can be written as

$$\mu' - \left(\frac{1}{\rho}\right)^2 + \left(\mu + \frac{1}{\rho}\right)^2 = 0\,.$$

In other words, if $\varkappa = \mu + \frac{1}{\rho}$, then

$$\varkappa' + \varkappa^2 = 0\,.$$

Assuming $\varkappa$ is not identically zero, this last equation has the general solution $\varkappa(\rho) = \frac{1}{\rho - \rho_s}$ for a constant distance ρ_s, which will turn out to be the Schwarzschild radius mentioned above with $v_0 = c$. We have chosen instead to subtract the radius, writing $\rho - \rho_s$ rather than the expression $\rho + \rho_s$ that occurred in our "toy" metric. We did so anticipating what we shall learn about the sign of ρ_s below. The difference is precisely the difference between the positive-definite metric of the Euclidean space $\mathbb{R}^4$—in general, such a metric is called *Riemannian*—and the four-dimensional space-time of special relativity, which has what is called a *pseudometric*, where ds^2 can be negative.

We now know that

$$\lambda'(\rho) = \frac{1}{\rho - \rho_s} - \frac{1}{\rho}\,,$$

so that

$$\lambda(\rho) = C + \ln\left(1 - \frac{\rho_s}{\rho}\right)\,.$$

Since $\lambda(\rho) \to 0$ as $\rho \to \infty$, it follows that $C = 0$. Summarizing, we find that the infinitesimal proper time increment ds is given by

$$(4.16) \qquad ds^2 = \frac{\rho - \rho_s}{\rho}\,dt^2 - \frac{\rho}{c^2(\rho - \rho_s)}\,d\rho^2 - \frac{\rho^2}{c^2}\,d\varphi^2 - \frac{\rho^2 \sin^2 \varphi}{c^2}\,d\theta^2\,.$$

What we have just done could be checked on a computer if necessary. In fact, the output from the second extension of *Mathematica* Notebook 6 (Volume 3) is

$$\{\{\lambda \to \text{Function}\,[\{\rho\}, C[2] - \text{Log}\,[\rho] + \text{Log}\,[1 - \rho C[1]]]\,,$$

$$\nu \to \text{Function}\,[\{\rho\}, C[3] + \text{Log}\,[\rho] - \text{Log}\,[\rho] - \text{Log}\,[1 - \rho C[1]]]\}\}\,.$$

The constants C[1], C[2], and C[3] can then be determined by the reasoning given above.

---

[28]Notice that dV appears to be an imaginary number. Einstein resolved this difficulty by *defining* dV to be $\sqrt{-\det(g)}$, where his g is the matrix of metric coefficients.

6.2. Geodesics in this metric. Now that we have the metric, we can write the four Euler equations for its geodesics. They are:

$$\frac{d}{ds}\left(\frac{2(\rho - \rho_s)}{\rho}\frac{dt}{ds}\right) = 0,$$

$$\frac{d}{ds}\left(\frac{2\rho}{c^2(\rho - \rho_s)}\frac{d\rho}{ds}\right) = \frac{\rho_s}{\rho^2}\left(\frac{dt}{ds}\right)^2 + \frac{\rho_s}{c^2(\rho - \rho_s)^2}\left(\frac{d\rho}{ds}\right)^2 - 2\rho\left(\frac{d\theta}{ds}\right)^2,$$

$$\frac{d}{ds}\left(2\rho^2\frac{d\varphi}{ds}\right) = 2\rho^2 \sin\varphi\cos\varphi\left(\frac{d\theta}{ds}\right)^2,$$

$$\frac{d}{ds}\left(2\rho^2 \sin^2\varphi\frac{d\theta}{ds}\right) = 0.$$

A suitable rotation will eliminate φ from consideration, just as happened in our earlier analysis. That is, we can assume that φ has an extreme value at $\varphi = \pi/2$, in which case $\varphi = \pi/2$ is the unique solution of the third of the four Euler equations. Since we know that the orbit lies in a plane, we switch to polar coordinates, replacing ρ with r and ρ_s with r_s. Also, just as in our previous analysis, the Euler equation for the variable θ once again gives Kepler's second law, with the difference that the independent variable is now proper time s rather than observed time t. (It is relativistic angular momentum that is conserved.)

With these reductions taken into account, we now need to deal with the following pair of differential equations:

$$\frac{d}{ds}\left(\frac{2(r - r_s)}{r}\frac{dt}{ds}\right) = 0,$$

$$\frac{d}{ds}\left(\frac{2r}{c^2(r - r_s)}\frac{dr}{ds}\right) = \frac{r_s}{r^2}\left(\frac{dt}{ds}\right)^2 + \frac{r_s}{c^2(r - r_s)^2}\left(\frac{dr}{ds}\right)^2 - 2r\left(\frac{d\theta}{ds}\right)^2.$$

That will be our task in the next section.

There is an obvious resemblance between Eqs. (4.7) and (4.16). We are going to see that ρ_s has essentially the same meaning in both equations. In the relativistic scheme, we know that the time coordinate and the spatial coordinate along the line of motion are entangled when two observers try to reconcile their observations. This entangling made two significant changes in the Newtonian scheme: (1) the purely conventional speed of information v_0, used to compute the "exchange rate" $dt = (1/v_0)\,dx$ between time and space, gets replaced by the speed of light c, and produces an absolute "exchange rate" $dt = (1/c)\,dx$; (2) the sign of the spatial portion of the metric is reversed. Therefore, we should expect a perturbation of the form $ds^2 = f(\rho)\,dt^2 - (g(\rho)/c^2)\,d\rho^2$ in spherical coordinates. And so it turned out.

We have mentioned that ρ_s is very small in comparison with the values of ρ typical in planetary orbits. Even for the orbit of Mercury, we always have $\rho_s/\rho < 10^{-7}$. When ρ_s is taken equal to zero, Eq. (4.16) becomes the flat space-time metric of special relativity, while Eq. (4.7) becomes the Euclidean metric on the spatial part of space-time. The importance of the Schwarzschild radius ρ_s was first recognized in connection with the relativistic law of gravity, although, as we saw above, it was "lurking" even in the Newtonian law and reveals itself when this law is formulated in geodesic terms. *All the "curvature" that causes the geometry of a gravitational field to differ from flat-space geometry is due solely to the presence of the radius ρ_s in the metrics.*

With those preliminaries taken care of, we now set out to study the motion of a particle, assuming it must be a geodesic in the modified space-time metric.

7. Computation of the Relativistic Orbit

If the gravitational field is computed to a degree of precision that is one order higher and the orbit of a material point infinitely small in comparison [to the size of the body creating the gravitational field] is computed to the same degree of precision, the following deviation from the Kepler–Newton laws of planetary motion arises. The elliptical orbit of a planet undergoes a slow rotation in the direction of its revolution by the amount

$$\varepsilon = 24\pi^3 \frac{a^2}{T^2 c^2 (1 - e^2)}$$

each time it is traversed. . . Calculation reveals a rotation of the orbit by 43″ per century for Mercury, corresponding precisely to the findings of the astronomers (Leverrier); to be specific, they found a precession of the perihelion of this planet that could not be accounted for by perturbations due to the other planets of exactly this amount.

Einstein, ([**21**], p. 822). My translation.

When we talk of computing the relativistic orbit of Mercury, we mean describing it in a frame of reference in which the Sun is fixed. In that frame, proper time s on Mercury, measured from some agreed-upon epochal event, is smaller than the time t the observers (we ourselves) are using. To compute this orbit, we need to find the geodesics for the space-time metric and the value of the constant ρ_s. In classical mechanics, this variational problem was stated as the brachistochrone problem, that is, the path that minimizes the time of transit. In analogy with the classical case, we choose to minimize the proper time s—that is, the four-dimensional space-time interval—rather than the laboratory time t. Thus, we want to know the paths that minimize the integral

$$\int \left\{ \left(\frac{\rho - \rho_s}{\rho} \right) (t')^2 - \left(\frac{\rho}{\rho - \rho_s} \right) (\rho'/c)^2 - (\rho/c)^2 (\varphi')^2 - (\rho/c)^2 \sin^2 \varphi \, (\theta')^2 \right\}^{\frac{1}{2}} du$$

$$= \int F(\rho, \varphi, t', \rho', \varphi', \theta') \, du \, .$$

Here the parameter u may be arbitrary, since the integral is invariant under changes of variable. Things become much simpler, however, if we use proper time s as parameter. The integrand is then equal to the constant 1, since s itself is merely the difference between the limits of integration:

$$F\big(\rho(s), \varphi(s), t'(s), \rho'(s), \varphi'(s), \theta'(s)\big) = 1 \, .$$

This fact simplifies the computation of the geodesics from the three Euler equations for the minimal path. (See Section 1 of Appendix 2 in Volume 2 for the statement and derivation of these equations. We shall not bother to verify that the critical path we find actually is a minimum rather than a maximum, since it is intuitively obvious that there can be no local maximum. One can always vary a path slightly and make it longer.)

Since our first reduction has already eliminated the co-latitude variable φ, and we have replaced the distance ρ by the variable r we customarily use in polar coordinates on the plane, and ρ_s by r_s at the same time, we have the proper-time length s of the geodesic in the form

$$s = \int ds = \int \sqrt{\left(1 - \frac{r_s}{r}\right)(t')^2 - \left(\frac{r}{c^2(r - r_s)}\right)(r')^2 - (r^2/c^2)(\theta')^2}\, ds$$

$$= \int F(r, t', r', \theta')\, ds.$$

Here $F \equiv 1$ along the geodesic. This is the integral for which Schwarzschild found the exact geodesics in his first 1916 paper. His r, however, was actually a radius that he denoted R, related to the normal one by the equation $R = r(1 + r_s^3/r^3)^{1/3}$. Schwarzschild died on 11 May 1916. Two weeks after his death, the dissertation of Johannes Droste (1886–1963) at the University of Leiden was presented (by Droste's advisor Lorentz) at the 27 May meeting of the Royal Netherlands Academy of Arts and Sciences [15]. It contained what we now know as the Schwarzschild solution in its present form.[29]

Schwarzschild remarked that for circular orbits, $d\theta/dt$—he wrote φ where we have written θ—satisfies $(d\theta/dt)^2 = r_s/(2R^3)$, so that Kepler's third law holds exactly for such orbits. Droste noted that the same was true for the normal polar coordinate r of Euclidean geometry, and we shall demonstrate this below.

7.1. Conservation of angular momentum. Let us now look at the Euler equations for the variables t and θ:

$$\frac{d}{ds}\left(\frac{(1 - \frac{r_s}{r})t'}{F}\right) = 0,$$

$$\frac{d}{ds}\left(\frac{-r^2\theta'}{c^2 F}\right) = 0.$$

The second of these equations says, after it is divided by $-r(s)^2$, that

$$\frac{d^2\theta}{ds^2} = \frac{2}{r}\frac{dr}{ds}\frac{d\theta}{ds}.$$

and we take the time to notice that in the "Newtonian limit" when $c \to \infty$ and $r_s \to 0$, so that $s \to t$, this is precisely the same as the Newtonian equation.

In the relativistic case, this equation implies that

$$r^2\frac{d\theta}{ds} = l,$$

for some constant l, which is precisely the statement of *conservation of relativistic angular momentum per unit rest mass*. This principle was a consequence of having a central force in the classical case, but it was also deduced from Euler's equation on θ, just as we have now done in the relativistic case. The difference now is that the derivative is with respect to proper time s rather than observed time t. To express it in terms of observer time t, we need the derivative of t with respect to s. The Euler equation gives us that, at the price of introducing another constant of integration that we will have to eliminate somehow.

---

[29]Droste is slighted in many accounts of relativity theory; his contributions were many and brilliant. He had earlier filled out the mathematical details of the 1913 "draft" of general relativity [24] by Einstein and Marcel Grossmann (1878–1936), and this paper represented his attempt to keep current, now that Einstein had found a fully invariant formulation of the theory. It was through a 1917 study by Hilbert that the world came to call this solution the Schwarzschild solution.

On the geodesic we have $t' = kr/(r - r_s)$ for some dimensionless constant[30] k and $\theta' = l/r^2$. Although we cannot immediately determine the constant k, s is the proper time on Mercury that is calculated by an observer in the heliocentric system; it is the time that an observer fixed in that coordinate system would actually observe on an accurate clock located on Mercury. It thus follows that $ds < dt$, since to the heliocentric observer, a clock on Mercury is losing time. As a consequence $t' = dt/ds > 1$. This result is consistent with the fact that $r - r_s < r$, but does not rule out the possibility that $k > 1$. Taking the ratio of the two derivatives, we find

$$\frac{d\theta}{dt} = \frac{\theta'}{t'} = \frac{l}{r^2} \frac{r - r_s}{kr} = \frac{l(r - r_s)}{kr^3}.$$

We can rewrite this last relation in two different ways:

(4.17) $$\frac{1}{2} r^2 \frac{d\theta}{dt} = \frac{l}{2k} \frac{r - r_s}{r}, \quad \text{or} \quad \frac{dt}{d\theta} = \frac{kr^3}{l(r - r_s)}.$$

The left-hand side in the first relation of Eq. (4.17) is the rate at which area is swept out by the radius vector, as measured by an observer at the origin. As we saw, in the Newtonian model Kepler's second law asserts that this rate is constant, numerically equal to half of the angular momentum per unit mass of the orbiting body. The result of that constancy is an orbit that remains an ellipse, year after year. We now see that in relativity this constant rate applies only to orbits with constant r, that is, circular orbits. The rate *is* constant, however, when measured in proper time, as we saw above. Hence, Kepler's second law continues to hold if we use proper time rather than laboratory time. We should not expect the total angular momentum to be constant in the general case, since the increase in speed when r decreases—which exactly balances the loss in momentum due to the decrease in r in the classical case—causes an increase in total mass.

Since Kepler's second law holds with extreme precision for any observer, we conclude that the distance r_s is very small. We would expect any variation from it to show up only in orbits of very small radius. As a result, we expect that the differences between the Newtonian predictions and the relativistic predictions will show up most clearly in the case of the innermost planet, Mercury.

7.2. The shape of the orbit. Rather than deal with the Euler variational equation on r, we find it simpler now to use the equation $F(r, r', t', \theta') = 1$ (which may be regarded as the integrated form of the Euler equation) together with Eq. (4.17). The equation $F(r, r', t', \theta') = 1$ says

$$\frac{r - r_s}{r} \left(\frac{dt}{ds}\right)^2 - \frac{r}{c^2(r - r_s)} \left(\frac{dr}{ds}\right)^2 - \frac{r^2}{c^2} \left(\frac{d\theta}{ds}\right)^2 = 1.$$

Dividing by $\left(\frac{d\theta}{ds}\right)^2$, which is l^2/r^4, we get

$$\frac{r - r_s}{r} \left(\frac{dt}{d\theta}\right)^2 - \frac{r}{c^2(r - r_s)} \left(\frac{dr}{d\theta}\right)^2 - \frac{r^2}{c^2} = \frac{1}{\left(\frac{d\theta}{ds}\right)^2} = \frac{r^4}{l^2}.$$

When we insert the value of $\frac{dt}{d\theta}$ and cancel a factor of r, we get

$$\frac{k^2 r^4}{l^2(r - r_s)} - \frac{1}{c^2(r - r_s)} \left(\frac{dr}{d\theta}\right)^2 - \frac{r}{c^2} = \frac{r^3}{l^2}.$$

---

[30]The constant k is not absolute. It depends on the particular geodesic.

When we solve this equation for $\left(\frac{dr}{d\theta}\right)^2$, we find

$$\frac{1}{r^4}\left(\frac{dr}{d\theta}\right)^2 = \frac{c^2 k^2}{l^2} - \frac{c^2(r - r_s)}{l^2 r} - \frac{r - r_s}{r^3}.$$

We can rewrite this equation as

$$\left(\frac{dr}{d\theta}\right)^2 = r\left(\frac{c^2(k^2 - 1)}{l^2}r^3 + \frac{c^2 r_s}{l^2}r^2 - r + r_s\right).$$

Recall that our attempt to geometrize Newton's astronomy failed at this point, since the right-hand side of the corresponding equation had to be nonnegative, but could assume the value 0 only for one positive value of r. That restriction excluded the possibility of elliptical orbits. Now, however, depending on the values of k, r_s, and l, the right-hand side of this equation may have up to three positive zeros, and hence the set of r for which it is positive *may* contain a finite interval of positive numbers, enabling us to consider bounded noncircular orbits. We don't know that it *does* contain such an interval, but at least such orbits are not mathematically excluded. If, for example, $k = 1$, there will be two positive zeros provided $l > 2cr_s$. The precise requirement is that the cubic equation

$$\frac{c^2(k^2 - 1)}{l^2}r^3 + \frac{c^2 r_s}{l^2}r^2 - r + r_s = 0$$

have at least two positive zeros between which the left-hand side is positive. This cannot happen if $k^2 > 1$, since the left-hand side is positive at $r = 0$ and tends to $-\infty$ as $r \to -\infty$. That means the equation has a negative root. By Descartes' rule of signs, it has only one negative root. But if it has two positive roots, it must be negative in between them, which means the equation $dr/d\theta = 0$ can be satisfied at only one of the two roots, since r cannot cross the interval between the two roots. Hence $k^2 \le 1$. If $k^2 = 1$, the equation is a quadratic equation and the left-hand side is once again negative between the two roots. Hence it must be that $k^2 < 1$. The left-hand side is still positive at $r = 0$, but goes to $-\infty$ as $r \to +\infty$. Again, by Descartes' rule of signs, the equation has no negative roots and either 1 or 3 positive roots. If it has three positive roots, it will be positive between the two larger roots, which is what we must have.

This equation resembles the equation for the orbit in Newtonian mechanics. As we have now done three times, we rewrite the equation in terms of $u = 1/r$, so that $du/d\theta = -(1/r^2)(dr/d\theta)$. The result is

$$\left(\frac{du}{d\theta}\right)^2 = \frac{c^2(k^2 - 1)}{l^2} + \frac{r_s c^2}{l^2}u - u^2 + r_s u^3,$$

which we rearrange as

(4.18) $$\left(\frac{du}{d\theta}\right)^2 + u^2 = \frac{c^2(k^2 - 1)}{l^2} + \frac{r_s c^2}{l^2}u + r_s u^3.$$

The presence of three as-yet-undetermined constants is still annoying, but since k occurs only in the constant term, we can get rid of it by differentiating and then dividing out the factor $2\frac{du}{d\theta}$:

(4.19) $$\frac{d^2 u}{d\theta^2} + u = \frac{r_s c^2}{2l^2} + \frac{3}{2}r_s u^2.$$

This equation is merely a perturbed version of the classical equation of Newtonian mechanics. It differs from that equation in only two ways: (1) the constant term GM/l^2 is replaced by $r_s c^2/(2l^2)$; (2) it contains the perturbing term $\frac{3}{2} r_s u^2$. It differs from the equation we obtained in our attempted geometrization of Newtonian mechanics only in the sign of the perturbation.

Now, if nature has any aesthetic sense at all, Eq. (4.19) must be telling us that the relativistic solution of this problem is just the Newtonian solution of the same problem, slightly perturbed due to the added small term $\frac{3}{2} r_s u^2$. Guided by this faith, we shall confidently assert, on the basis of aesthetics alone[31] that

$$r_s = \frac{2GM}{c^2} = 2953.80\,\mathrm{m}\,.$$

We can be even more confident of this assumption since this is precisely the length ρ_s that we introduced into Eq. (4.10) in the classical case if we take $v_0 = c$. If we needed any further justification beyond that point, we note that the only remaining perturbation of the classical equation is the term $\frac{3}{2} r_s u^2$, and the coefficient is exactly the negative of the coefficient of the perturbing term that we got when we attempted to reduce the Newtonian equations in the classical case to Euler's equations. Finally, we note that, if we take $r_s c^2 \equiv 2GM$, then letting $c \to \infty$ (the Newtonian limit) forces $r_s \to 0$, and the equation is precisely the Newtonian equation once again, as the quadratic term in u drops out. Einstein considered it to be very important that his equations become the Newtonian equations when $c \to \infty$.

We now have one of the constants we have been needing. It turns out to be a distance one could easily walk in a little more than half an hour. The average distance of Mercury from the Sun, which is the semi-major axis of its orbit (about 58 million kilometers), is about 20 million times as large as ρ_s. That is why there is little danger that r will get small enough to cross a critical point in the space-time metric. In particular, the value of u is always at least 14 million times as large as the perturbing term $r_s u^2$. We remarked above that a sphere of radius r_s could not be seen from the Earth, even with the Hubble telescope. We have idealized the Sun as a point-particle, which therefore lies entirely inside the Schwarzschild radius, and we are studying only the orbits of idealized particles outside the body of the Sun. Thus, the Schwarzschild radius is a kind of fiction, a purely mathematical entity that enables us to explain a physical phenomenon. In particular, it does *not* represent any singularity in the space-time metric. If through some miracle, we were able to penetrate to within that radius of the center of the Sun, we would of course be subject to unimaginable pressure, but otherwise nothing special would happen, because only the mass between us and the center would produce any gravitational force. That mass is much less than the mass of the Sun, and hence has a still smaller Schwarzschild radius, outside which we would remain. Only when the entire mass of a body is concentrated inside its Schwarzschild radius do we actually get a black

---

[31] It seems appropriate to quote Einstein himself at this point: "Raffiniert ist der Herrgott, aber boshaft ist Er nicht." ("The Lord God is subtle, but not malicious.") Einstein is said to have made this remark during a visit to Princeton University in 1921, in response to a recent announcement by Dayton Miller (1866–1941) that he had detected some difference in the speed of light relative to the motion of the Earth. (See p. 390 of the book of Ronald W. Clark [**8**].) The difference in speed was still much smaller than classical predictions made on the basis of an "ether drift," and Miller had to make more than 5 million measurements with his sophisticated interferometer in order to reveal it. The large number of measurements, analogous to turning up the volume on a radio full of static in order to hear the signal, suggests that the effect really was just "noise."

hole. In all other cases, the physical behavior of matter at that radius from the center is not different from its behavior anywhere else.

REMARK 4.9. If Kepler's third law held in this situation, we would have $GM = 4\pi^2 a^3/T^2$, where a is the semi-major axis of the orbit and T its period, and that would give us another expression for r_s, namely

$$r_s = \frac{8\pi^2 a^3}{T^2 c^2}.$$

But in relativity, the orbit isn't an ellipse when expressed in ordinary polar coordinates, and Kepler's third law doesn't hold in general. On the other hand, it does hold for *circular* orbits.

From now on, we shall work with the equation

$$\frac{d^2 u}{d\theta^2} + u = \alpha + \beta u^2,$$

where $\alpha = r_s c^2/(2l^2)$ and $\beta = 3r_s/2$.

Since u' does not occur explicitly in this equation, we can once again use the familiar technique of writing $u'' = p(dp/du)$, where $p = u'$. If we integrate this last equation from u_0 (aphelion) to u, taking $\theta = 0$ at aphelion, we get (after multiplying by 2) the equation

$$\left(\frac{du}{d\theta}\right)^2 = r_s\left(\frac{c^2}{l^2}(u - u_0) - (u^2 - u_0^2)/r_s + u^3 - u_0^3\right)$$

$$= r_s(u - u_0)\left(\frac{c^2}{l^2} - (u_0 + u)/r_s + u_0^2 + u_0 u + u^2\right).$$

Here, on the left-hand side, we use the fact that $du/d\theta = 0$ at aphelion. From the physical model, we know that $du/d\theta = 0$ also at perihelion (u_1), and so the constants r_s and l must be such that

$$\frac{c^2}{l^2} = (u_0 + u_1)/r_s - (u_0^2 + u_0 u_1 + u_1^2).$$

With this explicit expression for c^2/l^2, we can now write $(du/d\theta)^2$ in terms of u_0 and u_1:

$$\left(\frac{du}{d\theta}\right)^2 = (u - u_0)(u_1 - u)\big(1 - r_s(u + u_0 + u_1)\big).$$

REMARK 4.10. The expression for c^2/l^2 also allows us to write the angular momentum per unit rest mass in a form that involves only the parameters of the orbit and the Schwarzschild radius:

$$l = \frac{ca(1 - e^2)\sqrt{r_s}}{\sqrt{2a(1 - e^2) - r_s(3 + e^2)}}.$$

The fact that $du/d\theta$ is the square root of a cubic polynomial in u implies that u is an elliptic function of θ. To get the equation into a form in which we can compute this function, we need to make some changes of variable. The first is a simple linear transformation on u. Let $u = \gamma v + \delta$ with γ and δ chosen so that $u_0 = -\gamma + \delta$ and $u_1 = \gamma + \delta$, that is, v ranges over $[-1, 1]$. This is easily done: $\gamma = (u_1 - u_0)/2$, $\delta = (u_1 + u_0)/2$. In relation to the geometry of a planetary orbit in the Newtonian model, we have $\gamma = e/(a(1 - e^2))$ and $\delta = 1/(a(1 - e^2))$. When written in terms of v rather than u, the equation becomes

$$\left(\frac{dv}{d\theta}\right)^2 = (1 - v^2)\left(1 - \frac{3r_s}{(a(1 - e^2))} - \frac{r_s e}{(a(1 - e^2))}v\right).$$

Since we have eliminated the co-latitude φ from consideration, we are now free to use that symbol to represent a different angle, which will turn out to be a "phantom" longitude angle. Specifically, if we now let $v = \sin(\varphi)$, so that $dv/d\theta = \cos(\varphi)d\varphi/d\theta$, we get the very simple equation

$$\left(\frac{d\varphi}{d\theta}\right)^2 = 1 - \frac{3r_s}{a(1 - e^2)} - \frac{r_s e}{a(1 - e^2)}\sin(\varphi).$$

In this equation, since aphelion corresponds to $v = -1$, we need to take $\varphi = -\pi/2$ at this point.

Summarizing the changes of variable we have made so far, we have an equation for $u = 1/r$ that resembles the classical Newtonian solution given above:

$$u = \frac{e}{a(1 - e^2)}\sin(\varphi) + \frac{1}{a(1 - e^2)},$$

which becomes

$$r(1 - e\cos\varphi) = a(1 - e^2)$$

if, instead of taking $\varphi = -\pi/2$ at aphelion, we take $\varphi = 0$ at this point. We note that the eccentricity and semi-axis of this orbit are exactly the same as in the Newtonian model.

REMARK 4.11. This equation for the orbit in terms of the fictitious angle φ shows how to construct a kinematical model of the orbit based on the Keplerian orbit. We simply imagine a "phantom" Mercury traveling around the old Keplerian (elliptic) orbit predicted by the Newtonian theory. The radius vector from the Sun to the observable Mercury at right ascension θ equals the radius vector to the phantom Mercury at right ascension φ. (Again, this φ is *not* the co-latitude angle that we eliminated earlier!) The result is illustrated in Fig. 4.5.

Thus, in terms of the angle φ, the relativistic equation of the orbit is *exactly* the same as the classical equation. If we had $\varphi = \theta$, there would be no difference between them. As there *is* a difference, we need to explore how φ is related to θ.

Let us return to our original assumptions, whereby $\theta = 0$ and $\varphi = -\pi/2$ at aphelion. Thus

$$\theta = \int_0^\theta dt = \int_{-\pi/2}^\varphi \frac{ds}{\sqrt{1 - \frac{3r_s}{a(1-e^2)} - \frac{r_s e}{a(1-e^2)}\sin(s)}}$$

$$= \int_0^{\varphi+\pi/2} \frac{dt}{\sqrt{1 - \frac{3r_s}{a(1-e^2)} + \frac{r_s e}{a(1-e^2)}\cos(t)}},$$

where we replaced s by $t - \pi/2$ to get the last equation ($\sin(t - \pi/2) = -\cos(t)$).

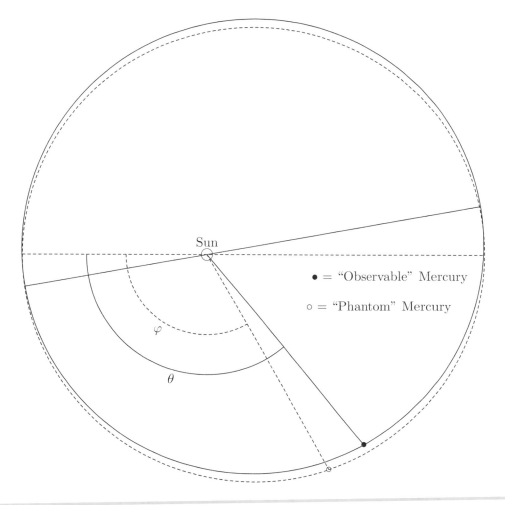

FIGURE 4.5. Kinematic model of the relativistic orbit of Mercury. The Newtonian orbit is shown in dashed lines as a "phantom" Mercury that traverses the same elliptical path year after year. The relativistic orbit, shown as a solid curve, has the same radius vector at right ascension θ that the Newtonian orbit has at right ascension φ. Since θ increases faster than φ (because the time in the heliocentric frame of reference of an event occurring on Mercury is larger than the proper time of that same event in the Mercury-centered frame of reference), the effect is that the ellipse rotates slowly in the direction of travel along the orbit.

We can reduce this last integral to a standard-form elliptic integral by using the identity $\cos t = 1 - 2\sin^2(t/2)$. From that identity we get, this time setting $s = t/2$,

$$\theta = 2\int_0^{\frac{\varphi}{2}+\frac{\pi}{4}} \frac{ds}{\sqrt{1 - \frac{r_s(3-e)}{a(1-e^2)} - \frac{2r_s e}{a(1-e^2)}\sin^2(s)}} = \frac{2}{C}\int_0^{\frac{\varphi}{2}+\frac{\pi}{4}} \frac{ds}{\sqrt{1 - m\sin^2(s)}},$$

where

$$C = \sqrt{1 - \frac{r_s(3-e)}{a(1-e^2)}} \quad \text{and} \quad m = \frac{2r_s e}{a(1-e^2) - r_s(3-e)}.$$

Thus, we find that

$$\theta = \frac{2}{C} \, \text{EllipticF}\left(\frac{\varphi}{2} + \frac{\pi}{4}, m\right).$$

By definition of the function EllipticF, this means that

$$\frac{\varphi}{2} + \frac{\pi}{4} = \text{am}\,(C\theta/2, m),$$

that is,

$$\varphi = 2\,\text{am}\,(C\theta/2, m) - \frac{\pi}{2}.$$

As in our toy Newtonian model, $\text{am}\,(x, m)$ is the *Jacobi amplitude* function corresponding to modulus m, that is, the inverse of the function $\text{EllipticF}\,(x, m)$. Here once again we find an echo of our unsuccessful attempt to geometrize Newton's analysis. The same two angles θ and φ have the same relation to each other here as in the earlier case. The difference is that the equation connecting r and φ is now the equation of an ellipse rather than a straight line.

We recall that $\varphi = -\pi/2$ at aphelion. Remeasuring now, so that $\varphi = 0$ at this point (that is, relabeling φ itself), we find the simple relation

$$\varphi = 2\,\text{am}\,(C\theta/2, m).$$

From the form of the equation of the orbit in terms of φ, with φ being set equal to 0 at *one particular* aphelion, we saw above that

$$r(1 - e\cos\varphi) = a(1-e^2).$$

We now see that the actual polar equation in terms of the observable angle θ looks almost the same, and we have the following elegant result:

THEOREM 4.4. *The observed polar coordinates (r, θ) of the orbiting particle satisfy the equation*

(4.20) $$r\big(1 - e\cos\big(2\,\text{am}\,(C\theta/2, m)\big)\big) = a(1-e^2).$$

For the rest of the argument, we need to insert some numerical values and do some precise calculations. The reader is referred to *Mathematica* Notebook 7 in Volume 3 for the necessary details, including the solution of the differential equation. We now take those results as known.

With these data, we quickly get numerical values for the constant C and the elliptic modulus m:

$$C = 1 - 7.441360 \times 10^{-8},$$
$$m = 2.1903594 \times 10^{-8}.$$

We can now see how very nearly equal φ and θ are. As a function of φ, r has period 2π. But an increment of 2π in the value of φ corresponds to a slightly *larger* increment in the polar angle θ. The actual increment in the polar angle θ for this increment in φ is

$$\Delta\theta = \frac{2}{C} \int_0^\pi \frac{ds}{\sqrt{1 - m\sin^2 s}}.$$

When we insert the numerical values of C and m into this equation and evaluate the integral, we find that

$$\Delta\theta = 2\pi + 5.02233126 \times 10^{-7}\,.$$

Another way of putting this is to say that the orbit of Mercury can be described by the usual equation with right ascension $\varphi(t)$ and radius $r(t)$ related by

$$r(t)\big(1 - e\cos\varphi(t)\big) = a(1 - e^2)\,,$$

except that $\varphi(t)$ differs slightly from the observed right ascension $\theta(t)$ and in particular, has a slightly longer period than the period T of the right ascension. Thus, we express the orbit in terms of a product of two observable, measurable periodic functions $r(t)$ and $\theta(t)$, which now have different periods. In the Newtonian elliptic model, these two functions have the same period T. In the Newtonian limit, in which the speed of information v_0 is infinite and consequently the Newtonian Schwarzschild radius r_s is 0, we get $m = 0$, $C = 1$, $\varphi(t) = \theta(t)$, and so the relativistic solution reduces to the Newtonian.

That small discrepancy, whereby the increment in θ between successive perihelions exceeds 2π by one two-millionth of a radian, accounts completely for the previously unexplained portion of the precession of the perihelion of Mercury. In fact, if we convert it into seconds—there are 3600 seconds in one degree—and find the accumulated discrepancy over a century, when Mercury makes some 415 complete revolutions, we learn that it is

$$5.02233126 \times 10^{-7} \times \frac{180}{\pi} \times 3600 \times 415 = 42.9678\,.$$

As we began by saying that we were trying to explain the *already observed* fact that the perihelion of Mercury was precessing by about 43 seconds of arc per century more than could be explained as the result of the gravitational action of the other planets, this very precise "fit" between theory and observation provides powerful evidence that the theory is more precise than the Newtonian theory.

REMARK 4.12. Taking the Sun as origin, we can regard the plane of Mercury's orbit as the complex plane and write the location of Mercury in it at time t as a complex-valued function of t, namely

$$r(t)e^{i\theta(t)}\,.$$

In Newtonian mechanics, the two functions $r(t)$ and $e^{i\theta(t)}$ have the same period, and therefore their product also has that period. In relativity, due to the different FitzGerald–Lorentz contractions in the radial and tangential directions, the two periods are different. As noted above, there is a constant $k = 2K/\pi$ such that $\theta(t) - kt$ has period 2π, and thus the location of the planet at time t is given as a product of two periodic functions:

$$r(t)e^{i\theta(t)} = \Big(r(t)e^{i\theta(t)-ikt}\Big)e^{ikt}\,,$$

whose periods are 2π and $2\pi/k = \pi^2/K$. The function is now periodic if and only if the ratio of the two periods $2K/\pi$ is a rational number. In any case, however, it is a *uniformly (Bohr-) almost-periodic* function,[32] and as such has a Fourier series

$$r(t)e^{i\theta(t)} = \sum_{n=1}^{\infty} c_n e^{2\pi i \lambda_n t},$$

where the frequencies λ_n are not necessarily integers. Thus, even the relativistic orbits confirm that Ptolemy's epicycle approach to astronomy, as formulated by Sternberg ([**77**], pp. 1–14), is feasible *in theory.*

7.3. Kepler's third law*. Kepler's first law has similar forms in Newtonian and relativistic physics, namely

$$\begin{aligned} r(1 + e\cos\theta) &= a(1 - e^2) \\ r(1 + e\cos\varphi) &= a(1 - e^2), \end{aligned}$$

where φ is an elliptic function of θ (measured, in this formula, from perihelion, in contrast to the formula given above, where it was measured from aphelion).

Similarly, Kepler's second law merely expresses conservation of angular momentum per unit rest mass:

$$\begin{aligned} l &= r^2 \frac{d\theta}{dt} \\ l &= r^2 \frac{d\theta}{ds}. \end{aligned}$$

The precession of perihelion means that r is no longer known to be a periodic function of θ, so that it would not appear to make sense to speak of its period. Still, relativity has not changed the aphelion and perihelion distances, so that r has a definite period as a function of time, and the average of the perihelion and aphelion distances is still defined. Those are the ingredients of Kepler's third law. The period T can be taken as the time elapsed between successive perihelia, and it might be measured in either "laboratory" (heliocentric) time or in proper time on the orbiting planet. Thus, we have already two possible forms for Kepler's third law, depending on which time we wish to use. We might get two more forms by considering the time required for θ to increase by 2π. In only one case, namely the case of circular orbits, would this period also be a period of the actual position of the planet. In that one case, we do indeed find that Kepler's third law holds. Replacing r' by 0 makes F independent of r' and thereby changes the Euler equation.

For a circular orbit of radius a, we have

$$1 = \sqrt{\left(1 - \frac{r_s}{r}\right)\left(\frac{dt}{ds}\right)^2 - \frac{r^2}{c^2}\left(\frac{d\theta}{ds}\right)^2} = F(t, r, \theta, t', r', \theta').$$

By Euler's equation on the variable r, we thus get (since F is independent of r', which means $\partial F/\partial r' = 0$)

$$0 = \frac{d}{ds}\left(\frac{\partial F}{\partial r'}\right) = \frac{\partial F}{\partial r} = \frac{r_s}{r^2}\left(\frac{dt}{ds}\right)^2 - \frac{2r}{c^2}\left(\frac{d\theta}{ds}\right)^2.$$

---

[32]The theory of these functions was developed single-handedly by Harald Bohr (1887–1951) during the second decade of the twentieth century in an attempt to settle the Riemann hypothesis (the still-unproved conjecture that all the nontrivial zeros of the Riemann zeta function have real part equal to 1/2). Harald Bohr was the brother of the famous physicist Niels Bohr(1885–1962).

Dividing this equation by $(dt/ds)^2$, we get for the period T

$$\frac{4\pi^2}{T^2} = \left(\frac{d\theta}{dt}\right)^2 = \frac{c^2}{2r}\frac{r_s}{r^2} = \frac{c^2 r_s}{2}\frac{1}{r^3} = \frac{GM}{r^3},$$

which is Kepler's third law in exactly the form it has in the Newtonian system.

8. The Speed of Light

Since proper time is constant for a photon traveling at the speed of light, we can get an expression for the speed of light in the gravitational field of a particle by setting $ds = 0$ in Eq. (4.16):

$$0 = \frac{\rho - \rho_s}{\rho}\,dt^2 - \frac{\rho}{c^2(\rho - \rho_s)}\,d\rho^2 - \frac{\rho^2}{c^2}\,d\varphi^2 - \frac{\rho^2 \sin^2\varphi}{c^2}\,d\theta^2 .$$

For a ray of light moving radially, we have $d\theta = 0 = d\varphi$, and so the *radial* speed of light c_r in this gravitational field is

$$c_r = \frac{d\rho}{dt} = c\left(1 - \frac{\rho_s}{\rho}\right).$$

In particular, light slows down in a gravitational field outside the Schwarzschild radius, and the closer to the radius, the more it slows down. (It comes to a halt at that radius.) This slowing down is barely noticeable at the distances of planetary orbits, since, as mentioned many times, ρ_s/ρ is of the order 10^{-7} for all the planets in the solar system.

The speed c_t is the same in every direction tangential to a sphere about the origin and can be obtained by setting $\varphi = \pi/2$, $d\varphi = 0 = dr$:

$$c_t = \rho\,\frac{d\theta}{dt} = c\sqrt{1 - \frac{\rho_s}{\rho}}.$$

As a consequence, the speed of light is *not* independent of direction, at least when we measure directions as if space-time were flat, that is, using the Euclidean metric on spheres centered at the gravitating particle. In other words, we appear to have contradicted the basic principle of special relativity, that the speed of light in empty space is constant. We need to add a proviso to that basic principle, so that it holds only *in the absence of a gravitational field*. It turns out that we can modify the metric slightly without changing any of the argument given above and obtain the result that the speed of light is the same in every direction from each given point, but does depend on the strength of the gravitational field at that point.

8.1. Isotropic coordinates. For some computations, it is useful to introduce what are called *isotropic* coordinates on the portion of space outside the Schwarzschild radius. The procedure for doing so is to replace the radial variable ρ by a slight modification of it, which we shall denote r. The two are related by the equations

$$\rho = \left(1 + \frac{\rho_s}{4r}\right)^2 r = \left(1 + \frac{r_s}{r}\right)^2 r$$

$$r = -\frac{1}{4}\rho_s + \rho\left(\frac{1 + \sqrt{1 - \rho_s/\rho}}{2}\right),$$

where $r_s = \rho_s/4$ is the value of r corresponding to $\rho = \rho_s$ (*not* the value given to it in the previous section, where it was merely another notation for ρ_s). These relations show that the ratio r/ρ increases from $1/4$ to 1 as ρ increases from ρ_s

to ∞. The increase is sufficiently fast that even at the visible surface of the Sun, which corresponds to $\rho = 2.3 \times 10^5 \rho_s$, we have

$$\frac{r}{\rho} \approx 0.999988\,.$$

The error in replacing our measured ρ by r is therefore about 0.001%, and therefore not much to worry about. The difference $\rho - r$ is very small indeed, since we have the formula

$$\rho - r = \rho_1 \left(\frac{1}{4} + \frac{1}{2\left(1 + \sqrt{1 - \rho_s/\rho}\right)} \right) < \frac{3\rho_s}{4}\,.$$

Outside the radius of the Sun, this difference has the nearly constant value of $\rho_s/2$, which is less than 2 km. Given that we have no empirical proof that the spatial portion of space-time actually *is* Euclidean, we could not prove anyone wrong who chose to use r instead of ρ in measuring interplanetary distances. The two sets of coordinates would give results that are experimentally indistinguishable. For example, even the very precise values we used for the perihelion and aphelion distances of Mercury were given with a precision only up to ± 50 km, which is nearly 30 times as large as this difference.

We introduce isotropic spherical coordinates and isotropic rectangular coordinates by replacing ρ with r while retaining the co-latitude (φ) and longitude (θ):

$$\begin{aligned} x &= r\sin\varphi\cos\theta\,,\\ y &= r\sin\varphi\sin\theta\,,\\ z &= r\cos\varphi\,. \end{aligned}$$

Naturally, these rectangular coordinates are not exactly equal to the ones we have been using, but, as already noted, the difference is too small to measure. With this modification, we then alter the metric so that Eq. (4.16) is replaced by

$$\begin{aligned} ds^2 &= \left(\frac{r - r_s}{r + r_s}\right)^2 dt^2 - \frac{1}{c^2}\left(1 + \frac{r_s}{r}\right)^4 \left(dr^2 + r^2\,d\varphi^2 + r^2\sin^2\varphi\,d\theta^2\right)\\ &= \left(\frac{r - r_s}{r + r_s}\right)^2 dt^2 - \frac{1}{c^2}\left(1 + \frac{r_s}{r}\right)^4 \left(dx^2 + dy^2 + dz^2\right)\,. \end{aligned}$$

The speed of light c_r at distance r from the attracting particle is now the same in every direction, namely

$$c_r = \sqrt{\left(\frac{dx}{dt}\right)^2 + \left(\frac{dy}{dt}\right)^2 + \left(\frac{dz}{dt}\right)^2} = \frac{1 - \frac{r_s}{r}}{\left(1 + \frac{r_s}{r}\right)^3}\,c\,.$$

The independence of the speed of light from direction is the reason for the name *isotropic* applied to these coordinates. When we add the gravitational fields of a number of particles, isotropic coordinates simplify the computations; and we shall use them in Chapter 7.

REMARK 4.13. The speed of light approaches zero as r decreases to the Schwarzschild radius r_s, but increases rapidly to its value in "empty" space c as $r \to \infty$. At Mercury's orbit, it is already $0.9999997c$.

9. Deflection of Light Near the Sun

If the acceleration of gravity at the surface of the Sun is substituted into the formula for $\tan\omega$*, taking the radius of the Sun as unity, we find* $\omega = 0.84''$*. If it were possible to observe fixed stars very near to the Sun, one would probably have to take this into account. However, as this is known not to be the case, the perturbation due to the Sun is to be ignored. For light rays coming from Venus, which Vidal has observed only two seconds from the edge of the Sun... the amount is much smaller, since one cannot take the distances from Venus and the Earth to be infinite.*

I trust no one finds it strange that I treat a light ray as if it were material. For one can see that light rays possess all the absolute properties of matter by considering the phenomenon of aberration, which is possible only if light rays really are material. Moreover, one cannot conceive of anything that exists and affects our senses without having mass.

> Nihil est, quod possis dicere ab omni
> Corpore seiunctum, secretumque esse ab inani:
> Quod quasi tertia sit rerum natura reperta.
> (Lucretius, *De rerum natura*, I, 431).

[There is nothing you could describe as being distinct from matter and yet not a void, nothing that would be a sort of third nature in this account.]

In any case, I see no need to apologize for publishing the present work, even though it shows that all the perturbations are unobservable. For we have an equal duty to learn what theory can tell us.

Johann Georg von Soldner ([**76**], pp. 171–172). My translation.

A light ray passing by the Sun thereby undergoes a deflection in the amount of $4 \cdot 10^{-5} = 0.83''$*. This is the amount by which the angle subtended by stars in the neighborhood of the Sun is increased by the curving of the rays. Since the fixed stars near the Sun become visible during a total solar eclipse, this consequence of the theory can be compared with observation. For the planet Jupiter the expected shift is some 1/100 of this amount. It would be very desirable for astronomers to investigate this problem.*

Einstein ([**18**], p. 908). My translation.

One of the two great early triumphs of general relativity was the explanation of the precession of Mercury's perihelion, which we have just discussed. The other was the prediction of the amount of deflection a ray of light would undergo when passing near to the Sun. There is a certain irony in the history of this second triumph of general relativity, as one can see by comparing the computation reported by Soldner in 1801, which was based on Newtonian mechanics, and the one Einstein reported in 1911, which was based on relativity. The predicted deflections are

identical!! Einstein was not aware of what Soldner had done a century earlier.[33]
He was very lucky that no one took up his urgent request for an investigation in
1911. If his prediction of that year had been observed, it would have confirmed
Newtonian and relativistic mechanics equally well. As it was, the war intervened,
during which Einstein published his general theory and Schwarzschild produced his
elegant solutions of the field equations in free space, under which the predicted
deflection is approximately doubled. It was not until the eclipse of 29 May 1919
that two expeditions were dispatched to make the measurements Einstein had been
urging, one to the island of Principe, a Portuguese possession off the west coast of
Africa, the other to Sobral in northern Brazil. Despite intermittent cloudy, rainy
weather, both were able to take a number of photographs, which could then be
studied and used to measure the deflection of light from stars near the Sun. We
now take up the comparison of Soldner's original result with Einstein's revised
result.

9.1. Soldner's result. Since the speed of light is not a barrier in Newtonian
mechanics, we can, like Soldner, take note of the fact that in the Newtonian scheme
all bodies, of whatever mass, undergo the same acceleration under Newton's law
of gravity, and just extend that fact to include massless particles like photons. At
the perihelion distance $r = r_{\mathrm{peri}}$, the velocity is perpendicular to the radius vector
and hence the speed is $c = r_{\mathrm{peri}}\theta'$. Since the angular momentum l per unit mass
is conserved, it follows that $l = r_{\mathrm{peri}}c$ at all times. Since a particle of light moves
faster than many observed material particles that describe hyperbolic orbits, and
theory guarantees that the orbit must be a conic section, we conclude that the path
of a light ray is hyperbolic and has a polar equation of the form

$$r(1 + e\cos\theta) = a(e^2 - 1)\,,$$

where $e > 1$ and the origin of these polar coordinates is at the center of the Sun.

We take the amount of deflection in passing near the Sun to be the angle
between the asymptotes of this hyperbola, that is, the angle that does *not* contain
the hyperbola. In rectangular coordinates the equation of the hyperbola is

$$\frac{(x + c)^2}{a^2} - \frac{y^2}{b^2} = 1\,,$$

where $c = \sqrt{a^2 + b^2}$. The foci of this hyperbola are at $x = 0$ and $x = -2c$. We shall
assume that the Sun is at $x = 0$ and that the path of the light ray is the branch
of the hyperbola that crosses the x-axis at $x = a - c$ (which, as shown in Fig. 4.6,
is a negative number). The asymptotes have the equations $y = \pm(b/a)(x + c)$.
The limiting value of y/x as x goes to infinity on the hyperbola is b/a, and the
minimum value of the polar angle θ is the arctangent of this value; we denote it
θ_∞, that is, $\theta_\infty = \arctan(b/a)$. The angle φ between the asymptotes is $\pi - 2\theta_\infty$.

---

[33]One wonders why he didn't undertake a Newtonian computation for comparison. Soldner
had no idea that photons were massless, but Einstein knew this had to be the case because of the
relativistic increase of mass with velocity. If Lucretius and Soldner had known about photons, or
what is now called *dark matter*, they would probably not have asserted that everything is either
matter or a void. Perhaps Einstein assumed that on Newtonian principles gravity could not affect
a massless particle. In the quotation, he seems quite confident that *any* bending of light around
the Sun would tell in favor of relativistic mechanics vis-à-vis Newtonian.

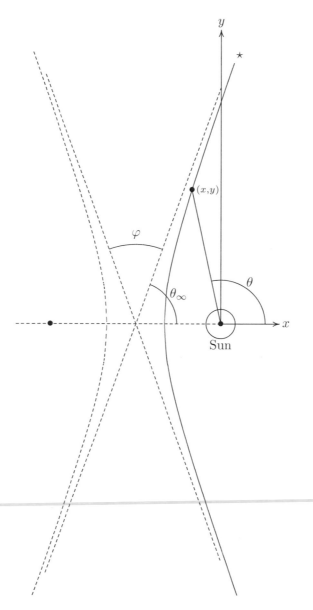

FIGURE 4.6. Deflection of a ray of light from a remote star passing close to the Sun. In the Newtonian case, the ray follows one branch of a hyperbola, shown here as a solid curve. The other branch is shown as a dashed curve. The deflection—greatly exaggerated in the figure—is the angle φ. In the relativistic case, the actual path is no longer a hyperbola, but does not differ qualitatively from what is shown here.

Since $e = \sqrt{1 + (b/a)^2}$, we have $b/a = \sqrt{e^2 - 1}$, and so the angle of deflection is

$$\varphi = \pi - 2\arctan\left(\sqrt{e^2 - 1}\right).$$

We now need to express the eccentricity of the hyperbola as a function of its perihelion point. Since a light ray follows a nearly straight line, the hyperbolic path must have an enormous eccentricity. In fact, when $r(1 + e\cos\theta) = a(e^2 - 1)$, we have $u = 1/r = (1 + e\cos\theta)/\big(a(e^2 - 1)\big)$, and it is easy to compute that

$$\frac{GM}{l^2} = u'' + u = \frac{1}{a(e^2 - 1)},$$

so that $l^2 = GMa(e^2 - 1)$.

We also have $r_{\text{peri}} = a(e-1)$, and $r_{\text{peri}}c = l = \sqrt{GMa(e^2 - 1)}$, as we saw earlier. Eliminating a between these two equations, we find $r_{\text{peri}}c = \sqrt{GMr_{\text{peri}}(e + 1)}$, so that $r_{\text{peri}} = GM(e + 1)^2/c^2$. That is, $e = 2r_{\text{peri}}/\rho_s - 1$, where once again ρ_s is the Schwarzschild radius.[34] Any light ray that does not actually fall into the Sun, must pass by it at a considerable distance. The minimum distance—the "limb" of the Sun, visible only during a total eclipse—is about $697,000$ kilometers. Since ρ_s is about three kilometers, we see first of all that, without any sensible error, we can replace $\sqrt{e^2 - 1}$ with $e + 1 = 2r_{\text{peri}}/\rho_s$, and that $e \approx 464,000$. Accordingly,

$$\varphi = \pi - 2\arctan(464000) \approx 4.31034 \times 10^{-6} = 0.89''.$$

9.2. Einstein's revised solution.
To get a relativistic solution of this problem for comparison, we note that the relativistic angular momentum per unit rest mass is $l = \alpha r^2\theta' = r^2 d\theta/ds$, where s is proper time. For a particle traveling at speed $v = c$, the factor $\alpha = (1 - v^2/c^2)^{-1/2}$ becomes *infinite!* Thus, a photon has infinite angular momentum "per (fictitious) unit mass." What this means is that the constant term GM/l^2 drops out of Eq. 4.19, and we get the simpler equation

$$u'' + u = \frac{3\rho_s}{2}u^2.$$

When we apply our equation-solving technique of letting $p = du/d\theta$, this equation can be integrated with u as independent variable starting from perihelion as a base point. At that point, $u = u_{\text{peri}}$ has its unique maximum value, and $p = 0$, so that, at a general point u,

$$p^2 = \rho_s(u^3 - u_{\text{peri}}^3) - (u^2 - u_{\text{peri}}^2).$$

Since θ increases from θ_∞ to π as u increases from 0 (that is $r = \infty$) to u_{peri}, we have the equation

$$\pi - \theta_\infty = \int_0^{u_{\text{peri}}} \frac{du}{\sqrt{\rho_s(u^3 - u_{\text{peri}}^3) - (u^2 - u_{\text{peri}}^2)}},$$

from which it follows that

$$\varphi = \pi - 2\theta_\infty = 2\int_0^{u_{\text{peri}}} \frac{du}{\sqrt{\rho_s(u^3 - u_{\text{peri}}^3) - (u^2 - u_{\text{peri}}^2)}} - \pi.$$

Once again, we would need elliptic functions to express the right-hand side of this equation with infinite precision; the path of the light ray is not a hyperbola. Nevertheless, we can be confident that it has the general shape of a hyperbola, and we don't need the exact solution. All we need to do is evaluate this integral numerically with great precision. Before we attempt to do so, we ought to verify

---

[34]Notice that the relativistic Schwarzschild radius arises naturally here, even though the analysis is purely within the context of Newtonian mechanics.

that the equation makes sense for $0 \leq u \leq u_{\text{peri}}$, that is, that the expression giving p^2 in terms of u is nonnegative on this interval. That fact follows from elementary algebra. By Descartes' rule of signs, the cubic polynomial in u on the right has precisely one negative root, say $u = -\varepsilon$, where $\varepsilon > 0$. It has the positive root $u = u_{\text{peri}}$, and the sum of its roots is the negative of the ratio of the coefficient of u^2 to the coefficient of u^3. That is, the sum of the roots is $1/\rho_s$. This means that the other positive root r is $1/\rho_s + \varepsilon - u_{\text{peri}}$, which is larger than $1/\rho_s - u_{\text{peri}}$. This number is certainly larger than u_{peri}, since $1/\rho_s$ is larger than $100,000u_{\text{peri}}$. It follows that the polynomial is positive over the entire interval $0 \leq u < u_{\text{peri}}$.

For the given values $\rho_s = 3$, $u_{\text{peri}} = 1/(697,000)$, *Mathematica* reveals that

$$\varphi = 8.61313 \times 10^{-6} \, \text{radians} = 1.77659''.$$

Thus, general relativity predicts a deflection almost exactly twice what Newtonian mechanics predicts. Here then is a test case to see which of the two works better, and it was (barely) within the limits of precision of the available instruments a century ago to measure the difference. Of course, because the paths are so nearly straight, it was necessary to get a ray of light that passed as close as possible to the Sun, and thus such a ray would be detectible on Earth only during a total solar eclipse.

9.3. The results of observation. The raw data obtained by a camera photographing the heavens do not give a direct read-out of the amount of deflection. These data have to be massaged and averaged, and reported in terms of a mean and standard deviation. On that basis, it appears that Einstein was right. The massaged data from Sobral gave a deflection of 1.87 ± 0.13 seconds, while the data from Principe yielded 1.98 ± 18 seconds. Both were large enough to rule out the Newtonian model and close enough to fit the relativistic model.

There is more to be said, however, and discussion of these results has continued. Sir Arthur Eddington was the most prominent member of the Principe expedition, and he was known to be an enthusiastic proponent of the theory of relativity, and also a man eager to reconcile with the Germans after the recent war. The suggestion was made that, consciously or unconsciously, he manipulated his data. It seems very unlikely that he would do so deliberately, since there were two independent expeditions, and he could manipulate the data of only one of them. He would have nothing to gain and everything to lose by such a risky gambit. For an account of this controversy, see the paper of Harvey [37].

10. Problems

PROBLEM 4.1. Verify Eq. (4.16).

PROBLEM 4.2. Let us explore the "punctured disk" of nonzero relativistic velocities as a two-dimensional manifold (see Appendix 2), with a metric $ds^2 = g_{11}dx^2 + g_{12}dx\,dy + g_{21}dy\,dx + g_{22}dy^2$. We will find it easier to do this using polar coordinates (r, θ), where r ranges over the real numbers in the interval $(0, c)$ and θ is the numerical measure of an angle, two angles being identified as usual if they differ by 2π. To get the squared element of arc length, consider an infinitesimal triangle with vertex at the origin and two sides equal to r and $r + dr$ enclosing an angle $d\theta$. The third side will be ds. On the infinitesimal level, the squared element of arc length is

$$ds^2 = \frac{r^2 + (r + dr)^2 - 2r\,(r + dr)\cos d\theta - r^2(r + dr)^2\sin^2 d\theta/c^2}{(1 - r\,(r + dr)\cos d\theta/c^2)^2}\,.$$

Expand every term in the numerator in a Maclaurin series in dr and $d\theta$, using the well-known expansions $\cos x = 1 - \frac{1}{2}x^2 + \frac{1}{24}x^4 - \cdots$ and $\sin^2 x = \frac{1}{2} - \frac{1}{2}\cos 2x = x^2 - \frac{1}{3}x^4 + \cdots$, retaining only terms of degree 2 or less in these infinitesimals. Then expand the denominator as a geometric series—that is, using the expansion $(1 - x)^{-2} = \frac{d}{dx}(1 - x)^{-1} = \frac{d}{dx}(1 + x + x^2 + x^3 + \cdots) = 1 + 2x + 3x^2 + \cdots$, and multiply the two expansions together to show that

$$ds^2 = \left(1 - \frac{r^2}{c^2}\right)^{-2}dr^2 + r^2\left(1 - \frac{r^2}{c^2}\right)^{-1}d\theta^2\,.$$

Then compute the length of the radius from the origin to the point with polar coordinates (r, θ) and the circumference of the circle through that point with center at the origin. Here, r is the *coordinate* of the endpoint of the radius, not its *length*. Denote the length $R(r)$, and express the circumference of the circle as a function of R.

PROBLEM 4.3. Compute the eight Christoffel symbols Γ_{ij}^k and the Ricci tensor Ric_{ab} for the metric of the previous problem.

PROBLEM 4.4. The expression for ds^2 in Problem 4.2 was known classically as the *first fundamental form*. When the metric on a two-dimensional surface in $\mathbb{R}^3$ has symmetry ($g_{12} = g_{21}$), the element of arc length can be written as[35]

$$ds^2 = E\,dr^2 + 2F\,dr\,d\theta + G\,d\theta^2\,.$$

In that case, the element of area on the surface is

$$dA = \sqrt{EG - F^2}\,dr\,d\theta\,.$$

Compute this element of area. Then use the expression for dA to compute the area enclosed by the (punctured) circle centered at the origin passing through the point $(r, 0)$. Finally, use the formula for dA to express the area of a triangle having sides u, v with included angle η.

PROBLEM 4.5. A finite piece of the hyperbolic plane can be represented accurately as part of a pseudo-sphere in $\mathbb{R}^3$ (see Appendix 1). The portion in question can be conveniently represented as the graph of a function in polar coordinates in an annulus, $0 < k_0 < r < k$, where k_0 may be an arbitrarily small positive number:

$$
\begin{aligned}
z(r, \theta) &= k\left(\ln\left(\frac{k}{r} + \sqrt{\left(\frac{k}{r}\right)^2 - 1}\right) - \sqrt{1 - \left(\frac{r}{k}\right)^2}\right) \\
&= k\left(\operatorname{arcsech}\left(\frac{r}{k}\right) - \sqrt{1 - \left(\frac{r}{k}\right)^2}\right).
\end{aligned}
$$

With only a small amount of tedium one can compute that

$$\frac{dz}{dr} = -\sqrt{\left(\frac{k}{r}\right)^2 - 1}\,,$$

---

[35]We do apologize for the abuse of the letter G, which we have made strenuous efforts to avoid in our notation for the Ricci tensor. It appears here as a metric coefficient. The notation is due to Gauss, and seems too venerable to change.

so that the element of arc length on this surface is very simple:

$$
\begin{aligned}
ds^2 &= dr^2 + r^2 \, d\theta^2 + dz^2 \\
&= \left(1 + \left(\frac{dz}{dr} \right)^2 \right) dr^2 + r^2 \, d\theta^2 \\
&= \frac{k^2}{r^2} \, dr^2 + r^2 \, d\theta^2 \, .
\end{aligned}
$$

Considering curves on the pseudo-sphere that are parameterized by arc length, show that a geodesic on which the point closest to the z-axis is $\left(r_0, \theta_0, z(r_0, \theta_0) \right)$ (assuming $k_0 < r_0$) must satisfy the system of Euler equations

$$
\theta' = \frac{r_0}{r^2}
$$

$$
rr'' - r'^2 = \left(\frac{r_0}{k} \right)^2 .
$$

Then show that the curve in the annulus whose polar equation is

$$
r = \frac{k r_0}{\sqrt{k^2 - r_0^2 (\theta - \theta_0)^2}}
$$

maps to a geodesic on the pseudo-sphere.

PROBLEM 4.6. Show that there is one other class of geodesics on the pseudo-sphere not included in the family of curves given in the previous problem, namely the curves whose parameterizations are $\left(r(s), \theta(s), z(r(s), \theta(s)) \right)$, where

$$
r(s) = r_0 e^{-s/k}, \quad \theta = \theta_0, \quad k \ln \left(\frac{r_0}{k} \right) < s < k \ln \left(\frac{r_0}{k_0} \right), \quad k_0 < r_0 < k .
$$

These are the hyperbolic analogs of lines of longitude on a sphere, and the parameter s is arc length.

PROBLEM 4.7. Show that $\mathrm{Ric}_{ab} = \mathrm{Ric}_{ba}$ for the space-time metric of general relativity and that the equation $\mathrm{Ric}_{ab} = 0$ is an identity when $a \neq b$.

PROBLEM 4.8. Show that the equations $\mathrm{Ric}_{33} = 0$ and $\mathrm{Ric}_{44} = 0$ are consequences of the equations $\mathrm{Ric}_{11} = 0 = \mathrm{Ric}_{22}$.

PROBLEM 4.9. In the Newtonian orbital computation, use the fact that $dr/dt = 0$ at both aphelion and perihelion to express l^2 in terms of the universal constants G and M and the planet-specific constants a and e. That is, show that $l^2 = GMa(1 - e^2)$.

PROBLEM 4.10. Prove that $\rho^2 \, (d\theta/ds)$ is constant for any radial space-time metric, that is, any metric of the form

$$
ds = \sqrt{ f(\rho) \, dt^2 - \frac{1}{c^2} \left(g(\rho) \, d\rho^2 - \rho^2 \, d\varphi^2 - \rho^2 \sin^2 \varphi \, d\theta^2 \right) } \, .
$$

Here, ρ is again the radial space coordinate, not charge density. *Hint:* Show first that a suitable choice of coordinates allows us to take $\varphi \equiv \pi/2$, $d\varphi/ds = 0$.

PROBLEM 4.11. Note the following series expansions

$$\frac{2}{C} = 2\Big(1 - \frac{r_s(3-e)}{a(1-e^2)}\Big)^{-\frac{1}{2}} = 2\Big(1 - \frac{3r_s}{a}\frac{1-e/3}{1-e^2}\Big)^{-\frac{1}{2}}$$

$$= 2\Big(1 - \frac{3r_s}{a}\Big(1 - \frac{e}{3} + e^2 - \frac{e^3}{3} + \cdots\Big)\Big)^{-\frac{1}{2}}$$

$$= 2\Big(1 + \frac{3r_s}{2a} - \frac{r_s e}{2a} + \cdots\Big)$$

$$\int_0^\pi (1 - m\sin^2 s)^{-\frac{1}{2}} = \int_0^\pi \Big(1 - \frac{m}{2} + \frac{m}{2}\cos(2s)\Big)^{-\frac{1}{2}}$$

$$= \int_0^\pi 1 + \frac{m}{4} - \frac{m}{4}\cos(2s)\cdots ds$$

$$m = \frac{2r_s e}{a(1-e^2) - r_s(3-e)} = \frac{2r_s e}{a(1-e^2)} + \cdots .$$

By neglecting terms that are not larger than $r_s e/a$—since r_s/a is already very small, as is the the eccentricity $e = 0.205$, so that $e/3$ is less than 7%—show that the increment in the polar angle θ when the angle φ increases by 2π is approximately

$$\Delta\theta = 2\pi + \frac{3\pi r_s}{a(1-e^2)} = 2\pi + \frac{24\pi^3 a^2}{T^2 c^2(1-e^2)},$$

as Einstein asserted.

PROBLEM 4.12. For a metric that has symmetry, that is, $g_{ij} = g_{ji}$, it is trivial to show that the Christoffel symbols are also symmetric in the two subscripts. Show that in this case, the metric coefficients g_{ij} satisfy the system of first-order linear partial differential equations

$$\frac{\partial g_{ij}}{\partial x^k} = \Gamma_{ik}^l g_{lj} + \Gamma_{jk}^l g_{il}.$$

It follows that the metric coefficients can be determined from the values they have at any one point, provided the Christoffel symbols are given.

PROBLEM 4.13. Use the result of the last problem to establish the dual differential equation

$$\frac{\partial g^{ij}}{\partial x^k} = -\Big(\Gamma_{mk}^j g^{im} + \Gamma_{mk}^i g^{jm}\Big).$$

PROBLEM 4.14. Again assuming symmetry of the metric coefficients, show that if $s \mapsto (x^1(s),\dots,x^n(s)) = \gamma(s)$ is the parameterization of a path of minimal total length, using arc length s as a parameter, then γ satisfies the system of differential equations

$$(x^m)''(s) + \Gamma_{jk}^m(x^1(s),\dots,x^n(s))(x^j)'(s)(x^k)'(s) = 0,$$

for $m = 1, 2, \dots, n$.

PROBLEM 4.15. Show that if the arc length s in the previous problem is replaced by an arbitrary parameter t for which $ds/dt > 0$ at all points, the equation satisfied is

$$(x^m)''(t) - \frac{d^2 s}{dt^2}\frac{ds}{dt}(x^m)'(t) + \Gamma_{jk}^m(x^1(t),\dots,x^n(t))(x^j)'(t)(x^k)'(t) = 0.$$

Notice that this equation is the same as the one derived in the previous problem if s is a *linear* function of t.

PROBLEM 4.16. Show that the Christoffel symbols are not altered (and hence neither is the Ricci tensor) if each of the metric coefficients g_{ij} is multiplied by the same constant k. Thus, curvature is independent of the scale by which distances are measured, as we should hope if it is to be the same number for all observers.

PROBLEM 4.17. Show that there is no surface $z = f(r, \theta)$ in $\mathbb{R}^3$ for which the metric coefficients induced from the metric on $\mathbb{R}^3$ are given by the diagonal matrix

$$\begin{pmatrix} \frac{r}{r+r_s} & 0 \\ 0 & r^2 \end{pmatrix}.$$

By the phrase "induced from the metric on $\mathbb{R}^3$," we mean that $ds^2 = dr^2 + r^2\, d\theta^2 + dz^2 = dr^2 + r^2\, d\theta^2 + \left((\partial f/\partial r)\, dr + (\partial f/\partial \theta)\, d\theta \right)^2 = \left(1 + (\partial f/\partial r)^2 \right) dr^2 + 2(\partial f/\partial r)(\partial f/\partial \theta)\, dr\, d\theta + \left((r^2 + (\partial f/\partial \theta)^2) \right) d\theta^2.$

PROBLEM 4.18. Show that a nonconstant uniformly almost-periodic function of time has more than one local extreme value. The definition of uniform almost-periodicity is as follows: A complex-valued function $f(t)$ of a real-variable is uniformly almost-periodic if for every $\varepsilon > 0$ there is a length L_ε such that every interval of length at least L_ε contains an ε-translate, which is a number T such that

$$|f(t + T) - f(t)| < \varepsilon$$

for all real numbers t. The fundamental theorem of almost-periodic functions says that if $f(t)$ is such a function, then for every $\varepsilon > 0$ there is a finite generalized trigonometric polynomial

$$p(t) = \sum_{j=1}^{n} c_{\lambda_j} e^{i\lambda_j t}$$

such that $|f(t) - p(t)| < \varepsilon$ for all t. The terms in this finite sum represent Ptolemy's epicycles. The frequencies λ_j are arbitrary, so that these polynomials are generally *not* periodic.

PROBLEM 4.19. Finish the proof of Theorem 4.3.

PROBLEM 4.20. Is it possible to endow space-time with a metric of the form

$$ds^2 = f(\rho)\, dt^2 - \frac{1}{c^2}\left(g(\rho)\, d\rho^2 + \rho^2\, d\varphi^2 + \rho^2 \sin^2 \varphi\, d\theta^2 \right)$$

such that the following conditions are met?
  (1) $f(\rho) \to 1$ and $g(\rho) \to 1$ as $\rho \to \infty$. (That is, the metric approaches the flat-space metric at infinity.)
  (2) The equations of a geodesic given by the Euler equations are Newton's equations of motion (1.5) and (4.6).

Hint: Notice the quotation from Eddington at the beginning of Section 4.

PROBLEM 4.21. Derive Eqs. (4.2) and (4.3) by using a moving frame of reference $\boldsymbol{\omega}_1(\theta) = \cos\theta\, \boldsymbol{i} + \sin\theta, \boldsymbol{j}$, $\boldsymbol{\omega}_2 = -\sin\theta\, \boldsymbol{i} + \cos\theta\, \boldsymbol{j}$ and setting the dot products of $(\alpha r')' + (GM/r^3)\boldsymbol{r}$ with $\boldsymbol{\omega}_1$ and $\boldsymbol{\omega}_2$ equal to zero. Solve the second of these equations for $\theta''(t)$ and substitute that value in the first of them. If you try to do this without the assistance of *Mathematica*, keep in mind that a fraction vanishes if and only if its numerator vanishes.

PROBLEM 4.22. Show that in Newtonian mechanics, the constant angular momentum per unit mass of a planet is $l = \sqrt{GMa(1 - e^2)}$, where a is the average distance of the planet from the Sun and e is the eccentricity of the elliptic orbit.

PROBLEM 4.23. Show that the solution of Eq. (4.3) with $u(0) = u_0$, $u'(0) = 0$ is the inverse of the function $\theta = \theta(u)$ given by

$$\theta = \int_{u_0}^{u} \frac{dx}{\sqrt{q(u_0)e^{2s(x-u_0)} - q(x)}},$$

where $q(x) = (x^2 - 2p/r_s)$, $p = GM/l^2$, and $r_s = 2GM/c^2$.

Show also that in this case we have

$$u'' + u = \frac{1}{2}q(u_0)e^{2r_s(u-u_0)}.$$

Thus, the study of Eq. (4.3) is subsumed in the general study of equations of the form

$$u'' + u = ae^{bu},$$

with constants a and b.

PROBLEM 4.24. Assuming Newtonian mechanics, imagine that a photon "falls" from infinity with initial speed c, use conservation of energy to show that its speed v at distance r from the Sun satisfies

$$v^2 = c^2 + 2GM/r,$$

that is, $v = c\sqrt{1 + r_s/r}$. (In contrast to what we found about the speed of light in a relativistic gravitational field, Newtonian mechanics predicts that light slows down as r increases. At the Schwarzchild radius, the speed of light would be $\sqrt{2}c$. In the case of the Sun, however, where $r_s/r < 10^{-5}$ for any photon that does not actually fall into the Sun, the speed would be nearly constant, and the resulting path nearly straight.)

CHAPTER 5

Concepts of Curvature, 1700–1850

In the preceding chapter, we left some important parts of the general theory of relativity unexplained. To mention only two, we did not explain why the word *curvature* is applied to describe what gravity is, and we did not explain why the formula that we gave for computing the curvature really does represent something that is intuitively a curvature. These two problems are the basic ones that lie at the foundation of an understanding of general relativity. The path to that understanding is a long and arduous one, even when it is stripped down to just the problem of gravitation in empty space.

The connection between curvature and physics will be discussed in more detail in Chapter 7. For now, it suffices to point out that the informal principle *forces produce curvature* describes a great deal of classical physics. The main example lies at the very heart of Newtonian mechanics in the form of the law of inertia: A body subject to no forces will move in a straight line at constant speed. In other words, to produce any curving of its path, forces are required. And conversely, when elastic forces are considered, *curvature produces force*. For example, the earliest attempt to analyze the vibrating string used the principle that the restoring force at each point of a stretched string is proportional to the curvature of the string at that point. Later, Euler showed that a particle subject to forces that confine it to a given surface but having no acceleration tangential to that surface will move along a geodesic of the surface, that is, a path of minimal curvature (see Appendix 2 of Volume 2).

Since the key concept is the curvature of space-time, we'll start much farther back and give some details of the classical subject of differential geometry, as developed in the eighteenth and nineteenth centuries. The twentieth-century concept of a differentiable manifold provides sufficient generality for our purposes. The theory of these manifolds is discussed in Appendix 4 of Volume 2. The evolution of the concept of curvature can be divided chronologically into four phases, associated with the names of Euler, Gauss, Riemann, and Ricci. The first two of these phases are the subject of the present chapter.

Our first goal is to present a streamlined version of the eighteenth-century work of Euler and the nineteenth-century work of Gauss on curvature. The context is two-dimensional surfaces in $\mathbb{R}^3$. Our aim is to see how the curvature of a parameterized surface can be computed from knowledge of its metric coefficients alone, independently of any embedding in $\mathbb{R}^3$. The notion of curvature that is needed for relativity involves some rather sophisticated multilinear algebra, especially the concept of a tensor and the differential geometry in which tensors are the natural language. But we approach the subject gently. Neither Euler nor Gauss had these concepts; and, although we have foreshadowed them in Chapter 2 with our discussion of covariant and contravariant objects and even computed with them in

Chapter 4, we need to build a bridge to them starting from the basic calculus that Euler and Gauss possessed (as, we presume, the reader does also). It is known (see Section 3) below that Gauss was able to get the curvature formula we are seeking, but he did so in a way that now seems unnecessarily cumbersome. What would have streamlined his work, had he known about it, is the notion of a Christoffel symbol. We have given an algebraic definition of these in Chapter 4. In the present chapter, we shall give a geometric definition and verify that it agrees with the algebraic definition we gave earlier. The Christoffel symbols are the essential bridge between the classical multi-variable calculus used by Euler and Gauss and the tensor analysis that now pervades both geometry and physics. After we obtain our formula for curvature in terms of the metric coefficients, we can devote the following two chapters to a discussion of the use of tensors in differential geometry and physics, bringing the technical part of this book to a close.

1. Differential Geometry

Differential geometry is the application of the differential and (to a lesser extent) integral calculus to study the geometry of curves and surfaces. In ancient Greek geometry, things that we now do easily and in great generality using algebra and infinitesimal methods had to be done on the macro-level. It was possible in this way to find the tangents to circles and conic sections, but more complicated curves could not be handled.[1] We begin by looking at material the reader has no doubt seen before, but emphasizing certain points that may have escaped notice earlier.

1.1. Derivatives and differentials. The elementary material we wish to review is the basic principle of differential calculus. Once you have the idea of graphing a distance-against-time relation, expressing the distance y as a function $y = f(x)$ of time x, you get a very easy geometric interpretation of average velocity over a time interval. In terms of Fig. 5.1, the average velocity is $\overline{BC}/\overline{AC} = \Delta y/\Delta x$, which is geometrically the slope of the secant AB. If the time interval Δx is very small, we expect that the average velocity will be close to what we intuitively think of as the *instantaneous velocity* at the time x_0. Intuitively, as the time interval gets very short (B slides down the curve toward A), that secant should approximate a small section of the tangent at A, and hence its slope should approximate the slope of the tangent at P. Both geometrically and algebraically, there is a difficulty in making sense of this, since we cannot make any sense out of $\Delta y/\Delta x$ when $\Delta x = 0$. (You cannot divide by zero.) We get lucky in the case of polynomial functions, however. For example, with a falling body, $y = y_0 + v_0 x + (g/2)x^2$, where y_0, v_0, and g are constants, and $\Delta y = y(x) - y(x_0) = v_0(x - x_0) + (g/2)(x^2 - x_0^2) = v_0(x - x_0) + (g/2)(x + x_0)(x - x_0) = \big(v_0 + g/2(x + x_0)\big)\Delta x$. Thus, for any value of x *except* $x = x_0$, we find that $\Delta y/\Delta x = v_0 + (g/2)(x + x_0)$. Although the left-hand side of this equation still does not make any sense when $x = x_0$, the right-hand side does: it is $v_0 + gx_0$. One would naturally be inclined to adopt that value as the *definition* of the instantaneous velocity at time $x = x_0$. This algebraic kindness

---

[1] Of all the ancients, only Archimedes came close to discovering the secret of the derivative, when he found that the tangent to an Archimedean spiral at the end of its first turn, the line from the origin of the spiral to the point in question, and the perpendicular to that line at the origin formed a right triangle whose area was equal to that of the circle through the point in question with center at the origin. In other words, he established what we now recognize as a connection between tangents and areas, dimly prefiguring the fundamental theorem of calculus.

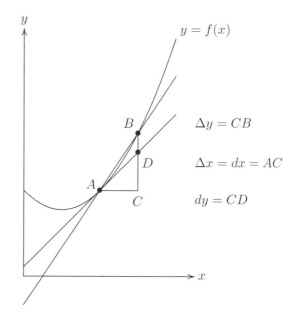

FIGURE 5.1. Geometric interpretation of increments Δx and Δy at a point A on a curve $y = f(x)$ and the differentials dx and dy at the same point.

of nature seems to have inspired a number of seventeenth-century mathematical physicists to take precisely that point of view. Newton tried to invoke another intuition to justify this, saying that the instantaneous velocity of a moving particle is the velocity it has exactly at the instant it arrives at a given location. That definition now appears circular.[2]

Our point of view on this subject is based on the concept of a best linear approximation to a given function. The tangent line is the best linear approximation to a curve near the point of tangency, and we can say in a quantitative sense just how good it is. Here's how. If two variables are related by a functional equation $y = f(x)$, their infinitesimal increments dy and dx are related as $dy = f'(x_0)\, dx$, where $f'(x_0)$ is the derivative at a point x_0 that will be held fixed. There is actually no reason why dy and dx in this equation need to be "small" if they are thought of as the increments of y and x on the tangent line whose equation is $y - f(x_0) = f'(x_0)(x - x_0)$. They can be any size, and their ratio dy/dx will always equal $f'(x_0)$. The reason we think of them as "small" is that they are approximated by the increments Δx and Δy, defined by the equation

$$\Delta y = f(x_0 + \Delta x) - f(x_0)\,,$$

---

[2]Logical purists like the philosopher George Berkeley (1685–1753) raised objections. Writing in *The Analyst* in 1736, about a decade after the death of Newton, he pointed out (using the example of $y(x) = x^3$) what the fallacy was: You can't set $\Delta x = 0$ in an equation derived under the assumption that $\Delta x \neq 0$. But he didn't wish to overthrow Newton's calculus. He merely pointed out that Newton's use of this kind of argument was only heuristic, and that Newton was happy to dispense with the infinitesimals once he had finite lines bearing the same ratio (CD and AC in Fig. 5.1).

when Δx is small. What makes calculus work is the *quantitative* fact that, when we take $dx = \Delta x$—the differential of an *independent* variable equals its finite increment—the approximation $dy = f'(x_0)\,dx = f'(x_0)\,\Delta x \approx \Delta y$ is good in the sense that $(\Delta y - dy)/\Delta x$ is very small when Δx is very small. In other words, the quantity $\Delta y - dy \approx 0$ is not merely small; it is small *even in comparison* with the small quantity Δx, and the smaller Δx is taken, the smaller the ratio $(\Delta y - dy)/\Delta x$ becomes. Despite the fact that the denominator is becoming small, the numerator is becoming small so much faster that this ratio gets arbitrarily small.

To summarize, there are three essential points to keep in mind: (1) For an independent variable, the infinitesimal increment dx can be identified with the finite increment Δx; (2) the differential dy is a linear function (constant multiple) of the differential dx; (3) the linear mapping $\Delta x = dx \mapsto dy$ is the best linear approximation to the mapping $\Delta x \mapsto \Delta y = f(x_0 + \Delta x) - f(x_0)$.

These three points are the essence of differential calculus. In practice, we often switch from the language of the derivative $f'(x)$ to the language of the differentials dy and dx and vice versa. For example, we may write a differential equation either in the form $f'(x) = af(x)$ or $dy = ay\,dx$. It is the same equation either way. Solving it means finding the function $f(x)$ such that the relation $y = f(x)$ is equivalent to the relation $dy = f'(x)\,dx$ together with, say, an initial condition $y = y_0$ when $x = x_0$, that is, $f(x_0) = y_0$.

Despite the near-equivalence of derivatives and differentials in practice, differential geometry makes a distinction between them. That difference shows up best when we have two sets of parameters that can be used to describe a geometrical object. Suppose these are $\boldsymbol{x} = (x^1, x^2, \ldots, x^n)$ and $\boldsymbol{y} = (y^1, y^2, \ldots, y^n)$. (It is essential that n be the same in both cases.) Assuming each ordered set of parameters is in one-to-one correspondence with the points P of the same geometrical object, we get a natural association

$$(x^1, \ldots, x^n) \leftrightarrow P \leftrightarrow (y^1, \ldots, y^n).$$

Thus, P can be regarded as a function of either set of parameters, and theoretically, linking these two associations defines each set of parameters as functions of the other, so that we get two sets of functions $\left(x^1(y^1, \ldots, y^n), \ldots, x^n(y^1, \ldots, y^n)\right)$ and $\left(y^1(x^1, \ldots, x^n), \ldots, y^n(x^1, \ldots, x^n)\right)$. If $z = f(P)$ is a function defined on the geometric object, then z can be regarded as a function of either set of parameters. As such, it has partial derivatives:

$$\frac{\partial z}{\partial x^i}, \quad \text{and} \quad \frac{\partial z}{\partial y^j}, \quad i = 1, \ldots, n, \ \ j = 1, \ldots, n.$$

It also has a total differential in each coordinate system:

$$dz = \frac{\partial z}{\partial x^1}\,dx^1 + \cdots + \frac{\partial z}{\partial x^n}\,dx^n \quad \text{and} \quad dz = \frac{\partial z}{\partial y^1}\,dy^1 + \cdots + \frac{\partial z}{\partial y^n}\,dy^n.$$

As in the case of a function of one variable, we can identify the differentials $dx^1, \ldots, dx^n$ of the independent variables $x^1, \ldots, x^n$ with their finite increments $\Delta x^1, \ldots, \Delta x^n$. When we do, dz (with the partial derivatives evaluated at a given point $\boldsymbol{x}_0 = (x_0^1, \ldots, x_0^n)$ that is held constant) is a linear function of $dx^1, \ldots, dx^n$, and the mapping

$$(\Delta x^1, \ldots, \Delta x^n) = (dx^1, \ldots, dx^n) \mapsto dz$$

is the best linear approximation to the mapping of the increments in the x-variables to the increment in z: $(\Delta x^1, \ldots, \Delta x^n) \mapsto \Delta z = z(x_0^1 + \Delta x^1, \ldots, x_0^n + \Delta x^n) - z(x_0^1, \ldots, x_0^n)$.

We need to know how to translate the partial derivatives and the differentials from one set of parameters to the other, since the two sets may not be equally convenient in every situation. The important infinitesimal relation that allows us to do this, and also contains the secret to understanding a huge amount of differential geometry, is the chain rule, which has both a derivative form and a differential form:

$$\frac{\partial z}{\partial y^j} = \frac{\partial z}{\partial x^1}\frac{\partial x^1}{\partial y^j} + \cdots + \frac{\partial z}{\partial x^n}\frac{\partial x^n}{\partial y^j}, \quad j = 1, \ldots, n,$$

$$dy^j = \frac{\partial y^j}{\partial x^1}dx^1 + \cdots + \frac{\partial y^j}{\partial x^n}dx^n, \quad j = 1, \ldots, n.$$

These two sets of relations can be written as matrix equations, with the partial derivatives of z written as single-rowed matrices and the differentials of the variables as single-column matrices:[3]

$$\nabla_{\boldsymbol{y}} z = \begin{pmatrix} \frac{\partial z}{\partial y^1} & \frac{\partial z}{\partial y^2} & \cdots & \frac{\partial z}{\partial y^n} \end{pmatrix}$$

$$= \begin{pmatrix} \frac{\partial z}{\partial x^1} & \frac{\partial z}{\partial x^2} & \cdots & \frac{\partial z}{\partial x^n} \end{pmatrix} \begin{pmatrix} \frac{\partial x^1}{\partial y^1} & \frac{\partial x^1}{\partial y^2} & \cdots & \frac{\partial x^1}{\partial y^n} \\ \frac{\partial x^2}{\partial y^1} & \frac{\partial x^2}{\partial y^2} & \cdots & \frac{\partial x^2}{\partial y^n} \\ \vdots & \vdots & \ddots & \vdots \\ \frac{\partial x^n}{\partial y^1} & \frac{\partial x^n}{\partial y^2} & \cdots & \frac{\partial x^n}{\partial y^n} \end{pmatrix} = \nabla_{\boldsymbol{x}} z \, \frac{\partial(x^1, \ldots, x^n)}{\partial(y^1, \ldots, \partial y^n)}$$

and

$$d\boldsymbol{y} = \begin{pmatrix} dy^1 \\ \vdots \\ dy^n \end{pmatrix} = \begin{pmatrix} \frac{\partial y^1}{\partial x^1} & \cdots & \frac{\partial y^1}{\partial x^n} \\ \vdots & & \vdots \\ \frac{\partial y^n}{\partial x^1} & \cdots & \frac{\partial y^n}{\partial x^n} \end{pmatrix} \begin{pmatrix} dx^1 \\ \vdots \\ dx^n \end{pmatrix} = \frac{\partial(y^1, \ldots, y^n)}{\partial(x^1, \ldots, x^n)} \, d\boldsymbol{x}.$$

These formulas contain the two mutually inverse Jacobian matrices that we introduced in Section 10 of Chapter 2.

$$\frac{\partial(x^1, \ldots, x^n)}{\partial(y^1, \ldots y^n)} = \begin{pmatrix} \frac{\partial x^1}{\partial y^1} & \cdots & \frac{\partial x^1}{\partial y^n} \\ \vdots & \ddots & \vdots \\ \frac{\partial x^n}{\partial y^1} & \cdots & \frac{\partial x^n}{\partial y^n} \end{pmatrix}$$

and

$$\frac{\partial(y^1, \ldots, y^n)}{\partial(x^1, \ldots x^n)} = \begin{pmatrix} \frac{\partial y^1}{\partial x^1} & \cdots & \frac{\partial y^1}{\partial x^n} \\ \vdots & \ddots & \vdots \\ \frac{\partial y^n}{\partial x^1} & \cdots & \frac{\partial y^n}{\partial x^n} \end{pmatrix}.$$

EXAMPLE 5.1. Consider the change from rectangular coordinates (x, y) to polar coordinates (r, θ) in the right half-plane where $x > 0$. We have

$$\begin{aligned} r &= \sqrt{x^2 + y^2} & x &= r\cos\theta \\ \theta &= \arctan(y/x) \, , & y &= r\sin\theta \, . \end{aligned}$$

---

[3]The nabla-symbols $\nabla_{\boldsymbol{x}}$ and $\nabla_{\boldsymbol{y}}$ are a useful notation for the *gradient* operator that we shall use frequently in what follows. The symbols are inserted here just to show how they are expressed algebraically.

Then

$$\frac{\partial(r,\theta)}{\partial(x,y)} = \begin{pmatrix} \frac{x}{\sqrt{x^2+y^2}} & \frac{y}{\sqrt{x^2+y^2}} \\ \frac{-y}{x^2+y^2} & \frac{x}{x^2+y^2} \end{pmatrix}, \qquad \frac{\partial(x,y)}{\partial(r,\theta)} = \begin{pmatrix} \cos\theta & -r\sin\theta \\ \sin\theta & r\cos\theta \end{pmatrix}.$$

It is easy to verify that the product of these two matrices is the identity matrix by replacing either set of variables in the product with the expressions they have in the equations of transformation.

1.2. Covariance and contravariance revisited. Although we discussed covariance and contravariance in Chapter 2, it will do no harm to review those concepts at this point from a slightly different point of view. By consideration of the individual components of the two mappings $\boldsymbol{x} \mapsto \boldsymbol{y}$ and $\boldsymbol{y} \mapsto \boldsymbol{x}$, we see that the Jacobian

$$\frac{\partial(y^1,\dots,y^n)}{\partial(x^1,\dots,x^n)}$$

is the matrix of the linear transformation $d\boldsymbol{x} \mapsto d\boldsymbol{y}$ that best approximates the increment mapping $d\boldsymbol{x} = \Delta\boldsymbol{x} \mapsto \Delta\boldsymbol{y} = \left(y^1(x_0^1 + \Delta x^1,\dots,x_0^n + \Delta x^n),\dots, y^n(x_0^1 + \Delta x^1,\dots x_0^n + \Delta x^n)\right) - \left(y^1(x_0^1,\dots,x_0^n),\dots,y^n(x_0^1,\dots,x_0^n)\right)$. As this matrix is obtained by regarding $y^1,\dots,y^n$ as functions of $x^1,\dots,x^n$, it goes with the mapping $\boldsymbol{x} \mapsto \boldsymbol{y}$. Thus, the differentials of the variables transform "in the same direction" as the variables themselves, and are said to be *covariant* quantities. In contrast, transforming the partial derivatives with respect to these variables requires the inverse Jacobian matrix $\frac{\partial(x^1,\dots,x^n)}{\partial(y^1,\dots,y^n)}$ and hence the partial derivatives are called *contravariant* quantities.[4] In a suggestive heuristic notation,

$$\frac{\partial\boldsymbol{y}}{\partial\boldsymbol{x}} = \frac{\partial(y^1,\dots,y^n)}{\partial(x^1,\dots,x^n)} \quad \text{and} \quad \frac{\partial}{\partial\boldsymbol{y}} = \frac{\partial}{\partial\boldsymbol{x}}\frac{\partial(x^1,\dots,x^n)}{\partial(y^1,\dots,y^n)}.$$

The distinction between the two kinds of vectors is signaled by the convention of indexing the coordinates of a contravariant object as superscripts and those of a covariant object as subscripts. The most important examples of these two concepts are contravariant tangent vectors (directional derivatives)

$$\boldsymbol{u} = u^1 \frac{\partial}{\partial x^1} + \cdots + u^n \frac{\partial}{\partial x^n},$$

and covariant vectors (covectors) (directional differentials, we might call them)

$$\boldsymbol{v} = u_1\, dx^1 + \cdots + u_n\, dx^n.$$

To keep the distinction between covariant and contravariant clear, superscripts on the coordinates indicate a contravariant vector and subscripts a covariant vector. What makes this convention more confusing is that the coordinates themselves are linear functionals acting on the vectors they define, and thus the coordinates of a covariant vector are contravariant, and vice versa, as will be shown below.

Matrices are used in differential geometry to denote three different kinds of "tensors of rank two," that is, bilinear forms acting on (1) a pair of vectors, (2) a pair consisting of a vector and a covector, or (3) a pair of covectors. These are

[4]The adjective *contravariant* is often omitted, and we speak of vectors and co-vectors rather than contravariant and covariant vectors. The covariant/contravariant terminology reappears when we consider *multi*-linear mappings (tensors), which are functions of several vectors and/or covectors that are linear in each argument when the other arguments are held constant.

called tensors of type $(0,2)$, $(1,1)$, and $(2,0)$ respectively. For the first type, a tensor $T(\boldsymbol{u}, \boldsymbol{v})$ of covariant rank 2, we write the matrix as

$$\begin{pmatrix} a_{11} & \cdots & a_{1n} \\ \vdots & & \vdots \\ a_{n1} & \cdots & a_{nn} \end{pmatrix},$$

meaning that the value of the tensor on a pair of (contravariant) vectors $\boldsymbol{u} = u^i \partial/\partial x^i$, $\boldsymbol{v} = v^j \partial/\partial x^j$ is

$$T(\boldsymbol{u}, \boldsymbol{v}) = \sum_{i,j=1}^{n} a_{ij} u^i v^j,$$

The tensor itself is written as $\sum a_{ij}\, dx^i\, dx^j$ to indicate its type. The metric tensor ds^2 introduced in Section 10 of Chapter 2 is an example of a tensor of this type, and it acts on a pair of contravariant vectors (which in physical applications will be velocities or four-velocities).

For a tensor of type $(1,1)$, say $T(\boldsymbol{u}, \boldsymbol{\omega})$, acting on a vector $\boldsymbol{u} = u^i \partial/\partial x^i$ and a covector $\boldsymbol{\omega} = v_j\, dx^j$ we write the matrix as

$$\begin{pmatrix} a_1^1 & \cdots & a_n^1 \\ \vdots & & \vdots \\ a_1^n & \cdots & a_n^n \end{pmatrix},$$

meaning that

$$T(\boldsymbol{u}, \boldsymbol{\omega}) = \sum_{i,j=1}^{n} a_i^j u^i v_j.$$

We leave it to the reader to work out that the matrix representation of a tensor of contravariant rank 2, say $T(\boldsymbol{v}, \boldsymbol{\omega})$, operating on a pair of covectors $\boldsymbol{v} = u_i\, dx^i$ and $\boldsymbol{\omega} = v_j\, dx^j$ would be

$$T(\boldsymbol{v}, \boldsymbol{\omega}) = \sum_{i,j=1}^{n} a^{ij} u_i v_j.$$

REMARK 5.1. One advantage of using subscripts and superscripts is that they naturally encode the Einstein summation convention: *If an index appears in a product as both subscript and superscript, the terms containing that index are to be summed over the appropriate range.* In this context, the superscript on a variable x^i counts as a subscript when it appears in the denominator of a partial derivative, that is, in the expression $\partial/\partial x^i$, the i is considered a subscript. In that notation, we have

$$\boldsymbol{u} = u^i \frac{\partial}{\partial x^i} = u^1 \frac{\partial}{\partial x^1} + \cdots + u^n \frac{\partial}{\partial x^n} \text{ and } \boldsymbol{v} = u_i\, dx^i = u_1\, dx^1 + \cdots + u_n\, dx^n.$$

By convention, the variables themselves are written with superscripts. That notation does *not* mean that they are either covariant or contravariant vectors.

Given the three kinds of tensors of rank 2, we see that our definition of the Kronecker delta in Section 10 of Chapter 2 may have been unnecessarily restrictive. The identity matrix there was assumed to result from multiplying a square contravariant matrix by a square covariant matrix, resulting in a square mixed matrix. But it very well might represent the identity (covariant) matrix of metric coefficients g_{ij} or its inverse g^{ij}, in which case we could take all three symbols to have the same literal meaning (denotation): $\delta_{ij} = \delta^i_j = \delta^{ij}$, the three cases being distinguished by context (connotation). We shall mostly need the mixed symbol that we have already introduced, however.

Just to maximize our confusion, the *coordinates* u^i of a contravariant vector $\boldsymbol{u}$, being the result of applying a linear functional to $\boldsymbol{u}$, transform in a covariant way. If we switch to coordinates $(y^1, \ldots, y^n)$ in which the components of $\boldsymbol{u}$ are $(v^1, \ldots, v^n)$, then

$$u^i \frac{\partial}{\partial x^i} = u^i \left(\frac{\partial y^j}{\partial x^i} \frac{\partial}{\partial y^j} \right) = \left(u^i \frac{\partial y^j}{\partial x^i} \right) \frac{\partial}{\partial y^j} = v^j \frac{\partial}{\partial y^j} \,,$$

that is

$$\begin{pmatrix} v^1 \\ \vdots \\ v^n \end{pmatrix} = \begin{pmatrix} \frac{\partial y^1}{\partial x^1} & \cdots & \frac{\partial y^1}{\partial x^n} \\ \vdots & & \vdots \\ \frac{\partial y^n}{\partial x^1} & \cdots & \frac{\partial y^n}{\partial x^n} \end{pmatrix} \begin{pmatrix} u^1 \\ \vdots \\ u^n \end{pmatrix}.$$

REMARK 5.2. The reader is warned that the notation

$$\frac{\partial(x^1, \ldots, x^n)}{\partial(y^1, \ldots, y^n)}$$

usually denotes the *determinant* of this matrix. Since we have need of the matrix itself, we shall explicitly write

$$\det \left(\frac{\partial(x^1, \ldots, x^n)}{\partial(y^1, \ldots, y^n)} \right) \quad \text{or} \quad \left| \frac{\partial(x^1, \ldots, x^n)}{\partial(y^1, \ldots, y^n)} \right|$$

when we need to mention this determinant. The important fact about the Jacobian, easily proved, is that it *must* be an invertible matrix if the correspondence $(x^1, \ldots, x^n) \leftrightarrow (y^1, \ldots, y^n)$ is to be differentiable in both directions, and that is a requirement we always impose. As matrices, we have

$$\frac{\partial(x^1, \ldots, x^n)}{\partial(y^1, \ldots y^n)} \frac{\partial(y^1, \ldots, y^n)}{\partial(x^1, \ldots x^n)} = \begin{pmatrix} 1 & 0 & \cdots & 0 & 0 \\ 0 & 1 & \cdots & 0 & 0 \\ \vdots & \vdots & \ddots & \vdots & \vdots \\ 0 & 0 & \cdots & 1 & 0 \\ 0 & 0 & \cdots & 0 & 1 \end{pmatrix},$$

while as determinants the relation is simply

$$\det \left(\frac{\partial(x^1, \ldots, x^n)}{\partial(y^1, \ldots y^n)} \right) \det \left(\frac{\partial(y^1, \ldots, y^n)}{\partial(x^1, \ldots x^n)} \right) = 1 \,,$$

as illustrated above in Example 5.1.

1.3. The duality of differentials and partial derivatives. The essence of differential calculus is to get *linear* functions whose increments approximate those of the more complicated functions that are necessary in the description of the world. That is shown most clearly in the fundamental formula of differential calculus, which we write using the Einstein summation convention:

$$dz = \frac{\partial z}{\partial x^i} \, dx^i \,.$$

In this formula, each partial derivative interacts with the corresponding differential in the simplest possible way (through multiplication) to produce the all-important linear approximation dz of the dependent variable z. As remarked above, the differentials $dx^1, \ldots, dx^n$ are really the same as the increments $\Delta x^1, \ldots, \Delta x^n$, and amount to merely a new set of variables. In view of that fact, we feel free to alter our point of view on these differentials and regard them as operators that interact with partial derivatives according to the formula

$$dx^i \left(\frac{\partial}{\partial x^j} \right) = \delta^i_j .$$

The entry in row i, column j of the identity matrix is δ^i_j. In the language of Appendix 4 of Volume 2, $\{dx^1, \ldots, dx^n\}$ is the basis of the cotangent space *dual* to the basis $\{\partial/\partial x^1, \ldots, \partial/\partial x^n\}$ of the tangent space. The tangent and cotangent spaces act on each other and can be treated symmetrically.

Parametrizations, vectors, and covectors are the basic tools of differential geometry. Our object in this chapter is to explore curved spaces using these tools, and specifically to investigate the concept of curvature. Several different definitions of curvature have been found useful in physics and geometry, but we shall start with the classical notion of curvature of curves and surfaces in $\mathbb{R}^3$ or, more generally, $\mathbb{R}^n$, and generalize from there. Surfaces in $\mathbb{R}^3$ provide a very good visualizable model for understanding the general situation. Although we live in a three-dimensional world, we can imagine ourselves observing beings who are confined to a two-dimensional surface embedded in that space. From our three-dimensional perspective, we can visualize the surface, but they cannot; it is their whole universe and they cannot imagine anything outside it. From our point of view, the metric, the geodesics, and the curvature of that surface are determined by its embedding in $\mathbb{R}^3$, which serves as a kind of scaffolding to be used when building a condominium. It is attached to the "condominium," which is the surface itself, but does not form an intrinsic part of that structure. We visualize the "scaffolding" (tangent plane) as dangling in a space that does not exist as far as the residents of the condominium know. Eventually, by removing the scaffolding, we—and the residents of the condominium—hope to be able to study the condominium/surface intrinsically, replacing concepts defined by use of the tangent plane and the third dimension with concepts that can be defined in terms of the parameterization alone, without reference to anything outside the surface. After we see how to do that for a surface in $\mathbb{R}^3$, we will be in a position to do it for our own four-dimensional space-time, which is not, as far as we know or need to know, embedded in any space of higher dimension. We shall look at this imaginary world again at the end of the next section, after we have introduced the general notion of curvature.

2. Curvature, Phase 1: Euler

Out of the welter of visible phenomena, ancient people all over the world focused their attention on the two simplest kinds of planar figures, namely polygons and circles. Using just these simple figures, the ancient Greeks constructed an elaborate theory of geometric proportion and area-preserving transformations of figures, with full proofs. When this theory led them to more complicated problems for which straight lines and circles (ruler-and-compass constructions) were not adequate—the famous problems of squaring a circle, trisecting an arbitrary angle and doubling a cube—they introduced motion and allowed their circles and lines to generate cones, planes, spheres, and cylinders, by means of which the last two of these three problems could be solved. By combining rotational motion with straight-line motion of a point, Archimedes generated a spiral, and a mathematician named

Hippias[5] combined rotational motion with the straight-line motion of a line to generate the *quadratrix*. Both of these curves solved the problem of squaring the circle. Archimedes even began to introduce infinitesimal arguments, which could have enriched geometry immensely. One tool, however, the ancient Greeks lacked: algebra, specifically the introduction of symbols to stand for unknown or unspecified quantities of a general type—what we now call variables. Although there are many examples of such symbolism in early mathematics in many parts of the world, these did not come into general use until the early seventeenth century. When they did, the result was the calculus, with its myriad applications in geometry and mechanics. One of those uses was to fit complicated curves as closely as possible with the straight lines and circles that are easy to understand.

2.1. Curvature of a plane curve. The switch from the macroscopic point of view of the Greek geometers to the modern application of differential calculus to solve problems in geometry has had profound consequences. It would have been impossible without two key ingredients: the algebraic apparatus that Descartes brought to bear on geometry, and the infinitesimal reasoning of Descartes, Fermat, Pascal, Newton, Leibniz, and other seventeenth-century mathematicians. Newton's *Principia*, although it contains the elements of differential calculus, still reflects a preference for finitistic reasoning. Newton liked to express himself in the classical language of Euclidean geometry. It took more than two centuries for the "action-at-a-distance" view of gravity in Newtonian mechanics to be supplanted by reasoning about a metric given in infinitesimal form.

In the previous section, we pointed out that differential calculus is useful because a large class of curves can be fitted closely with straight lines via the differential approximation described above. In accordance with that discussion, we note that, for a fixed base point x_0, the straight line whose equation is

$$y = f(x_0) + f'(x_0)(x - x_0)$$

approximates the curve $y = f(x)$ to first order. In the language of infinitesimals, on the first-order infinitesimal level, a smooth curve *is* a straight line. On that infinitesimal level, we replace $\Delta y = y - f(x_0)$ by dy and write $dy = f'(x_0)\, dx$. We remind the reader that for an independent variable x, we have $\Delta x = dx = x - x_0$ on both the finite and infinitesimal levels.)

Linear functions are easily handled without infinitesimals, and that is because in this case Δy is not just approximated by dy, but actually equal to it. Linear motion at constant speed amounts to the simple "distance = rate × time" formula taught in high-school algebra. Given the Descartes/Newton law of inertia, we see that this motion is in a sense the natural state of things. In Newtonian physics, it is *departures* from this simple formula that need to be explained, and this is done by positing *forces* that act on the irregularly moving object. For that reason, we need to go beyond the level of first-order approximation.

Approximating with straight lines, we can't normally do any better than first-order approximation unless the curve we are approximating is itself "flat" at a point. That is, the point is a point of inflection, where the curve is locally very close to being a straight line. Such points are exceptional. If the curve does not

---

[5]Which of several people named Hippias did this is disputed among historians of ancient Greek mathematics.

actually bend, we will find that the quantity

$$\frac{y - f(x_0) - f'(x_0)(x - x_0)}{(x - x_0)^2}$$

is close to 0 when x is close to x_0. But again, points where this relation holds (points of inflection, for example) are rare. In general, on the infinitesimal level of second order, a curve does bend, and the purpose of passing to the second derivative, as we are about to do, is to get a quantitative measure of how much it bends per unit of arc length. That bending is measured by the amount of rotation of the tangent line, the unit of curvature being one radian per unit of arc length.

To take the simplest example, for a plane curve that is the graph of a function, described by an equation $y = f(x)$, the length of arc measured from the point $(0, f(0))$, is

$$s = \int_0^x \sqrt{1 + \left(f'(t)\right)^2}\, dt\,,$$

a relation obtained by integrating the infinitesimal version of the Pythagorean theorem: $ds^2 = dx^2 + dy^2$. That is,

$$\frac{ds}{dx} = \sqrt{1 + \left(f'(x)\right)^2} \text{ and } \frac{dx}{ds} = \frac{1}{\sqrt{1 + \left(f'(x)\right)^2}}\,.$$

The tangent line to the curve at the point $(x, f(x))$ makes an angle $\alpha(x)$ with the positive x-axis, where

$$\alpha(x) = \arctan\left(f'(x)\right).$$

DEFINITION 5.1. The *curvature* of the graph of the function $y = f(x)$ at the point $(x, f(x))$ is the number

$$\frac{d\alpha}{ds} = \frac{d\alpha}{dx}\frac{dx}{ds} = \frac{f''(x)}{1 + \left(f'(x)\right)^2}\frac{1}{\sqrt{1 + \left(f'(x)\right)^2}} = \frac{f''(x)}{\left(1 + \left(f'(x)\right)^2\right)^{3/2}}\,.$$

If $f''(x) > 0$, the curvature is positive, which means the local portion of the plane just *above* the graph is convex; if $f''(x) < 0$, the portion just below the graph is convex.

2.2. Curves in $\mathbb{R}^n$. The notion of curvature generalizes easily to curves in $\mathbb{R}^n$, but in this case we will have to settle for the absolute value of the curvature, since what appears as a clockwise rotation from one side of a plane looks like a counterclockwise rotation from the other side. The case $n = 3$ is sufficiently general to convey the idea. Suppose the curve is given parametrically as the graph of a vector-valued function $\boldsymbol{r}(t)$. Then arc length measured from the point $\boldsymbol{r}(t_0)$ in the direction of increasing t is

$$s = \int_{t_0}^t \sqrt{\boldsymbol{r}'(x) \cdot \boldsymbol{r}'(x)}\, dx\,,$$

which is to say

$$\frac{ds}{dt} = \sqrt{\boldsymbol{r}'(t) \cdot \boldsymbol{r}'(t)} \text{ and } \frac{dt}{ds} = \frac{1}{\sqrt{\boldsymbol{r}'(t) \cdot \boldsymbol{r}'(t)}}\,,$$

and we assume for the sake of simplicity that the point "keeps moving," that is, $\boldsymbol{r}'(t)$ is never the zero vector.

The unit tangent vector pointing in the direction of increasing parameter t is

$$\boldsymbol{\tau}(t) = \frac{1}{\sqrt{\boldsymbol{r}'(t) \cdot \boldsymbol{r}'(t)}} \boldsymbol{r}'(t).$$

DEFINITION 5.2. The *vector curvature* $\boldsymbol{\kappa}(t)$ of a parameterized curve $t \mapsto \boldsymbol{r}(t)$ is the *vector* $\boldsymbol{\kappa}$ perpendicular to $\boldsymbol{\tau}$ given by

$$\boldsymbol{\kappa}(t) = \frac{d\boldsymbol{\tau}}{ds} = \frac{d\boldsymbol{\tau}}{dt}\frac{dt}{ds} = \frac{1}{\boldsymbol{r}' \cdot \boldsymbol{r}'}\left(\boldsymbol{r}'' - \frac{(\boldsymbol{r}' \cdot \boldsymbol{r}'')}{\boldsymbol{r}' \cdot \boldsymbol{r}'}\boldsymbol{r}'\right).$$

In geometric terms, $\boldsymbol{\kappa}(t)$ is the projection of $\boldsymbol{r}''$ perpendicular to $\boldsymbol{r}'$ divided by the square of the absolute value of $\boldsymbol{r}'$.

DEFINITION 5.3. The *numerical curvature* $\kappa(t)$ is defined to be the absolute value (norm) of the vector curvature. (Notice that we distinguish the two by using boldface for the vector curvature and normal font face for the numerical curvature.)

In symbols, we have

$$\kappa^2(t) = \frac{|\boldsymbol{r}'(t)|^2|\boldsymbol{r}''(t)|^2 - (\boldsymbol{r}'(t) \cdot \boldsymbol{r}''(t))^2}{|\boldsymbol{r}'(t)|^6}.$$

By the Schwarz inequality, $\kappa^2(t)$ is a nonnegative number. If $\boldsymbol{r}'' = \boldsymbol{0}$, it is zero. Thus, its square root $\kappa(t)$ is a nonnegative number representing the absolute rate of rotation of the tangent line to the curve. The rotation is counterclockwise when looked at from the position of the vector $\boldsymbol{\tau} \times \boldsymbol{\kappa}$, where the origin is taken at $\boldsymbol{r}(t)$.

There are two particular choices of parameter in which the curvature has an interesting expression. One of them is arc length s, since $\boldsymbol{\kappa}(s)$ is then simply the vector $\boldsymbol{r}''(s)$. (It is actually, the projection of that vector perpendicular to $\boldsymbol{r}'(s)$, but $\boldsymbol{r}'(s)$ is a unit vector and hence its derivative is already perpendicular to it and doesn't need to be projected.) The other is the case when the curve is the graph of a function of one of the coordinates, say $\boldsymbol{r}(x) = (x, y(x), z(x))$. In that case, a bit of algebra shows that

$$\kappa^2 = \frac{(y'')^2 + (z'')^2 + (y'z'' - y''z')^2}{\left(1 + (y')^2 + (z')^2\right)^3}.$$

The two-dimensional version of this case, when $z \equiv 0$ and $\boldsymbol{r}(x) = (x, f(x))$, is the case already discussed:

$$\kappa(x) = \frac{|f''(x)|}{\left(1 + (f'(x))^2\right)^{3/2}}.$$

The vector curvature $\boldsymbol{\kappa}$ is by definition normal to the curve. As an example, a circle of radius R can be parameterized as $\boldsymbol{r}(t) = R\cos t\, \boldsymbol{i} + R\sin t\, \boldsymbol{j}$. Then $ds = |\boldsymbol{r}'(t)|\, dt = R\, dt$, and so the parameterization by arc length (for which we inconsistently retain the letter $\boldsymbol{r}$) is $\boldsymbol{r}(s) = R\cos(s/R)\boldsymbol{i} + R\sin(s/R)\boldsymbol{j}$. It is obvious that the tangent vector $\boldsymbol{\tau}(s) = \boldsymbol{r}'(s)$ is $-\sin(s/R)\boldsymbol{i} + \cos(s/R)\boldsymbol{j}$, and $\boldsymbol{\kappa}(s) = \boldsymbol{r}''(s) = (-1/R^2)\boldsymbol{r}(s)$. Since $|\boldsymbol{r}(s)| = R$, $\kappa(s)$ has the constant value $1/R$ (see Problem 5.1 below).

Two important principles to be kept in mind as we now progress to surfaces in $\mathbb{R}^3$ are the following:

(1) The vector curvature $\boldsymbol{\kappa}(t)$ of a parameterized curve $\boldsymbol{r}(t)$ is determined by the first two derivatives of $\boldsymbol{r}(t)$.

(2) The vector curvature $\boldsymbol{\kappa}(t)$ is the projection of $\boldsymbol{r}''(t)$ perpendicular to the tangent vector $\boldsymbol{r}'(t)$ divided by the square of the magnitude of $\boldsymbol{r}'(t)$. If $\boldsymbol{r}$ is assumed to have physical dimension of length—and we normally take that for granted—then the curvature has the physical dimension length^{-1}, *whatever physical dimension is ascribed to the parameter*[6] *t.*

2.3. Surfaces in $\mathbb{R}^3$. Near any given "base" point $(x_0, y_0, z_0) \in \mathbb{R}^3$, the equation of the most general smooth (differentiable) surface can always be written in the form $F(x, y, z) = 0$. Except in degenerate cases, this equation can be solved locally for one of the variables, which without any loss of generality we may take to be z, and rewritten as $z = f(x, y)$. Such a solution exists by the implicit function theorem, near any base point (x_0, y_0, z_0) where $\frac{\partial F}{\partial z}$ is not zero. The function $f(x, y)$ is determined by the condition $f(x_0, y_0) = z_0$ and the formulas for its partial derivatives, obtained by differentiating the identity $F\big(x, y, f(x, y)\big) \equiv 0$:

$$\frac{\partial f}{\partial x} = -\frac{\frac{\partial F}{\partial x}}{\frac{\partial F}{\partial z}},$$

$$\frac{\partial f}{\partial y} = -\frac{\frac{\partial F}{\partial y}}{\frac{\partial F}{\partial z}}.$$

When y is held fixed, the first of these is a differential equation for a function $\varphi_y(x) = f(x, y)$ that has a unique solution up to a "constant" that actually depends on y. That is, $f(x, y) = \varphi_y(x) + c(y)$. When $\varphi_y(x) + c(y)$ is substituted into the second equation, the result is a differential equation that determines $c(y)$ (that is to say, $f(x, y)$) up to a constant term k. The constant term is then determined from the condition $f(x_0, y_0) = z_0$. (See Appendix 5 for justification of this argument.)

EXAMPLE 5.2. The simplest example of this process is provided by the sphere whose equation is $F(x, y, z) = x^2 + y^2 + (z - r_0)^2 - r_0^2 = 0$. Near the point $(0, 0, 0)$, we can solve explicitly for z and write

$$z = r_0 - \sqrt{r_0^2 - x^2 - y^2}.$$

This equation is valid for all points (x, y, z) in the "Southern Hemisphere," where $0 \le z < r_0$. If we had found it necessary to resort to the differential equation to determine this function, we would have used the equations

$$\frac{\partial z}{\partial x} = \frac{\partial f}{\partial x} = -\frac{2x}{2(z - r_0)} = -\frac{x}{z - r_0},$$

$$\frac{\partial z}{\partial y} = \frac{\partial f}{\partial y} = -\frac{2y}{2(z - r_0)} = -\frac{y}{z - r_0}.$$

The first of these would yield, via the differential equation $x\, dx + (z - r_0)\, dz = 0$ that holds when y is constant (that is, $dy = 0$), the relation

$$z = r_0 - \sqrt{c(y) - x^2},$$

(the negative sign is needed since $z < r_0$), and when this equation is differentiated with respect to y, holding x fixed, we find that the second equation says $c'(y)/2(z - r_0) = \partial z/\partial y = -y/(z - r_0)$, which yields $c(y) = -y^2 + k$ for a constant

---

[6]In the example of the circle just given, t appears as the argument of trigonometric functions; it must therefore be dimensionless.

k. Then, given that $z = 0$ when $x = y = 0$, we find $k = r_0^2$, and thereby arrive at the solution already obtained.

2.4. Curvature of a surface in $\mathbb{R}^3$: I. We use the simplified assumption of an explicit equation $z = f(x, y)$ in discussing geodesics in Appendix 2, since it provides a horizontal plane of reference (the xy-plane) over which a local portion of the surface is spread. In that case, for a direction in the xy-plane making a fixed angle θ with the positive x-axis we can consider the curve on the surface whose parameterization is $\boldsymbol{r}(\theta, t) = (x + t\cos\theta)\boldsymbol{i} + (y + t\sin\theta)\boldsymbol{j} + f(x + t\cos\theta, y + t\sin\theta)\boldsymbol{k}$. This curve is the intersection of the surface with the vertical plane $y = x\tan\theta$. We are interested in computing the curvature of this curve at the parameter value $t = 0$. We are assuming that the reader knows what is meant by the *gradient* of a function $f(x, y)$ of two variables, that is, is the vector

$$\nabla f = \frac{\partial f}{\partial x}\boldsymbol{i} + \frac{\partial f}{\partial y}\boldsymbol{j}.$$

We have already used this notation more than once. More generally, the gradient of a function $f(x^1, \ldots, x^n)$ is the vector

$$\nabla f = \left(\frac{\partial f}{\partial x^1}, \ldots, \frac{\partial f}{\partial x^n} \right).$$

DEFINITION 5.4. The *Hessian* matrix[7] of $f(x, y)$ is

$$H_f = \begin{pmatrix} \frac{\partial^2 f}{\partial x^2} & \frac{\partial^2 f}{\partial x\,\partial y} \\ \frac{\partial^2 f}{\partial y\,\partial x} & \frac{\partial^2 f}{\partial y^2} \end{pmatrix},$$

We regard the Hessian as a tensor of type $(0, 2)$, so that its entries will be denoted with a pair of subscripts as h_{ij}. As such, its effect on a pair of contravariant vectors $\boldsymbol{u} = (u^1\boldsymbol{i} + u^2\boldsymbol{j})$ and $\boldsymbol{v} = v^1\boldsymbol{i} + v^2\boldsymbol{j}$ is written

$$H_f(\boldsymbol{u}, \boldsymbol{v}) = h_{ij}u^iv^j = \frac{\partial^2}{\partial x^2}u^1v^1 + \frac{\partial^2 f}{\partial x\,\partial y}u^1v^2 + \frac{\partial^2 f}{\partial y\,\partial x}u^2v^1 + \frac{\partial^2 f}{\partial y^2}u^2v^2.$$

More generally, we can define the Hessian $H_f(\boldsymbol{u}, \boldsymbol{v})$ of a function $f(x^1, \ldots, x^n)$ on $\mathbb{R}^n$ as an $n \times n$ matrix, the matrix of the bilinear function (tensor of covariant rank 2, that is, type $(0, 2)$) whose action on a pair of vectors $\boldsymbol{u} = (u^1, \ldots, u^n)$ and $\boldsymbol{v} = (v^1, \ldots v^n)$ is given by

$$H_f(\boldsymbol{u}, \boldsymbol{v}) = \sum_{i,j=1}^{n} \frac{\partial^2 f}{\partial x^i\,\partial x^j}u^iv^j.$$

Notice that H_f varies from one point (x, y) to another, but we are suppressing that argument to avoid getting formulas that look too cumbersome. But keep in mind that there is always a "base point" (x, y) where the partial derivatives are to be evaluated.

Because the mixed second-order partial derivatives are equal, the Hessian matrix is symmetric. As the reader likely knows already, the gradient and the Hessian on $\mathbb{R}^n$ play the roles of first and second derivatives in the discussion of series expansions of functions of several variables. Thus, potential extreme values of a function $f(x^1, \ldots, x^n)$ are found by looking for points where the gradient is zero (all the first-order partial derivatives vanish). When those points are found, a sufficient

---

[7]Named after Ludwig Otto Hesse (1811–1874).

condition for a minimum is that the Hessian be a positive-definite matrix and a sufficient condition for a maximum is that it be negative-definite. We shall use the Hessian to compute the curvature of a surface in $\mathbb{R}^3$ just as we used the second derivative to compute the curvature of a curve.

We now fix a base point (x, y) and a direction corresponding to a fixed angle θ. Just to remind ourselves that we are doing so, we write $\boldsymbol{r}_\theta(t)$ as an abbreviation for $\boldsymbol{r}(t) = \big(x + t\cos\theta, y + t\sin\theta, f(x + t\cos\theta, y + t\sin\theta)\big)$. We find that

$$
\begin{aligned}
\boldsymbol{r}'_\theta(t) &= \cos\theta\,\boldsymbol{i} + \sin\theta\,\boldsymbol{j} + \Big(\cos\theta\frac{\partial f}{\partial x} + \sin\theta\frac{\partial f}{\partial y}\Big)\boldsymbol{k} \\
&= \cos\theta\,\boldsymbol{i} + \sin\theta\,\boldsymbol{j} + A_\theta(t)\boldsymbol{k}, \\
|\boldsymbol{r}'_\theta(t)| &= \sqrt{1 + \Big(\cos\theta\frac{\partial f}{\partial x} + \sin\theta\frac{\partial f}{\partial y}\Big)^2} = \sqrt{1 + \big((\cos\theta\,\boldsymbol{i} + \sin\theta\,\boldsymbol{j})\cdot\nabla f\big)^2}. \\
\boldsymbol{r}''_\theta(t) &= \Big(\cos^2\theta\frac{\partial^2 f}{\partial x^2} + 2\cos\theta\sin\theta\frac{\partial^2 f}{\partial x\,\partial y} + \sin^2\theta\frac{\partial^2 f}{\partial y^2}\Big)\boldsymbol{k} \\
&= H_\theta(t)\,\boldsymbol{k},
\end{aligned}
$$

where this last equation is simply the definition of the symbols $A_\theta(t) = \big((\cos\theta\,\boldsymbol{i} + \sin\theta\,\boldsymbol{j})\cdot\nabla f\big)$ and $H_\theta(t) = H_f\big(\cos\theta\,\boldsymbol{i} + \sin\theta\,\boldsymbol{j}, \cos\theta\,\boldsymbol{i} + \sin\theta\,\boldsymbol{j}\big)$. In both of these, it is understood that the partial derivatives of f that occur are to be evaluated at $(x + t\cos\theta, y + t\sin\theta)$.

To take a simple case for example, say $f(x, y) = xy$, we have $A_\theta(t) = x\sin\theta + y\cos\theta + 2t\cos\theta\sin\theta$ and $H_\theta(t) = 2\cos\theta\sin\theta$.

Although we really would like to use arc length s as a parameter, we cannot assume that $t = s$ unless

$$
\cos\theta\frac{\partial f}{\partial x} + \sin\theta\frac{\partial f}{\partial y} = 0,
$$

since the assumption that $t = s$ implies that $|\boldsymbol{r}'_\theta(t)| = 1$. In general this is true only for one direction, but if the xy-plane is tangent to the surface at the point $(x, y, f(x, y))$, then the two first-order partial derivatives of f will vanish at that point, and this equation will hold in all directions θ. It will *not* hold in general at nearby points where $t \neq 0$, and we need those points in order to compute the second-order partial derivatives. Hence, we do not assume that $t = s$.

The relation $A'_\theta(t) = H_\theta(t)$ is immediate. From the expression above, we have

$$
\frac{dt}{ds} = \frac{1}{|\boldsymbol{r}'_\theta(t)|} = \frac{1}{\sqrt{1 + A_\theta(t)^2}},
$$

and the unit tangent vector is, when we suppress the argument t,

$$
\boldsymbol{\tau} = \frac{1}{\sqrt{1 + A_\theta^2}}\big(\cos\theta\,\boldsymbol{i} + \sin\theta\,\boldsymbol{j} + A_\theta\,\boldsymbol{k}\big).
$$

The vector curvature of the parameterized curve $\boldsymbol{r}_\theta(t)$ (see Definition 5.2 above) works out to be

$$
\boldsymbol{\kappa}_\theta(t) = \frac{-A_\theta H_\theta}{(1 + A_\theta^2)^2}\big(\cos\theta\,\boldsymbol{i} + \sin\theta\,\boldsymbol{j}\big) + \frac{H_\theta}{(1 + A_\theta^2)^2}\boldsymbol{k},
$$

from which it follows that the numerical curvature at $(x, y, f(x,y))$ in the direction given by θ is

$$\kappa_\theta(0) = \frac{|H_\theta|}{(1 + A_\theta^2)^{3/2}},$$

where H_θ and A_θ are evaluated at $t = 0$. In our example $z = xy$, we find $\kappa_\theta(t) = 2\sin\theta\cos\theta/(1 + x^2\sin^2\theta + 2xy\sin\theta\cos\theta + y^2\cos^2\theta)^{(3/2)}$. At the point $x = y = 0$, we thus get $\kappa_\theta(0) = \sin(2\theta)$, and, as we would expect, the curvature is 0 along the axes and ±1 at points making a 45° angle with the axes.

Notice the analogy with the case of a plane curve, where

$$\kappa_\theta(0) = \frac{f''(x)}{\left(1 + \left(f'(x)\right)^2\right)^{3/2}}.$$

Because of this convenient analogy, we can now relax our definition of the numerical curvature. That is, we can allow $\kappa_\theta(0)$ to be negative, defining it by the equation

$$\kappa_\theta(0) = \frac{H_\theta}{(1 + A_\theta^2)^{3/2}}.$$

The complication here is the directionality of this curvature. It depends on θ. We shall now explore this dependence in more detail.

Since we are now finished with the variable t, we set it equal to zero and write

$$\kappa_\theta(x, y) = \frac{H_\theta}{(1 + A_\theta^2)^{3/2}},$$

where now the partial derivatives that occur in H_θ and A_θ are to be evaluated at (x, y), both coordinates of which are still being held fixed. Our next task is to get an expression that will measure the curvature of the *surface* at the point (x, y) by selecting information from the numerical curvatures in various directions. The algebra involved will become prohibitively complicated at this point unless we make an intuitive "leap of faith" and believe that the curvature is intrinsic to the surface and would be the same in any coordinates. We explore the algebraic basis of that faith below (Problem 5.2), and a full proof of its correctness can be found in Appendix 6 of Volume 2. Right now we simply make the leap.

We first translate our coordinate axes so that the origin is at $(x, y, f(x, y))$, that is, $(x, y) = (0, 0)$ and $f(0, 0) = 0$. Then, by a rotation of axes, we force the tangent plane at this point to be the xy-plane. Both of these changes of variable are rigid motions of $\mathbb{R}^3$, and hence should not affect the curvature if it is what we picture it as being. The fact that the z-axis is perpendicular to the tangent plane provides a notion of "above" and "below" relative to that plane, and hence a notion of positive and negative curvature for sections of the surface by planes containing the z-axis (perpendicular to the tangent plane). By this choice of coordinates, we gain a great deal of simplicity, since now the first partial derivatives of $f(x, y)$ at $(0, 0)$ are both equal to zero. Thus $A_\theta = 0$ at that point, and our curvature κ_θ is simply

$$\kappa_\theta = H_\theta.$$

By using trigonometric identities, we can write

$$H_\theta = a + b\cos(2\theta) + c\sin(2\theta),$$

where

$$a = \frac{1}{2}\left(\frac{\partial^2 f}{\partial x^2} + \frac{\partial^2 f}{\partial y^2}\right),$$

$$b = \frac{1}{2}\left(\frac{\partial^2 f}{\partial x^2} - \frac{\partial^2 f}{\partial y^2}\right),$$

$$c = \frac{\partial^2 f}{\partial x \partial y},$$

and then H_θ is given as

$$H_\theta = a + d\cos(2\theta - \varphi),$$

where

$$d = \sqrt{b^2 + c^2} = \frac{1}{2}\sqrt{\left(\frac{\partial^2 f}{\partial x^2} - \frac{\partial^2 f}{\partial y^2}\right)^2 + 4\left(\frac{\partial^2 f}{\partial x \partial y}\right)^2},$$

$$\cos\varphi = \frac{b}{d},$$

$$\sin\varphi = \frac{c}{d}.$$

We now need to study the way κ_θ varies as a function of θ. Being of period π it necessarily has a maximum and minimum value in each period, of course. From the equations just written, we can see that the values of θ at which these extrema occur are those for which $\cos(2\theta - \varphi) = +1$ and $\cos(2\theta - \varphi) = -1$ respectively. The numerical curvatures in these directions are called the *principal curvatures* of the surface at that point. The result is the following theorem, due to Euler, and illustrated in Fig. 5.2:

THEOREM 5.1. *The directions of maximum and minimum curvature of planar sections perpendicular to the tangent plane are given by the angles*

$$\theta = \frac{\varphi}{2} \text{ and } \frac{\pi + \varphi}{2}.$$

In particular, they are perpendicular to each other.

These maximum and minimum values are $a \pm d$.

The two curvatures just described are called the *principal curvatures* of the surface at the base point. If the surface has points $(x, y, f(x,y))$ arbitrarily near the base point $(x_0, y_0, f(x_0, y_0))$ where $f(x,y) \geq f(x_0, y_0)$, then the maximum curvature of such sections is nonnegative. Likewise, if there are points arbitrarily near $(x_0, y_0, f(x_0, y_0))$ where $f(x,y) \leq f(x_0, y_0)$, then the minimum curvature is nonpositive.

The preceding considerations motivate Euler's definition of curvature for a surface.

DEFINITION 5.5. The *curvature* κ at (x,y) of the surface $z = f(x,y)$ is the product of the minimum and maximum numerical values of the curvatures of planar sections of the surface through the point $(x, y, f(x,y))$ perpendicular to the tangent plane at that point.

In terms of the angle φ defined above and the constants a and d defined above, this means

$$\kappa = \kappa_{\varphi/2}\kappa_{(\varphi+\pi)/2} = a^2 - d^2 = \left(\frac{\partial^2 f}{\partial x^2}\right)^2\left(\frac{\partial^2 f}{\partial y^2}\right)^2 - \left(\frac{\partial^2 f}{\partial x \partial y}\right)^2.$$

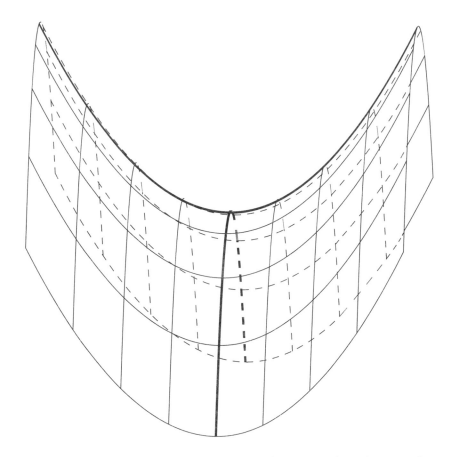

FIGURE 5.2. Euler's theorem: The two directions of maximum and minimum curvature of planar sections perpendicular to the tangent plane are perpendicular to each other. This principle is shown here with the hyperbolic paraboloid. At the saddle point, the surface intersects one plane in an upward-opening parabola (maximum curvature), and a plane perpendicular to it in a downward-opening parabola (minimum curvature).

The curvature of a surface must have physical dimension length^{-2}, that is, the reciprocal of area, if $\mathbb{R}^3$ is regarded as the product of three physical lines.

REMARK 5.3. One may well ask: Why the product? Surely there are other combinations of the two directional curvatures that would tell us something about the shape of the surface at a given point. For example, the sum of the two principal curvatures was used by Sophie Germain (1776–1831) in a paper on the theory of elasticity. Half of this sum, which is their arithmetic mean, is called the *mean curvature*. In Chapter 6, we shall also encounter the *sectional curvature* of a manifold, which is, for example $1/R^2$ for an n-dimensional sphere of radius R in $\mathbb{R}^{n+1}$. Also in Chapter 6, we shall introduce the *scalar curvature*, which for a two-dimensional manifold embedded in $\mathbb{R}^3$ is twice the curvature defined by Euler. Thus, we have a variety of notions of curvature that we might deal with. But the product of the two principal curvatures has one outstanding advantage, which will be developed below: It allows the curvature of a surface in $\mathbb{R}^3$ to be defined using the curvature of a sphere as a standard, just as the circle is the standard of curvature for curves. That approach was developed by Gauss, who revamped this whole subject nearly a century later. The curvature that Euler defined has come to be called *Gaussian curvature*.

REMARK 5.4. It follows from the definition of curvature that if all the points on the the surface near the base point lie on the same side of the tangent plane, the curvature at the base point is nonnegative, whereas if there are pairs of points arbitrarily near the base point that are on opposite sides of the tangent plane, then the curvature is nonpositive. It is easy to see intuitively that near a point of negative curvature, the surface must be shaped rather like a "saddle." Indeed, the Hessian determinant of $f(x,y)$ must be negative at such a point, so that the point is neither a maximum nor a minimum, even though both of its first partial derivatives vanish there.

2.5. Curvature of a surface in $\mathbb{R}^3$: II. With these preliminaries out of the way, we can now state Euler's main result, which motivates the definition of curvature of a surface just given .

THEOREM 5.2. *In terms of the quantities a, b, c, and d introduced above, the curvature of the surface $z = f(x,y)$ at a point where the xy-plane is tangent to that surface is given by*

$$\kappa(x,y) = a^2 - d^2 = a^2 - b^2 - c^2 = \frac{\partial^2 f}{\partial x^2}\frac{\partial^2 f}{\partial y^2} - \left(\frac{\partial^2 f}{\partial x \partial y}\right)^2.$$

PROOF. This proof is self-working and is merely a matter of inserting the values of a, b, c, and d into the definition just given for κ. $\qquad\square$

The curvature turns out to be the determinant of the Hessian matrix, but only in coordinates where the xy-plane is tangent to the surface at the point where curvature is being computed. It is the analogy between the Hessian and the second derivative that motivates our definition of curvature.

It is important to keep in mind the special configuration of coordinates that makes this simple expression possible. When general coordinates are used, even if the equation of the surface has the form $z = f(x,y)$, the formula for the curvature is a little more complicated. We state this result as a separate theorem.

THEOREM 5.3. *For a surface that is the graph of a function $z = f(u,v)$, the curvature at a point $\bigl(u, v, f(u,v)\bigr)$ is*

$$\kappa(u,v) = \frac{\frac{\partial^2 f}{\partial u^2}\frac{\partial^2 f}{\partial v^2} - \left(\frac{\partial^2 f}{\partial u \partial v}\right)^2}{\left(1 + \left(\frac{\partial f}{\partial u}\right)^2 + \left(\frac{\partial f}{\partial v}\right)^2\right)^2}.$$

PROOF. We assume a coordinate mapping $(u,v) \mapsto \boldsymbol{r}(u,v) = u\boldsymbol{i} + v\boldsymbol{j} + f(u,v)\boldsymbol{k}$, and we make the substitution $u = u_0 + x$, $v = v_0 + y$, thus getting a new mapping $(x,y) \mapsto \boldsymbol{s}(x,y) = u_0\boldsymbol{i} + v_0\boldsymbol{j} + f(u_0,v_0)\boldsymbol{k} + \bigl((x\boldsymbol{i} + y\boldsymbol{j} + (f(u_0 + x, v_0 + y) - f(u_0,v_0))\boldsymbol{k}\bigr) = u_0\boldsymbol{i} + v_0\boldsymbol{j} + (x\boldsymbol{i} + y\boldsymbol{j} + g(x,y)\boldsymbol{k}$, where $g(x,y) = f(u_0 + x, v_0 + y) - f(u_0,v_0)$. We then have $g(0,0) = 0$. This change of variables is "transparent" to the partial derivative operators, that is, $\partial/\partial x = \partial/\partial u$ and $\partial/\partial y = \partial/\partial v$. The formula given in the theorem therefore holds for f if it holds when $\big(u, v, f(u, v)\big)$ is replaced by $\big(x, y, g(x, y)\big)$. Thus without any loss of generality, we can assume that $u_0 = 0$ and $v_0 = 0$. We have only to perform a rotation of the coordinate system in order to get the tangent plane to be the xy-plane.

We are assuming that the surface is the graph of a function $f(u, v)$, and this means that the tangent vectors

$$\frac{\partial \boldsymbol{r}}{\partial u} = \boldsymbol{i} + \frac{\partial f}{\partial u}\boldsymbol{k} \text{ and } \frac{\partial \boldsymbol{r}}{\partial v} = \boldsymbol{j} + \frac{\partial f}{\partial v}\boldsymbol{k}$$

are linearly independent. In particular, the normal vector $\boldsymbol{n}$, which is the cross product of these tangent vectors, is perpendicular to the tangent plane and not the zero vector.

To keep our computations straight, we now introduce some abbreviations:

$$f_1 = \frac{\partial f}{\partial u}, \quad f_2 = \frac{\partial f}{\partial v}, \quad a = \sqrt{1 + f_1^2}, \quad b = \sqrt{1 + f_1^2 + f_2^2},$$

where the two partial derivatives are to be evaluated $(u_0, v_0) = (0, 0)$. All four of the quantities f_1, f_2, a, and b are constants, not functions of u and v.

We now introduce a new orthonormal basis of $\mathbb{R}^3$, as follows:

$$\begin{aligned}
\boldsymbol{i}' &= \frac{1}{a}\boldsymbol{i} + \frac{f_1}{a}\boldsymbol{k}, \\
\boldsymbol{j}' &= \frac{-f_1 f_2}{ab}\boldsymbol{i} + \frac{1 + f_1^2}{ab}\boldsymbol{j} + \frac{f_2}{ab}\boldsymbol{k}, \\
\boldsymbol{k}' &= \frac{-f_1}{b}\boldsymbol{i} + \frac{-f_2}{b}\boldsymbol{j} + \frac{1}{b}\boldsymbol{k}.
\end{aligned}$$

(This change of coordinates is not unique; all that really matters is that $\boldsymbol{k}'$ be a unit vector normal to the surface.)

All the partial derivatives in these expressions are evaluated at $(0, 0)$, and hence are constants. They are not to be differentiated when we differentiate expressions in which they occur.

This being a transformation of one right-handed orthonormal system into another, its inverse has coefficients that are merely the transpose of these:

$$\begin{aligned}
\boldsymbol{i} &= \frac{1}{a}\boldsymbol{i}' + \frac{-f_1 f_2}{ab}\boldsymbol{j}' + \frac{-f_1}{b}\boldsymbol{k}', \\
\boldsymbol{j} &= \frac{1 + f_1^2}{ab}\boldsymbol{j}' + \frac{-f_2}{b}\boldsymbol{k}', \\
\boldsymbol{k} &= \frac{f_1}{a}\boldsymbol{i}' + \frac{f_2}{ab}\boldsymbol{j}' + \frac{1}{b}\boldsymbol{k}'.
\end{aligned}$$

We now suppose that the function $g(x, y)$ is such that

$$u\boldsymbol{i} + v\boldsymbol{j} + f(u, v)\boldsymbol{k} = x\boldsymbol{i}' + y\boldsymbol{j}' + g(x, y)\boldsymbol{k}'.$$

The equations of transformation show that

$$u = \frac{x}{a} - \frac{f_1 f_2 y}{ab} - \frac{f_1 g(x,y)}{b},$$

$$v = \frac{a y}{b} - \frac{f_2 g(x,y)}{b},$$

$$g(x,y) = \frac{-(u f_1 + v f_2) + f(u,v)}{b},$$

$$\frac{\partial u}{\partial x} = \frac{1}{a} - \frac{f_1}{b}\frac{\partial g}{\partial x},$$

$$\frac{\partial u}{\partial y} = \frac{-f_1 f_2}{ab} - \frac{f_1}{b}\frac{\partial g}{\partial y},$$

$$\frac{\partial v}{\partial x} = \frac{-f_2}{b}\frac{\partial g}{\partial x},$$

$$\frac{\partial v}{\partial y} = \frac{a}{b} - \frac{f_2}{b}\frac{\partial g}{\partial y}.$$

From the third of these equations, we easily deduce that at a point (u,v) we have

$$\frac{\partial g}{\partial x} = -\frac{1}{b}\left(f_1 \frac{\partial u}{\partial x} + f_2 \frac{\partial v}{\partial x}\right) + \frac{1}{b}\left(\frac{\partial f}{\partial u}\frac{\partial u}{\partial x} + \frac{\partial f}{\partial v}\frac{\partial v}{\partial x}\right).$$

It follows from this expression that $\partial g/\partial x = 0$ at the origin, and similarly $\partial g/\partial y = 0$. Thus the tangent plane to the surface is indeed the xy-plane. In other words, we have now made the reduction that Euler made in defining curvature. As a consequence, we find that at this point

$$\frac{\partial u}{\partial x} = \frac{1}{a},$$

$$\frac{\partial u}{\partial y} = \frac{-f_1 f_2}{ab},$$

$$\frac{\partial v}{\partial x} = 0,$$

$$\frac{\partial v}{\partial y} = \frac{a}{b}.$$

Moreover, by extending the use of the chain rule in the formula for $g(x,y)$, we find, again at the point $(0,0)$, that the second-order partial derivatives of u and v with respect to x and y cancel out, and we are left with

$$\frac{\partial^2 g}{\partial x^2} = \frac{1}{b}\left(\frac{\partial^2 f}{\partial u^2}\left(\frac{\partial u}{\partial x}\right)^2 + 2\frac{\partial^2 f}{\partial u \partial v}\frac{\partial u}{\partial x}\frac{\partial v}{\partial x} + \frac{\partial^2 f}{\partial v^2}\left(\frac{\partial v}{\partial x}\right)^2\right),$$

$$\frac{\partial^2 g}{\partial x \partial y} = \frac{1}{b}\left(\frac{\partial^2 f}{\partial u^2}\frac{\partial u}{\partial x}\frac{\partial u}{\partial y} + \frac{\partial^2 f}{\partial u \partial v}\left(\frac{\partial u}{\partial x}\frac{\partial v}{\partial y} + \frac{\partial u}{\partial y}\frac{\partial v}{\partial x}\right) + \frac{\partial^2 f}{\partial v^2}\frac{\partial v}{\partial x}\frac{\partial v}{\partial y}\right),$$

$$\frac{\partial^2 g}{\partial y^2} = \frac{1}{b}\left(\frac{\partial^2 f}{\partial u^2}\left(\frac{\partial u}{\partial y}\right)^2 + 2\frac{\partial^2 f}{\partial u \partial v}\frac{\partial u}{\partial y}\frac{\partial v}{\partial y} + \frac{\partial^2 f}{\partial v^2}\left(\frac{\partial v}{\partial y}\right)^2\right).$$

From this point, a slightly messy computation (see Problems 5.2 and 5.4 below) reveals that the curvature is

$$\kappa(u,v) = \frac{\frac{\partial^2 g}{\partial x^2}\frac{\partial^2 g}{\partial y^2} - \left(\frac{\partial^2 g}{\partial x\,\partial y}\right)^2}{} ,$$

$$= \frac{\frac{\partial^2 f}{\partial u^2}\frac{\partial^2 f}{\partial v^2} - \left(\frac{\partial^2 f}{\partial u\,\partial v}\right)^2}{\left(1 + \left(\frac{\partial f}{\partial u}\right)^2 + \left(\frac{\partial f}{\partial v}\right)^2\right)^2} .$$

(Four terms cancel in pairs in the expanded version of the right-hand side.) □

Theorem 5.3 is an analog of the formula for the curvature of a plane curve that is the graph of a function $y = f(x)$. Recall that this curvature is

$$\kappa(x) = \frac{f''(x)}{\left(1 + (f'(x))^2\right)^{3/2}} .$$

By comparing with Theorem 5.3, we can see that the second derivative is replaced by the determinant of the Hessian, the square of the first-order derivative by the sum of the squares of the two first-order partial derivatives, and the exponent $3/2$ by $2 = 4/2$.

DEFINITION 5.6. The *radius of curvature* at a point of the surface is the reciprocal of the mean proportional between the maximum and minimum radii of curvature among all planar sections of the surface passing through that point perpendicular to the tangent plane. In other words, it is the square root of the reciprocal of the curvature.

For a sphere of radius R, the radius of curvature is R, but at a point where a surface has negative curvature, the radius of curvature is an imaginary number. An important example of a surface of constant negative curvature is the *pseudo-sphere*, which plays a large role as a model of the hyperbolic plane and will frequently serve us as an example, in this chapter and the next.

EXAMPLE 5.3. Consider the lower hemisphere of radius r_0, for which $f(x,y) = r_0 - \sqrt{r_0^2 - x^2 - y^2}$. We first compute the Hessian at a general point:

$$H(x,y) = \frac{\partial^2 f}{\partial x^2}\frac{\partial^2 f}{\partial y^2} - \left(\frac{\partial^2 f}{\partial x\,\partial y}\right)^2$$

$$= \frac{(r_0^2 - x^2)(r_0^2 - y^2) - x^2 y^2}{(r_0^2 - x^2 - y^2)^3}$$

$$= \frac{r_0^4 - r_0^2(x^2 + y^2)}{(r_0^2 - x^2 - y^2)^3}$$

$$= \frac{r_0^2}{(r_0^2 - x^2 - y^2)^2} .$$

Since

$$1 + \left(\frac{\partial f}{\partial x}\right)^2 + \left(\frac{\partial f}{\partial y}\right)^2 = \frac{r_0^2}{r_0^2 - x^2 - y^2} ,$$

we find that the curvature is

$$\kappa(x,y) = \frac{r_0^2}{r_0^4} = \frac{1}{r_0^2} .$$

Since the equation of this surface will be exactly the same at every point when the tangent plane is taken as the xy-plane, we infer that the curvature of a sphere is constant and equal to $1/r_0^2$. For a sphere, the radius of curvature is precisely the radius of the sphere.

At this point, *in theory*, we know exactly what to do to compute the curvature of a surface in $\mathbb{R}^3$. If it is a proper surface, it is the graph of an equation $F(x, y, z) \equiv 0$ such that the gradient $\nabla F(x, y, z)$ never vanishes on the surface itself. For example, the sphere of radius r_0 with center at the origin has the equation $F(x, y, z) = x^2 + y^2 + z^2 - r_0^2 \equiv 0$, and the gradient $\nabla F(x, y, z) = 2x\boldsymbol{i} + 2y\boldsymbol{j} + 2z\boldsymbol{k}$ does not vanish at any point of the surface. (It vanishes only at the origin, which is not on the surface.) Given that fact, the implicit function theorem tells us that every "proper" surface in $\mathbb{R}^3$ is the graph of a function, say $z = f(x, y)$. Granted, it may be difficult to find the function $f(x, y)$ analytically, but that is a separate issue. If $f(x, y)$ can be found, the curvature can be computed.

What more could we want? For one thing, given that a surface may have more than one analytic equation, we want assurance that the curvature is determined by the surface itself, and not by any particular analytic representation of it. Second, we'd rather *not* have to find the function $f(x, y)$. For example, we might like to parameterize the sphere of radius r_0 with longitude-latitude coordinates $(\theta, \varphi) \mapsto r_0 \cos\theta \cos\varphi\, \boldsymbol{i} + r_0 \sin\theta \cos\varphi\, \boldsymbol{j} + r_0 \sin\varphi\, \boldsymbol{k}$. It would be less messy to work directly with these trigonometric functions than to have to manipulate the square root of a sum of squares. What we hope to get, ultimately, is a set of key functions derivable directly from the parameterization of the surface in terms of which the curvature can be expressed. More than anything else, we want to know how to generalize this notion of shape to higher-dimensional objects. The crucial steps in that direction were taken by Gauss and will be studied in the next section.

2.6. A two-dimensional universe. The space that formed the background of this discussion was the three-dimensional space $\mathbb{R}^3$ of Euclidean geometry. We used that three-dimensional model to generate an expression for the element of arc length along a curve in terms of the parametric function that defines the curve, and we then extended that discussion to define the curvature of the surface using Euclidean circles and finally the sphere as our standard of curvature. As a simple model of this surface, we might imagine (x, y) to be coordinates on a small portion of a flat plane and $f(x, y)$ the elevation of the point of the Earth's surface directly above or below that flat plane. We used the three-dimensional Pythagorean theorem to define the element of arc length on this surface.

Imagine now a different species of earthlings whose visual perception is handicapped, so that they cannot *see* hills and valleys. As far as they are concerned, the pair (x, y) (longitude and latitude, perhaps) is the complete set of coordinates of a point, and altitude does not exist for them as a thing they can perceive. They may nevertheless be able to form some concept of it. True, our three-dimensional Pythagorean theorem is of no use to them, since they cannot directly perceive any third dimension; their geometry is that of a two-dimensional world. Because of gravity, however, they *can* detect and deal with the third dimension that we perceive and explain it to themselves in terms of the gravitational field. Even though they can't see themselves going uphill or downhill, they know when they are doing either by the amount of effort involved in moving with or against the gravitational field. They might define the distance from point A to point B as a suitably dimensioned multiple of the amount of work required to move a standard unit mass from A to B, that is, the difference in gravitational potential between the two points. That "distance" is negative if B is downhill of A in the three-dimensional model and zero if the two points are at the same altitude. What we, from our three-dimensional perspective, picture as a hill is, for the inhabitants of this two-dimensional world, a region containing a point of maximum gravitational potential (a point that it requires the maximum amount of work to reach), surrounded by a family of closed level curves of the potential.

This trichotomy of distances recalls the space-like and time-like intervals between two events and the light cone of an event in the four-dimensional universe that we actually do inhabit. Relative to that four-dimensional universe, we are in the position the inhabitants of the two-dimensional world find themselves in according to the scenario we just imagined. The proper-time interval ds given by $ds^2 = dt^2 - (dx^2 + dy^2 + dz^2)/c^2$ has some of the properties of the metric we just defined using the difference in gravitational potential. Just as our imaginary denizens lacking three-dimensional vision will still perceive a difference between uphill and downhill, so are we able to perceive a difference between time and space even though we cannot perceive or imagine a world of four-dimensional *spatial* coordinates.

In the next section, we shall recast what we have just done in $\mathbb{R}^3$ in a way that will provide a generalization to n-dimensional spaces.

3. Curvature, Phase 2: Gauss

As remarked above, in all *local* analysis, including in particular the computation of curvature, there is no loss of generality in assuming that a surface in $\mathbb{R}^3$ is the graph of a function. By translating and rotating coordinates, one can assume that the point in question is the origin $(0, 0, 0)$ and that the tangent plane at that point is horizontal. This applies in particular to surfaces given parametrically by a mapping

$$\boldsymbol{r}(u, v) = \big(x(u, v), y(u, v), z(u, v)\big).$$

There is a difficulty, however, since such a surface is presented in terms of the parameters (u, v), which may range over an arbitrary region of $\mathbb{R}^2$. The function-graph model we used requires an orthonormal coordinate system (x, y, z). Although such a system can be constructed in general by orthogonalizing the pair of tangent vectors $\big(\frac{\partial \boldsymbol{r}}{\partial u}, \frac{\partial \boldsymbol{r}}{\partial v}\big)$ to get unit vectors that can be taken as $\boldsymbol{i}$ and $\boldsymbol{j}$, the computation involved in converting from (u, v)-coordinates to (x, y)-coordinates and back again, is complicated. Gauss found another way to look at this problem.[8] He expounded his ideas during the years 1825–1827 in two treatises bearing the title *Disquisitiones generales circa superficies curvas* (*General Investigations of Curved Surfaces*). In discussing Gauss's ideas, we are going to update some of his notation, making use of the simplification of vector notation, which Gauss did not have, and especially the concept of Christoffel symbols.

The new perspective that Gauss brought to the problem contained many profound ideas, the two most important of which are:

(1) The notion of metric coefficients, which Gauss denoted E, F, and G, and used to construct the *first fundamental form*, in terms of which arc length and area can be computed, leading eventually to an expression for Euler's

---

[8] He was inspired to do so by his participation in the geodetic survey of the region around Hannover in Lower Saxony.

curvature that depends only on these coefficients and avoiding the geometric transformations needed to compute curvature in Euler's approach. This line of reasoning leads to the Riemann curvature tensor mentioned in the previous chapter. That tensor will be the final outcome of the present chapter, but it will be written in the language of calculus. (In the next chapter, we shall give it a more geometric form, better adapted to purely abstract spaces.)

(2) The notion of geodesic polar coordinates in a neighborhood of a given point. Gauss showed that the points lying at geodesic distance r from a given point on the surface lie on a curve perpendicular to each of the geodesics through the given point. This line of reasoning leads to the exponential mapping and normal coordinates, and meshes well with the Ricci tensor that provides Einstein's law of gravity. All of that must await the reformulation of the Riemann curvature tensor in the next chapter.

3.1. The spherical mapping. Intuitively, our notion of the curvature of a one-dimensional manifold, such as the graph of a function $y = f(x)$, is based on a computation of the amount that its unit tangent vector $\boldsymbol{\tau}$ "wiggles" as it slides along the curve. The limit of the oscillations of this vector over a small piece of the curve, divided by the length of that small piece, as the piece shrinks down on a point $\big(x, f(x)\big)$ is precisely what we call the curvature $\kappa(x)$. For a surface in $\mathbb{R}^3$ there is an obvious generalization of this notion. Just take the unit *normal* vector at a point P on the surface, measure the area that a unit vector from the origin and parallel to it "scratches out" on the unit sphere as the point P ranges over a small patch of surface, and divide that area by the area of the small patch. That should give some measure of the "average curvature" over the patch, compared with the constant curvature of the unit sphere. If the surface is then shrunk down on a point P, the limit of that average curvature should be a measure of the curvature of the surface at the point P. That is what Gauss did, and it is easily provable that for surfaces of the form $z = f(x, y)$, this new definition of curvature coincides with the definition given by Euler. The details now follow.

Along with the mapping $\boldsymbol{r}(u, v)$ that parameterizes the surface, Gauss introduced the mapping $\boldsymbol{n}(u, v)$ from the parameter space into the unit sphere $\mathbb{S}^2$, as just described. Here $\boldsymbol{n}(u, v)$ is the unit normal to the surface at the point $\boldsymbol{r}(u, v)$ and can be expressed as

$$\boldsymbol{n}(u, v) = \frac{\frac{\partial \boldsymbol{r}}{\partial u} \times \frac{\partial \boldsymbol{r}}{\partial v}}{\left| \frac{\partial \boldsymbol{r}}{\partial u} \times \frac{\partial \boldsymbol{r}}{\partial v} \right|}.$$

This mapping is shown in Fig. 5.3.

What makes this unit normal useful is that it can be pictured as a flagpole anchored in a vehicle rolling over the surface. The flatter the surface, the less the flagpole will oscillate and the more nearly it will remain parallel to itself. For example, if the surface is a plane, the unit normal will always remain parallel to its original position and will not "scratch out" any area on the unit sphere at all. If the surface is a sphere of radius r_0 with center at the origin, the unit normal will always point directly away from the center of the sphere. In that case, the image of a region U under $\boldsymbol{r}$ will be simply r_0 times the image of the region under $\boldsymbol{n}$, and hence the ratio of the area of the latter to that of the former will be $1/r_0^2$, which is precisely the curvature of the sphere. Thus one can see that the ratio of the two

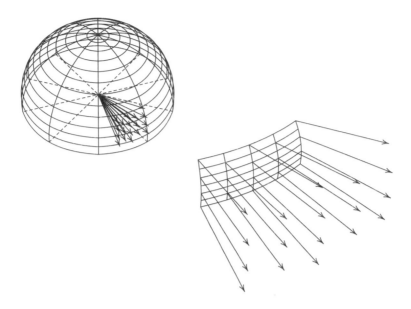

FIGURE 5.3. Unit normals to a curved surface with their tails translated to the origin, so that their heads lie on the upper unit hemisphere.

areas provides a measure of the curvature of the surface. Letting the region U in the parameter space shrink down on a point (u_0, v_0), we can take the limit of the ratio of the area of its image under n to the area of its image under r and call that limit the curvature at the point $r(u_0, v_0)$. That is the theory behind the approach used by Gauss. What remains is the actual algebraic computation of the curvature using these principles and verifying that it is the same as the curvature defined by Euler.

We shall retain much of the notation used by Gauss, who defined six functions on the surface. The first three, which Gauss denoted E, F, and G, are

$$
\begin{aligned}
E &= \frac{\partial r}{\partial u} \cdot \frac{\partial r}{\partial u} = \left(\frac{\partial x}{\partial u}\right)^2 + \left(\frac{\partial y}{\partial u}\right)^2 + \left(\frac{\partial z}{\partial u}\right)^2, \\
F &= \frac{\partial r}{\partial u} \cdot \frac{\partial r}{\partial v} = \frac{\partial x}{\partial u}\frac{\partial x}{\partial v} + \frac{\partial y}{\partial u}\frac{\partial y}{\partial v} + \frac{\partial z}{\partial u}\frac{\partial z}{\partial v}, \\
G &= \frac{\partial r}{\partial v} \cdot \frac{\partial r}{\partial v} = \left(\frac{\partial x}{\partial v}\right)^2 + \left(\frac{\partial y}{\partial v}\right)^2 + \left(\frac{\partial z}{\partial v}\right)^2.
\end{aligned}
$$

The three functions E, F, and G are called the *metric coefficients* of the surface, and it is useful to put them into a "metric matrix" M:

$$
M = \begin{pmatrix} E & F \\ F & G \end{pmatrix}.
$$

Later, when we wish to be more systematic and generalize this work, we shall change the notation to $E = g_{11}$, $F = g_{12} = g_{21}$, $G = g_{22}$.

The metric coefficients can be used to compute the length (s) of a parameterized curve $t \mapsto r\big(u(t), v(t)\big)$ on the surface and the area S of a portion of the surface via the formulas

$$ds^2 = \left(E\left(u'(t)\right)^2 + 2F\,u'(t)v'(t) + G\left(v'(t)\right)^2 \right) dt^2\,,$$
$$dS = \sqrt{EG - F^2}\,du\,dv\,.$$

The first of these relations is usually written in invariant differential form, independent of the parameterization of the specific curve, but depending on the parameterization of the surface:

$$ds^2 = E\,du^2 + 2F\,du\,dv + G\,dv^2\,.$$

The quadratic form ds^2 is called the *first fundamental form* of the surface. Notice that the element of area on the surface is

$$dS = \sqrt{\det(M)}\,du\,dv\,.$$

The coefficients E and G are obviously nonnegative, as (by the Schwarz inequality) is the determinant $\det(M)$. We shall assume that $\det(M) > 0$, which implies that $E > 0$ and $G > 0$ as well, so that the matrix is *positive-definite*.[9] The metric on a parameterized surface is such that ds^2 is always nonnegative, and the matrix M is symmetric. Such a metric is now called a *Riemannian* metric.

The analogous expressions for the mapping $\boldsymbol{n}$ lead to the *second fundamental form*, which (up to a constant, since definitions of this concept vary) we write—again following Gauss—as

$$D\,du^2 + 2D'\,du\,dv + D''\,dv^2\,.$$

REMARK 5.5. Because the cross product of the two partial derivatives of $\boldsymbol{r}$ that defines the normal vector is anti-symmetric, the mapping $\boldsymbol{n}$ has a direction determined by the order of the parameters. If u and v are interchanged, then $\boldsymbol{n}$ becomes $-\boldsymbol{n}$. This is not important in itself, but it does call attention to the two-sided ambiguity of the unit normal vector. When u and v are interchanged, however, the cross product of the two partial derivatives $\boldsymbol{n}$ also reverses its sign. As a result, *the vector triple product* $\boldsymbol{n} \cdot \frac{\partial \boldsymbol{n}}{\partial u} \times \frac{\partial \boldsymbol{n}}{\partial v}$ *does not change when the parameters* u *and* v *are interchanged.* If the sign of that triple product is positive, we call the curvature positive, and if negative, we call it negative.

With the second fundamental form, we can compute what Gauss called the curvature, namely the ratio of the infinitesimal areas.

DEFINITION 5.7. The *Gaussian curvature* of the surface is

$$\kappa = \pm\sqrt{\frac{DD'' - (D')^2}{EG - F^2}}\,,$$

where the ambiguous sign is the sign of the vector triple product $\boldsymbol{n} \cdot \frac{\partial \boldsymbol{n}}{\partial u} \times \frac{\partial \boldsymbol{n}}{\partial v}$.

EXAMPLE 5.4. Consider the hyperbolic paraboloid whose equation is

$$z = (x^2 - y^2)/a\,.$$

---

[9]At points where $\det(M) = 0$, the parameterization has a singularity, and we need to use a different set of parameters. For example, co-latitude and longitude coordinates on a sphere have such a singularity along the equator, and so we generally use them only to study the punctured upper hemisphere with the north pole (where longitude is not defined) removed.

(We include the constant a, which is intended to have the geometric dimension of length, so that both sides of the equation will represent a length.) We shall use x and y as parameters so that $\boldsymbol{r}(x,y) = \big(x, y, (x^2 - y^2)/a\big)$. We then have

$$\frac{\partial \boldsymbol{r}}{\partial x} = (1, 0, 2x/a),$$

$$\frac{\partial \boldsymbol{r}}{\partial y} = (0, 1, -2y/a).$$

Thus the normal vector to this surface is

$$\boldsymbol{N}(x,y) = \frac{\partial \boldsymbol{r}}{\partial x} \times \frac{\partial \boldsymbol{r}}{\partial y} = (1, 0, 2x/a) \times (0, 1, -2y/a) = (-2x/a, 2y/a, 1),$$

and so the *unit* normal is

$$\boldsymbol{n}(x,y) = \left(\frac{-2x}{\sqrt{a^2 + 4x^2 + 4y^2}}, \frac{2y}{\sqrt{a^2 + 4x^2 + 4y^2}}, \frac{a}{\sqrt{a^2 + 4x^2 + 4y^2}} \right).$$

From these expressions, we find

$$E = 1 + 4x^2/a^2,$$
$$F = -4xy/a^2,$$
$$G = 1 + 4y^2/a^2,$$
$$D = \frac{4(a^2 + 4y^2)}{(a^2 + 4x^2 + 4y^2)^2},$$
$$D' = \frac{-16xy}{(a^2 + 4x^2 + 4y^2)^2},$$
$$D'' = \frac{4(a^2 + 4x^2)}{(a^2 + 4x^2 + 4y^2)^2}.$$

Notice that E, F, and G are dimensionless, while D, D', and D'' have the dimension of length raised to power -2. Thus, curvature will also have this dimension, as it should have.

Computation reveals that

$$\frac{\partial \boldsymbol{r}}{\partial x} \times \frac{\partial \boldsymbol{r}}{\partial y} \cdot \frac{\partial \boldsymbol{n}}{\partial x} \times \frac{\partial \boldsymbol{n}}{\partial y} = \frac{-4}{a^2 + 4x^2 + 4y^2},$$

so that the curvature of this surface at every point is negative. That is no surprise, since everyone knows that a hyperbolic paraboloid is saddle-shaped. The computation shows that the curvature is

$$\kappa(x,y) = \frac{-4a^2}{(a^2 + 4x^2 + 4y^2)^2}.$$

The curvature at the origin, where the surface is tangent to the xy-plane, is $\kappa(0,0) = -4/a^2$, which is exactly the value computed by Euler's method.

The definition of curvature given by Gauss for a parameterized surface is consistent with Euler's definition for the curvature of the graph of a function of two variables, as we can see by applying Gauss's definition to the parameterization $\boldsymbol{r}(x,y) = \big(x, y, h(x,y)\big)$. *Mathematica* Notebook 8 in Volume 3 verifies this assertion. Apart from the annoying fact that *Mathematica* prefers working with the negative of the Hessian rather than the Hessian itself and stubbornly insists on expressing an explicitly real number a as $\mathrm{Abs}\,[-a]\,\mathrm{Sign}\,[a]$, one can see from the output of Notebook 8 that this is exactly Euler's result, derived in detail in Problem 5.4 below.

3.2. Curvature in terms of the metric coefficients. In § 11 of his first treatise on differential geometry, Gauss gave the following formula[10] for curvature,[11] except for using p and q as parameters where we are going to be using u and v and writing k instead of κ for the curvature.

$$
(5.1) \quad 4(EG - F^2)^2 \kappa = E\left(\frac{\partial E}{\partial v}\frac{\partial G}{\partial v} - 2\frac{\partial F}{\partial u}\frac{\partial G}{\partial v} + \left(\frac{\partial G}{\partial u}\right)^2\right)
$$
$$
+ F\left(\frac{\partial E}{\partial u}\frac{\partial G}{\partial v} - \frac{\partial E}{\partial v}\frac{\partial G}{\partial u} - 2\frac{\partial E}{\partial v}\frac{\partial F}{\partial v} + 4\frac{\partial F}{\partial u}\frac{\partial F}{\partial v} - 2\frac{\partial F}{\partial u}\frac{\partial G}{\partial u}\right)
$$
$$
+ G\left(\frac{\partial E}{\partial u}\frac{\partial G}{\partial u} - 2\frac{\partial E}{\partial u}\frac{\partial F}{\partial v} + \left(\frac{\partial E}{\partial v}\right)^2\right) - 2(EG - F^2)\left(\frac{\partial^2 E}{\partial v^2} - 2\frac{\partial^2 F}{\partial u\,\partial v} + \frac{\partial^2 G}{\partial u^2}\right).
$$

Given this formula, it becomes possible to begin the discussion of a surface with the metric coefficients E, F, and G, ignoring their origin as dot products of tangent vectors to the surface. The ambient space in which the surface is embedded then becomes superfluous. The shape of the surface is determined by its parameters, and the conversion from one set of parameters to another is governed by the Jacobian matrix. In other words, algebra turns the study of the geometric object into the study of a set of equations, and the foundation is laid for generalization to geometry in any number of dimensions.

Although this formula applies only to two-dimensional surfaces in $\mathbb{R}^3$, its analytic form naturally inspires the question how it might be generalized to give a definition of curvature in higher-dimensional manifolds. Our aim is to study the "shape" of such manifolds as functions of parameters without the need to refer to any space in which the manifold is embedded. The key to doing so—defining a "direction" in a completely abstract space—is described in Appendix 4 of Volume 2, where it is shown how the tangent plane can be generalized as the space of *derivations* of locally C^∞ mappings. A *derivation* is a linear operator X on locally differentiable functions f having the multiplicative property: $X(fg) = fX(g) + gX(f)$. In Euclidean space, such an operator is necessarily a directional derivative, so that these operators provide an algebraic encoding of the concept of a direction. This fact makes up for the ambiguity in the notion of direction when a surface is given parametrically. Each parameter domain can be regarded as a Euclidean space made up of points whose analytic expressions as ordered sets of real numbers provide a natural set of directions. But the fact that a given surface can be parameterized in infinitely many ways means that any such choice of "direction" is arbitrary. The derivation operators X, however, operate on real-valued functions defined on the surface itself, independently of any parameterization. Each such operator therefore "stands in" for a direction on the surface away from a given point, thereby providing the abovementioned algebraic encoding of the concept.

The generalization of this work of Gauss involves considerable difficulty. The form of Eq. (5.1) is sufficiently asymmetric that it is not obvious what the analogous

---

[10]Felix Klein ([**46**], p. 155) commented that Gauss obtained this formula "rather laboriously, by a lengthy computation" (*nicht ohne Mühe, auf Grund längerer Rechnung*).

[11]The standard English translation of this work by James Caddall Morehead and Adam Miller Hiltbeitel, published by the Princeton University Library in 1902, has a misprint on the left-hand side of the formula, where $(EG - F^2)$ appears instead of the correct expression $(EG - F^2)^2$.

formula would be for, say, a three-dimensional manifold embedded in $\mathbb{R}^4$. For that reason, we will need to do some work to construct the proper generalization of it. In the course of that work, we will replace the formula that Gauss gave with one having more geometric transparency. The key to doing so will be the Christoffel symbols, introduced below.

Our goal is to get a definition of curvature that is (1) intrinsic to a manifold and (2) identical with Gaussian curvature when applied to surfaces in $\mathbb{R}^3$. Up to now, we have needed an explicit mapping $(u, v) \mapsto \boldsymbol{r}(u, v) = \big(x(u, v), y(u, v), z(u, v)\big)$ to define the metric coefficients of the surface and then we needed to repeat that operation with the mapping $(u, v) \mapsto \boldsymbol{n}(u, v)$ in order to find its curvature. The latter process has involved the vector cross product, which does not generalize to $\mathbb{R}^n$ in the same form. (Its generalization to $\mathbb{R}^n$ is a multilinear function of $n - 1$ vectors, albeit an important one, since it includes the volume element on $(n - 1)$-dimensional hypersurfaces.) Moreover, we would like to cut the ambient space $\mathbb{R}^n$ entirely out of the conversation and work only with the parameterization. We shall keep Gauss's formula in mind as a guide to tell us whether we are on the right track or not.

Our first step will be to derive a formula for curvature that is equivalent to the one given by Gauss, but has a form that generalizes in an obvious way to n-dimensional manifolds. All that we know right now—and we have not proved even that—is the following obvious consequence of Gauss's formula (Eq. (5.1)):

THEOREM 5.4. *The curvature of a surface can be expressed as a function of the three metric coefficients E, F, and G and their first- and second-order partial derivatives.*

To add some heuristic support for this (unproved) assertion, we note that the curvature is defined as the square root of the quotient of the first and second fundamental forms. Now the first fundamental form *is* essentially the three metric coefficients. As for the second, it is obtained from the unit normal vector $\boldsymbol{n}$ just as the first fundamental form is computed from $\boldsymbol{r}$, that is, by taking dot products of partial derivatives. Since $\boldsymbol{n}$ is a unit vector, its partial derivatives are perpendicular to it, that is, they are vectors in the tangent plane, and therefore expressible as linear combinations of the two first derivatives of $\boldsymbol{r}$. Their dot products can therefore be expressed as linear combinations of E, F, and G.

The weak point in this reasoning is that, when the second fundamental form is expressed in terms of E, F, and G, the coefficients involve derivatives of the cross product of the two tangent vectors. That is the piece of the computation that specifically requires three dimensions and so needs to be cut out and replaced with something intrinsic to the parameterization.

One can see easily see that Theorem 5.4 holds when the surface is the graph of a function $z = f(x, y)$ (Problems 5.2 and 5.4).

3.3. Geometric definition of the Christoffel symbols. In Chapter 4 we introduced the Christoffel symbols, giving a purely algebraic definition of them. The discussion at that point implied that they have an intimate connection with the curvature of a manifold. We are now going to explore the connection between curvature and the Christoffel symbols in the context of surfaces in $\mathbb{R}^3$. Our first job is to give a geometric definition of these symbols in that context and then show

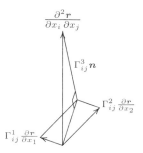

FIGURE 5.4. The standard Christoffel symbols Γ_{ij}^1 and Γ_{ij}^2 are the coefficients of the tangential component of a second-order partial derivative, that is, the components of the projection of the partial derivative into the tangent plane. The nonstandard symbol Γ_{ij}^3 is the coefficient of its normal component.

that it is the same as the algebraic definition given in Chapter 4. That task is the purpose of the present subsection.

In the previous subsection, we remarked that when a surface is described in absolute terms as the graph of a vector-valued function $r(u, v)$—say it consists of the points $\big(x(u,v), y(u,v), z(u,v)\big)$—the curvature can be expressed in terms of the first and second derivatives of the vector-valued function $r(u,v) = x(u,v)\,\boldsymbol{i} + y(u,v)\,\boldsymbol{j} + z(u,v)\,\boldsymbol{k}$. The difficulty we are trying to overcome is the fact that the second-order derivatives, regarded as vectors in $\mathbb{R}^3$, are not tangential to the surface. We have no way, as yet, of expressing them that is intrinsic to the surface and independent of its embedding in $\mathbb{R}^3$ and hence applicable to any parameterization of the surface. To deal with this difficulty, we separate the second-order partial derivatives of the mapping r into components tangential and normal to the surface. We write

$$\frac{\partial^2 r}{\partial u^2} = \Gamma_{11}^1 \frac{\partial r}{\partial u} + \Gamma_{11}^2 \frac{\partial r}{\partial v} + \Gamma_{11}^3 \boldsymbol{n},$$

$$\frac{\partial^2 r}{\partial u\, \partial v} = \Gamma_{12}^1 \frac{\partial r}{\partial u} + \Gamma_{12}^2 \frac{\partial r}{\partial v} + \Gamma_{12}^3 \boldsymbol{n},$$

$$\frac{\partial^2 r}{\partial v^2} = \Gamma_{22}^1 \frac{\partial r}{\partial u} + \Gamma_{22}^2 \frac{\partial r}{\partial v} + \Gamma_{22}^3 \boldsymbol{n},$$

where the vector $\boldsymbol{n}$ is the unit normal to the surface, so that its dot products with both $\frac{\partial r}{\partial u}$ and $\frac{\partial r}{\partial v}$ are zero.

The symbols Γ_{kj}^i will turn out to denote the same objects we denoted by these symbols in Chapter 4. As mentioned there, they are called the *Christoffel symbols*, after Elwin Bruno Christoffel (1829–1900). They are intimately involved with the differential geometry of vector fields on manifolds. The important Christoffel symbols are the coefficients of the two components in the tangent plane, that is, Γ_{ij}^1 and Γ_{ij}^2. They are the two components of the projection of the second-order partial derivative into the tangent plane, as illustrated in Fig. 5.4.

The subscript and superscript notation we have introduced here will gradually come to dominate all of our computations, allowing us to ignore any embedding of a manifold in a higher-dimensional ambient space and work directly with the parameters and the metric coefficients. For the moment, however, we need the vector notation of $\mathbb{R}^3$ to instill an intuition about the meaning of the algebraic formulas we are going to be deriving. Without that intuition, the basic formulas of differential geometry would simply be learned by rote, the way we all once learned the multiplication table. The route by which these formulas were originally discovered would remain a mystery, and the fact that they encode something useful in the description of nature would be an even deeper mystery. Also, without the intuitive geometric picture of their meaning they would not be of much help in suggesting physical hypotheses. Anyone can manipulate the formulas; but in order to be of any use, that manipulation must be guided by intuition.[12]

Notice that the two subscripts j and k correspond to the two parameters u and v; thus the subscript pair 11 corresponds to taking the derivative with respect to u both times, 12 and 21 correspond to the mixed derivative, and 22 corresponds to two differentiations with respect to v. Since the mixed partial derivative is the same in either order, it follows that $\Gamma^i_{21} = \Gamma^i_{12}$. The superscript $i = 1$ or $i = 2$ corresponds to the first-order partial derivative vector whose coefficient is the Christoffel symbol in question.

REMARK 5.6. Normally, on a manifold of dimension n, there are n^3 Christoffel symbols, all defined by the formula that was given in Chapter 4. On a two-dimensional surface in $\mathbb{R}^3$ the four Christoffel symbols with superscript 3 would not normally be used, and we shall occasionally rewrite them as what they are, namely:

$$\Gamma^3_{11} = \frac{\partial^2 \boldsymbol{r}}{\partial u^2} \cdot \boldsymbol{n},$$

$$\Gamma^3_{12} = \Gamma^3_{21} = \frac{\partial^2 \boldsymbol{r}}{\partial u \, \partial v} \cdot \boldsymbol{n},$$

$$\Gamma^3_{22} = \frac{\partial^2 \boldsymbol{r}}{\partial v^2} \cdot \boldsymbol{n}.$$

For the time being, we shall distinguish these four Christoffel symbols, which are defined only for surfaces in $\mathbb{R}^3$ (and perhaps only in the present book!), by calling them the *nonstandard* Christoffel symbols, although of course we wouldn't have introduced them unless we had some use for them. As a matter of fact, we shall find it easy to standardize them by introducing a third parameter. That will allow the subscripts to assume the value 3 as well, and we shall then have a 3×3 matrix of metric coefficients and a triply-indexed tableau of 27 Christoffel symbols. This seeming complication will save us some labor even in the two-variable case we are now studying, since it will allow us to write formulas in a uniform notation that

---

[12]What we are trying to do here is raise the level of abstract visualization. The goal is to harmonize intuitive but not rigorous geometric pictures with analytic expressions that contain unimpeachable rigor but lack intuitive content. All higher-level mathematics courses depend on the student's bridging this gap. Algebraists learn to *visualize* abstract groups, and analysts to *visualize* abstract topological spaces. In both cases, there is a basic substrate of concrete examples (small finite groups, or two- and three-dimensional Lie groups in the case of algebraists, ordinary Euclidean space in the case of analysts) that allows abstract definitions to seat themselves in the mind. Conjectures about abstract objects are suggested intuitively, but have to be checked against logic, since an abstract object does not have *all* of the properties found in the intuitive stock of mental pictures. In 1962, I had the privilege of learning real analysis from the late Ralph P. Boas, Jr. (1912–1992). I remember vividly his posing the question whether the closure of a ball in a metric space is compact. (The answer is no. In finite-dimensional Euclidean spaces, it is compact; but the example of any infinite set equipped with the discrete metric shows that this is not true in general.) Thus, another staple in the diet of research mathematicians is a stock of *counterexamples* to restrain our potentially misleading intuition.

otherwise would contain anomalous terms. For the time being, the nonstandard Christoffel symbols are defined purely by convention, that is, from the equations displayed above.

As for the *standard* Christoffel symbols Γ^i_{kj}, for which the superscript has the same range ($i = 1, 2$) as the two subscripts, it is proved in Appendix 6 of Volume 2 that if we set $g_{11} = E$, $g_{12} = g_{21} = F$, and $g_{22} = G$, these symbols are just the two-dimensional version of the four-dimensional symbols in Eq. (4.11), that is, when $x^1 = u$ and $x^2 = v$,

$$\Gamma^i_{jk} = \frac{1}{2} \sum_l g^{il} \left(\frac{\partial g_{jl}}{\partial x^k} + \frac{\partial g_{lk}}{\partial x^j} - \frac{\partial g_{jk}}{\partial x^l} \right).$$

In the present case, the range of summation on the index l is $l = 1$ to $l = 2$, whereas in Chapter 4 it was $l = 1$ to $l = 4$. From now on, we are going to omit the summation sign in expressions of this type by invoking the Einstein summation convention. When that is done, this formula becomes formally identical to formula (4.11).

We remind the reader that all this machinery is being hauled out in order to get an efficient algebraic expression for curvature, one that is equivalent to the formula given by Gauss but amenable to generalization and application to manifolds of any dimension. As it turns out, the most obvious expression for the curvature does not involve the standard Christoffel symbols, but instead contains the nonstandard symbols Γ^3_{ij}. In order to get the standard Christoffel symbols into the expression for curvature and the nonstandard ones out of it, we need to differentiate the standard symbols. Doing so leads us to the central concept on which everything we are going to do will depend: the *covariant Riemann curvature tensor*, defined below. The argument that leads from Gauss's definition of curvature (Definition 5.7) to the covariant Riemann curvature tensor consists of nothing more complicated than trivial manipulations of the partial derivative operator and the vector dot and cross products. It is elementary from beginning to end. Unfortunately, it is long and winding and involves several combinatorial devices and a large number of functions; it is a messy argument. Such arguments command assent, but they do not give any insight into the meaning of the concepts under discussion, and they certainly do not inspire reverence for mathematical beauty. For that reason, we are relegating this essential proof to Appendix 6 of Volume 2 and devoting the remainder of the present chapter to a statement of the result and an exploration of its consequences.

THEOREM 5.5. *For indices i, j, k, and l each equal to 1 or 2, the Christoffel symbols satisfy the relation*

(5.2) $$\Gamma^3_{ki}\Gamma^3_{lj} - \Gamma^3_{li}\Gamma^3_{kj} = g_{mi} \left(\frac{\partial \Gamma^m_{lj}}{\partial x^k} - \frac{\partial \Gamma^m_{kj}}{\partial x^l} + \left(\Gamma^m_{kp}\Gamma^p_{lj} - \Gamma^m_{lp}\Gamma^p_{kj} \right) \right),$$

where the range of summation over the indices m and p on the right is from 1 to 2.

The right-hand side of Eq. (5.2), which we shall denote R_{ijkl}, consists of the coordinates of what is called the *covariant Riemann curvature tensor*. Physicists generally denote this tensor as $R_{\lambda\mu\rho\sigma}$, using Greek letters to signify that the symbol represents a quadrilinear operator on four contravariant vectors. We shall denote this quadrilinear operator (tensor of type $(0, 4)$) as Rie. Thus,

$$\text{Rie}\,(\boldsymbol{u}, \boldsymbol{v}, \boldsymbol{w}, \boldsymbol{z}) = R_{ijkl}u^i v^j w^k z^l.$$

It is shown in Appendix 6 of Volume 2 that the curvature κ is given by the nonstandard Christoffel symbols on the left-hand side of Eq. 5.2 via the formula

$$\kappa = (g^{11}g^{22} - g^{12}g^{21})\Gamma^3_{11}\Gamma^3_{22} - \Gamma^3_{12}\Gamma^3_{21}\,.$$

We can therefore express the Gaussian curvature of a surface in $\mathbb{R}^3$ in terms of its covariant Riemann curvature tensor and metric coefficients:

COROLLARY 5.1. *The curvature κ is*

$$\kappa = \frac{g_{m1}\left(\frac{\partial\Gamma^m_{12}}{\partial x^2} - \frac{\partial\Gamma^m_{21}}{\partial x^1} + \left(\Gamma^m_{2p}\Gamma^p_{12} - \Gamma^m_{1p}\Gamma^p_{21}\right)\right)}{g_{11}g_{22} - g_{12}g_{21}}\,,$$

(5.3)

where the range of summation on the indices m and p is from 1 to 2. (Thus, only the standard metric and Christoffel symbols, intrinsic to the surface, appear in this formula.)

It appears that we have here an embarrassment of riches, since the two sides of Eq. (5.2) can be made to assume the value $(g_{11}g_{22} - g_{12}g_{21})\kappa$ by two obvious choices of i, j, k, l: either $i = k = 1$, $j = l = 2$ or $i = k = 2$, $j = l = 1$. One of our choices for expressing the curvature κ can be

$$\kappa = \frac{\operatorname{Rie}\left(\frac{\partial r}{\partial u}, \frac{\partial r}{\partial v}, \frac{\partial r}{\partial u}, \frac{\partial r}{\partial v}\right)}{g_{11}g_{22} - g_{12}g_{21}}\,.$$

We have already encountered this kind of redundancy when we used the vanishing of the Ricci tensor to determine the metric coefficients for space-time (see Eq. (4.16) in Chapter 4). In that situation, on a manifold of dimension 4, the Ricci tensor has 16 components, 12 of which vanish automatically, just as happens here in Eq. (5.2).

REMARK 5.7. Although we have now succeeded in expressing the curvature in terms of the metric coefficients, those coefficients are not absolute. They are defined within a particular system of parameters. The proof that the same number is obtained for the curvature in any other set of parameters is the main purpose of Appendix 6 in Volume 2.

To see that Eq. (5.2) really does work, examine *Mathematica* Notebook 9 in Volume 3, which uses the equation to compute some curvatures directly from the metric coefficients. We have at last uncovered the secret of generalizing curvature in the covariant Riemann curvature tensor $\operatorname{Rie}(\boldsymbol{u}, \boldsymbol{v}, \boldsymbol{w}, \boldsymbol{z})$.

REMARK 5.8. The mathematical beneficence of the universe is wonderfully revealed in the mere fact that the Riemann curvature tensor *really is* a tensor. We should not expect this, since it is assembled from the Christoffel symbols and their partial derivatives, which are *not* tensors. (See Appendix 6 of Volume 2.) The Christoffel symbols fail to be a tensor in just exactly the right way to make the Riemann tensor transform correctly.

3.4. A look ahead. Although the expression given here for the covariant Riemann curvature tensor is still just a formula containing a lot of symbols, the subscript notation that allows us to regard it as a tensor of type $(0, 4)$ provides the foundation that will enable us to modify it in the next chapter. We shall first raise one of the indices to get the standard Riemann curvature tensor, usually denoted $R^\rho_{\sigma\mu\nu}$, a tensor of type $(1,3)$ that operates on one covector $\boldsymbol{v} = u_i\, dx^i$ and three contravariant vectors $\boldsymbol{v} = v^j\,\partial/\partial x^j$, $\boldsymbol{w} = w^k\,\partial/\partial x^k$, $\boldsymbol{z} = z^l\,\partial/\partial x^l$ to produce the number $R^\rho_{\sigma\mu\nu}u_\rho v^\sigma w^\mu z^\nu$. (The notation for partial derivative operators here will be explained in the next chapter.)

By contracting this tensor on the indices ρ and μ, we get the Ricci tensor of Chapter 4. Furthermore, by holding one of its arguments fixed, we can regard it as a bilinear mapping of two contravariant vectors whose image is a covector. This flexibility of interpretation will enable us to interpret the Ricci tensor geometrically and, we hope, provide insight into the reason for its effectiveness in explaining gravity as a geometric phenomenon. All that is reserved to the next chapter.

3.5. An application of curvature. Our final result before we take up the question of curvature in higher-dimensional manifolds is a famous result known as the *theorema egregium* (*elegant theorem*) of Gauss:

THEOREM 5.6. *Let U and V be open sets on surfaces $\Sigma \subset \mathbb{R}^3$ and $T \subset \mathbb{R}^3$ respectively, and let $\varphi : U \to V$ be an isometry (distance-preserving mapping) of U onto V such that φ and its inverse φ^{-1} are both C^3 mappings. Then Σ and T have the same curvature at corresponding points.*

PROOF. Because φ preserves distances, the two surfaces have equal metric coefficients at corresponding points. (That is $g_U(x,y) = g_V\big(\varphi(x,y)\big)$.) Since the metric coefficients determine the curvature, it follows that the surfaces have the same curvature at corresponding points. □

This theorem explains why it is possible to draw an accurate flat map of a cylindrical or conical surface, that is, one that depicts both distances and areas accurately, but this is not possible for a sphere or a torus. In mechanical terms, you can wrap a piece of paper around a cylinder or cone without wrinkling it, but you cannot do that with a sphere or torus. The cylinder and cone have curvature that is identically zero, just like a flat plane, but the sphere and torus do not.

4. Problems

PROBLEM 5.1. Verify that the curvature of a circle of radius r_0 is $1/r_0$.

PROBLEM 5.2. Assume that the tangent plane to the surface $z = f(x,y)$ (where $f(0,0) = 0$) is horizontal at the point $(0,0,0)$, so that $\partial f/\partial x = 0 = \partial f/\partial y$ at this point. It was shown in the text that the curvature at the point $(0,0,0)$ is the determinant of the Hessian, that is,

$$\kappa(0,0) = \frac{\partial^2 f}{\partial x^2}\frac{\partial^2 f}{\partial y^2} - \left(\frac{\partial^2 f}{\partial x\,\partial y}\right)^2.$$

Assume now that the coordinates (x,y,z) are changed by an orthogonal matrix

$$O = \begin{pmatrix} a_{11} & a_{12} & a_{13} \\ a_{21} & a_{22} & a_{23} \\ a_{31} & a_{32} & a_{33} \end{pmatrix}$$

into the coordinates (u,v,w), and that in the new coordinates the equation of the surface is $w = h(u,v)$. That is,

$$\begin{aligned} u &= a_{11}x + a_{12}y + a_{13}f(x,y), \\ v &= a_{21}x + a_{22}y + a_{23}f(x,y), \\ w = h(u,v) &= a_{31}x + a_{32}y + a_{33}f(x,y). \end{aligned}$$

Show that in the new coordinates the curvature is given as

$$\kappa(0,0) = \frac{\frac{\partial^2 f}{\partial x^2}\frac{\partial^2 f}{\partial y^2} - \left(\frac{\partial^2 f}{\partial x\,\partial y}\right)^2}{\left(1 + \left(\frac{\partial h}{\partial u}\right)^2 + \left(\frac{\partial h}{\partial v}\right)^2\right)^2}.$$

PROBLEM 5.3. Compute the Gaussian curvature of the upper pseudo-hemisphere (see Appendix 1) using parameterization by a length $u > 0$ representing latitude and an angle v representing longitude, as follows:

$$\boldsymbol{r}(u,v) = \left(a\operatorname{sech}\left(\frac{u}{a}\right)\cos(v), a\operatorname{sech}\left(\frac{u}{a}\right)\sin(v), u - a\tanh\left(\frac{u}{a}\right)\right).$$

PROBLEM 5.4. Show how to express the curvature of a surface $z = f(x,y)$ using only the metric coefficients $E = 1 + \left(\frac{\partial f}{\partial x}\right)^2$, $F = \frac{\partial f}{\partial x}\frac{\partial f}{\partial y}$, and $G = 1 + \left(\frac{\partial f}{\partial y}\right)^2$.

PROBLEM 5.5. Show that $\operatorname{Rie}\big((1,0),(0,1),(1,0),(0,1)\big)$ is the right-hand side of Gauss's expression for the curvature, that is, Eq. (5.1).

PROBLEM 5.6. Apply Gauss's formula for the curvature to the case of a parameterization $(u,v) \mapsto \big(u,v,f(u,v)\big)$, and show that the second-order derivatives of E, F, and G that occur in formula (5.1) are such that the third-order derivatives of $\boldsymbol{r}$ (which involve the derivatives of the Christoffel symbols), cancel one another out. Specifically, show that

$$\frac{\partial^2 E}{\partial v^2} - 2\frac{\partial^2 F}{\partial u\,\partial v} + \frac{\partial^2 G}{\partial u^2} = 2\left(\frac{\partial^2 f}{\partial u\,\partial v}\right)^2 - \frac{\partial^2 f}{\partial u^2}\frac{\partial^2 f}{\partial v^2}.$$

PROBLEM 5.7. Show that $\operatorname{Rie}(\boldsymbol{u},\boldsymbol{v},\boldsymbol{w},\boldsymbol{z}) \equiv -\operatorname{Rie}(\boldsymbol{v},\boldsymbol{u},\boldsymbol{w},\boldsymbol{z})$.

CHAPTER 6

Concepts of Curvature, 1850–1950

We now take up the story of curvature where we left it in the previous chapter. The torch was passed from one mathematical giant to another in 1854, when Riemann gave his inaugural lecture at the University of Göttingen with the aged Gauss in the audience. (The latter had only one year of life remaining.) Riemann took Gauss's concept of the first fundamental form and generalized it to what he called (and what is still called) the metric on an *n-dimensional manifold*. The idea of a manifold is an abstraction and generalization of the concept of a surface. The parameterizations introduced by Euler and Gauss are applicable only locally, in a neighborhood of a point. If their domain is enlarged too far, singularities are typically encountered. For example, polar coordinates are not applicable at the origin in the plane, since the polar angle θ is not defined at that point. Furthermore, if every point in the plane except the origin is to have unique coordinates, it is necessary to restrict θ to some half-open interval, usually $[-\pi, \pi)$, leaving a "barricade" at one end. There is no way of using these parameters in such a way that (1) every point except the origin has unique coordinates and (2) each coordinate can vary in some interval about its value at any given point. Using parameterizations, one can normally study only a local piece of a surface, while the surface itself is thought of as a single object, all of whose points are equally dignified and deserving of attention.

As originally introduced, manifolds had this same limitation, and the same was true in the theory of Lie groups (a way of extending the Galois theory of algebraic equations to differential equations). Originally, a Lie group was not actually a group, but only a piece of a group near its identity element. Nevertheless, the power of analytic symbolism (parameterization) to express deep geometric intuition had profound consequences, even at the earliest stage. In the present chapter, we work mostly in such an environment, using whatever local parameterization is convenient for a specific purpose. These methods sufficed for Einstein's work on general relativity.

It was only in the mid-twentieth century that the modern concepts of a *global* differentiable manifold and manifold-with-boundary were developed to remove the limitations on parameterizations mentioned above. As a consequence, a manifold once again became a unified object, and Lie groups became actual groups. The systematization and improved perspective provided by the concept of a manifold are well worth knowing, even though we can get along without it as far as the application to relativity is concerned. To avoid cluttering up the exposition with all this geometric machinery, we have relegated the modern theory of manifolds to Appendix 4, which shows the advantages and the cost of getting a theory that allows a manifold to be regarded as a single object, like a surface in $\mathbb{R}^3$. Even in Appendix 4, our discussion of this material is minimal, and a more systematic treatment can be found in, for example, the book by Sternberg [**78**].

1. Second-Order Derivations

Comparison of the formula (5.1) given by Gauss for curvature in terms of the metric coefficients with formula (5.3), which expresses it in terms of the Riemann curvature tensor shows the importance of the ingredient that Gauss was lacking: the Christoffel symbols. We used these symbols in an *ad hoc* way in the previous chapter, but their application in Chapter 4 shows how much greater still their potential is. The present chapter is devoted to developing that potential, extending what we have done on surfaces in $\mathbb{R}^3$ to a general abstract manifold. We begin by describing a way of replacing tangent vectors in $\mathbb{R}^3$ with a corresponding object that can be expressed in terms of parameters without the need for the ambient space $\mathbb{R}^3$.

A vector field on a manifold (see Appendix 4) is intuitively interpreted as a choice of a "directional derivative" at each point. More formally, a vector field $\boldsymbol{u}$ defined in a neighborhood of a point P of a manifold is a *derivation*, which is a linear mapping of the space of local C^∞ functions into itself satisfying the multiplicative condition $\boldsymbol{u}(fg) = \boldsymbol{u}(f)\,g + f\,\boldsymbol{u}(g)$. It is proved in Appendix 4 that such a functional can be expressed in terms of a local coordinate system near P as a linear combination of partial derivatives in which the coefficients are C^∞ functions. A covector field $\boldsymbol{v}$ is a linear mapping of these vector fields into the space of C^∞ functions. The reader is assumed to be familiar with the duality of vector and covector spaces in linear algebra, and the analogous fact holds here. Although a vector field is *defined* as operating on the space of locally C^∞ functions, it can equally well be regarded as acting on the space of covector fields, since the mapping $\boldsymbol{v}(\boldsymbol{u})$ can be "turned around" and regarded as a mapping $\boldsymbol{u}(\boldsymbol{v})$. What we really have is a bilinear function $\langle \boldsymbol{v}, \boldsymbol{u} \rangle = \boldsymbol{v}(\boldsymbol{u}) = \boldsymbol{u}(\boldsymbol{v})$.

Any smooth surface in $\mathbb{R}^3$ has—theoretically—a tangent plane, but to describe it without algebra is extremely difficult. Only for a few very simple surfaces were the ancient Greeks able to characterize the tangent plane to a surface in a comprehensible way. And without such a description, what does it mean to say that the surface even *has* a tangent plane? We may feel intuitively that it ought to, but to express what it is in the language of Greek geometry is a fool's errand. Fortunately, analytic geometry solves this problem. We can retain the absolute character of the surface and still say precisely what its tangent plane is by defining the surface as the points (x, y, z) in a universal Euclidean space $\mathbb{R}^3$ that satisfy an equation $F(x, y, z) = 0$. In that case, we can characterize the tangent plane at a point (x_0, y_0, z_0) as the set of points (x, y, z) such that the vector $(x - x_0)\boldsymbol{i} + (y - y_0)\boldsymbol{j} + (z - z_0)\boldsymbol{k}$ is perpendicular to the gradient $\nabla F(x_0, y_0, z_0)$. This characterization amounts to the vector equation

$$\nabla F(x_0, y_0, z_0) \cdot \big((x - x_0)\boldsymbol{i} + (y - y_0)\boldsymbol{j} + (z - z_0)\boldsymbol{k}\big) = 0\,,$$

that is,

$$(x - x_0)\frac{\partial F}{\partial x} + (y - y_0)\frac{\partial F}{\partial y} + (z - z_0)\frac{\partial F}{\partial z} = 0\,,$$

where the partial derivatives are evaluated at (x_0, y_0, z_0).

When the surface is parameterized by two real variables s and t as a vector-valued function $\boldsymbol{r}(s, t) = \big(x(s, t), y(s, t), z(s, t)\big)$ defined on an open set U in $\mathbb{R}^2$ containing (x_0, y_0), the tangent plane coincides with the set of all points $\boldsymbol{u}$ in $\mathbb{R}^3$

of the form

$$u = x_0 i + y_0 j + z_0 k + u^1 \frac{\partial r}{\partial s} + u^2 \frac{\partial r}{\partial t},$$

where u^1 and u^2 both range over the set of real numbers and again the partial derivatives are evaluated at (x_0, y_0). In this way, an observer using the coordinates s and t and the parameterization r has an interpretation of the tangent plane and the notion of a tangent vector expressed in absolute terms.

A field of tangent vectors is a vector-valued function

$$u(s, t) = u^1(s, t) \frac{\partial r}{\partial s}(s, t) + u^2(s, t) \frac{\partial r}{\partial t}(s, t).$$

The tangent vector $u(s, t)$ here is identified with the point $u(s, t) + r(s, t)$ in $\mathbb{R}^3$, which is a point in the tangent plane to the surface at the point $r(s, t)$. We have included the arguments (s, t) to emphasize that a vector field is a collection of tangent vectors. If we wish to free ourselves from the ambient space and talk about the surface as it intrinsically is, independent of any such embedding, we must get rid of the vector-valued function r in this expression. One way to do this is to associate with the vector field u whose expression in (s, t)-coordinates is the one just exhibited the derivation $u_d : C^\infty(U) \to C^\infty(U)$ whose expression in (s, t)-coordinates is

$$u_d(f) = u^1 \frac{\partial f}{\partial s} + u^2 \frac{\partial f}{\partial t}$$

for all smooth functions f. It is shown in Appendix 4 that derivations are completely characterized by the properties of being linear functions of the argument f having the multiplicative property $u_d(fg) = u_d(f) g + f u_d(g)$.

Although the identification $u \leftrightarrow u_d$ seems formally an obvious one to make, it requires some explanation. If we assume that $f(x, y, z)$ is a C^∞-function defined on an open set containing the surface we are dealing with, it coincides on the surface itself with the function $\tilde{f}(s, t) = f(r(s, t)) = f(x(s, t), y(s, t), z(s, t))$, and we have the equation

$$u_d(\tilde{f}) = u \cdot \nabla f.$$

Notice that the abstract vector (derivation) operator $u_d = u^1 \partial/\partial s + u^2 \partial/\partial t$ is obtained by simply erasing the symbol r. From now on, we shall work mostly with the derivation u_d, omitting the subscript d. We shall keep the identification of the derivation $\partial/\partial s$ with the derivative $\partial r/\partial s$ for use whenever we wish to make an argument geometrically intuitive, but most of our discussion will be carried out in the language of algebra and will be simpler in the former notation, without the specific parameterization r in view. When we discuss the Lie bracket $[u, v]$ and the covariant derivative $\nabla_u u$ below, it will be essential to think of a vector v as a directional derivative operator $v^i \partial/\partial x^i$ rather than an "arrow" $v^i \partial r/\partial x^i$. The symbol r here only gets in the way of the computation.

1.1. Repeated derivations and the Lie bracket. Second-order partial derivatives are essential for defining curvature, and since vector fields are the version of first-order derivatives used as a tool in differential geometry, we need to see the result of repeated application of them. If u is a vector field, and f a C^∞ function, then $u(f)$ is a C^∞ function, and hence a vector field v can be applied to it. Unfortunately, if we do that, we lose the all-important property of tangency, since the result is not a vector field. It fails the multiplicative test, as we easily see:

$$
\begin{aligned}
\boldsymbol{v}\big(\boldsymbol{u}(fg)\big) &= \boldsymbol{v}\big(g\,\boldsymbol{u}(f)+f\,\boldsymbol{u}(g)\big) \\
&= \boldsymbol{v}\big(g\,\boldsymbol{u}(f)\big)+\boldsymbol{v}\big(f\,\boldsymbol{u}(g)\big) \\
&= g\,\boldsymbol{v}\big(\boldsymbol{u}(f)\big)+\boldsymbol{u}(f)\,\boldsymbol{v}(g)+\boldsymbol{v}(f)\,\boldsymbol{u}(g)+f\,\boldsymbol{v}\big(\boldsymbol{u}(g)\big)\,.
\end{aligned}
$$

If $\boldsymbol{vu}$ were a vector field, we would have had the relation

$$
\boldsymbol{v}\big(\boldsymbol{u}(fg)\big)=g\,\boldsymbol{v}\big(\boldsymbol{u}(f)\big)+f\,\boldsymbol{v}\big(\boldsymbol{u}(g)\big)\,.
$$

Instead, we have this expression augmented by two extra terms:

$$
\boldsymbol{u}(f)\,\boldsymbol{v}(g)+\boldsymbol{v}(f)\,\boldsymbol{u}(g)\,.
$$

In a coordinate system in which $\boldsymbol{u}$ has the representation $u^1\,\partial/\partial s+u^2\,\partial/\partial t$ and $\boldsymbol{v}$ the representation $v^1\,\partial/\partial s+v^2\,\partial/\partial t$, the fully expanded expression for $\boldsymbol{v}\big(\boldsymbol{u}(fg)\big)$ consists of 24 terms. At the price of a slight inconsistency—regarding f and g as functions of s and t rather than functions of $\boldsymbol{r}(s,t)$, which they actually are[1]—we can group these terms as follows:

$$
\begin{aligned}
(6.1)\quad \boldsymbol{v}\big(\boldsymbol{u}(fg)\big) &= \left(v^1\frac{\partial}{\partial s}+v^2\frac{\partial}{\partial t}\right)\left(u^1\frac{\partial(fg)}{\partial s}+u^2\frac{\partial(fg)}{\partial t}\right) \\
&= \left(v^1\frac{\partial u^1}{\partial s}+v^2\frac{\partial u^1}{\partial t}\right)\left(f\frac{\partial g}{\partial s}+g\frac{\partial f}{\partial s}\right)+\left(v^1\frac{\partial u^2}{\partial s}+v^2\frac{\partial u^2}{\partial t}\right)\left(f\frac{\partial g}{\partial t}+g\frac{\partial f}{\partial t}\right) \\
&+ \left(v^1\frac{\partial g}{\partial s}+v^2\frac{\partial g}{\partial t}\right)\left(u^1\frac{\partial f}{\partial s}+u^2\frac{\partial f}{\partial t}\right)+\left(v^1\frac{\partial f}{\partial s}+v^2\frac{\partial f}{\partial t}\right)\left(u^1\frac{\partial g}{\partial s}+u^2\frac{\partial g}{\partial t}\right) \\
&+ v^1 u^1\left(g\frac{\partial^2 f}{\partial s^2}+f\frac{\partial^2 g}{\partial s^2}\right)+v^1 u^2\left(g\frac{\partial^2 f}{\partial s\,\partial t}+f\frac{\partial^2 g}{\partial s\,\partial t}\right) \\
&+ v^2 u^1\left(g\frac{\partial^2 f}{\partial t\,\partial s}+f\frac{\partial^2 g}{\partial t\,\partial s}\right)+v^2 u^2\left(g\frac{\partial^2 f}{\partial t^2}+f\frac{\partial^2 g}{\partial t^2}\right)\,.
\end{aligned}
$$

This is not the first indication we have had that matters rapidly get out of hand if we try to compute too much directly from definitions. We definitely need to streamline expressions of this type. Somehow, we must annihilate the extra terms here, that is, get rid of the nontangential parts of the iterated derivations. One seemingly procrustean way of adapting our tools to the job at hand is to note that the "unwanted" terms in the expression for $\boldsymbol{v}\big(\boldsymbol{u}(f)\big)$ (which is to say, the last three lines of Eq. (6.1), which contain second-order derivatives and products of first-order derivatives) occur symmetrically and can be erased by considering the expression known as the *Lie bracket*.[2]

DEFINITION 6.1. The *Lie bracket* of two tangent vectors $\boldsymbol{u}$ and $\boldsymbol{v}$ is the tangent vector $[\boldsymbol{v},\boldsymbol{u}]$ whose action on a C^∞ function f is given by

$$
[\boldsymbol{v},\boldsymbol{u}]f=\boldsymbol{v}(\boldsymbol{u}f)-\boldsymbol{u}(\boldsymbol{v}f)\,.
$$

From our previous expression for $\boldsymbol{vu}(fg)$, we see that the two extra terms $\boldsymbol{u}(f)\,\boldsymbol{v}(g)+\boldsymbol{v}(f)\,\boldsymbol{u}(g)$ cancel out of the expression for $[\boldsymbol{u},\boldsymbol{v}](fg)$.

---

[1]See the Feller principle, discussed in Appendix 6. We judge that nothing would be gained at this point by a pedantic insistence on using, say, the symbols $\tilde{f}$ and $\tilde{g}$ for the representation of these functions in terms of parameters, as we did above in order to clarify the meaning of the symbol $\boldsymbol{u}_d$.

[2]Named after the Norwegian mathematician Sophus Lie (1842–1899).

THEOREM 6.1. *The Lie bracket satisfies the* Jacobi identity[3]

$$\big[u,[v,w]\big] + \big[v,[w,u]\big] + \big[w,[u,v]\big] = 0.$$

The proof of this theorem is a routine computation. The Jacobi identity replaces the associative law, which does not hold in so-called *Lie algebras*. (Three-dimensional Euclidean space $\mathbb{R}^3$ is a Lie algebra if the cross product is used as multiplication, that is, if $[u,v]$ is defined as $u \times v$.)

THEOREM 6.2. *The Lie bracket of two derivations is a derivation, and hence is a contravariant vector field. This means that*

$$[v,u](fg) = g\,[v,u](f) + f\,[v,u](g).$$

When this expression is written out in terms of partial derivatives in a parameterization, as was done above for $v\big(u(fg)\big)$, it yields the explicit formula

$$[v,u]f = \left(v^1\frac{\partial u^1}{\partial s} + v^2\frac{\partial u^1}{\partial t} - u^1\frac{\partial v^1}{\partial s} - u^2\frac{\partial v^1}{\partial t}\right)\frac{\partial f}{\partial s}$$
$$+ \left(v^1\frac{\partial u^2}{\partial s} + v^2\frac{\partial u^2}{\partial t} - u^1\frac{\partial v^2}{\partial s} - u^2\frac{\partial v^2}{\partial t}\right)\frac{\partial f}{\partial t}.$$

The fact that the same number would be obtained for $[v,u]f$ in any other coordinate system—in other words, the tensor nature of the Lie bracket—is established in Appendix 6. Thus, this number depends only on the tangent vectors u and v, not on their coordinates in any particular set of parameters. It is, as mathematicians say, "well defined."

REMARK 6.1. For surfaces in $\mathbb{R}^3$, the Lie bracket is the zero vector if and only if

$$v^1\frac{\partial u^1}{\partial s} + v^2\frac{\partial u^1}{\partial t} = u^1\frac{\partial v^1}{\partial s} + u^2\frac{\partial v^1}{\partial t} \text{ and } v^1\frac{\partial u^2}{\partial s} + v^2\frac{\partial u^2}{\partial t} = u^1\frac{\partial v^2}{\partial s} + u^2\frac{\partial v^2}{\partial t}.$$

This condition is not trivial, but there is one obvious case in which it does hold, namely when the coefficients u^1, u^2, v^1, and v^2 are all constants. Then both sides of both equations vanish here. Of course, that condition merely expresses the fact that the derivative of a constant is zero.

1.2. The covariant derivative of a vector field. The Lie bracket is one of the two tools we need to discuss the abstract Riemann curvature tensor. The other, which we introduced in the previous chapter without giving it a name, is the covariant derivative of one vector field with respect to another. It provides the clue to a geometric interpretation of the Christoffel symbols. Recalling our geometric definition of these symbols in the previous chapter, we can see a second way to get a (tangential) vector field out of the second derivatives: Project them orthogonally into the tangent plane. That is, simply *erase* the last term in the equations

$$\frac{\partial^2 r}{\partial x^i\,\partial x^j} = \Gamma^1_{ij}\frac{\partial r}{\partial x^1} + \Gamma^2_{ij}\frac{\partial r}{\partial x^2} + \Gamma^3_{ij}n,$$

replacing the second-order partial derivative on the left with what we shall call the *covariant derivative* of the vector field $\partial/\partial x^i$ with respect to the vector field $\partial/\partial x^j$ and write as the abstract relation

$$\nabla_{\frac{\partial}{\partial x^j}}\frac{\partial}{\partial x^i} = \Gamma^1_{ij}\frac{\partial}{\partial x^1} + \Gamma^2_{ij}\frac{\partial}{\partial x^2}.$$

----
[3]Named after Carl Jacobi (1804–1851).

At first sight, that device appears to be just as procrustean an adaptation as the Lie bracket.[4] It retains only the coefficients Γ^1_{ij} and Γ^2_{ij} and annihilates the normal components $\Gamma^3_{ij}\boldsymbol{n}$. That is why the whole subject is as complicated as it is: These annihilated Christoffel symbols were exactly what was needed to get the curvature. It might appear that we have lost the game we were playing. But the tangential coefficients Γ^1_{ij} and Γ^2_{ij} in the two equations interact with one other and with the normal components in such a way that their *derivatives* can regenerate the information that was contained in the normal coefficients, and in that regeneration we found precisely the expression for curvature that we were seeking, namely the covariant Riemann curvature tensor Rie_{ijkl}. In general, a function of any type can be recovered, up to a constant, from its set of partial derivatives. Since the partial derivatives of the unit normal vector are tangential, they can be expressed in terms of any basis of the tangent space. That is the mechanism underlying the success of our project.

The directional derivative at P of a vector-valued function of parameters s and t, that is, a vector field $\boldsymbol{u}(s,t) = u^1(s,t)\frac{\partial \boldsymbol{r}}{\partial s} + u^2(s,t)\frac{\partial \boldsymbol{r}}{\partial t}$ in the direction of a vector $\boldsymbol{v} = v^1(s,t)\frac{\partial \boldsymbol{r}}{\partial s} + v^2(s,t)\frac{\partial \boldsymbol{r}}{\partial t}$ in the tangent space at $P = \boldsymbol{r}(s_0,t_0)$, is not generally tangent to the surface, and only its tangential component has intrinsic meaning in terms of the geometry of the surface. We now formally define what that tangential component is.

DEFINITION 6.2. The *covariant derivative of the tangent vector field* $\boldsymbol{u}$ *with respect to a tangent vector field* $\boldsymbol{v}$, denoted $\nabla_{\boldsymbol{v}}\boldsymbol{u}$, is the orthogonal projection on the tangent plane of the directional derivative of $\boldsymbol{u}$ in the direction of $\boldsymbol{v}$. (This directional derivative is *not* a derivation.) The covariant derivative is given in parameters (s,t) as follows:

$$\nabla_{\boldsymbol{v}}\boldsymbol{u} = \left(v^1\frac{\partial u^1}{\partial s} + v^2\frac{\partial u^1}{\partial t} + u^1 v^1\,\Gamma^1_{11} + (u^1 v^2 + u^2 v^1)\,\Gamma^1_{12} + u^2 v^2\,\Gamma^1_{22}\right)\frac{\partial \boldsymbol{r}}{\partial s}$$
$$+ \left(v^1\frac{\partial u^2}{\partial s} + v^2\frac{\partial u^2}{\partial t} + u^1 v^1\,\Gamma^2_{11} + (u^1 v^2 + u^2 v^1)\,\Gamma^2_{12} + u^2 v^2\,\Gamma^2_{22}\right)\frac{\partial \boldsymbol{r}}{\partial t}\,.$$

This covariant derivative *is* a derivation, being given as an explicit linear combination of tangent vectors. As an example, when $\boldsymbol{u} = \partial \boldsymbol{r}/\partial s$ and $\boldsymbol{v} = \partial \boldsymbol{r}/\partial t$, this definition becomes

$$\nabla_{\frac{\partial \boldsymbol{r}}{\partial t}}\frac{\partial \boldsymbol{r}}{\partial s} = \Gamma^1_{12}\frac{\partial \boldsymbol{r}}{\partial s} + \Gamma^2_{12}\frac{\partial \boldsymbol{r}}{\partial t}\,.$$

Since this definition is stated in terms of a particular set of parameters, it needs to be verified that the same tangent vector results in every parameterization. This work is carried out in Appendix 6. Since it is the orthogonal projection of a derivative, the covariant derivative is obviously linear as a function of $\boldsymbol{u}$ when $\boldsymbol{v}$ is held fixed. It is less obvious, but true, that it is linear as a function of $\boldsymbol{v}$ when $\boldsymbol{u}$ is held fixed. (The dependence on $\boldsymbol{v}$ is determined by the ways in which v^1 and v^2 occur in the formula, and the expression is linear in those variables when u^1 and u^2 are held fixed.)

We can also define the covariant derivative of a scalar-valued function f, that is, when $\boldsymbol{u}$ is replaced by f. By convention, it is just the usual directional derivative

---

[4]But, after all, the tangent space is all that we *can* define intrinsically. What else could we do?

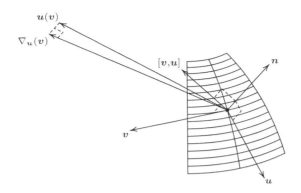

FIGURE 6.1. Small section of the sphere $x^2 + y^2 + z^2 = 1$ near the point $(1/2, 1/2, 1/\sqrt{2})$, showing two tangent vectors $\boldsymbol{u}$ and $\boldsymbol{v}$ and their Lie bracket $[\boldsymbol{v}, \boldsymbol{u}]$ all of which lie in the plane perpendicular to the normal vector $\boldsymbol{n}$ at that point. The vector $\boldsymbol{u}(\boldsymbol{v})$ does not lie in this plane, but its perpendicular projection into the plane is the covariant derivative $\nabla_{\boldsymbol{u}}(\boldsymbol{v})$. (All the vectors are drawn to scale except $\boldsymbol{u}$ and $[\boldsymbol{v}, \boldsymbol{u}]$, which are shown one-third of their actual size.)

of f in the direction of $\boldsymbol{v}$, which is to say, the usual result of applying the derivation $\boldsymbol{v} = v^1 \partial/\partial s + v^2 \partial/\partial t$ to f:

$$\nabla_{\boldsymbol{v}} f = v^1 \frac{\partial f}{\partial s} + v^2 \frac{\partial f}{\partial t} = \boldsymbol{v}(f).$$

In particular,

$$\nabla_{\frac{\partial}{\partial s}} f = \frac{\partial f}{\partial s} \text{ and } \nabla_{\frac{\partial}{\partial t}} f = \frac{\partial f}{\partial t}.$$

The covariant derivative has the usual property of a derivation when applied to the product of a scalar-valued function and a vector-valued function (Problem 6.18):

$$\nabla_{\boldsymbol{v}} (f \boldsymbol{u}) = (\nabla_{\boldsymbol{v}} f) \boldsymbol{u} + f (\nabla_{\boldsymbol{v}} \boldsymbol{u}).$$

As a bonus from the covariant derivative, we obtain another way of writing the Lie bracket, namely

$$[\boldsymbol{u}, \boldsymbol{v}] = \left(u^l \frac{\partial v^i}{\partial x^l} - v^l \frac{\partial u^i}{\partial x^l} \right) \frac{\partial}{\partial x^i} = \nabla_{\boldsymbol{u}} \boldsymbol{v} - \nabla_{\boldsymbol{v}} \boldsymbol{u},$$

and the interesting formula

$$\nabla_{[\boldsymbol{u}, \boldsymbol{v}]} = \nabla_{\nabla_{\boldsymbol{u}} \boldsymbol{v} - \nabla_{\boldsymbol{v}} \boldsymbol{u}} = \nabla_{\nabla_{\boldsymbol{u}} \boldsymbol{v}} - \nabla_{\nabla_{\boldsymbol{v}} \boldsymbol{u}}.$$

The proofs of these formulas are easy exercises and are left to the reader (Problem 6.6 below).

The concepts of tangent vectors, Lie bracket, and covariant derivative are illustrated in the case of the unit sphere in Fig. 1.2, which exhibits the vector fields $\boldsymbol{u} = (1/(4y))\partial/\partial x + (1/(4x^2))\partial/\partial y$ and $\boldsymbol{v} = x\partial/\partial x - y\partial/\partial y$ at the point $(1/2, 1/2, 1/\sqrt{2})$, along with the second-order field $\boldsymbol{u}(\boldsymbol{v})$, the Lie bracket $[\boldsymbol{v}, \boldsymbol{u}]$, and the covariant derivative $\nabla_{\boldsymbol{u}}(\boldsymbol{v})$. These are all coplanar (tangent) vectors, except $\boldsymbol{u}(\boldsymbol{v})$, whose projection into the tangent plane is the covariant derivative. The field $\boldsymbol{v}(\boldsymbol{u})$, not shown, also lies outside the tangent plane.

2. Curvature, Phase 3: Riemann

Although we produced the metric coefficients E, F, and G from a parameterization $(s,t) \mapsto \boldsymbol{r}(s,t)$ of a two-dimensional surface in $\mathbb{R}^3$, nothing prevents us from starting with four arbitrary functions of two variables s and t, say E, F_1, F_2, and G, which need not be regarded as the dot products of the partial derivatives of any vector-valued function, and considering an element of arc length dL given in terms of two parameters as

$$dL^2 = E\,ds^2 + F_1\,ds\,dt + F_2\,dt\,ds + G\,dt^2\,.$$

Of course, being the *square* of an infinitesimal distance, the expression on the right-hand side here needs to be nonnegative for all values of the infinitesimal increments ds and dt. Since we are not distinguishing between $ds\,dt$ and $dt\,ds$, we may as well assume that $F_1 = F_2 = F$. With the crucial exception of the metrics on space-time, we are usually going to assume that the matrix of metric coefficients is positive-definite. That means that $E > 0$ and $EG - F^2 > 0$. These inequalities are automatic when E, F, and G are given as the dot products of the partial derivatives of a nonsingular parameterization $\boldsymbol{r}$.

When we start from metric coefficients in this way, the surface in $\mathbb{R}^3$ that we were talking about is no longer needed. We can produce the Christoffel symbols and the curvature of a hypothetical two-dimensional manifold from the metric coefficients alone. We know how to convert the metric coefficients from one coordinate system to another; and so, as far as we are concerned, the surface might as well consist of just a collection of open sets in $\mathbb{R}^2$, with change-of-coordinate mappings between them that determine the conversion of the metric coefficients. Whether there actually is a subset of $\mathbb{R}^3$ (or indeed $\mathbb{R}^n$) from which these metric coefficients could be defined in absolute terms using the metric of the ambient space, as we have done up to now, is an interesting question, but one that we don't need to answer in order to discuss manifolds in general. (See the Whitney embedding theorem, a special case of which is proved in Appendix 4.) We are now in the realm of abstract manifolds.

The generalization to an n-dimensional manifold is immediate. We take a set of parameters $x^1, \ldots, x^n$—these are superscripted variables, not powers of a single variable x—and an $n \times n$ matrix of functions (g_{ij}) and write a squared element of arc length ds as

$$ds^2 = g_{ij}\,dx^i\,dx^j\,.$$

The essential ideas in this abstraction from the work of Gauss were introduced by Riemann. In his inaugural lecture[5] at the University of Göttingen in 1854, with Gauss in the audience, Riemann introduced what he called an *n-fold extended quantity*—what we now call an n-dimensional Riemannian manifold—and argued

---

[5]This lecture was published as a 14-page paper containing almost no mathematical symbols, under the title "Über die Hypothesen die der Geometrie zu Grunde liegen" (On the assumptions that underlie geometry). Later, in 1861, he developed this subject more mathematically in a paper on heat conduction submitted in a competition established by the Paris Academy of Sciences. The paper was entitled "Commentatio mathematica, qua respondere tentatur quaestioni ab Illma Academia Parisiensi propositae" (A mathematical note attempting to answer a question posed by the distinguished Paris Academy), but was not published until after his death. In the course of that paper, he noted that an expression of the type shown above could be regarded as the squared length of an infinitesimally short curve. Incidentally, in the same paper, he was led to study differential equations of third order, which had been rare in physics up to the time.

that the simplest definition of arc length ds on such an object would be

$$ds = \sqrt{a_{ij}(x^1,\dots,x^n)\,dx^i\,dx^j}\,,$$

where, he said, the expression under the square root must be nonnegative. Such a metric is now called a *Riemannian* metric. In the case of the Euclidean space $\mathbb{R}^n$, we have $a_{ii}=1$ and $a_{ij}=0$ if $i\neq j$. Riemann called a manifold *flat* if its metric coefficients were constant. The curvature at a point is a measure of the local departure from flatness.

Riemann went on to say that the next simplest case would be one in which the Maclaurin series for ds^2 as a function of the variables dx^i would begin with fourth-degree terms. Since this was a general lecture attended by nonmathematicians as well as mathematicians, Riemann kept the technical details to a minimum and did not discuss this case. As he said, "The investigation of this more general type, to be sure, would not require any essentially different principles, but it would be rather time-consuming, and cast relatively little new light on the theory of space; and moreover the results could not be expressed geometrically."

2.1. A reformulation of the Riemann curvature tensor. The expression for the covariant Riemann curvature tensor Rie given in the previous chapter as the right-hand side of Eq. (5.3) looks rather asymmetrical, consisting mostly of the Christoffel symbols and their partial derivatives, but with a small "impurity" in the form of the metric coefficients g_{im}. If the metric coefficients and the summation over m are omitted, we get a tensor called simply the *Riemann curvature tensor*, which operates on three vectors, say v, w, and z, and one covector, say v. To tell them apart, since we wrote the coordinates of the covariant Riemann curvature tensor as R_{ijkl}, we write those of the more basic Riemann curvature tensor as R^i_{jkl}. Thus, we have

(6.2)
$$R^i_{jkl}=\frac{\partial\Gamma^i_{lj}}{\partial x^k}-\frac{\partial\Gamma^i_{kj}}{\partial x^l}+\Gamma^i_{kp}\Gamma^p_{lj}-\Gamma^i_{lp}\Gamma^p_{kj}$$

whereas, by Eq. (5.3),

(6.3)
$$R_{ijkl}=g_{im}R^m_{jkl}=g_{mi}\left(\frac{\partial\Gamma^m_{lj}}{\partial x^k}-\frac{\partial\Gamma^m_{kj}}{\partial x^l}+\Gamma^m_{kp}\Gamma^p_{lj}-\Gamma^m_{lp}\Gamma^p_{kj}\right).$$

As explained in Appendix 6, multiplying R^m_{jkl} by g_{im} and summing over m is known as *lowering* the index m, to produce the purely covariant tensor whose components are R_{ijkl}. We thus get a vector $R=R(v,w,z)$ depending on three other vectors v, w, and z that operates on a covector v to produce the number

$$R(v,w,z)(v)=R^i_{jkl}v^jw^kz^lv_i\,.$$

To picture this all embedded in a Euclidean space, we can think of the Riemann curvature tensor as the mapping $(v,w,z)\mapsto R(v,w,z)$, where

$$R(v,w,z)=R^i_{jkl}v^jw^kz^l\frac{\partial r}{\partial x^i}\,.$$

In other words, the operator $R(\cdot,\cdot,\cdot)$ maps $\mathbb{R}^n\times\mathbb{R}^n\times\mathbb{R}^n$ into $\mathbb{R}^n$ in accordance with the formula just written. On a general manifold, a linear operator Λ that acts on covectors v is associated by duality with a vector, as follows: $\Lambda\leftrightarrows u_\Lambda$, where

$$u_\Lambda=\Lambda(dx^i)\frac{\partial}{\partial x^i}\,,$$

and so, instead of thinking of the Riemann curvature tensor as a quadrilinear mapping of three vectors and one covector into the real numbers, it makes much better sense to think of it as a trilinear mapping of three vectors into a fourth vector.

REMARK 6.2. Here once again we are defining an absolute object (a tangent vector) in terms of its expression in a particular set of parameters. The absoluteness of this vector depends on its yielding the same numbers as values when it is applied as a derivation to a given function in any parameterization. The proof that this one has that property is given below.

Equation (6.2) looks cleaner than Eq. (6.3), consisting entirely of Christoffel symbols and their partial derivatives. Just as our ability to compute curvature directly from the metric coefficients allows us to take those metric coefficients as starting points and forget about the way they were constructed from an embedding in Euclidean space, the Riemann curvature tensor whose coefficients are R^i_{jkl} suggests that we could start even farther in, taking the Christoffel symbols as our point of departure and not worrying about how they are computed from some hypothetical metric coefficients. It is possible to do this since the equations that define the Christoffel symbols in terms of the partial derivatives of the metric coefficients can be solved for the partial derivatives of the metric coefficients in terms of the Christoffel symbols.[6] The metric coefficients satisfy a homogeneous linear system of partial differential equations whose coefficients are Christoffel symbols (see Problems 4.12 and 4.13). We are not going to explore that possibility, however, since in the application we have in mind—relativistic gravitation—our approach in Chapter 4 was based on assuming a certain form for the metric coefficients, which in turn imposed a certain form on the Christoffel symbols. We proceeded by first positing that the vanishing of the Ricci tensor was the law of gravitation, then *computing* what the metric coefficients had to be, knowing the coefficients of that tensor, and finally computing the geodesics in the resulting metric to get a description of the motion of a particle.

REMARK 6.3. In his systematic exposition of general relativity in the fourth series of the *Annalen der Physik*, Einstein gave an exposition of these results, along with some more general information about tensors. Using another standard notation for the Christoffel symbols, namely

$$\Gamma^i_{jk} = \left\{ \begin{matrix} jk \\ i \end{matrix} \right\},$$

he referred to the tensor whose components are R^i_{jkl} as the *Riemann–Christoffel* tensor and wrote Eq. (6.2) as

$$B^\rho_{\mu\sigma\tau} = -\frac{\partial}{\partial x_\tau}\left\{\begin{matrix}\mu\sigma\\\rho\end{matrix}\right\} + \frac{\partial}{\partial x_\sigma}\left\{\begin{matrix}\mu\tau\\\rho\end{matrix}\right\} - \left\{\begin{matrix}\mu\sigma\\\alpha\end{matrix}\right\}\left\{\begin{matrix}\alpha\tau\\\rho\end{matrix}\right\} + \left\{\begin{matrix}\mu\tau\\\alpha\end{matrix}\right\}\left\{\begin{matrix}\alpha\sigma\\\rho\end{matrix}\right\}.$$

(See Eq. (43) on page 800 of [**21**].) Eddington's notation was similar, except that he wrote $\Gamma^k_{ij} = \{ij, k\}$. He also wrote ([**16**], p. 72) the covariant Riemann curvature tensor that we have denoted R_{ijkl} as B_{jkli}.)

---

[6] On a manifold of dimension n, there are n^3 Christoffel symbols defined as linear functions of the n^3 partial derivatives $\partial g_{ij}/\partial x^k$, with coefficients that are rational functions of the g_{ij}.

The geometric significance of lowering an index can be made clearer by introducing the natural inner product on the tangent space to a manifold:[7]

$$\langle \boldsymbol{u}, \boldsymbol{v} \rangle = g_{im} u^i v^m = g_{im} u^i v^m .$$

If we think of the Riemann curvature tensor as described above, that is, as the operator that maps the ordered triple of vectors $\boldsymbol{v}$, $\boldsymbol{w}$, $\boldsymbol{z}$ to the vector

$$\boldsymbol{R} = \boldsymbol{R}(\boldsymbol{v}, \boldsymbol{w}, \boldsymbol{z}) = R^i_{jkl} v^j w^k z^l \frac{\partial}{\partial x^i} ,$$

we have the following expression for the action of the covariant Riemann curvature tensor on the ordered quadruple $(\boldsymbol{u}, \boldsymbol{v}, \boldsymbol{w}, \boldsymbol{z})$:

$$\mathrm{Rie}(\boldsymbol{u}, \boldsymbol{v}, \boldsymbol{w}, \boldsymbol{z}) = \langle \boldsymbol{u}, \boldsymbol{R}(\boldsymbol{v}, \boldsymbol{w}, \boldsymbol{z}) \rangle .$$

It is also convenient to think of the vector $\boldsymbol{R}(\boldsymbol{v}, \boldsymbol{w}, \boldsymbol{z})$ as being the image of the vector $\boldsymbol{v}$ under a linear mapping $R(\boldsymbol{w}, \boldsymbol{z})$, so that $\boldsymbol{R}(\boldsymbol{u}, \boldsymbol{v}, \boldsymbol{w}) = R(\boldsymbol{w}, \boldsymbol{z})(\boldsymbol{v})$:

$$R(\boldsymbol{w}, \boldsymbol{z}) \boldsymbol{v} = R^i_{jkl} v^j w^k z^l \frac{\partial}{\partial x^i} .$$

We have thereby introduced a *bilinear operation* R that operates on a pair of vectors $\boldsymbol{w}$ and $\boldsymbol{z}$, to produce the *linear operator* $R(\boldsymbol{w}, \boldsymbol{z})$, which in turn operates on the vector $\boldsymbol{v}$, producing finally the vector $R(\boldsymbol{w}, \boldsymbol{z}) \boldsymbol{v} = \boldsymbol{R}(\boldsymbol{v}, \boldsymbol{w}, \boldsymbol{z})$.

The justification for this seeming complication is that the operator $R(\boldsymbol{w}, \boldsymbol{z})$ can be interpreted in terms of covariant derivatives with respect to $\boldsymbol{w}$ and $\boldsymbol{z}$.

To show how this works, we give a definition of the operator $R(\boldsymbol{w}, \boldsymbol{z})$ in terms of the covariant derivative and the bracket of two vector fields, which we shall then show is equivalent to the definition given above.

THEOREM 6.3.

$$R(\boldsymbol{w}, \boldsymbol{z}) \boldsymbol{v} = \nabla_{\boldsymbol{w}} \left(\nabla_{\boldsymbol{z}} \boldsymbol{v} \right) - \nabla_{\boldsymbol{z}} \left(\nabla_{\boldsymbol{w}} \boldsymbol{v} \right) - \nabla_{[\boldsymbol{w}, \boldsymbol{z}]} \boldsymbol{v} .$$

PROOF. We first establish that this new definition of $R(\boldsymbol{z}, \boldsymbol{w}) \boldsymbol{v}$ has the correct components, that is, those given by Eq. (6.2). We shall prove this in the context of an abstract n-dimensional manifold whose parameters are $(x^1, \ldots, x^n)$ and whose metric coefficients are $g_{ij}(x^1, \ldots, x^n)$. We shall adhere to the following notation

$$\boldsymbol{v} = v^i \frac{\partial}{\partial x^i} , \quad \boldsymbol{w} = w^i \frac{\partial}{\partial x^i} , \quad \boldsymbol{z} = z^i \frac{\partial}{\partial x^i} ,$$

where v^i, w^i, and z^i are functions of $(x^1, \ldots, x^n)$ for $i = 1, 2, \ldots, n$. With that notation, we have

$$\nabla_{\boldsymbol{z}} \boldsymbol{v} = z^l \left(\frac{\partial v^i}{\partial x^l} + v^j \Gamma^i_{lj} \right) \frac{\partial}{\partial x^i} .$$

Denoting the coefficient of $\partial/\partial x^i$ here by u^i, we have

$$u^p = z^l \left(\frac{\partial v^p}{\partial x^l} + v^j \Gamma^p_{lj} \right)$$

and

$$\frac{\partial u^i}{\partial x^k} = \frac{\partial z^l}{\partial x^k} \frac{\partial v^i}{\partial x^l} + z^l \frac{\partial^2 v^i}{\partial x^k \partial x^l} + \left(\frac{\partial z^l}{\partial x^k} v^p + z^l \frac{\partial v^p}{\partial x^k} \right) \Gamma^i_{lp} + z^l v^p \frac{\partial \Gamma^i_{lp}}{\partial x^k} .$$

---

[7]Observe that this inner product is not a pure number. In the notation of Subsection 5.2 of Chapter 4 it has the physical dimension $[d]^2$, that is, "length"-squared, as length is measured on the manifold. In particular, $\langle \boldsymbol{u}, \boldsymbol{u} \rangle$ is the squared "length" of the vector $\boldsymbol{u}$. That is exactly as it should be, in analogy with $\mathbb{R}^3$.

The coefficient of $\partial/\partial x^i$ in the expression for $\nabla_{\boldsymbol{w}}(\nabla_{\boldsymbol{z}}\boldsymbol{v})$, which is

$$w^k\left(\frac{\partial u^i}{\partial x^k} + u^j\Gamma^i_{kj}\right),$$

can then be written as a sum of seven terms, each of which involves summation on two, three, or four indices, namely

$$w^k\frac{\partial z^l}{\partial x^k}\frac{\partial v^i}{\partial x^l} + w^k z^l\frac{\partial^2 v^i}{\partial x^k\,\partial x^l} + w^k\frac{\partial z^l}{\partial x^k}v^p\Gamma^i_{lp} + w^k z^l\frac{\partial v^p}{\partial x^k}\Gamma^i_{lp}$$
$$+ w^k z^l v^j\frac{\partial\Gamma^i_{lj}}{\partial x^k} + w^k z^l\frac{\partial v^p}{\partial x^l}\Gamma^i_{kp} + w^k z^l v^j\Gamma^p_{lj}\Gamma^i_{kp}\,.$$

(Here, we replaced the dummy index of summation p in the fifth term by the index j. Doing so does not change the value of that term.)

If we interchange $\boldsymbol{w}$ with $\boldsymbol{z}$ and k with l, the second term remains unchanged here, and the fourth and sixth terms are interchanged. Now the interchange of k with l makes no difference to any of the terms, since both of these indices are mere dummy variables over which summation is performed. Therefore these three terms drop out of the expression $\nabla_{\boldsymbol{w}}(\nabla_{\boldsymbol{z}}\boldsymbol{v}) - \nabla_{\boldsymbol{z}}(\nabla_{\boldsymbol{w}}\boldsymbol{v})$, and what remains can be conveniently written as

$$\nabla_{\boldsymbol{w}}(\nabla_{\boldsymbol{z}}\boldsymbol{v}) - \nabla_{\boldsymbol{z}}(\nabla_{\boldsymbol{w}}\boldsymbol{v}) = \left(\left(w^k\frac{\partial z^l}{\partial x^k} - z^k\frac{\partial w^l}{\partial x^k}\right)\left(\frac{\partial v^i}{\partial x^l} + v^p\Gamma^i_{lp}\right)\right.$$
$$\left. + (w^k z^l - w^l z^k)v^j\left(\frac{\partial\Gamma^i_{lj}}{\partial x^k} + \Gamma^i_{kp}\Gamma^p_{lj}\right)\right)\frac{\partial}{\partial x^i}\,.$$

Since

$$[\boldsymbol{w},\boldsymbol{z}] = \left(w^k\frac{\partial z^l}{\partial x^k} - z^k\frac{\partial w^l}{\partial x^k}\right)\frac{\partial}{\partial x^l}\,,$$

computation reveals that

$$\nabla_{[\boldsymbol{w},\boldsymbol{z}]}\boldsymbol{v} = \left(w^k\frac{\partial z^l}{\partial x^k} - z^k\frac{\partial w^l}{\partial x^k}\right)\left(\frac{\partial v^i}{\partial x^l} + v^p\Gamma^i_{lp}\right)\frac{\partial}{\partial x^i}\,,$$

and so the vector

$$R(\boldsymbol{w},\boldsymbol{z})\boldsymbol{v} = \nabla_{\boldsymbol{w}}(\nabla_{\boldsymbol{z}}\boldsymbol{v}) - \nabla_{\boldsymbol{z}}(\nabla_{\boldsymbol{w}}\boldsymbol{v}) - \nabla_{[\boldsymbol{w},\boldsymbol{z}]}\boldsymbol{v}$$

consists of just the second half of the formula above, that is

$$\nabla_{\boldsymbol{w}}(\nabla_{\boldsymbol{z}}\boldsymbol{v}) - \nabla_{\boldsymbol{z}}(\nabla_{\boldsymbol{w}}\boldsymbol{v}) - \nabla_{[\boldsymbol{w},\boldsymbol{z}]}\boldsymbol{v} = (w^k z^l - w^l z^k)v^j\left(\frac{\partial\Gamma^i_{lj}}{\partial x^k} + \Gamma^i_{kp}\Gamma^p_{lj}\right)\frac{\partial}{\partial x^i}\,.$$

By interchanging the summation indices k and l in the subtracted terms and then rearranging, we find that

$$\nabla_{\boldsymbol{w}}(\nabla_{\boldsymbol{z}}\boldsymbol{v}) - \nabla_{\boldsymbol{z}}(\nabla_{\boldsymbol{w}}\boldsymbol{v}) - \nabla_{[\boldsymbol{w},\boldsymbol{z}]}\boldsymbol{v}$$
$$= \left(\frac{\partial\Gamma^i_{lj}}{\partial x^k} - \frac{\partial\Gamma^i_{kj}}{\partial x^l} + (\Gamma^i_{kp}\Gamma^p_{lj} - \Gamma^i_{lp}\Gamma^p_{kj})\right)v^j w^k z^l\frac{\partial}{\partial x^i}\,.$$

Comparison of this result with Eq. (6.2) shows that the coefficient of $\partial/\partial x^i$ in this expression is precisely

$$R^i_{jkl}v^j w^k z^l\,.$$

$\square$

REMARK 6.4. Using this alternative approach to the Riemann curvature tensor, one can immediately see that it really is a tensor. The work done in Appendix 6 shows that $\boldsymbol{v}$, $\boldsymbol{w}$, $\boldsymbol{z}$, $[\boldsymbol{w}, \boldsymbol{z}]$, $\nabla_{\boldsymbol{z}}\boldsymbol{v}$, and so forth, are all tensors, and hence $R(\boldsymbol{w}, \boldsymbol{z})\boldsymbol{v}$ must also be a tensor. That is, under a change of coordinates, it transforms correctly in terms of the Jacobian matrix.

REMARK 6.5. The Riemann curvature tensor can also be elegantly rewritten in terms of the so-called *second covariant derivative*, which is defined as

$$\nabla^2_{\boldsymbol{u},\boldsymbol{v}} = \nabla_{\boldsymbol{u}}(\nabla_{\boldsymbol{v}}) - \nabla_{\nabla_{\boldsymbol{u}}\boldsymbol{v}}.$$

Taking account of Problem 6.5, we have the formula

$$R(\boldsymbol{u}, \boldsymbol{v})\boldsymbol{w} = \nabla^2_{\boldsymbol{u},\boldsymbol{v}}\boldsymbol{w} - \nabla^2_{\boldsymbol{v},\boldsymbol{u}}\boldsymbol{w}.$$

In particular, it is not generally true that a mixed second covariant derivative is independent of the order in which the covariant derivatives are taken, although it is independent of that order when the two vector fields $\boldsymbol{u}$ and $\boldsymbol{v}$ have constant coefficients and it is applied to a function f rather than to a vector field $\boldsymbol{w}$. As we shall see below, this second covariant derivative is essentially the Hessian operator used by Euler to define the curvature of a surface at a point where the surface is tangent to the xy-plane.

Regarded as a tangent vector that operates on functions, the second covariant derivative is

$$\nabla^2_{\frac{\partial}{\partial x^i},\frac{\partial}{\partial x^j}} = \frac{\partial^2}{\partial x^i\,\partial x^j} - \Gamma^k_{ij}\frac{\partial}{\partial x^k}$$

(Note that the right-hand side of this relation is symmetric in i and j.) In particular, on a flat space like $\mathbb{R}^3$, where all the Christoffel symbols are zero, we have

$$\nabla^2_{\frac{\partial}{\partial x^i},\frac{\partial}{\partial x^j}} = \frac{\partial^2}{\partial x^i\,\partial x^j},$$

that is, the second covariant derivative with respect to a pair of basis vectors in a flat space is just the ordinary mixed second-order partial derivative.

2.2. A tensor formulation of curvature. If we lower the superscripts i in the components of the vector $R(\boldsymbol{w}, \boldsymbol{z})\boldsymbol{v}$ as described in Appendix 4, we get a covector $\boldsymbol{v}$ that operates on a vector $\boldsymbol{u}$ to produce the number

$$\mathrm{Rie}(\boldsymbol{u}, \boldsymbol{v}, \boldsymbol{w}, \boldsymbol{z}) = \langle \boldsymbol{u}, R(\boldsymbol{w}, \boldsymbol{z})\boldsymbol{v}\rangle = g_{im}R^m_{jkl}u^iv^jw^kz^l = R_{ijkl}u^iv^jw^kz^l.$$

With this connection established, we can now give a tensor expression for the curvature. To do so, we need a lemma:

LEMMA 6.1. *The covariant Riemann curvature tensor satisfies the relation* $R_{ijkl} = -R_{jikl}$. *In terms of vectors,*

$$\langle \boldsymbol{u}, R(\boldsymbol{w}, \boldsymbol{z})\boldsymbol{v}\rangle = -\langle \boldsymbol{v}, R(\boldsymbol{w}, \boldsymbol{z})\boldsymbol{u}\rangle.$$

Consequently,

$$\langle \boldsymbol{u}, R(\boldsymbol{w}, \boldsymbol{z})\boldsymbol{v}\rangle = \langle \boldsymbol{v}, R(\boldsymbol{z}, \boldsymbol{w})\boldsymbol{u}\rangle.$$

PROOF. By Eq. (5.3),

$$R_{ijkl} = \Gamma^3_{ki}\Gamma^3_{lj} - \Gamma^3_{li}\Gamma^3_{kj}.$$

It is obvious that interchanging i and j causes the expression on the right-hand side to reverse its sign.

Since it is obvious from the definition of the operator $R(\cdot,\cdot)$ that it is antisymmetric in its arguments, the last equation now follows as well.   □

REMARK 6.6. Since Eq. (5.3) was proved above only for surfaces in $\mathbb{R}^3$, this proof is not general. The lemma is true in general, however, as Problem 5.7 shows.

THEOREM 6.4. *For any two linearly independent tangent vectors $\boldsymbol{u}$ and $\boldsymbol{v}$, the Gaussian curvature κ is given by*

$$(6.4) \qquad \kappa = \frac{\langle \boldsymbol{u}, R(\boldsymbol{u},\boldsymbol{v})\boldsymbol{v}\rangle}{\langle \boldsymbol{u},\boldsymbol{u}\rangle \langle \boldsymbol{v},\boldsymbol{v}\rangle - \langle \boldsymbol{u},\boldsymbol{v}\rangle^2}\,.$$

PROOF. If we take $\boldsymbol{u} = \boldsymbol{w} = \partial/\partial x^1$ and $\boldsymbol{v} = \boldsymbol{z} = \partial/\partial x^2$, we get the situation described in the discussion following the proof of Theorem 6.5 of Appendix 6 in Volume 2, in which the left-hand side of Eq. (5.3) is the curvature multiplied by $(g_{11}g_{22} - g_{12}g_{21}) = \langle \boldsymbol{u},\boldsymbol{u}\rangle \langle \boldsymbol{v},\boldsymbol{v}\rangle - \langle \boldsymbol{u},\boldsymbol{v}\rangle \langle \boldsymbol{v},\boldsymbol{u}\rangle$. Since the right-hand side of Eq. (5.3) is the numerator in Eq. (6.4) in this case (that is, since $u^1 = 1 = w^1 = v^2 = z^2$ and $u^2 = 0 = w^2 = v^1 = z^1$, so that the only nonzero terms occur when $i = k = 1$ and $j = l = 2$), we see that Eq. (6.4) is true in this special case.

Let us temporarily denote the quotient that we are claiming is equal to the curvature by $\kappa(\boldsymbol{u},\boldsymbol{v})$. Since the denominator vanishes when the vectors $\boldsymbol{u}$ and $\boldsymbol{v}$ are linearly dependent, and therefore $\kappa(\boldsymbol{u},\boldsymbol{v})$ is not defined in this case, we fix a linearly independent pair $\boldsymbol{u}$, $\boldsymbol{v}$ to start with for which $\kappa(\boldsymbol{u},\boldsymbol{v})$ equals the curvature. We have just shown that there is one such pair $\boldsymbol{u}$, $\boldsymbol{v}$. All that remains is to show that this expression is unaltered if $\boldsymbol{u}$ and $\boldsymbol{v}$ are replaced by any linear combinations $a\boldsymbol{u} + b\boldsymbol{v}$, $c\boldsymbol{u} + d\boldsymbol{v}$ with $ad - bc \neq 0$. (This restriction is necessary to assure that the new pair is linearly independent.) The homogeneity of the expressions in the numerator and denominator implies that replacing $\boldsymbol{u}$ by $a\,\boldsymbol{u}$ and $\boldsymbol{v}$ by $b\,\boldsymbol{v}$, where a and b are both nonzero, leaves $\kappa(\boldsymbol{u},\boldsymbol{v})$ unaltered. (Both numerator and denominator are multiplied by a^2b^2.) But the numerator and denominator are also unaffected by a shear transformation $\boldsymbol{u} \mapsto \boldsymbol{u} + \boldsymbol{v}$, as one can easily see. Indeed, looking at the numerator, we find

$$\langle \boldsymbol{u}+\boldsymbol{v}, R(\boldsymbol{u}+\boldsymbol{v},\boldsymbol{v})\boldsymbol{v}\rangle = \langle \boldsymbol{u}, R(\boldsymbol{u},\boldsymbol{v})\boldsymbol{v}\rangle + \langle \boldsymbol{u}, R(\boldsymbol{v},\boldsymbol{v})\boldsymbol{v}\rangle + \langle \boldsymbol{v}, R(\boldsymbol{u},\boldsymbol{v})\boldsymbol{v}\rangle + \langle \boldsymbol{v}, R(\boldsymbol{v},\boldsymbol{v})\boldsymbol{v}\rangle\,.$$

The first term on the right-hand side is just the numerator of the expression in Eq. (6.4). The second and fourth terms vanish due to the antisymmetry of the operator $R(\cdot,\cdot)$, and the third term vanishes by Lemma 6.1. It follows that the numerator remains invariant. The invariance of the denominator under the same shear mapping is a routine computation. Hence, if a and b are both nonzero,

$$\kappa(\boldsymbol{u},\boldsymbol{v}) = \kappa(a\boldsymbol{u}, b\boldsymbol{v}) = \kappa(a\boldsymbol{u} + b\boldsymbol{v}, b\boldsymbol{v}) = \kappa(a\boldsymbol{u} + b\boldsymbol{v}, \boldsymbol{v})\,.$$

Since $\kappa(a\boldsymbol{u},\boldsymbol{v}) = \kappa(\boldsymbol{u},\boldsymbol{v})$ if $a \neq 0$, this equation also holds when $b = 0$. The symmetry relation $\kappa(\boldsymbol{u},\boldsymbol{v}) = \kappa(\boldsymbol{v},\boldsymbol{u})$ then implies that $\kappa(\boldsymbol{u},\boldsymbol{v}) = \kappa(\boldsymbol{u}, c\boldsymbol{u} + d\boldsymbol{v})$ provided $d \neq 0$.

If $d = 0$, we have $\kappa(a\boldsymbol{u} + b\boldsymbol{v}, c\boldsymbol{u} + d\boldsymbol{v}) = \kappa(a\boldsymbol{u} + b\boldsymbol{v}, c\boldsymbol{u}) = \kappa(b\boldsymbol{v} + a\boldsymbol{u}, \boldsymbol{u}) = \kappa(\boldsymbol{v},\boldsymbol{u}) = \kappa(\boldsymbol{u},\boldsymbol{v})$, provided $bc \neq 0$. Since we are assuming $ad - bc \neq 0$, we will not have both $bc = 0$ and $d = 0$. Without loss of generality, assume $d \neq 0$.

We then have $\kappa(\boldsymbol{u},\boldsymbol{v}) = \kappa(\boldsymbol{u}, c\boldsymbol{u} + d\boldsymbol{v}) = \kappa\big(s\boldsymbol{u} + t(c\boldsymbol{u} + d\boldsymbol{v}), c\boldsymbol{u} + d\boldsymbol{v}\big)$. Taking $s = a - bc/d = (ad - bc)/d$, which is nonzero, and $t = b/d$, we get

$$\kappa(\boldsymbol{u},\boldsymbol{v}) = \kappa\Big(\frac{ad - bc}{d}\boldsymbol{u} + \frac{b}{d}(c\boldsymbol{u} + d\boldsymbol{v}), c\boldsymbol{u} + d\boldsymbol{v}\Big) = \kappa(a\boldsymbol{u} + b\boldsymbol{v}, c\boldsymbol{u} + d\boldsymbol{v})\,. \qquad □$$

3. Parallel Transport

Now that we have a geometric interpretation of the Riemann curvature tensor, we can see why it is called a curvature tensor. We know already that it has a connection with curvature through its covariant form, in which the upper index is lowered. The key ingredient in producing this tensor, as we now see, is the covariant derivative, applied three times, with the term involving the covariant derivative with respect to the Lie bracket playing the role of a "corrective" term when the components of the vector fields u and v are not constant.

The covariant derivative makes it possible to speak of two vectors that are tangent to a manifold at different points as being, in some sense, "the same vector." The idea is quite intuitive. A vivid image of it can be conveyed by imagining an ancient warrior marching eastward along the equator carrying a spear that always points directly to the front. From the cosmic perspective, that spear, representing a tangent vector at each point where the warrior happens to be, does rotate. As a vector in $\mathbb{R}^3$, it does not remain the same. But, relative to the spherical Earth that it is on, it comes "as close as possible" to being the same. That is, it rotates by the minimal amount, and the warrior would have to exert some force on it to keep it *cosmically* parallel to itself. From a terrestrial point of view, the rest of the cosmos does not count, and parallelism is relative to the surface of the Earth. That is the picture we want to keep in mind as we describe parallel transport around a curve in any manifold, independently of any embedding that manifold may have in an ambient Euclidean space. The covariant derivative turns out to be the key to a precise encoding of this intuition: essentially, we require that the covariant derivative of a "parallel transported" vector with respect to the tangent vector to the curve vanish identically.

From now on, it will save a lot of writing and confusion if we refer to a point P on a manifold $\mathfrak{M}$ having coordinates $(x^1, \ldots, x^n)$ in some chart as "the point $P = (x^1, \ldots, x^n)$."

DEFINITION 6.3. Let $\gamma(t)$ be a parameterized curve on a manifold $\mathfrak{M}$ in a coordinate system $(x^1, x^2, \ldots, x^n)$. (We think of γ as a mapping into the parameter space, that is, $\gamma(t) = \big(x^1(t), \ldots, x^n(t)\big)$.) Let w_0 be a tangent vector at a point $\gamma(t_0) = (x_0^1, \ldots, x_0^n)$. The vector field

$$\boldsymbol{w}(t) = w^i(t) \frac{\partial}{\partial x^i}\bigg|_{\gamma(t)},$$

that is the (unique) solution of the initial-value problem

$$(w^i)'(t) + (\gamma^j)'(t) w^k(t) \Gamma^i_{jk}\big(\gamma(t)\big) = 0, \quad w^i(t_0) = w_0^i, \quad i, 1, 2, \ldots, n,$$

is the *parallel transport* of w_0 along the curve $\gamma(t)$.

REMARK 6.7. Although this definition appears unmotivated, we can provide the motivation through the concept of the covariant derivative. As we have given it, the vector field $\boldsymbol{w}(t)$ is defined only at points of the curve $\gamma(t)$. But suppose that $\boldsymbol{w}(x^1, \ldots, x^n)$ is a vector field defined on an open set U in the parameter space containing the curve γ and is such that $\boldsymbol{w}\big(\gamma(t)\big) = \boldsymbol{w}(t)$. Then we have

$$(w^i)'(t) = (\gamma^j)'(t) \frac{\partial w^i}{\partial x^j},$$

and the system of equations that defines the parallel transport of $\boldsymbol{w}_0$ along $\boldsymbol{\gamma}$ is equivalent to the vector-field equation

$$\nabla_{\boldsymbol{\gamma}'(t)}\boldsymbol{w}\big(\boldsymbol{\gamma}(t)\big) = \boldsymbol{0}, \quad \boldsymbol{w}\big(\boldsymbol{\gamma}(t_0)\big) = \boldsymbol{w}_0.$$

This covariant derivative makes sense only if $\boldsymbol{w}(x^1, \ldots, x^n)$ is defined on an open set, and that restriction is not necessary for our definition of parallel transport. If it happens to be satisfied, then the condition for parallel transport has the intuitive meaning that the directional derivative of the vector field $\boldsymbol{w}$ in the direction of the tangent vector to the curve, when projected back into the tangent plane, is zero. That means that, relative to the tangent line to the curve, the field is constant. In that sense, the parallel transport of $\boldsymbol{w}_0$ is "the same vector" as $\boldsymbol{w}_0$, only based at a different point.

Another way of describing the situation is to say that, while the definition of the covariant derivative $\nabla_{\boldsymbol{v}}\boldsymbol{u}$ at a point requires that $\boldsymbol{u}$ be defined as a vector field in a neighborhood of the point, we can expand this definition to cover one more important case, allowing $\nabla_{\boldsymbol{\gamma}'(t)}\boldsymbol{w}$ to be defined by this last equation if $\boldsymbol{w}\big(\boldsymbol{\gamma}(t)\big)$ is defined for t in an open interval.

It needs to be emphasized that this "sameness" of $\boldsymbol{w}_0$ and $\boldsymbol{w}(t)$ is relative to the curve $\boldsymbol{\gamma}$, and if one arrives at the same point over a different path, the parallel transport of $\boldsymbol{w}_0$ will very likely be a different vector. We thus have the paradoxical situation of two different tangent vectors at a given point both being "the same vector" as a third vector at a different point. This possibility is the very essence of what is meant by a curved space, and we shall investigate it thoroughly below when we given a geometric interpretation of the Riemann curvature tensor.

To summarize, our *motivation* for this definition comes from the notion of a covariant derivative, but parallel transport is *actually defined* using minimal principles, requiring only an initial vector and a smooth curve. The question of whether or not there exists a vector field that is defined on an open set and whose covariant derivative with respect to the tangent vector at each point vanishes is psychologically helpful as motivation, but not logically relevant.

EXAMPLE 6.1. The simplest example is parallel transport along meridians of longitude and parallels of latitude on a sphere. The geometry of one such parallel transport is shown in Fig. 6.2. We shall take the trouble here to give the algebraic details for the portion of the transport that goes along the equator. Let the sphere be parameterized by latitude and longitude coordinates (φ, θ), that is, $\boldsymbol{r}(\varphi, \theta) = (R\cos\varphi\cos\theta, R\cos\varphi\sin\theta, R\sin\varphi)$. We give the equatorial circle the parameterization $\boldsymbol{\gamma}(\theta) = (R\cos\theta, R\sin\theta, 0) = \boldsymbol{r}(0, \theta)$, so that

$$\boldsymbol{\gamma}'(\theta) = R(-\sin\theta, \cos\theta, 0) = \frac{\partial}{\partial\theta}.$$

and we shall take $\theta_0 = 0$, so that $\boldsymbol{\gamma}(\theta_0) = (R, 0, 0)$. Let $\boldsymbol{w}_0$ be the unit "eastward-pointing" vector at this point, that is, $\boldsymbol{w}_0 = (1/R)\partial/\partial\theta$ (which, at this particular point on the equator, would be the vector $\boldsymbol{j}$ in the usual vector basis of $\mathbb{R}^3$). Thus we get the vector $\boldsymbol{w}(\theta) = w^1(\theta)\partial/\partial\varphi + w^2\partial/\partial\theta$ as the solution of the initial-value problem

$$\nabla_{\frac{\partial}{\partial\theta}}\left(w^1(\theta)\frac{\partial}{\partial\varphi} + w^2(\theta)\frac{\partial}{\partial\theta}\right) = \boldsymbol{0} = 0\frac{\partial}{\partial\varphi} + 0\frac{\partial}{\partial\theta}; \quad \boldsymbol{w}(0) = \frac{\partial}{\partial\theta}.$$

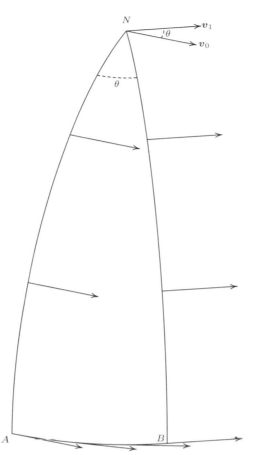

FIGURE 6.2. The vector v_0 is located at the north pole N of a hemisphere. It is then parallel-transported directly south along a meridian of longitude, always pointing east, until it reaches the equator at point A, still pointing east. It is then parallel transported eastward along the equator to B, which differs from A in longitude by the angle θ. From B it is again parallel-transported north along the meridian of longitude through B, still pointing east, until it returns to the north pole N. Despite having been kept strictly parallel to itself, when it returns to N, it has become the vector v_1, which makes angle θ with v_0. The rotation is due to the curvature of the hemisphere. For a surface of constant curvature such as the sphere, the angle between the initial and final positions is proportional to the area enclosed by the path.

Thus, in terms of Definition 6.2, we have $s = \varphi$, $t = \theta$, $v^1 = 0$, $v^2 = R(-\sin\theta + \cos\theta)$, $u^1 = w^1$, $u^2 = w^2$, and since w^1 and w^2 are functions of θ alone, their partial derivatives are full derivatives. Now the only nonzero Christoffel symbols here are $\Gamma_{22}^1 = \cos\varphi \sin\varphi$, and $\Gamma_{12}^2 = \Gamma_{21}^2 = -\tan(\varphi)$. (You can determine this fact algebraically, using the metric coefficients $g_{11} = R^2$, $g_{12} = g_{21} = 0$, $g_{22} = R^2 \cos^2 \varphi$; or you can take a geometric approach and create the unit normal vector $\boldsymbol{n} = -\cos\varphi \cos\theta \, \boldsymbol{i} - \cos\varphi \sin\theta \, \boldsymbol{j} - \sin\varphi \, \boldsymbol{k}$, then, for each $i = 1, 2, 3$, set up three systems of three linear equations in the coefficients Γ^i_{jk}, $j, k = 1, 2$. If you take the second approach, you will get the same Christoffel symbols, along with the unneeded fact that $\Gamma^3_{11} = \Gamma^3_{12} = 0$, $\Gamma^3_{22} = R \cos^2 \varphi$.) Knowing the Christoffel symbols, we set up two pairs of initial-value problems to determine w^1 and w^2, and these turn out to be

$$
\begin{aligned}
0 &= \frac{dw^1(\theta)}{d\theta} + (\sin\varphi \cos\varphi)\, w^2 \\
0 &= \frac{dw^2(\theta)}{d\theta} - (\tan\varphi)\, w^1 \\
0 &= w^1(0) \\
1/R &= w^2(0).
\end{aligned}
$$

Here we use the fact that $\varphi \equiv 0$ on the path, so that the obvious solution is $w^1 \equiv 0$, $w^2 \equiv 1/R$, and so the parallel transport of the vector $(1/R)\partial/\partial\theta$ remains $(1/R)\partial/\partial\theta$ over the whole path. In terms of our identification of directional derivatives with "arrows," this derivation is to be pictured as the vector $(1/R)\partial\boldsymbol{r}/\partial\theta = -\sin\theta \, \boldsymbol{i} + \cos\theta \, \boldsymbol{j}$ in $\mathbb{R}^3$, which is indeed what we usually think of as the unit eastward-pointing tangent vector at a point $(R\cos\theta, R\sin\theta, 0)$ along the equator.

3.1. Parallel transport and the Riemann curvature tensor.
Parallel transport is discussed further in Appendix 4, and it is shown in examples there that when a tangent vector is parallel-transported around a triangle on a surface of constant positive, zero, or negative curvature, the angle between its initial and final positions is directly proportional to the product of the curvature and the area of the triangle. This fact can be generalized from a triangle to any closed smooth curve. In that sense, parallel transport of a vector provides a sort of planimeter on surfaces of constant curvature. When a unit vector is transported around a simple closed curve, the angle between its initial and final positions is proportional to the area of the portion of the surface inside the closed curve.

We shall now see how the concept of parallel transport relates to the Riemann curvature tensor, showing first that if a vector returns to its original value after being parallel-transported around any closed curve on a surface, then the Riemann curvature tensor vanishes on that surface. Thus the Riemann curvature tensor of a "flat" surface (one having this parallel-transport property) vanishes identically. We shall then discuss the geometric interpretation of the Riemann tensor in the general case when parallel transport is *not* independent of the path. To carry out the first part of this program, for the sake of intuition, we once again return to our notation in $\mathbb{R}^3$, using the geometric vector $\partial\boldsymbol{r}/\partial x^j$ where in the abstract discussion we used the derivation $\partial/\partial x^j$.

THEOREM 6.5. *Fix a point P on a portion of the surface $\Sigma \subset \mathbb{R}^3$ that is the image of a connected open set $U \subset \mathbb{R}^2$ under a parameterization $(x^1, x^2) \mapsto \boldsymbol{r}(x^1, x^2) = \big(x^1, x^2, z(x^2, x^2)\big)$. Suppose that for any tangent vector $\boldsymbol{w}_0$ at P, the vector $\boldsymbol{w}\big(\boldsymbol{\gamma}(t)\big)$ obtained by parallel-transporting $\boldsymbol{w}_0$ over a path $\boldsymbol{\gamma}(s)$ in Σ, $0 \leq s \leq t$, depends only on the point $Q = \boldsymbol{\gamma}(t)$ and not on the particular path $\boldsymbol{\gamma}$ or the parameter value t. Denote the resulting vector field by $\boldsymbol{w}(Q)$. Then the Riemann curvature tensor $R(\boldsymbol{w}, \boldsymbol{z})\boldsymbol{v}$ vanishes identically on Σ. In terms of components,*

$$\frac{\partial \Gamma^i_{jk}}{\partial x^l} - \frac{\partial \Gamma^i_{lj}}{\partial x^k} + \big(\Gamma^i_{lp}\Gamma^p_{kj} - \Gamma^i_{kp}\Gamma^p_{lj}\big) \equiv 0$$

for all sixteen values of i, j, k, and l and all parameter values $(x^1, x^2) \in U$.

PROOF. Suppose that the tangent vector $\boldsymbol{w}_0$ at P has been parallel-transported to every point Q in a neighborhood $\mathfrak{S}$ of P on the surface Σ as a single-valued vector field $\boldsymbol{w}(Q)$. Then $\boldsymbol{w}(Q)$ is defined to be $\boldsymbol{w}\big(\boldsymbol{\gamma}(t)\big)$ when $Q = \boldsymbol{\gamma}(t)$, and this definition depends only on the point $Q \in \mathfrak{S}$, not on the particular path $\boldsymbol{\gamma}$ or the value of the parameter t. As a result, the vector field $\boldsymbol{w}\big(\boldsymbol{r}(x^1, x^2)\big) = w^1(x^1, x^2)\frac{\partial \boldsymbol{r}}{\partial x^1} + w^2(x^1, x^2)\frac{\partial \boldsymbol{r}}{\partial x^2}$ satisfies the system of differential equations

(6.5)
$$\frac{\partial w^i}{\partial x^k} = -w^j \Gamma^i_{kj}, \quad i = 1, 2, \quad k = 1, 2,$$

where the summation on j is from $j = 1$ to $j = 2$. (Note that we regard the coefficients of $\boldsymbol{w}$ as functions of x^1 and x^2 rather than what they properly are: functions of $\boldsymbol{r}(x^1, x^2) = \big(x^1, x^2, z(x^1, x^2)\big)$.) Consider a closed curve on $\mathfrak{S}$ parameterized as $\boldsymbol{\gamma}(t)$, say

$$\boldsymbol{\gamma}(t) = \boldsymbol{r}\big(\gamma^1(t), \gamma^2(t)\big) = \big(\gamma^1(t), \gamma^2(t), z(t)\big), \quad 0 \leq t \leq 1,$$

where $\boldsymbol{\gamma}(0) = \boldsymbol{\gamma}(1)$, and $z(t)$ is an abbreviation for $z\big(\gamma^1(t), \gamma^2(t)\big)$. Then

(6.6) $(\gamma^k)'(t)\dfrac{\partial w^i\big(\gamma^1(t), \gamma^2(t)\big)}{\partial x^k} = -(\gamma^k)'(t)w^j\big(\gamma^1(t), \gamma^2(t)\big)\Gamma^i_{kj}\big(\boldsymbol{\gamma}(t)\big), \quad i = 1, 2,$

where the summation indices j and k assume the values 1 and 2.

The left-hand side of Eq. (6.6) is

$$\frac{d}{dt}\Big(w^i\big(\gamma^1(t), \gamma^2(t)\big)\Big).$$

Therefore, if we integrate it from $t = 0$ to $t = 1$, the result is $w^i\big(\gamma^1(1), \gamma^2(1)\big) - w^i\big(\gamma^1(0), \gamma^2(0)\big) = 0$. We conclude that

$$\int_0^1 (\gamma^k)'(t)w^j\big(\gamma^1(t), \gamma^2(t)\big)\Gamma^i_{kj}\big(\boldsymbol{\gamma}(t)\big)\, dt = 0, \quad i = 1, 2.$$

Now this last integral can be interpreted as a line integral in the $x^1 x^2$-plane:

$$\int_C A(x^1, x^2)\, dx^1 + B(x^1, x^2)\, dx^2,$$

where C is the closed curve given as the set of values $\big(\gamma^1(t), \gamma^2(t)\big)$, $0 \leq t \leq 1$, $A(x^1, x^2) = w^j(x^1, x^2)\Gamma^i_{1j}\big(\boldsymbol{r}(x^1, x^2)\big)$, and $B = w^j(x^1, x^2)\Gamma^i_{2j}\big(\boldsymbol{r}(x^1, x^2)\big)$. Since we can vary the curve C to suit outselves, let us assume that it has no self-intersections except that the point P corresponds to both parameter values 0 and 1. By Green's Theorem,

$$0 = \iint\limits_{S} \left(\frac{\partial B}{\partial x^1} - \frac{\partial A}{\partial x^2} \right) dx^1 \, dx^2 = 0 \,,$$

where S is the portion of the plane enclosed by the curve C.[8] Because the region inside C can be varied at will, the integrand here must vanish identically.

It is a routine computation, using Eq. (6.5), to show that

$$\frac{\partial B}{\partial x^1} - \frac{\partial A}{\partial x^2} = w^j \left(\frac{\partial \Gamma^i_{2j}}{\partial x^1} - \frac{\partial \Gamma^i_{1j}}{\partial x^2} + \left(\Gamma^i_{1p} \Gamma^2_{2j} - \Gamma^i_{2p} \Gamma^p_{1j} \right) \right).$$

Again, since $\boldsymbol{w}$ is arbitrary, we can vary w^j at will. Thus we conclude for $i = 1, 2$ and $j = 1, 2$

$$\frac{\partial \Gamma^i_{2j}}{\partial x^1} - \frac{\partial \Gamma^i_{1j}}{\partial x^2} + \left(\Gamma^i_{1p} \Gamma^p_{2j} - \Gamma^i_{2p} \Gamma^p_{1j} \right) \equiv 0 \,.$$

Comparison with the Riemann curvature tensor shows that this equation says

$$R^i_{j21} \equiv 0 \,.$$

Since $R^i_{j12} = -R^i_{j21}$ while $R^i_{j11} = 0 = R^i_{j22}$, it follows that all sixteen components of the Riemann tensor vanish in this case. □

In the case of an n-dimensional manifold $\mathfrak{M}$, we can give a similar argument, but we need to consider the covariant (invariant) form of Stokes's theorem in order to do so. We shall not give the details, but merely note that, just as above, we can consider a closed curve $\boldsymbol{\gamma}$, parameterized by a real variable $t \in [0, 1]$ on the submanifold $\mathfrak{S}$. This curve encloses an open region S of $\mathfrak{S}$, of which it is the boundary ∂S. Since this curve is also contained in the manifold $\mathfrak{M}$, the differential equations that characterize the components of the transported vector field $\boldsymbol{w}(Q)$ in the n-dimensional parameter space are

$$\frac{\partial w^i}{\partial x^j} = -w^k \Gamma^i_{kj} \,, \quad i = 1, 2, \ldots, n \,, \quad j = 1, 2, \ldots, n \,.$$

The parallel transport of $\boldsymbol{w}$ around $\boldsymbol{\gamma}$ leads to the equations

$$0 = \int_{\partial S} \omega^i \,, \quad i = 1, 2, \ldots, n \,,$$

where ω^i is the one-form

$$\omega^i = w^k \Gamma^i_{kj} \, dx^j \,.$$

By Stokes's theorem in its covariant form, we then have the equations

$$0 = \int_{S} d\omega^i = \iint_{U} \frac{\partial (w^k \Gamma^i_{kj})}{\partial x^l} \, dx^l \wedge dx^j \,, \quad i = 1, 2, \ldots, n \,.$$

Here $dx^l \wedge dx^j$ stands for the differential

$$\det \left(\frac{\partial (x^l, x^j)}{\partial (x^1, x^2)} \right) dx^1 \, dx^2 \,,$$

---

[8]Because we assume C is nonintersecting, the Jordan Curve Theorem implies that it has a well-defined inside.

and U is the region of the parameter space whose image under the parameterization is S.

Again, since this holds for any region $S \subseteq \mathfrak{S}$ bounded by a curve γ contained in the given two-dimensional submanifold of $\mathfrak{M}$, the integrand must vanish identically, and then the equations of parallel transport imply that the n-dimensional version of the Riemann curvature tensor must vanish identically.

Conversely, the vanishing of the Riemann curvature tensor implies that the one-form ω^i is closed. On a *simply* connected region, a closed form is exact, and hence the vanishing of the Riemann curvature tensor on such a region implies that parallel transport of a vector from one point to another within the region is independent of the path followed.

3.2. Geometric significance of the Riemann curvature tensor*.

As we have seen, if the parallel transport of every vector around every closed loop in a local set of coordinates leads to a final value of the vector equal to its initial value, then the Riemann curvature tensor vanishes identically. That is good evidence that the Riemann curvature tensor has something to do with curvature, since we know that the angle between the initial and final values of such a transported vector is proportional to the product of the area enclosed by the loop and the curvature, when the curvature is constant.[9] What we have illustrated is the case of curvature zero.

In general, if the final value of a transported vector differs from its initial value, we cannot even *define* the vector field to which we just applied Stokes's theorem, and this whole argument falls apart. In that case, we need to look more closely at the Riemann curvature tensor in order to coax a geometric interpretation out of it. The case of a surface parameterized by an open subset of the plane will suffice to make the point. We consider a tangent vector $\boldsymbol{w} = w^1 \partial/\partial x^1 + w^2 \partial/\partial x^2$ at a fixed point A which, without loss of generality, we take to be $(0,0)$ and a variable point $B = (r, s)$. We transport the vector $\boldsymbol{w}$ to points inside the rectangle whose opposite corners are A and B in two different ways. At the point (x, y) we define $\boldsymbol{u}_y(x) = u_y^1 \partial/\partial x^1 + u_y^2 \partial/\partial x^2$ to be the result of transporting $\boldsymbol{w}$ first to $(0, y)$ along the vertical left edge of the rectangle, then to (x, y) along the horizontal line whose second coordinate is y. Similarly, we define $\boldsymbol{v}_x(y) = v_x^1 \partial/\partial x^1 + v_x^2 \partial/\partial x^2$ to be the result of parallel translation to $(x, 0)$ along the bottom horizontal edge, then along the vertical line whose first coordinate is x. The crucial point that we shall need below is that both $\boldsymbol{v}_x(y)$ and $\boldsymbol{u}_y(x)$ are differentiable functions of both x and y, and as a result, there is an absolute constant such that $|u_{y_1}^i(x) - u_{y_2}^i(x)| \leq K|y_1 - y_2|$, and $v_{x_1}(y) - v_{x_2}(y)| \leq K|x_1 - x_2|$ for all x, y, x_1, x_2, y_1, y_2.

We then transport the vector $\boldsymbol{w}$ parallel to itself from A to B along two different paths ACB and ADB, as shown in Fig. 6.3, where $C = (r, 0)$ and $D = (0, s)$. For this purpose, we parameterize the four edges of the rectangle $ACBD$ as paths γ_m,

---

[9]In spherical and hyperbolic geometry, the area of a triangle is proportional to its angle excess or defect—that is, the amount by which its angle sum differs from π radians—and this is also the angle between the initial and final positions of a vector parallel-transported around the triangle.

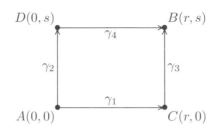

FIGURE 6.3. Parallel transport around a rectangle

$m = 1, 2, 3, 4$, defined as follows:

$$\begin{aligned}
\boldsymbol{\gamma}_1(x) &= (x, 0), & 0 \le x \le r, \\
\boldsymbol{\gamma}_2(y) &= (0, y), & 0 \le y \le s, \\
\boldsymbol{\gamma}_3(y) &= (r, y), & 0 \le y \le s, \\
\boldsymbol{\gamma}_4(x) &= (x, s), & 0 \le x \le r.
\end{aligned}$$

If we transport $\boldsymbol{w}$ from A to B via the paths $\boldsymbol{\gamma}_1 + \boldsymbol{\gamma}_3$, the result will be what we called $\boldsymbol{v}_r(s)$, while if we go over the path $\gamma_2 + \gamma_4$, the result will be $\boldsymbol{u}_s(r)$. We are interested in what difference there may be between these two vectors. Notice that if $\boldsymbol{u}_s(r)$ is parallel-transported from B around the edge of the rectangle in the positive orientation (counterclockwise), it will be $\boldsymbol{w}$ when it reaches A and hence $\boldsymbol{v}_r(s)$ when it reaches B again.

THEOREM 6.6. *Under the conditions just described, the difference $\boldsymbol{u}_s(r) - \boldsymbol{v}_r(s)$ satisfies the limiting relation*

$$\lim_{(r,s) \to (0,0)} \frac{1}{rs}(\boldsymbol{u}_s(r) - \boldsymbol{v}_r(s)) = R\Big(\boldsymbol{w}, \frac{\partial}{\partial x^1}, \frac{\partial}{\partial x^2}\Big) = R\Big(\frac{\partial}{\partial x^1}, \frac{\partial}{\partial x^2}\Big)\boldsymbol{w}\,.$$

This theorem gives us the geometric interpretation we have been wanting for the Riemann curvature tensor $R(\boldsymbol{w}, \boldsymbol{u}, \boldsymbol{v})$: Its value when $\boldsymbol{u}$ and $\boldsymbol{v}$ are infinitesimal increments in two mutually perpendicular directions and $\boldsymbol{w}$ is any vector in the tangent space is, up to higher-order vanishing, equal to the discrepancy when $\boldsymbol{w}$ is parallel-transported around the infinitesimal parallelogram spanned by $\boldsymbol{u}$ and $\boldsymbol{v}$, divided by the area of the pre-image of that parallelogram in the parameter space.

PROOF. The only thing required in the proof other than the definition of parallel transport is a rather abusive use of the mean-value theorem. Since $\boldsymbol{u}_0(r)$ is the parallel transport of $\boldsymbol{w}$ over γ_1 from A to C, we have

$$u_0^i(r) = w^i - \int_0^r u_0^j(x)\Gamma^i_{j1}(x, 0)\,dx$$

for $i = 1, 2$. Here the summation index j runs from 1 to 2, as will the other summation index p that will occur below.

Since $\boldsymbol{v}_r(s)$ is the parallel transport of $\boldsymbol{u}_0(r)$ over γ_3 from C to B, we get

$$
\begin{aligned}
v_r^i(s) &= u_0^i(r) - \int_0^s v_r^j(y)\Gamma_{j2}^i(r,y)\,dy \\
&= w^i - \int_0^r u_0^j(x)\Gamma_{j1}^i(x,0)\,dx - \int_0^s v_r^j(y)\Gamma_{j2}^i(r,y)\,dy\,.
\end{aligned}
$$

Similarly, since $\boldsymbol{u}_s(r)$ is the parallel transport of $\boldsymbol{w}$ over $\gamma_2 + \gamma_4$, we have

$$
u_s^i(r) = w^i - \int_0^r u_s^j(x)\Gamma_{j1}^i(x,s)\,dx - \int_0^s v_0^j(y)\Gamma_{j2}^i(0,y)\,dy\,.
$$

We therefore get

$$
u_s^i(r) - v_r^i(s) = \int_0^s \left(v_r^j(y)\Gamma_{j2}^i(r,y) - v_0^j(y)\Gamma_{j2}^i(0,y)\right)dy
$$

$$
- \int_0^r \left(u_s^j(x)\Gamma_{j1}^i(x,s) - u_0^j(x)\Gamma_{j1}^i(x,0)\right)dx\,.
$$

By the mean-value theorem, there are values x^* of x and y^* of y such that $0 < x^* < r$, $0 < y^* < s$, and

$$
u_s^i(r) - v_r^i(s) = s\left(v_r^j(y^*)\Gamma_{j2}^i(r,y^*) - v_0^j(y^*)\Gamma_{j2}^i(0,y^*)\right)
$$

$$
- r\left(u_s^j(x^*)\Gamma_{j1}^i(x^*,s) - u_0^j(x^*)\Gamma_{j1}^i(x^*,0)\right)\,.
$$

We now rearrange the coefficient of $-r$ on the right-hand side of this last equality, writing it as

$$
u_0^j(x^*)\left(\Gamma_{j1}^i(x^*,s) - \Gamma_{j1}^i(x^*,0)\right) + \left(u_s^j(x^*) - u_0^j(x^*)\right)\Gamma_{j1}^i(x^*,s)\,.
$$

Again, by the mean-value theorem, the first term in this expression has the representation

$$
su_0^j(x^*)\frac{\partial\Gamma_{j1}^i}{\partial x^2}(x^*,y^{**})
$$

for some y^{**} between 0 and s.

The expression for $u_s^i(r) - v_r^i(s)$ therefore contains a sum

$$
-rsu_0^j(x^*)\frac{\partial\Gamma_{j1}^i}{\partial x^2}(x^*,y^{**})\,,
$$

and the coefficient of rs in this last expression tends to

$$
-w^j\frac{\partial\Gamma_{j1}^i}{\partial x^2}(0,0)
$$

as $(r,s) \to (0,0)$.

As for the second term, we note that

$$
u_s^j(x^*) \quad - \quad v_0^j(s) - \int_0^{x^*} u_s^p(x)\Gamma_{p1}^j(x,s)\,dx\,,
$$

$$
u_0^l(x*) \quad = \quad w^j - \int_0^{x^*} u_0^p(x)\Gamma_{p1}^j(x,0)\,dx\,,
$$

We thus get two more terms to be multiplied by $-r$, namely the terms $\left(v_0^j(s) - w^j\right)\Gamma_{j1}^i(x^*,0)$ and

$$
\left(\int_0^{x^*} u_s^p(x)\Gamma_{p1}^j(x,s) - u_0^p(x)\Gamma_{p1}^j(x,0)\,dx\right)\Gamma_{j1}^i(x^*,0)\,.
$$

For some y^{***} between 0 and s, we have

$$v_0^j(s) - w^j = -\int_0^s v_0^p(y)\Gamma_{p2}^j(0,y)\,dy = -sv_0^p(y^{***})\Gamma_{p2}^j(0,y^{***})\,.$$

We thus find in our expression for $u_s^i(r) - v_r^i(s)$ a sum

$$rsv_0^p(y^{***})\Gamma_{p2}^j(0,y^{***})\Gamma_{j1}^i(x^*,0)\,,$$

and as $(r,s) \to (0,0)$, the coefficient of rs here tends to $w^p\Gamma_{p2}^j(0,0)\Gamma_{j1}^i(0,0)$, which is the same as $w^j\Gamma_{j2}^p(0,0)\Gamma_{p1}^i(0,0)$, since both j and p are dummy indices of summation. We are nearly done discussing the terms contributed by the coefficient of r. It remains only to consider the integral

$$\int_0^{x^*} u_s^p(x)\Gamma_{p1}^j(x,s) - u_0^p(x)\Gamma_{p1}^j(x,0)\,dx\,.$$

Fortunately, we don't need the exact value of this integral, and it suffices to estimate it. As we noted before the statement of the theorem, we have an inequality

$$|u_s^p(x) - u_0^p(x)| \le Ks$$

with some absolute constant K for all values of s, p, and x. A similar inequality holds for the Christoffel coefficient, that is, $|\Gamma_{p1}^j(x,s) - \Gamma_{p1}^j(x,0)| \le Ks$, and thus we find, since $x^* < r$, that this integral is at most Krs. In the expression for $u_s^i(r) - v_r^i(s)$, this term gets multiplied by r. Thus, the term containing r as a factor contributes to the expression for $u_s^i(r) - v_r^i(s)$ an amount

$$-rsw^j\left(\frac{\partial \Gamma_{j1}^i}{\partial x^2} - \Gamma_{j2}^p\Gamma_{p1}^i + E\right),$$

where from now on, all Christoffel symbols are to be evaluated at $(0,0)$, $|E| \le Kr + o(1)$, and $o(1)$ is a quantity that tends to zero. (This last term is the difference in the value of the Christoffel symbols and their derivatives at points near $(0,0)$ and the point $(0,0)$ itself.)

A similar analysis of the term containing a factor of s yields exactly analogous results, and when we subtract the latter from the former, we get the relation

$$\lim_{(r,s)\to(0,0)} \frac{1}{rs}\left(v_r^i(s) - u_s^i(r)\right) = \left(\frac{\partial \Gamma_{j2}^i}{\partial x^1} - \frac{\partial \Gamma_{j1}^i}{\partial x^2} + \Gamma_{p1}^i\Gamma_{j2}^p - \Gamma_{j1}^p\Gamma_{p2}^i\right)w^j$$

$$= \left(\frac{\partial \Gamma_{jl}^i}{\partial x^k} - \frac{\partial \Gamma_{jk}^i}{\partial x^l} + \Gamma_{pk}^i\Gamma_{jl}^p - \Gamma_{jk}^p\Gamma_{pl}^i\right)w^j\delta_1^k\delta_2^l$$

$$= R_{jkl}^i w^j u^k v^l\,,$$

where $\boldsymbol{u} = \partial/\partial x^1$ and $\boldsymbol{v} = \partial/\partial x^2$, so that $u^k = \delta_1^k$ and $v^l = \delta_2^l$. Thus we see that

$$\lim_{(r,s)\to(0,0)} \frac{1}{rs}\left(\boldsymbol{v}_r(s) - \boldsymbol{u}_s(r)\right) = R(\boldsymbol{w},\boldsymbol{u},\boldsymbol{v}) = R(\boldsymbol{u},\boldsymbol{v})\boldsymbol{w}\,. \qquad \square$$

4. The Exponential Mapping and Normal Coordinates

In this section, we shall construct a "canonical" system of coordinates about each point of a manifold. The reader unfamiliar with manifolds is referred to Appendix 4 in Volume 2 for basic information and examples. Some important relations show up clearly in this system of coordinates that are obscured when arbitrary systems are used. Of course, given that to us a manifold *is* a huge collection of coordinate mappings—we don't discuss what $\mathfrak{M}$ *intrinsically is*—the word *canonical* must be taken with reservations. The basis of this system of coordinates is found in what is called the *exponential map*, which we are about to explain. The use of the term *exponential* is better understood in the context of Lie groups, which are analytic manifolds that also happen to be topological groups. The tangent space at the identity of a Lie group can be made into a Lie algebra using the Lie bracket as multiplication, and then there is a natural map from that Lie algebra into the group $X \mapsto \exp(X)$ that has the basic exponential property $\exp((s+t)X) = \exp(sX)\exp(tX)$, where the multiplication is the group operation. For a fixed tangent vector X, the image of the line $\{tX : -\infty < t < \infty\}$ is a one-parameter subgroup (line or circle) that is a geodesic. An example is given in Problem 6.22 below.

For the sake of intuitive clarity, we appeal to an absolute n-dimensional C^∞-manifold $\mathfrak{M}$ embedded in a Euclidean space $\mathbb{R}^m$. (This is the last time we shall use this aid to visualization. After this, we shall discuss our ideas in the language of algebra and calculus.) Let $\boldsymbol{r} : U \to \mathbb{R}^m$ be an embedding of a connected open set $U \subset \mathbb{R}^n$ in this manifold (in other words, a parameterization of its image under $\boldsymbol{r}$). Thus, we have $\boldsymbol{r}(x^1,\ldots,x^n) = \left(y^1(x^1,\ldots,x^n),\ldots,y^m(x^1,\ldots,x^n)\right)$. Let $P_0 = (y_0^1,\ldots,y_0^m) = \boldsymbol{r}(x_0^1,\ldots,x_0^n)$ be a given point of $\mathfrak{M}$ in the image of $\boldsymbol{r}$, that is, $(x_0^1,\ldots,x_0^n) \in U$.

We require that $(x_0^1,\ldots,x_0^n)$ be a regular point of the parameterization. This means that the Jacobian matrix

$$J(x^1,\ldots,x^n) = \frac{\partial(y^1,\ldots,y^m)}{\partial(x^1,\ldots,x^n)} = \begin{pmatrix} \frac{\partial y^1}{\partial x^1} & \cdots & \frac{\partial y^1}{\partial x^n} \\ \vdots & \ddots & \vdots \\ \frac{\partial y^m}{\partial x^1} & \cdots & \frac{\partial y^m}{\partial x^n} \end{pmatrix}$$

is of full rank n at the point $(x_0^1,\ldots,x_0^n)$. (Naturally, this requirement entails that $m \geq n$.) The columns of this matrix, regarded as m-tuples in $\mathbb{R}^m$ are just the partial derivatives of the mapping $\boldsymbol{r}$, that is, the ith column is $\partial\boldsymbol{r}/\partial x^i$. Our requirement then is that the columns of this matrix be linearly independent.

The metric coefficients of $\mathfrak{M}$ near P are given by the matrix $g(x^1,\ldots,x^n) = (g_{ij})$, $1 \leq i,j \leq n$, where

$$g_{ij}(x^1,\ldots,x^n) = \frac{\partial\boldsymbol{r}}{\partial x^i} \cdot \frac{\partial\boldsymbol{r}}{\partial x^j} = g_{ji}(x^1,\ldots,x^n).$$

From what was just said, it is obvious that g_{ij} is the entry in row i, column j of the matrix $J^t J$. We claim that because J has full rank n, so does the matrix $g(x^1,\ldots,x^n)$. In fact, suppose $c^1,\ldots,c^n$ are constants such that $c^j g_{ij} = 0$ for $i = 1,2,\ldots,n$. This means

$$c^j \frac{\partial\boldsymbol{r}}{\partial x^i} \cdot \frac{\partial\boldsymbol{r}}{\partial x^j} = 0$$

for $i = 1,2,\ldots,n$. That says the vector $\boldsymbol{v} = c^j(\partial\boldsymbol{r}/\partial x^j)$ satisfies $\boldsymbol{v} \cdot (\partial\boldsymbol{r}/\partial x^i) = 0$ for all i and hence

$$\boldsymbol{v} \cdot \boldsymbol{v} = c^i \boldsymbol{v} \cdot (\partial\boldsymbol{r}/\partial x^i) = 0.$$

But this implies that $\boldsymbol{v} = \boldsymbol{0}$, and since the columns of J are linearly independent, that $c^i = 0$ for all i. Thus the matrix $g(x^1,\ldots,x^n)$ has an inverse $g^{-1}(x^1,\ldots,x^n)$ defined near $(x_0^1,\ldots,x_0^n)$. We denote the entries in the inverse matrix, as usual, by $g^{ij}(x^1\ldots,x^n)$.

4.1. Geodesics. We have been preoccupied with curvature in this chapter, but up to now we have not discussed geodesics, which were the main piece of differential geometry that we needed in the preceding chapter to discuss relativistic gravitation. We are now about to close this gap by using geodesics to construct a canonical parameterization in a neighborhood of the point P. For convenience we shall assume that $P = r(0, \ldots, 0)$.

Fix a nonzero tangent vector u at point P on $\mathfrak{M}$; let the coordinate expression of this vector be

$$u = u^i \frac{\partial r}{\partial x^i}.$$

There is a unique geodesic $\gamma_u(s)$, defined for s in a neighborhood of 0 and determined by the two conditions $\gamma_u(0) = P = r(x_0^1, \ldots, x_0^n)$ and $\gamma_u'(0) = u$, together with the system of Euler equations of the calculus of variations (see Appendix 2 in Volume 2 and Problem 4.15 above). If the parameter s is the arc length given by the metric corresponding to the matrix g, the tangent vector at each point will be a unit vector, and the Euler conditions will hold:

$$
\begin{aligned}
0 &= (x^i)''(s) + \Gamma^i_{kj}\big(x^1(s), \ldots, x^n(s)\big)(x^j)'(s)\,(x^k)'(s), \\
x^i(0) &= x_0^i, \\
(x^i)'(0) &= u^i/|u|.
\end{aligned}
$$

For technical reasons that will become obvious below, we find it better to change variables in these equations by the linear substitution $s = t|u|$, where[10]

$$|u| = \sqrt{u \cdot u} = \sqrt{g_{ij}\,u^i\,u^j}.$$

If $\gamma_u(t) = r\big(\tilde{x}^1(t), \ldots, \tilde{x}^n(t)\big)$, then $\tilde{x}^i(t) = x^i(s)$ and $(\tilde{x}^i)'(t) = (x^i)'(s)(ds/dt) = (x^i)'(s)|u|$. In particular $(\tilde{x}^i)'(0) = u^i$. The parameter is the "most natural" parameter to use for the particular geodesic whose tangent vector at P is u. Further, $(\tilde{x}^i)''(t) = |u|^2 (x^i)''(s)$, and so the factor $|u|^2$ cancels out of the second-order Euler differential equation when it is written in terms of t. Thus, we can now drop the tilde signs and rewrite the conditions that define $\gamma_u(t) = r\big(x^1(t), \ldots, x^n(t)\big)$ in terms of the parameter $t = s/|u|$:

$$
\begin{aligned}
0 &= (x^i)''(t) + \Gamma^i_{kj}\big(x^1(t), \ldots, x^n(t)\big)(x^j)'(t)\,(x^k)'(t), \\
x^i(0) &= x_0^i, \\
(x^i)'(0) &= u^i.
\end{aligned}
$$

The length of this path from the point $P = \gamma_u(0)$ to $Q = \gamma_u(t)$ is just the value of s, corresponding to t. That is, the length is $t|u|$. What we wish to show is that, if $|u|$ is sufficiently small, the domain of γ_u contains the whole interval $[0, 1]$. That gives us one parameter value $t = 1$ for which $\gamma_u(t)$ is defined in every direction u, and hence defined on a ball containing the origin in the tangent space.

As shown in Appendix 5, this initial-value problem has a unique solution defined for $|t| < \delta/2nM$, where M and δ depend only on the magnitude of the partial derivatives of the coefficients in the equation. Thus there are positive numbers ε

---

[10] At this point, the reader can probably see that the mapping r is superfluous here. What we really need amounts to just the mapping into the parameter space $s \mapsto \big(x^1(s), \ldots, x^n(s)\big)$ and the metric coefficients g_{ij}. The absolute vector u could perfectly well be replaced by the vector in the parameterized tangent space $u^i \partial/\partial x^i$, provided the Euclidean dot product is replaced by the natural inner product on the tangent space: $\langle u, v \rangle = g_{ij} u^i v^j = u^i \frac{\partial r}{\partial x^i} \cdot v^j \frac{\partial r}{\partial x^j} = u \cdot v$.

and η such that the solution is defined for $|t| < \eta$ whenever $|\boldsymbol{u}| < \varepsilon$. If $\boldsymbol{\gamma_u}(t)$ is this path, then $\boldsymbol{\delta_r}(t) = \boldsymbol{\gamma_u}(rt)$ is a solution of the same system of differential equations, passes through the same point at parameter value $t = 0$, and satisfies $\boldsymbol{\delta'_r}(0) = r\boldsymbol{u}$. By the uniqueness of the solution to this initial-value problem (proved in Appendix 5), this means that $\boldsymbol{\gamma_u}(rt) = \boldsymbol{\gamma_{ru}}(t)$. If $r < \eta$ and $|\boldsymbol{u}| < \varepsilon$, then $\boldsymbol{\gamma_u}(t)$ is defined at $t = r$, that is, $\boldsymbol{\gamma_{ru}}(t)$ is defined at $t = 1$. We have now achieved our goal:

DEFINITION 6.4. The *exponential mapping* $\exp(\boldsymbol{u})$ is the mapping $\boldsymbol{u} \mapsto \boldsymbol{\gamma_u}(1)$. In other words, if $0 < |\boldsymbol{u}| < \varepsilon\eta$, then $\exp(\boldsymbol{u}) = \boldsymbol{\gamma_u}(1) = \boldsymbol{r}(x^1(1), \dots, x^n(1))$.

4.2. Coordinates of the exponential mapping. Having started with an arbitrary (nonsingular) parameterization of a patch of the manifold $\mathfrak{M}$ near P, we have now produced a new parameterization of this patch that looks somehow "more natural." We made no assumptions as to the form of the initial parameterization. Since we started with an absolute embedding of $\mathfrak{M}$ in $\mathbb{R}^m$, which may be pictured as the solution of a system of $m - n$ independent equations $F_1(x^1, \dots, x^m) = 0, \dots,$ $F_{m-n}(x^1, \dots, x^m) = 0$, we can use the implicit function theorem to solve these equations, say—reordering the variables if necessary—for $x^{n+1}, \dots, x^m$ in terms of $x^1, \dots, x^n$ and thereby get an absolute parameterization as our starting point, that is, a mapping $\boldsymbol{r}(x^1, \dots, x^n) = (x^1, \dots, x^n, x^{n+1}(x^1, \dots, x^n), \dots, x^m(x^1, \dots, x^n))$. We mention this representation to show that what we are doing can still be anchored in absolute geometric language. After we prove Theorem 6.7 below, the parameterization $\boldsymbol{r}$ will have served its purpose, and we shall henceforth dispense with it, identifying a point P on an n-dimensional manifold $\mathfrak{M}$ with its coordinates $(x^1, \dots, x^n)$ in some coordinate chart $\psi : \mathfrak{M} \to \mathbb{R}^n$ and writing vectors as linear combinations of the derivation operators $\partial/\partial x^i$. To that end, the following sequence of computations is obvious:

$$
\begin{aligned}
(x^i)'(s) &= (x^i)'(0) + \int_0^s (x^i)''(r)\,dr \\[2mm]
&= u^i - \int_0^s \Gamma^i_{jk}(x^1(r), \dots, x^n(r))\,(x^j)'(r)\,(x^k)'(r)\,dr\,, \\[2mm]
x^i(t) &= x^i_0 + \int_0^t (x^i)'(s)\,ds \\[2mm]
&= x^i_0 + u^i t - \int_0^t \int_0^s \Gamma^i_{jk}(x^1(r), \dots, x^n(r))\,(x^j)'(r)\,(x^k)'(r)\,dr\,ds \\[2mm]
&= x^i_0 + u^i t - \int_0^t \int_r^t \Gamma^i_{jk}(x^1(r), \dots, x^n(r))\,(x^j)'(r)\,(x^k)'(r)\,ds\,dr \\[2mm]
&= x^i_0 + u^i t - \int_0^t (t - r)\Gamma^i_{jk}(x^1(r), \dots, x^n(r))\,(x^j)'(r)\,(x^k)'(r)\,dr\,.
\end{aligned}
$$

Since $x^i(1)$ is the value of the solution of the initial-value problem given above, it is determined by the initial conditions of the problem, one of which—the initial point x^i_0—is of no importance, being held fixed throughout. Thus we are justified in regarding $x^i(1)$ as a function of $\boldsymbol{u}$ and (following the Feller principle discussed in Appendix 6) writing it as $x^i(u^1, \dots, u^n)$. We then have

$$
\exp(u^1, \dots, u^n) = \boldsymbol{r}(x^1(u^1, \dots, u^n), \dots, x^n(u^1, \dots, u^n))\,,
$$

where
$$x^i(u^1, \ldots, u^n) = x_0^i + u^i - v^i(u^1, \ldots, u^n),$$
and, regarding x^j and $(x^k)'$ once again as functions of a real variable r,

$$v^i(u^1, \ldots, u^n) = \int_0^1 (1-r)\Gamma_{jk}^i\big(x^1(r), \ldots, x^n(r)\big)\, (x^j)'(r)\, (x^k)'(r)\, dr.$$

The exponential mapping $\boldsymbol{u} \mapsto \boldsymbol{\gamma_u}(1) = \exp(\boldsymbol{u})$ takes a tangent vector $\boldsymbol{u}$ satisfying $|\boldsymbol{u}| < \eta\varepsilon$ to a point of the manifold. Moreover $\boldsymbol{\gamma_u}(t) = \exp(t\boldsymbol{u})$. It is obvious from the smooth dependence of solutions of differential equations on the initial conditions that this is a C^∞ mapping. In this way, the exponential mapping allows us to use the tangent space as the natural source of a parameterization of a local piece of the manifold. To verify that, we need to show that the mapping $\exp(\boldsymbol{u})$ is nonsingular at the point $\boldsymbol{0}$.

THEOREM 6.7. *The exponential mapping is of full rank n in some neighborhood of the origin.*

PROOF. The exponential mapping is the composite mapping
$$\exp(u^1, \ldots, u^n) = \boldsymbol{r}\big(x^1(u^1, \ldots, u^n), \ldots, x^n(u^1, \ldots, u^n)\big).$$

It follows that its Jacobian is the product of the Jacobian of $\boldsymbol{r}$ at $(x_0^1, \ldots, x_0^n)$ and the Jacobian $\partial(x^1, \ldots, x^n)/\partial(u^1, \ldots, u^n)$. Thus, it suffices to show that the latter is nonsingular at $(0, \ldots, 0)$. In fact, it is the identity matrix, as we now show. At the point $(0, \ldots, 0)$, we have, for any sufficiently small *fixed* value of r

$$
\begin{aligned}
\frac{\partial x^i}{\partial u^j} &= \lim_{s \to 0} \frac{1}{s} \gamma^i_{s\,\frac{\partial \boldsymbol{r}}{\partial x^j}}(1) \\
&= \lim_{s \to 0} \frac{1}{s} \gamma^i_{\frac{s}{r}\, r\,\frac{\partial \boldsymbol{r}}{\partial x^j}}(1) \\
&= \lim_{s \to 0} \frac{1}{s} \gamma^i_{r\,\frac{\partial \boldsymbol{r}}{\partial x^j}}\Big(\frac{s}{r}\Big) \\
&= \frac{1}{r}\big(\gamma^i_{r\,\frac{\partial \boldsymbol{r}}{\partial x^j}}\big)'(0) = \frac{1}{r}\, r\Big(\frac{\partial \boldsymbol{r}}{\partial x^j}\Big)^i = \delta^i_j,
\end{aligned}
$$

since $\boldsymbol{r}(x^1, \ldots, x^n) = \big(x^1, \ldots, x^n, x^{n+1}(x^1, \ldots, x^n), \ldots, x^m(x^1, \ldots, x^n)\big)$.

Thus the exponential mapping is of full rank at the origin and by continuity also in some neighborhood of the origin. It therefore provides a coordinate chart on the manifold $\mathfrak{M}$. $\square$

This result implies that, given any basis of the tangent space, say $\boldsymbol{u}_1, \ldots, \boldsymbol{u}_n$, the mapping $\boldsymbol{r}(x^1, \ldots, x^n) = \exp(x^1\boldsymbol{u}_1 + \cdots + x^n\boldsymbol{u}_n)$ is a parameterization whose inverse $\psi : U \to \mathbb{R}^n$ is a local chart at P. This chart is the preferred one for some applications. Its advantage is that it is an isometry along radial lines in the tangent plane from the point of tangency, so that the image of each such line can be regarded as a "straight line" in the surface. All the non-Euclidean features of the manifold $\mathfrak{M}$ are then concentrated in the hypersurfaces of dimension $n-1$ orthogonal to these radial "lines." Moreover, being the mapping from the tangent space into the manifold that is an isometry along radial lines near the origin and "preserves directions" at that point (in the sense that for each *unit* vector $\boldsymbol{u}$, the derivative of the mapping $t \mapsto \exp(t\boldsymbol{u})$ at $t = 0$ is $\boldsymbol{u}$), the exponential mapping is the unique solution to the corresponding initial-value problem. That point has important implications, which we shall explore after looking at some examples.

EXAMPLE 6.2. The simplest nontrivial example is the sphere of radius r_0. We take the point P as $(0, 0, r_0)$. The geodesics through P are the intersections of the sphere with planes containing the z-axis. In cylindrical coordinates (r, θ, z), the equation of the sphere is $r^2 + z^2 = r_0^2$, and the length of the geodesic (great-circle) arc between $(0, \theta, r_0)$ and $\left(r, \theta, \sqrt{r_0^2 - r^2}\right)$ is $r_0 \arcsin(r/r_0)$. Thus the exponential map is given by $(r, \theta, r_0) \mapsto \left(r_0 \sin(r/r_0), \theta, r_0 \cos(r/r_0)\right)$.

We shall want to know what the metric coefficients are for this mapping, and we cannot compute them in cylindrical coordinates, since the mapping needs to be C^∞ throughout a neighborhood of the origin in the tangent space. That is not the case with cylindrical coordinates, since the polar angle θ is not even defined at the origin. In fact, when we use cylindrical coordinates, the surface is parameterized by the mapping $(r, \theta) \mapsto \left(r \cos \theta, r \sin \theta, \sqrt{r_0^2 - r^2}\right)$, and the metric coefficients are $g_{11} = 1 + r^2/(r_0^2 - r^2)$, $g_{12} = g_{21} = 0$, $g_{22} = r^2$. Since $g_{22} = 0$ at $r = 0$, the parameterization has a singularity at the north pole. For that reason, we state the exponential mapping in rectangular coordinates:

$$\exp(x, y) = \frac{r_0 x}{\sqrt{x^2 + y^2}} \sin\left(\frac{\sqrt{x^2 + y^2}}{r_0}\right) \boldsymbol{i}$$

$$+ \frac{r_0 y}{\sqrt{x^2 + y^2}} \sin\left(\frac{\sqrt{x^2 + y^2}}{r_0}\right) \boldsymbol{j} + r_0 \cos\left(\frac{\sqrt{x^2 + y^2}}{r_0}\right) \boldsymbol{k}.$$

This expression also appears to have a singularity at $x = 0 = y$, but actually doesn't. It can be rewritten as

$$\exp(x, y) = \left(x\varphi\left(\frac{x^2 + y^2}{r_0^2}\right), y\varphi\left(\frac{x^2 + y^2}{r_0^2}\right), r_0\psi\left(\frac{x^2 + y^2}{r_0^2}\right)\right),$$

where φ and ψ are analytic functions:

$$\varphi(t) = 1 - \frac{t}{3!} + \frac{t^2}{5!} - \frac{t^3}{7!} + \cdots$$

$$= \sum_{k=0}^{\infty} (-1)^k \frac{t^k}{(2k+1)!},$$

$$\psi(t) = 1 - \frac{t}{2!} + \frac{t^2}{4!} - \frac{t^3}{6!} + \cdots$$

$$= \sum_{k=0}^{\infty} (-1)^k \frac{t^k}{(2k)!}.$$

For nonnegative values of t, and in particular for the only value we need, namely $t = (x^2 + y^2)/r_0^2$, we have $\varphi(t) = \sin(\sqrt{t})/\sqrt{t}$ and $\psi(t) = \cos(\sqrt{t})$. As this very simple example shows, the computational aspect of the exponential mapping can be messy. In Problem 6.10 below you are invited to work out the metric coefficients in this parameterization of the upper hemisphere and show that, in polar coordinates

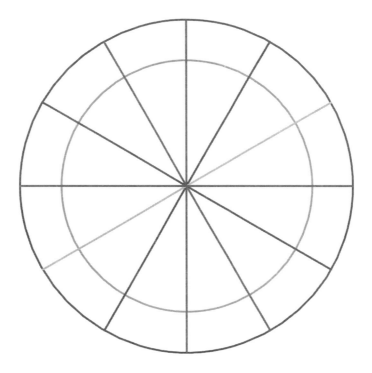

FIGURE 6.4. The exponential mapping at the north pole of a sphere. Top view.

(r, θ),

$$g_{11} = \cos^2 \theta + \frac{r_0^2}{r^2} \sin^2 \theta \sin^2 \left(\frac{r}{r_0}\right),$$

$$g_{12} = \cos \theta \sin \theta - \frac{r_0^2}{r^2} \cos \theta \sin \theta \sin^2 \left(\frac{r}{r_0}\right),$$

$$g_{22} = \sin^2 \theta + \frac{r_0^2}{r^2} \cos^2 \theta \sin^2 \left(\frac{r}{r_0}\right).$$

It follows that $g_{11} = 1 = g_{22}$ and $g_{12} = 0$ at $x = 0 = y$ ($r = 0$, θ arbitrary). In particular, for each fixed angle θ_0, along the radial path $\gamma(t) = \exp(t \cos \theta, t \sin \theta)$ we have $x = t \cos \theta_0$, $y = t \sin \theta_0$, that is, $r = t$ and $\theta = \theta_0$, from which it follows that

$$ds^2 = \left(g_{11}(x, y)(x'(t))^2 + 2g_{12}(x, y)\, x'(t)y'(t) + g_{22}(x, y)\, (y'(t))^2\right) dt^2 = dt^2,$$

so that the parameter $t = r$ is indeed arc length on γ.

The exponential mapping in this case is very transparent, being a wrapping of each tangent line along the line of longitude in whose plane the tangent line lies. The top view is shown in Fig. 6.4, and an oblique view in Fig. 6.5.

EXAMPLE 6.3. Our second example is the pseudo-hemisphere of radius[11] c (see Subsection 1.2 of Appendix 1), obtained by revolving the tractrix about the z-axis.

---

[11]We like to use the letter c for distances that have some application to relativity. Think of this c as the distance that light travels in unit time.

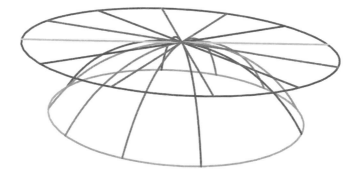

FIGURE 6.5. The exponential mapping at the north pole of a sphere. Oblique view.

Its equation in cylindrical coordinates (r, θ, z), $0 < r < c$, is

$$z = c\left(\operatorname{arcsech}\left(\frac{r}{c}\right) - \sqrt{1 - \left(\frac{r}{c}\right)^2}\right)$$

$$= c\ln\left(\frac{c + \sqrt{c^2 - r^2}}{r}\right) - \sqrt{c^2 - r^2}\,.$$

Note that this equation does not contain θ. Since we are not using the origin, we can use polar coordinates in the punctured disk and the mapping $(r, \theta) \mapsto r\cos\theta\,\boldsymbol{i} + r\sin\theta\,\boldsymbol{j} + z(r)\,\boldsymbol{k}$ as a parameterization. In those coordinates, the first fundamental form is

$$ds^2 = \frac{c^2}{r^2}\,dr^2 + r^2\,d\theta^2\,.$$

In rectangular coordinates, the matrix of metric coefficients is

$$g(x, y) = \begin{pmatrix} \frac{c^2 x^2 + r^2 y^2}{r^4} & \frac{(c^2 - r^2)xy}{r^4} \\ \frac{(c^2 - r^2)xy}{r^4} & \frac{c^2 y^2 + r^2 x^2}{r^4} \end{pmatrix},$$

where $r^2 = x^2 + y^2$, and the determinant of the matrix $g(x, y)$ works out to be simply c^2/r^2 at every point. Thus, it is a nonsingular matrix, as required.

The equations of a geodesic joining points with coordinates (r_1, θ_1) and (r_2, θ_2) are derived in Appendix 1, and it is shown that a geodesic on this surface is either a meridian with equation $\theta = \theta_0$ (when $\theta_1 = \theta_2 = \theta_0$) or a curve whose projection into the equatorial (punctured) disk has an equation of the form

(6.7) $$r(\theta) = \frac{cr_0}{\sqrt{c^2 - r_0^2(\theta - \theta_0)^2}}\,.$$

The two kinds of geodesics intersect at right angles at the point $\big(r_0, \theta_0, z(r_0)\big)$. We now wish to consider the entire family of geodesics whose projections pass through a given point (r_1, θ_1), which will not in general be the projection of the point $\big(r_0, \theta_0, z(r_0)\big)$ just mentioned. We leave it to the reader to verify that if $|a| < c$, the curve given by Eqs. (6.8) and (6.9) below can be transformed into Eq. (6.7) with $r_0 = \sqrt{c^2 - a^2}$ and $\theta_0 = \theta_1 + c\sqrt{a^2 - c^2 + r_1^2}/(r_1\sqrt{c^2 - a^2}) = \theta_1 + c\sqrt{r_1^2 - r_0^2}/(r_0 r_1)$.

To minimize the messiness of the computations, we shall assume $\theta_1 = 0$, and we shall consider only those geodesics whose tangents at the point $P = r_1\,\boldsymbol{i} + z(r_1)\,\boldsymbol{k}$ project into the portion of the quadrant of the xy-plane where $y > 0$ and $x < r_1$.

This family of curves is indexed by a parameter a satisfying $\sqrt{c^2 - r_1^2} \leq |a| \leq c$, and the projections of these curves into the punctured disk satisfy Eqs. (6.8)–(6.9). In general (not yet putting $\theta_1 = 0$), the equations of the geodesics that pass through $(r_1, \theta_1, z(r_1, \theta_1))$ are

$$(6.8) \quad r(t) = r_1 \cosh\left(\frac{t}{c}\right) - \sqrt{a^2 - (c^2 - r_1^2)} \sinh\left(\frac{t}{c}\right),$$

$$c \ln\left(\frac{c - a}{r_1 - \sqrt{a^2 - (c^2 - r_1^2)}}\right) < t < c \ln\left(\frac{c + a}{r_1 - \sqrt{a^2 - (c^2 - r_1^2)}}\right).$$

$$(6.9) \quad \theta(t) = \theta_1 - \frac{(c^2 - r_1^2)\sqrt{c^2 - a^2}}{r_1(ar_1 + c\sqrt{a^2 - (c^2 - r_1^2)})}$$

$$+ \frac{(c^2 - r(t)^2)\sqrt{c^2 - a^2}}{r(t)(ar(t) + c\sqrt{a^2 - (c^2 - r(t)^2)})} = \theta_0 + c\sqrt{\frac{1}{r_0^2} - \frac{1}{r^2}}.$$

In Eq. (6.8), it is assumed that $|a| < c$. (When $a = c$, the range on t is from $t = c \ln(r_1/c)$ to $t = +\infty$.) Equation (6.8) shows that $r(0) = r_1$, and then Eq. (6.9) shows that $\theta(0) = \theta_1$. The meridian geodesic $\theta = \theta_1$ corresponds to $a = c$. The geodesic that passes through the point $r_1 \cos\theta_1\, \boldsymbol{i} + r_1 \sin\theta_1\, \boldsymbol{j} + z(r_1, \theta_1)\, \boldsymbol{k}$ perpendicular to the meridian geodesic corresponds to $a = \sqrt{c^2 - r_1^2}$. In general, the geodesic whose tangent makes angle φ with the meridian corresponds to $a = \sqrt{c^2 - (r_1 \sin\varphi)^2}$.

From the last expression for $\theta(t)$, we deduce that

$$\left(\frac{d\theta}{dt}\right)^2 = \frac{c^2(c^2 - a^2)}{r^4(r^2 - c^2 + a^2)}\left(\frac{dr}{dt}\right)^2.$$

Standard identities satisfied by hyperbolic functions yield the identity

$$\left(\frac{r}{c}\right)^2 - \left(\frac{dr}{dt}\right)^2 = \frac{c^2 - a^2}{c^2},$$

so that

$$\left(\frac{dr}{dt}\right)^2 = \frac{a^2 - c^2 + r^2}{c^2}.$$

Combining the expression for the first fundamental form given above with the expressions just derived for the derivatives of r and θ, we find that

$$\left(\frac{ds}{dt}\right)^2 \equiv 1,$$

that is, the parameter t is arc length.

Figure 6.6 shows a view from directly above the point of tangency on the pseudo-hemisphere with four selected geodesics superimposed on their pre-images in the tangent plane. Figure 6.7 shows the same geodesics and is rotated to show how the surface is oriented with respect to the tangent plane.

4.3. Normal metric coefficients. We defined the exponential mapping as the mapping $(u^1, \ldots, u^n) \mapsto \boldsymbol{r}(x^1(1), \ldots, x^n(1))$, where the function $x^i(t)$ satisfies the Euler equation of the calculus of variations and two initial conditions. The Euler equation contains a Christoffel symbol that has to come from the initial parameterization $\boldsymbol{r}(x^1, \ldots, x^n)$, which, however, was arbitrary. So far as two vectors can be identified with each other by the usual change-of-variable formula involving

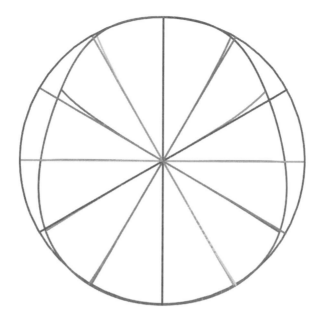

FIGURE 6.6. The exponential mapping on the pseudo-hemisphere. Top view.

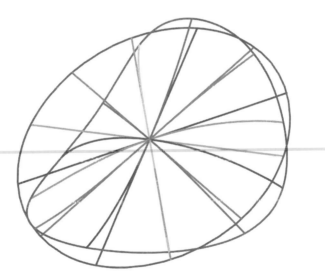

FIGURE 6.7. The exponential mapping on the pseudo-hemisphere. Oblique view.

the Jacobian, the exponential mapping is independent of the choice of initial parameterization. The exponential mapping is uniquely determined as the mapping that is an isometry on radial lines through the origin and preserves directions at that point. Thus, it will be the same mapping no matter what initial parameterization is chosen, provided only that the mapping is of full rank. But the exponential mapping itself is of full rank at $(0, \ldots, 0)$, and so, if we knew what it was—and we sometimes do, as the two preceding examples have shown—we could use the exponential mapping itself as the initial mapping to produce itself. This fact has consequences that are more profound than one might expect from such a tautological beginning. When the initial mapping r is the exponential mapping, we have $x^i(t) = tu^i$, and as a result:

$$
\begin{aligned}
x^i_0 &= 0, \\
(x^i)'(t) &\equiv u^i, \\
(x^i)''(t) &\equiv 0.
\end{aligned}
$$

It follows from the last two of these and the Euler equation that in normal coordinates $\Gamma^i_{jk}(tu^1, \ldots, tu^n)\, u^j\, u^k \equiv 0$ for all $t \in [0,1]$, all n-tuples $(u^1, \ldots, u^n)$ that are sufficiently small, and all $i = 1, \ldots, n$. One might think that this last identity makes all the Christoffel symbols vanish identically near the origin. That, of course, would be absurd, since these coordinates exist at every point of every manifold, while the vanishing of all the Christoffel symbols on an open set implies that the manifold is flat on that open set. What is true, however, is that all the Christoffel symbols vanish at the origin itself, that is, $\Gamma^i_{jk}(0, \ldots, 0) = 0$ for all i, j, k. This equality is not difficult to prove. Just take $t = 0$, and you find $\Gamma^i_{jk}(0, \ldots, 0)\, u^j\, u^k = 0$ for all $(u^1, \ldots, u^n)$. Taking $u^j = u \neq 0$, $u^k = 0$ for $k \neq j$ we find that $\Gamma^i_{jj}(0, \ldots, 0) = 0$. Then, if $j \neq k$, take $u^j = u^k = u \neq 0$ and $u^l = 0$ for $l \neq j$ and $l \neq k$. The result is that $0 = \Gamma^i_{jk}(0, \ldots, 0) + \Gamma^i_{kj}(0, \ldots, 0) = 2\Gamma^i_{jk}(0, \ldots, 0)$.

We can employ similar reasoning to get another fundamental fact about the Christoffel symbols.

THEOREM 6.8. *in normal coordinates:*

(6.10)
$$
\frac{\partial \Gamma^i_{jk}}{\partial u^l} + \frac{\partial \Gamma^i_{lj}}{\partial u^k} + \frac{\partial \Gamma^i_{kl}}{\partial u^j} = 0
$$

at the point $(0, \ldots, 0)$.

PROOF. This is a little more complicated than the previous fact, but still not difficult. We differentiate the equation $\Gamma^i_{jk}(tu^1, \ldots, tu^n) \equiv 0$ with respect to t and then set $t = 0$, getting the equality

$$
\frac{\partial \Gamma^i_{jk}}{\partial u^l}\, u^j\, u^k\, u^l \equiv 0
$$

for all sufficiently small values of u^j, u^k, and u^l. (Be careful to distinguish the two uses of the symbol u^l in this equation. As the variable of differentiation, it could be called anything. It is not a number. But in the argument of this function, it *is* a number.)

First, fix an index j and take $u^j = u \neq 0$ and $u^k = 0 = u^l$ for $k \neq j$ and $l \neq j$. The result is

$$
\frac{\partial \Gamma^i_{jj}}{\partial u^j} = 0,
$$

which implies Eq. (6.10) when $j = k = l$. Next, fix j and k with $j \neq k$, let $u^j = u \neq 0$ and $u^k = v \neq 0$, where u and v are distinct small (fixed) positive numbers, and let $u^l = 0$ for $l \neq j$ and $l \neq k$. The result is the equality

$$
\left(\frac{\partial \Gamma^i_{jk}}{\partial u^j} + \frac{\partial \Gamma^i_{jj}}{\partial u^k} + \frac{\partial \Gamma^i_{kj}}{\partial u^j} \right) u^2 v + \left(\frac{\partial \Gamma^i_{jk}}{\partial u^k} + \frac{\partial \Gamma^i_{kj}}{\partial u^k} + \frac{\partial \Gamma^i_{kk}}{\partial u^j} \right) uv^2 = 0.
$$

Since $uv \neq 0$, we cancel it and find

$$\left(\frac{\partial \Gamma^i_{jk}}{\partial u^j} + \frac{\partial \Gamma^i_{jj}}{\partial u^k} + \frac{\partial \Gamma^i_{kj}}{\partial u^j}\right)u + \left(\frac{\partial \Gamma^i_{jk}}{\partial u^k} + \frac{\partial \Gamma^i_{kj}}{\partial u^k} + \frac{\partial \Gamma^i_{kk}}{\partial u^j}\right)v = 0\,.$$

Since this equation must hold for all $u \neq 0$ and $v \neq 0$, the coefficients of u and v must both be zero. Thus, we have

$$\frac{\partial \Gamma^i_{jk}}{\partial u^j} + \frac{\partial \Gamma^i_{jj}}{\partial u^k} + \frac{\partial \Gamma^i_{kj}}{\partial u^j} = 0\,,$$

which is Eq. (6.10) for $l = j$, and obviously the same holds by symmetry if $l = k$ or $j = k$.

We are now left with just the case in which no two of the indices k, j, l are equal. In that case, we fix three distinct indices j, k, and l, and take $u^j = u^k = u^l = u \neq 0$. Using the symmetry of the Christoffel symbols in the two subscripts, and canceling $2u^3$, we get Eq. (6.10). $\square$

A useful consequence of Eq. (6.10) is the following identity.

COROLLARY 6.1. *In normal coordinates* $(u^1, \ldots, u^n)$, *the following equality holds at the point* $(0, \ldots, 0)$ *for all indices* j, k, l, r:

$$(6.11) \qquad \frac{\partial}{\partial u^r}\left(\frac{\partial g_{jk}}{\partial u^l} + \frac{\partial g_{lj}}{\partial u^k} + \frac{\partial g_{kl}}{\partial u^j}\right) = 0\,.$$

PROOF. Denote the left-hand side of Eq. (6.11) by L. According to Problem 4.12, we have

$$\frac{\partial g_{ij}}{\partial u^k} = \Gamma^m_{jk}g_{im} + \Gamma^m_{ik}g_{jm}\,;$$

and since all the derivatives of the metric coefficients therefore vanish at $(0, \ldots, 0)$, we have at that point

$$L = 2\left(\frac{\partial \Gamma^m_{jl}}{\partial u^r}g_{mk} + \frac{\partial \Gamma^m_{lk}}{\partial u^r}g_{mj} + \frac{\partial \Gamma^m_{kj}}{\partial u^r}g_{ml}\right),$$

which, by Eq. (6.10), yields

$$L = -2\left(\left(\frac{\partial \Gamma^m_{rj}}{\partial u^l} + \frac{\partial \Gamma^m_{lr}}{\partial u^j}\right)g_{mk} + \left(\frac{\partial \Gamma^m_{rl}}{\partial u^k} + \frac{\partial \Gamma^m_{kr}}{\partial u^l}\right)g_{mj} + \left(\frac{\partial \Gamma^m_{rk}}{\partial u^j} + \frac{\partial \Gamma^m_{jr}}{\partial u^k}\right)g_{ml}\right),$$

This last relation can be rearranged to yield

$$L = -2\left(\left(\frac{\partial \Gamma^m_{rj}}{\partial u^l}g_{mk} + \frac{\partial \Gamma^m_{rk}}{\partial u^l}g_{mj}\right) + \right.$$
$$\left. \left(\frac{\partial \Gamma^m_{rl}}{\partial u^j}g_{mk} + \frac{\partial \Gamma^m_{rk}}{\partial u^j}g_{ml}\right) + \left(\frac{\partial \Gamma^m_{rk}}{\partial u^l}g_{mj} + \frac{\partial \Gamma^m_{rj}}{\partial u^k}g_{ml}\right)\right).$$

Again differentiating the equality in Problem 4.12, we find that

$$L = -2L\,.$$

It therefore follows that $L = 0$, as asserted. $\square$

COROLLARY 6.2. *In normal coordinates* $(u^1, \ldots, u^n)$ *at the origin*

$$\frac{\partial \Gamma^i_{jk}}{\partial u^l} = -\frac{\partial^2 g_{kj}}{\partial u^l\, \partial u^i}\,.$$

and consequently the Riemann curvature tensor has components

$$
\begin{aligned}
R^i_{jkl} &= \frac{\partial \Gamma^i_{lj}}{\partial u^k} - \frac{\partial \Gamma^i_{kj}}{\partial u^l} \\
&= \frac{\partial^2 g_{kj}}{\partial u^l \, \partial u^i} - \frac{\partial^2 g_{lj}}{\partial u^k \, \partial u^i} \\
&= \frac{\partial}{\partial u^i} \left(\frac{\partial g_{kj}}{\partial u^l} - \frac{\partial g_{lj}}{\partial u^k} \right).
\end{aligned}
$$

PROOF. In fact, we have

$$
\frac{\partial \Gamma^i_{kj}}{\partial u^l} = \frac{1}{2} \frac{\partial g^{im}}{\partial u^l} \left(\frac{\partial g_{jm}}{\partial u^k} + \frac{\partial g_{mk}}{\partial u^j} - \frac{\partial g_{jk}}{\partial u^m} \right) + \frac{1}{2} g^{im} \left(\frac{\partial^2 g_{jm}}{\partial u^l \partial u^k} + \frac{\partial^2 g_{mk}}{\partial u^l \partial u^j} - \frac{\partial^2 g_{jk}}{\partial u^l \partial u^m} \right).
$$

At the origin, the first summation here vanishes, since all the partial derivatives of the metric coefficients vanish there. We therefore have only to deal with the second sum. Applying Eq. (6.11), we see that the sum of the first two terms in the second set of parentheses is

$$
-\frac{\partial^2 g_{kj}}{\partial u^l \, \partial u^m},
$$

so that

$$
\frac{\partial \Gamma^i_{kj}}{\partial u^l} = -g^{im} \frac{\partial^2 g_{kj}}{\partial u^l \, \partial u^m} = -\frac{\partial^2 g_{kj}}{\partial u^l \, \partial u^i}
$$

since $g^{im} = \delta^m_i$ at the origin.

Since Γ^i_{jk} also vanishes at the origin, there is very little left of the Riemann curvature tensor at this point, and it reduces to the expression given in the Corollary. □

REMARK 6.8. The coordinates $(u^1, \ldots, u^n)$ used by default for the analytic expression of the exponential mapping are those "inherited" from the original parameterization. The implied basis of the tangent space for which these are the coordinates of a point is $\{\partial/\partial x^1, \ldots, \partial/\partial x^n\}$. As a result, the matrix of metric coefficients at the point P in these coordinates is the same as it was in the original coordinates. But once we have the exponential mapping, we don't need the original parameterization any more. It therefore makes sense to change the basis of the tangent space into an orthonormal basis, in which case the matrix of metric coefficients at the point P becomes the identity matrix.

Because all the Christoffel symbols vanish at $(0, \ldots, 0)$ and the partial derivatives of the metric coefficients are linear combinations of those Christoffel symbols, it follows that the Maclaurin series of the metric coefficient g_{ij} in normal coordinates is

$$
g_{ij}(u^1, \ldots, u^n) = g_{ij}(0, \ldots, 0) + \frac{1}{2} \frac{\partial g_{ij}}{\partial u^p \, \partial u^q} u^p \, u^q + E,
$$

where the "approximation error" E is not larger than some absolute constant times $|\boldsymbol{u}|^3$. This result will be of importance below when we discuss the Laplace–Beltrami operator and the Ricci tensor.

5. Sectional Curvature

At this point, since we are are dispensing entirely with the embedding of a manifold in $\mathbb{R}^n$, it is useful to review the evolution of our notation. We began by considering parameterizations $\boldsymbol{r} : U \to \mathbb{R}^3$, where U is an open connected set of points (s, t) in $\mathbb{R}^2$. We found that arc length on the surface that is the image of this parameterization could be conveniently given by the first fundamental form $E \, ds^2 + 2F \, ds \, dt + G \, dt^2$, where

$$E = \frac{\partial \boldsymbol{r}}{\partial s} \cdot \frac{\partial \boldsymbol{r}}{\partial s} \qquad F = \frac{\partial \boldsymbol{r}}{\partial s} \cdot \frac{\partial \boldsymbol{r}}{\partial t} \qquad G = \frac{\partial \boldsymbol{r}}{\partial s} \cdot \frac{\partial \boldsymbol{r}}{\partial t},$$

When it came to finding the Gaussian curvature of such a surface (defined earlier by Euler and reformulated by Gauss using the spherical mapping), we needed second-order partial derivatives of $\boldsymbol{r}$, and that involved introducing the unit normal vector

$$\boldsymbol{n} = \frac{\frac{\partial \boldsymbol{r}}{\partial s} \times \frac{\partial \boldsymbol{r}}{\partial t}}{\left| \frac{\partial \boldsymbol{r}}{\partial s} \times \frac{\partial \boldsymbol{r}}{\partial t} \right|}.$$

That normal vector allowed us to define the standard Christoffel symbols Γ^i_{jk}, $i = 1, 2$ and the nonstandard Γ^3_{jk}, which we soon standardized by introducing a third parameter. Letting $s = x^1$ and $t = x^2$, we defined the Christoffel symbols by the equations

$$\frac{\partial^2 \boldsymbol{r}}{\partial x^j \, \partial x^k} = \Gamma^1_{jk} \frac{\partial \boldsymbol{r}}{\partial x^1} + \Gamma^2_{jk} \frac{\partial \boldsymbol{r}}{\partial x^2} + \Gamma^3_{jk} \boldsymbol{n}.$$

By introducing the systematic notation $E = g_{11}$, $F = g_{12} = g_{21}$, $G = g_{22}$, and the matrix

$$g = \begin{pmatrix} g_{11} & g_{12} \\ g_{21} & g_{22} \end{pmatrix},$$

with inverse

$$g^{-1} = \begin{pmatrix} g^{11} & g^{12} \\ g^{21} & g^{22} \end{pmatrix}$$

we were able to get a unified algebraic formula for the standard Christoffel symbols

$$\Gamma^i_{jk} = \frac{1}{2} g^{il} \left(\frac{\partial g_{jl}}{\partial x^k} + \frac{\partial g_{kl}}{\partial x^j} - \frac{\partial g_{jk}}{\partial x^l} \right),$$

where by the Einstein convention, the terms were summed on l over the range $l = 1$ to $l = 2$.

The curvature was at first expressed as a simple formula in the nonstandard Christoffel symbols Γ^3_{jk}, and accordingly, we standardized them by introducing a third variable x^3 and the modified mapping $\tilde{\boldsymbol{r}}(x^1, x^2, x^3) = \boldsymbol{r}(x^1, x^2) + x^3 \boldsymbol{n}(x^1, x^2)$. When we did so, we found that the 19 new Christoffel symbols could all be obtained by the simple device of allowing the indices i, j, k to range from 1 to 3 and the summation on l to extend from $l = 1$ to $l = 3$ in the algebraic formula already given for the Christoffel symbols.

Then, through intricate combinatorial work, we were able to express the curvature in terms of the standard Christoffel symbols and their derivatives, dispensing with the third variable x^3. All this was done to lay down groundwork for the more abstract treatment of the subject that we are now engaged in. We have repeatedly stressed that the true starting point of all this geometry is the set of metric coefficients g_{ij}. We don't actually need to know the mapping $\boldsymbol{r}$ that produces them as dot products of its derivatives. We have already seen that the ambient space $\mathbb{R}^3$ can be replaced by any Euclidean space $\mathbb{R}^n$, and this makes no change whatever in the definition of the curvature of a two-dimensional manifold, provided only that the Christoffel symbols continue to be given by the same formula, with the index of summation now running from 1 to n. (On the other hand, as the Whitney Embedding Theorem proved in Appendix 4 for compact manifolds makes clear, any compact two-dimensional manifold is equivalent in the manifold sense to a surface in $\mathbb{R}^5$.)

In accordance with these considerations, we shall henceforth leave the hypothetical parameterization r entirely out of the picture and work only with its domain, which is to be a connected open set U in $\mathbb{R}^n$, consisting of points $(x^1, \ldots, x^n)$. We *assume* that we know the $n \times n$ matrix g whose entries are the metric coefficients g_{ij}, along with its inverse g^{-1}, whose entries are g^{ij}, and we define vectors as directional derivative operators applying to real-valued C^∞-functions defined locally on U. We also define the n^3 Christoffel symbols Γ^i_{jk} by the formula above, letting all indices i, j, k, l range from 1 to n. Then, once again, we can define the Riemann curvature tensor $R(w, u, v) = R(u, v)w$ as above. Our intuitive anchor for all this is Theorem 6.6 above, which we proved in an abstract setting, and which shows that on the infinitesimal level (when u and v are small), $R(u, v)w$ is obtained by dividing the difference between the initial and final values obtained when w is parallel-transported around a small rectangle of sides u and v by the area of that rectangle in the parameter space.

In our original two-dimensional model, we found that curvature was expressed as $\langle R(u, v)v, u \rangle / (\langle u, u \rangle \langle v, v \rangle - (\langle u, v \rangle)^2)$, where $\langle u, v \rangle = g_{ij} u^i v^j$ is the natural inner product on the tangent space (equal to $u \cdot v$ if u and v are regarded as directional derivatives of the parameterization r in $\mathbb{R}^n$). Our aim in the present section is to see what geometric interpretation is to be given to this curvature when there are n parameters instead of only 2.

Although one might expect curvature on an n-dimensional manifold to be an "n-dimensional" entity—for example, given that the curvature of a circle of radius r is $1/r$ and the Gaussian curvature of a two-sphere $\mathbb{S}^2_r$ of radius r is $1/r^2$, it is natural to assume that the curvature of a three-sphere $\mathbb{S}^3_r$ would be $1/r^3$—it turns out that we can discuss curvature in general using two-dimensional *sections* of a manifold (two-dimensional geodesic submanifolds), and that the *sectional curvature* of the n-sphere $\mathbb{S}^n_r$ of radius r is $1/r^2$ for all $n \geq 2$. We are now going to define this sectional curvature and show that it is precisely what is given by the Riemann curvature tensor in accordance with the formula just written.

5.1. Two-dimensional submanifolds of a manifold.
We found the curvature of a surface in $\mathbb{R}^3$ in terms of the Christoffel coefficients by temporarily introducing a third parameter, which we subsequently set equal to zero. Geometrically, this meant regarding the surface we were interested in as a two-dimensional submanifold of a three-dimensional manifold. Te geometric interpretation of the curvature in the case of n parameters is to be obtained in precisely this way: Given two linearly independent tangent vectors u and v at a point, it will be the curvature at that point of some two-dimensional submanifold passing through that point, whose tangent space is spanned by u and v. Since there are infinitely many such submanifolds, the question we need to answer is, "Which one?" A one-dimensional analogy that illustrates this problem is obtained if we consider a fixed line tangent to a sphere at a point P. If the plane tangent to the sphere at P is rotated about that line, it intersects the sphere in a family of circles (one-dimensional submanifolds of the sphere), all having the given line as a tangent and whose radii range from 0 to the radius of the sphere. These circles have curvatures ranging from the curvature of a great circle to infinity. If we had to pick out one of these as

"canonical," we would undoubtedly choose the great circle, that is, the geodesic, which has minimal curvature.

Our task would be difficult if not for the canonical normal coordinates provided by the exponential mapping. Since the Riemann curvature tensor *is* a tensor, the sectional curvature we are going to define is the same number in any coordinate system, and we are now fortunate in having this canonical parameterization near any point P, in which the domain is an open ball about the origin in $\mathbb{R}^n$ and the origin maps to P. We can also assume that the matrix g is the identity at the origin, that the variables $(u^1, \ldots, u^n)$ are coordinates of an orthonormal system in $\mathbb{R}^n$, that all the Christoffel symbols vanish at the origin (so that all partial derivatives of the metric coefficients g_{ij} also vanish there), and that the image of each line through the origin is a geodesic. It is this canonical set of coordinates that makes it easy, given a pair of linearly independent tangent vectors $\boldsymbol{u}$ and $\boldsymbol{v}$ at P, to construct a two-dimensional submanifold through the point P whose tangent plane contains $\boldsymbol{u}$ and $\boldsymbol{v}$ (is the subspace of the tangent space spanned by $\boldsymbol{u}$ and $\boldsymbol{v}$) and whose curvature at P is given by the curvature formula we derived in terms of the Riemann curvature tensor. This submanifold is in fact the set of geodesics through P that are tangent to the plane spanned by $\boldsymbol{u}$ and $\boldsymbol{v}$.

THEOREM 6.9. *Let $\boldsymbol{u}$ and $\boldsymbol{v}$ be two linearly independent tangent vectors at a point P of a manifold $\mathfrak{M}$ of dimension larger than 2. The set of geodesics through the point P whose tangents at P lie in the subspace of the tangent space to $\mathfrak{M}$ spanned by $\boldsymbol{u}$ and $\boldsymbol{v}$ is a two-dimensional submanifold $\widetilde{\mathfrak{M}} \subset \mathfrak{M}$ whose curvature at P is*

$$\kappa(\boldsymbol{u}, \boldsymbol{v}) = \frac{\langle \boldsymbol{u}, R(\boldsymbol{u}, \boldsymbol{v})\boldsymbol{v} \rangle}{\langle \boldsymbol{u}, \boldsymbol{u} \rangle \langle \boldsymbol{v}, \boldsymbol{v} \rangle - \langle \boldsymbol{u}, \boldsymbol{v} \rangle^2}.$$

PROOF. In canonical coordinates at P, the description of $\widetilde{\mathfrak{M}}$ is the simplest possible. It consists of the points whose coordinates are of the form $(u^1, u^2, 0, \ldots, 0)$. This two-dimensional submanifold has its own metric coefficients $\tilde{g}_{ij}$, Christoffel symbols $\widetilde{\Gamma}^i_{jk}$, and Riemann curvature tensor $\widetilde{R}(\boldsymbol{u}, \boldsymbol{v})\boldsymbol{w}$ in terms of which its curvature is

$$\tilde{\kappa}(\boldsymbol{u}, \boldsymbol{v}) = \frac{\langle \boldsymbol{u}, \widetilde{R}(\boldsymbol{u}, \boldsymbol{v})\boldsymbol{v} \rangle}{\langle \boldsymbol{u}, \boldsymbol{u} \rangle \langle \boldsymbol{v}, \boldsymbol{v} \rangle - \langle \boldsymbol{u}, \boldsymbol{v} \rangle^2}.$$

The proof of this theorem amounts to showing that $\tilde{\kappa}(\boldsymbol{u}, \boldsymbol{v}) = \kappa(\boldsymbol{u}, \boldsymbol{v})$. Intuitively, the theorem says that when the Riemann curvature tensor is restricted to $\widetilde{\mathfrak{M}}$, we have $\langle \boldsymbol{u}, R(\boldsymbol{u}, \boldsymbol{v})\boldsymbol{v} \rangle = \langle \boldsymbol{u}, \widetilde{R}(\boldsymbol{u}, \boldsymbol{v})\boldsymbol{v} \rangle$. Without any loss of generality, since we are using canonical coordinates, we can assume that $\boldsymbol{u} = \partial/\partial u^1$ and $\boldsymbol{v} = \partial/\partial u^2$. We need to show that $R^1_{212} = \widetilde{R}^1_{212}$ at the origins in $\mathbb{R}^n$ and $\mathbb{R}^2$ respectively. We shall show more than that, namely that $R^i_{jkl} - \widetilde{R}^i_{jkl}$ at the origin for all $i, j, k, l = 1, 2$.

The proof is very short. First of all, the metric coefficients $\tilde{g}_{ij}(u^1, u^2)$ on $\widetilde{\mathfrak{M}}$, $i, j = 1, 2$, are inherited from the corresponding coefficients g_{ij} on $\mathfrak{M}$ via the equation

$$\tilde{g}_{ij}(u^1, u^2) = g_{ij}(u^1, u^2, 0, \ldots, 0)$$

The proof of this fact is straightforward. If $\tilde{\boldsymbol{\gamma}}(t) = \left(\tilde{u}^1(t), \tilde{u}^2(t)\right)$ is a path in $\widetilde{\mathfrak{M}}$, then it coincides with the path $\boldsymbol{\gamma}(t) = \left(u^1(t), u^2(t), 0, \ldots, 0\right)$ in $\mathfrak{M}$, where $u^i(t) = \tilde{u}^i(t)$, $i = 1, 2$, and hence the element of arc length along it is $ds^2 = g_{ij}\, du^i\, du^j$, where the summation on i and j runs from 1 to n. But, since $u^i \equiv 0$ for $i > 2$, it follows that $du^i\, du^j = 0$ if either i or j is larger than 2, and thus the sum actually extends only over $i, j = 1, 2$. We thus have $g_{ij}\, du^i\, du^j = ds^2 = \tilde{g}_{ij} d\tilde{u}^i\, d\tilde{u}^j = \tilde{g}_{ij}\, du^i\, du^j$, and therefore the stated equality must hold. Thus, for i and j equal to 1 or 2, the metric coefficients $\tilde{g}_{ij}(u^1, u^2)$ *and all their partial derivatives with respect to u^1 and u^2* coincide with the corresponding coefficients $g_{ij}(u^1, u^2, 0, \ldots, 0)$ and their partial derivatives with respect to the same variables.

It is obvious that the coordinates (u^1, u^2) are normal coordinates on $\widetilde{\mathfrak{M}}$, and in particular $\widetilde{\Gamma}^i_{jk}(0,0) = 0$ for all $i, j, k = 1, 2$, just as $\Gamma^i_{jk}(0, 0, \ldots, 0) = 0$ for all $i, j, k = 1, 2, \ldots, n$. That leaves very little of the Riemann curvature tensor to be accounted for, and it now suffices to show that

$$\frac{\partial \widetilde{\Gamma}^i_{jk}}{\partial u^l} = \frac{\partial \Gamma^i_{jk}}{\partial u^l},$$

where the left-hand side is evaluated at $(0,0)$ and the right-hand side at $(0, 0, \ldots, 0)$.

In fact, we have

$$\frac{\partial \widetilde{\Gamma}^i_{kj}}{\partial u^l} = \frac{1}{2}\frac{\partial \tilde{g}^{im}}{\partial u^l}\left(\frac{\partial \tilde{g}_{jm}}{\partial u^k} + \frac{\partial \tilde{g}_{mk}}{\partial u^j} - \frac{\partial \tilde{g}_{jk}}{\partial u^m}\right) + \frac{1}{2}\tilde{g}^{im}\left(\frac{\partial^2 \tilde{g}_{jm}}{\partial u^l\, \partial u^k} + \frac{\partial^2 \tilde{g}_{mk}}{\partial u^l\, \partial u^j} - \frac{\partial^2 \tilde{g}_{jk}}{\partial u^l\, \partial u^m}\right),$$

where the summation extends from $m = 1$ to $m = 2$, and

$$\frac{\partial \Gamma^i_{kj}}{\partial u^l} = \frac{1}{2}\frac{\partial g^{im}}{\partial u^l}\left(\frac{\partial g_{jm}}{\partial u^k} + \frac{\partial g_{mk}}{\partial u^j} - \frac{\partial g_{jk}}{\partial u^m}\right) + \frac{1}{2}g^{im}\left(\frac{\partial^2 g_{jm}}{\partial u \cdot \partial u^k} + \frac{\partial^2 g_{mk}}{\partial u^l\, \partial u^j} - \frac{\partial^2 g_{jk}}{\partial u^l\, \partial u^m}\right),$$

where the summation extends from $m = 1$ to $m = n$. Now all we have to do is note that (1) the first term on the right-hand side of both expressions vanishes at the origin of $\mathbb{R}^2$ or $\mathbb{R}^n$ respectively, since the Christoffel symbols vanish there in both cases, and (2) the summation on m in the second term amounts to just the single term where $m = i$, since $g^{im} = \delta^m_i = \tilde{g}^{im}$ at the origin. Thus, as long as i, j, k, l assume only the values 1 and 2, the two expressions coincide at the origin.

It then follows that $\widetilde{R}^i_{jkl} = R^i_{jkl}$ for that range of indices, and we are done. $\square$

DEFINITION 6.5. Let $\boldsymbol{u}$ and $\boldsymbol{v}$ be linearly independent tangent vectors to a manifold $\mathfrak{M}$ at a point P. The *sectional curvature* of $\mathfrak{M}$ determined by P and the subspace of the tangent space at P spanned by $\boldsymbol{u}$, and $\boldsymbol{v}$ is the number

$$\kappa(\boldsymbol{u}, \boldsymbol{v}) = \frac{\langle \boldsymbol{u}, R(\boldsymbol{u}, \boldsymbol{v})\boldsymbol{v}\rangle}{\langle \boldsymbol{u}, \boldsymbol{u}\rangle\langle \boldsymbol{v}, \boldsymbol{v}\rangle - \langle \boldsymbol{u}, \boldsymbol{v}\rangle^2}.$$

As a general rule, the computations involved in finding sectional curvature directly from the definition are lengthy. We shall illustrate this concept for just one simple case, that of the three-dimensional sphere $\mathbb{S}^3_{r_0}$ of radius r_0 in $\mathbb{R}^4$. *Mathematica* Notebook 10 in Volume 3 carries out the computations. The output from this notebook shows that the sectional curvature is $1/r_0^2$. Thus, the sectional curvature of the 3-sphere of radius r_0 has that constant value over any section, at any point.

6. The Laplace–Beltrami Operator

We are now almost finished creating the tiles from which the mosaic of general relativity can be assembled. Three points remain to be addressed. The first is that, while generalizing so much of differential geometry from surfaces in $\mathbb{R}^3$ to abstract manifolds, we have not brought along with us the fundamental object that Euler invoked to study the curvature of surfaces $z = f(x, y)$ tangent to the xy-plane, namely the Hessian matrix

$$H_f(x, y) = \begin{pmatrix} \frac{\partial^2 f}{\partial x^2} & \frac{\partial^2 f}{\partial y \, \partial x} \\ \frac{\partial^2 f}{\partial x \, \partial y} & \frac{\partial^2 f}{\partial y^2} \end{pmatrix}.$$

The second missing piece is the Laplacian operator, which plays a central role in classical physics. Since the metric coefficients are going to be playing a role analogous to potential energy in our geometrized theory of gravity in the next chapter, we need to explore the analog of the Laplacian on a general manifold and see what it means for the Laplacian of the metric coefficients to vanish, as the ordinary Laplacian does for Newtonian potential functions.

These two lacunae will be filled in the present section. The third and final missing piece is the actual nexus between geometry and physics. We shall discuss the role of the Laplacian in classical physics at the end of the present section and the application of the generalized Laplacian in relativity in the next chapter.

6.1. The Hessian and the Laplacian in $\mathbb{R}^3$. As Euler showed, the determinant of the Hessian matrix is the curvature of the surface $z = f(x, y)$ at a point where it is tangent to the xy-plane if the matrix of metric coefficients at that point is the identity, which we now temporarily assume. For 2×2 matrices M, the determinant is the constant term in its characteristic polynomial $\chi_M(\lambda)$, given by the equality

$$\chi_M(\lambda) = \det(\lambda I - M) = \lambda^2 - \mathrm{Tr}\,(M)\lambda + \det(M)\,,$$

where I is the 2×2 identity matrix.

The linear term in this relation is of independent interest. We have denoted its coefficient Tr, since it is generally called the *trace* of M. When M is the Hessian matrix $H_f(x, y)$, the trace is the well-known *Laplacian*:

$$\mathrm{Tr}\,\big(H(x, y)\big) = \frac{\partial^2 f}{\partial x^2} + \frac{\partial^2 f}{\partial y^2} = \nabla^2 f\,.$$

The eigenvalues of the Hessian matrix can then be expressed in terms of the Laplacian on the parameter space and the curvature as

$$\frac{\nabla^2 f \pm \sqrt{\big(\nabla^2 f + 2\sqrt{\kappa}\,\big)\big(\nabla^2 f - 2\sqrt{\kappa}\,\big)}}{2}\,.$$

The Laplacian can be decomposed into a sequence of two applications of a single formal vector operator ∇: $\nabla^2 f(x, y) = \nabla \cdot (\nabla f) = \mathrm{div}\,\mathrm{grad}\,\big(f(x, y)\big)$. It is named after Pierre-Simon Laplace (1749–1827), who introduced it in cylindrical coordinates in a work devoted to the study of gravity. It is a linear second-order differential operator, and so we might suspect that it has some connection with the covariant derivative of a vector field. It is sometimes denoted by the symbol Δ, and sometimes prefixed by a negative sign, since geometers prefer a positive-definite operator to a negative-definite one. One can easily verify that the Laplacian as we have defined it is negative-definite as an operator on a function space where the inner product of two continuous functions of compact support on $\mathbb{R}^n$ is

$$\langle f, g \rangle = \int_{\mathbb{R}^n} f(\boldsymbol{x}) g(\boldsymbol{x}) \, d\boldsymbol{x}\,,$$

that is, $\langle \nabla^2 f, f \rangle < 0$ for all nonzero C^2-functions f of compact support. (See Problem 6.13.)

The Hessian and the Laplacian are also connected with the second covariant derivative defined earlier. In fact, the second covariant derivative provides the natural definition of the Hessian in generalized coordinates. When that definition is made (see Subsection 6.6 below), it turns out that the double contraction of the Hessian is the correct generalization of the Laplacian to these coordinates. In order to bring out these points fully, we need to explore the operator ∇ whose iteration is the Laplacian. This is a part of vector analysis that is fairly well known, but we include a summary of it now so that we can efficiently discuss the generalization of the Laplacian to an arbitrary manifold.

6.2. The operators ∇ and d and the wedge product. The gradient-curl-divergence operator on $\mathbb{R}^3$ is

$$\nabla = \frac{\partial}{\partial x}\, \boldsymbol{i} + \frac{\partial}{\partial y}\, \boldsymbol{j} + \frac{\partial}{\partial z}\, \boldsymbol{k}\,.$$

It operates on scalar fields (functions) f and—in two ways—on vector fields $\boldsymbol{u} = P\,\boldsymbol{i} + Q\,\boldsymbol{j} + R\,\boldsymbol{k}$, as follows:

$$
\begin{aligned}
\nabla f &= \operatorname{grad} f = \frac{\partial f}{\partial x}\,\boldsymbol{i} + \frac{\partial f}{\partial y}\,\boldsymbol{j} + \frac{\partial f}{\partial z}\,\boldsymbol{k}\,, \\[4pt]
\nabla \times \boldsymbol{u} &= \operatorname{curl} \boldsymbol{u} = \left(\frac{\partial R}{\partial y} - \frac{\partial Q}{\partial z}\right)\boldsymbol{i} + \left(\frac{\partial P}{\partial z} - \frac{\partial R}{\partial x}\right)\boldsymbol{j} + \left(\frac{\partial Q}{\partial x} - \frac{\partial P}{\partial y}\right)\boldsymbol{k}\,, \\[4pt]
\nabla \cdot \boldsymbol{u} &= \operatorname{div} \boldsymbol{u} = \frac{\partial P}{\partial x} + \frac{\partial Q}{\partial y} + \frac{\partial R}{\partial z}\,.
\end{aligned}
$$

We could naturally identify the unit vectors $\boldsymbol{i}$, $\boldsymbol{j}$, and $\boldsymbol{k}$ with the partial derivative operators $\partial/\partial x$, $\partial/\partial y$, and $\partial/\partial z$. With that identification, only the curl has any possibility of representing a covariant derivative, mapping vectors to vectors. Since the space $\mathbb{R}^3$ is flat, all of its Christoffel symbols are zero, and it is easy to compute that

$$\nabla_{\boldsymbol{u}}\left(P\,\frac{\partial}{\partial x} + Q\,\frac{\partial}{\partial y} + R\,\frac{\partial}{\partial z}\right) = (\boldsymbol{u} \cdot \nabla P)\,\frac{\partial}{\partial x} + (\boldsymbol{u} \cdot \nabla Q)\,\frac{\partial}{\partial y} + (\boldsymbol{u} \cdot \nabla R)\,\frac{\partial}{\partial z}\,,$$

which is nothing like the curl of $\boldsymbol{u}$ as defined above.

We have used the notation $\nabla_{\boldsymbol{u}}$ for the covariant derivative with respect to a vector $\boldsymbol{u}$, and $\nabla^2_{\boldsymbol{u},\boldsymbol{v}}$ for the second covariant derivative

$$\nabla^2_{\boldsymbol{u},\boldsymbol{v}} = \nabla_{\boldsymbol{u}}\big(\nabla_{\boldsymbol{v}}\big) - \nabla_{\nabla_{\boldsymbol{u}}\boldsymbol{v}}\,.$$

As we shall see below (Theorem 6.10), the Laplace–Beltrami operator that we are going to introduce is related to the second covariant derivative, justifying to some extent the slight risk of confusing them.

On real-valued functions f defined on an open set in $\mathbb{R}^3$ the Laplacian is given in rectangular, cylindrical, and spherical (longitude and co-latitude) coordinates respectively by the formulas

$$\nabla^2 f = \frac{\partial^2 f}{\partial x^2} + \frac{\partial^2 f}{\partial y^2} + \frac{\partial^2 f}{\partial z^2},$$

$$\nabla^2 f = \frac{1}{r}\frac{\partial}{\partial r}\left(r\frac{\partial f}{\partial r}\right) + \frac{1}{r^2}\frac{\partial^2 f}{\partial\theta^2} + \frac{\partial^2 f}{\partial z^2},$$

$$\nabla^2 f = \frac{1}{\rho^2\sin\varphi}\left(\sin\varphi\frac{\partial}{\partial\rho}\left(\rho^2\frac{\partial f}{\partial\rho}\right) + \frac{\partial}{\partial\varphi}\left(\sin\varphi\frac{\partial f}{\partial\varphi}\right) + \frac{1}{\sin\varphi}\frac{\partial^2 f}{\partial\varphi^2}\right).$$

The three coordinate systems are related by the usual transformations, for example:

$$x = r\cos\theta = \rho\cos\theta\,\sin\varphi,$$
$$y = r\sin\theta = \rho\sin\theta\,\sin\varphi,$$
$$z = \rho\cos\varphi.$$

The polar distances r and ρ vary in the open interval between 0 and $+\infty$, the longitude angle θ in the open interval from $-\pi$ to π, and the co-latitude angle[12] φ in the open interval from 0 to $\pi/2$. These coordinate transformations are undefined on the closed half-line (half-plane in $\mathbb{R}^3$) where $y = 0$ and $x \leq 0$.

REMARK 6.9. This classical vector notation models both the tangent space and the cotangent spaces, which it is convenient to distinguish when discussing general manifolds. The basis vectors $\boldsymbol{i}$, $\boldsymbol{j}$, and $\boldsymbol{k}$ can be associated with the basis vectors $\partial/\partial x$, $\partial/\partial y$, and $\partial/\partial z$ of the tangent space, or with the dual basis dx, dy, and dz of the cotangent space (the space of 1-forms) and also with the basis $dy \wedge dz$, $dz \wedge dx$, $dx \wedge dy$ of the space of 2-forms (defined below). The identification with differential forms is more useful, because of the concept of *duality* on k-forms, which we are about to define. By using differential forms instead of partial derivative operators, we shall be able to define an operator like the Laplacian on any manifold.

DEFINITION 6.6. A *k-form* ω in coordinates $\{x^1, \ldots, x^n\}$ is a k-linear functional of ordered sets of k vector fields $(\boldsymbol{u}_1, \ldots, \boldsymbol{u}_k)$ that is *alternating*, which means that

$$\omega(\boldsymbol{u}_1, \ldots, \boldsymbol{u}_{i-1}, \boldsymbol{u}_i, \boldsymbol{u}_{i+1} \ldots, \boldsymbol{u}_{j-1}, \boldsymbol{u}_j, \boldsymbol{u}_{j+1}, \ldots, \boldsymbol{u}_k)$$
$$= -\omega(\boldsymbol{u}_1, \ldots, \boldsymbol{u}_{i-1}, \boldsymbol{u}_j, \boldsymbol{u}_{i+1}, \ldots, \boldsymbol{u}_{j-1}, \boldsymbol{u}_i, \boldsymbol{u}_{j+1}, \ldots, \boldsymbol{u}_k).$$

In other words, it reverses sign if two of its arguments are interchanged.

In particular, any covector is a 1-form, since there is no possibility of interchanging any arguments when there is only one argument.

The *tensor product* of two one-forms (differentials) is denoted $dx^k \otimes dx^l$ and defined by the relation

$$(dx^k \otimes dx^l)(\boldsymbol{u}_i, \boldsymbol{u}_j) = dx^k(\boldsymbol{u}_i)dx^l(\boldsymbol{u}_j).$$

The tensor product is bilinear, but not generally alternating. It becomes alternating, however, if we replace the tensor product operation $\otimes$ with the *wedge product* operation $\wedge$, defined by the formula

$$(dx^k \wedge dx^l)(\boldsymbol{u}_i, \boldsymbol{u}_j) = dx^k \otimes dx^l(\boldsymbol{u}_i, \boldsymbol{u}_j) - dx^l \otimes dx^k(\boldsymbol{u}_i, \boldsymbol{u}_j).$$

---

[12]The use of co-latitude is a concession to the preference of physics textbooks. It introduces a completely unnecessary singularity at the equator, and thereby confines us to the northern hemisphere.

If $\boldsymbol{u}_i = u_i^m \partial/\partial x^m$ and $\boldsymbol{u}_j = u_j^n \partial/\partial x^n$, we have

$$(dx^k \wedge dx^l)(\boldsymbol{u}_i, \boldsymbol{u}_j) = u_i^k u_j^l - u_i^l u_j^k = \det \begin{pmatrix} u_i^k & u_j^k \\ u_i^l & u_j^l \end{pmatrix}.$$

If $(u^1, \ldots, u^n)$ are new coordinates in which $\boldsymbol{u}_i = \partial/\partial u^i$, then, as we know from the chain rule

$$\boldsymbol{u}_i = \frac{\partial}{\partial u^i} = \frac{\partial x^k}{\partial u^i} \frac{\partial}{\partial x^k},$$

so that in the coordinates $\{x^1, \ldots, x^n\}$ we have

$$u_i^k = \frac{\partial x^k}{\partial u^i},$$

and so

$$dx^k \wedge dx^l = \det \left(\frac{\partial(x^k, x^l)}{\partial(u^i, u^j)} \right) du^i \wedge du^j.$$

Because of this transformation law, the wedge product is a tensor of type $(0, 2)$. We can create alternating k-linear functionals by "antisymmetrizing" tensor products. The process is sufficiently straightforward that we give just the example of a 3-form: If $\boldsymbol{u}_l = u_l^m \partial/\partial x^m$, then

$$dx^i \wedge dx^j \wedge dx^k(\boldsymbol{u}_1, \boldsymbol{u}_2, \boldsymbol{u}_3) = \det \begin{pmatrix} u_1^i & u_1^j & u_1^k \\ u_2^i & u_2^j & u_2^k \\ u_3^i & u_3^j & u_3^k \end{pmatrix}.$$

6.3. Duality. When vectors on $\mathbb{R}^3$ are identified with 1-forms and 2-forms on $\mathbb{R}^3$, as illustrated above with the operator ∇, a function $f(x, y, z)$ corresponds to itself, regarded as a 0-form, and also to its *adjoint* three-form $*f = f\, dx \wedge dy \wedge dz$; a vector $P\boldsymbol{i} + Q\boldsymbol{j} + R\boldsymbol{k}$ corresponds to the 1-form $\omega = P\, dx + Q\, dy + R\, dz$ and also to its adjoint 2-form $*\omega = P\, dy \wedge dz + Q\, dz \wedge dx + R\, dx \wedge dy$. On 2-forms and 3-forms in three variables the adjoint operation $*$ (called the *Hodge star*[13]) is the inverse of what it is on 1-forms and 0-forms respectively, so that $*(*\omega) = \omega$ in all cases. In n variables, we shall find below that $*(*\omega) = (-1)^{k(n-k)}\omega$ for k-forms ω.

The differential operator d that maps k-forms into $(k+1)$-forms corresponds to the gradient when acting on 0-forms (functions), to the curl when acting on 1-forms, and to the divergence when acting on 2-forms. It is the zero mapping on 3-forms. It has the important property that $d(d\omega) = 0$ for any k-form ω. In vector language, the curl of the gradient of a function is the zero vector, and the divergence of the curl of a vector is the zero function; both of these properties follow from the equality of mixed partial derivatives. By definition, on k-forms in n variables $\{x^1, \ldots, x^n\}$, the differential is the linear operator d such that if

$$\omega = P\, dx^{i_1} \wedge \cdots \wedge dx^{i_k},$$

then

$$\begin{aligned} d\omega &= (dP) \wedge dx^{i_1} \wedge \cdots \wedge dx^{i_k} \\ &= \frac{\partial P}{\partial x^j} dx^j \wedge dx^{i_1} \wedge \cdots \wedge dx^{i_k}. \end{aligned}$$

---

[13]Named after William Vallance Douglas Hodge (1903–1975).

We can define an operator δ that goes on the opposite direction from d—that is, it maps k-forms into $(k-1)$-forms—by letting $\delta(\omega) = *(d(*\omega))$. (Some textbooks take this as the definition of $-\delta$.) Thus, in three variables,

$$\delta(f) = 0 \quad \text{for 0-forms } f,$$

$$\delta(P\,dx + Q\,dy + R\,dz) = *d\big(P\,dy \wedge dz + Q\,dz \wedge dx + R\,dx \wedge dy\big)$$

$$= *\Big(\frac{\partial P}{\partial x} + \frac{\partial Q}{\partial y} + \frac{\partial R}{\partial z}\Big)\,dx \wedge dy \wedge dz$$

$$= \frac{\partial P}{\partial x} + \frac{\partial Q}{\partial y} + \frac{\partial R}{\partial z}\,.$$

$$\delta(P\,dy \wedge dz + Q\,dz \wedge dx + R\,dx \wedge dy) = *d(P\,dx + Q\,dy + R\,dz)$$

$$= *\Big(\Big(\frac{\partial R}{\partial y} - \frac{\partial Q}{\partial z}\Big)dy \wedge dz + \Big(\frac{\partial P}{\partial dz} - \frac{\partial R}{\partial x}\Big)dz \wedge dx + \Big(\frac{\partial Q}{\partial x} - \frac{\partial P}{\partial y}\Big)dx \wedge dy\Big)$$

$$= \Big(\frac{\partial R}{\partial y} - \frac{\partial Q}{\partial z}\Big)dx + \Big(\frac{\partial P}{\partial z} - \frac{\partial R}{\partial x}\Big)dy + \Big(\frac{\partial Q}{\partial x} - \frac{\partial P}{\partial y}\Big)dz\,,$$

$$\delta(f\,dx \wedge dy \wedge dz) = *df$$

$$= *\Big(\frac{\partial f}{\partial x}\,dx + \frac{\partial f}{\partial y}\,dy + \frac{\partial f}{\partial z}\,dz\Big)$$

$$= \frac{\partial f}{\partial x}\,dy \wedge dz + \frac{\partial f}{\partial y}\,dz \wedge dx + \frac{\partial f}{\partial z}\,dx \wedge dy\,.$$

Like the differential operator d, the operator δ satisfies $\delta(\delta\omega) = 0$, again because of the equality of mixed partial derivatives.

The Laplacian $\nabla^2 f$ for a function (0-form) f then has a succinct expression: $\nabla^2 f = \delta(df)$. In fact, we can define the Laplacian of any k-form on $\mathbb{R}^3$ by setting $\nabla^2(\omega) = d(\delta\omega) + \delta(d\omega)$, and the reader can verify that this definition is the same as the one already given for 0-forms.[14]

All of this generalizes to any manifold, but there are some subtleties involved, since what has been done up to here applies only in the case when the matrix of metric coefficients g_{ij} is an orthogonal matrix at each point. This hypothesis is trivially satisfied in $\mathbb{R}^3$. We are now about to generalize it to an arbitrary coordinate system.

A generalization of the operator ∇^2 was introduced for a manifold having metric coefficients $g_{ij}(x^1,\ldots,x^n)$ by Eugenio Beltrami (1835–1900) in an 1868 paper on spaces of constant curvature. The generalization is obvious for a flat space such as $\mathbb{R}^n$, where, in rectangular coordinates, the operator ∇^2 is the standard Laplacian:

$$\nabla^2 f = \sum_{i=1}^{n} \frac{\partial^2 f}{\partial(x^i)^2}\,,$$

while in cylindrical and spherical (longitude and co-latitude) coordinates on $\mathbb{R}^3$ it has the forms given above (Problem 6.12).

---

[14]In this form, the operator with a negative sign attached is called the *Laplace–de Rham* operator, after Georges de Rham (1903–1990).

6.4. Invariant definition. It is not obvious that the Laplacian ∇^2 just defined is a tensor, since in general second-order differential operators are not tensors. But if we take the definition $\nabla^2 = d\delta + \delta d$, we can see that such is the case, since both d and δ are tensor operations. This is already clear in the case of d, which maps tensors of type $(0, k)$ to tensors of type $(0, k + 1)$. (Since an alternating tensor retains its alternating character when coordinates are changed, restricting d to alternating tensors of type $(0, k)$—that is, to k-forms—does not change its tensor character.) It remains to be shown that δ is also a tensor. In view of the definition $\delta = *d*$, it suffices to show that the adjoint operation $*$ is "tensorial," and we have not yet given the general definition of this operation. We shall now do so. After we give that definition and verify that it is tensorial, we still need to show (Theorem 6.11 below) that ∇^2 as defined by Beltrami coincides with the restriction of δd to 0-forms.

Here is how the operation $*$ works in general coordinates. First, we choose, in a smooth way, an orthonormal basis $\{\boldsymbol{u}_1, \ldots, \boldsymbol{u}_n\}$ of the tangent space at each point P—orthonormal, that is, relative to the standard inner product $\langle \boldsymbol{u}, \boldsymbol{v} \rangle = g_{pq} u^p v^q$— with dual basis $\{\boldsymbol{v}_1, \ldots, \boldsymbol{v}_n\}$ of the cotangent space. This is easily done in a number of ways. We shall demonstrate below how it can be done by changing coordinates in a local chart. For now, just assume that $\boldsymbol{u}_i = u_i^j (\partial/\partial x^j)$, where $(x^1, \ldots, x^n)$ are the local coordinates. Then let $\boldsymbol{v}_1, \ldots, \boldsymbol{v}_n$ be the dual base of the co-tangent space, that is, $\boldsymbol{v}_i(\boldsymbol{u}_j) = \delta_j^i$. If the matrix[15] (u_i^j) has inverse (v_j^i), then $\boldsymbol{v}_i = v_j^i \, dx^j$. The definition of $*$ is as follows.

Let $\omega = \boldsymbol{v}_{i_1} \wedge \cdots \wedge \boldsymbol{v}_{i_k}$, where $1 \leq i_1 < \cdots < i_k \leq n$. Let $j_1 < \cdots < j_{n-k}$ be the remaining indices, that is $\{j_1, \ldots, j_{n-k}\} = \{1, \ldots, n\} \setminus \{i_1, \ldots, i_k\}$. (Important: Note that the sets of indices $\{i_l\}$ and $\{j_l\}$ are both arranged in *ascending* order.) We define

$$*\omega = \varepsilon(i_1, \ldots, i_k, j_1, \ldots, j_{n-k}) \, \boldsymbol{v}_{j_1} \wedge \cdots \wedge \boldsymbol{v}_{j_{n-k}},$$

where $\varepsilon(i_1, \ldots, i_k, j_1, \ldots, j_{n-k})$ is $+1$ if $(i_1, \ldots, i_k, j_1, \ldots, j_{n-k})$ is an even permutation of $(1, 2, \ldots, n)$ and -1 if this permutation is odd. It is easy to prove that $\varepsilon(m_1, \ldots, m_n) = (-1)^N$ for any permutation $(m_1, \ldots, m_n)$ of $(1, \ldots, n)$, where N is the number of inversions in the permutation (the number of ordered pairs (m_i, m_j) with $i < j$ and $m_i > m_j$).

Since the reader can easily verify that the parity of $(i_1, \ldots, i_k, j_1, \ldots, j_{n-k})$ differs from that of $(j_1, \ldots, j_{n-k}, i_1, \ldots, i_k)$ by the factor $(-1)^{k(n-k)}$ (Problem 6.9), it follows that $*(*\omega) = (-1)^{k(n-k)}\omega$. It is important to keep the signs straight here. The sign of each term is the parity of the permutation of the n-form $\boldsymbol{v}_1 \wedge \cdots \wedge \boldsymbol{v}_n$ obtained by putting the indices of the *image* wedge product $\boldsymbol{v}_{j_1} \wedge \cdots \wedge \boldsymbol{v}_{j_{n-k}}$ in order, *after* those of the pre-image $\boldsymbol{v}_{i_1} \wedge \cdots \wedge \boldsymbol{v}_{i_k}$, that is, by looking at the n-form

$$\boldsymbol{v}_{i_1} \wedge \cdots \wedge \boldsymbol{v}_{i_k} \wedge \boldsymbol{v}_{j_1} \wedge \cdots \wedge \boldsymbol{v}_{j_{n-k}}.$$

We note that if the number n of variables is odd, then either k or $n - k$ is even, so that $*(*\omega) = \omega$. If n is even, $*(*\omega) = (-1)^k \omega$ for each k-form ω.

---

[15]There seem to be advantages and disadvantages in all ways of writing matrices. In these change-of-coordinate formulas, it seems advisable to use a subscript to indicate the row of an entry and a superscript to indicate the column. Thus u_i^j denotes the entry in row i, column j. On the other hand, the matrix of metric coefficients $g = (g_{ij})$, since it represents a tensor of type $(0, 2)$, really needs to have both of its indices subscripted, and this is particularly convenient as a way of denoting its inverse $g^{-1} = (g^{ij})$. For much of what we are doing in the present section, the matrices involved are symmetric, so that the distinction between rows and columns is not crucial.

To define the operation $*\omega$ in general coordinates $\{x^1, \ldots, x^n\}$, once we have done so with a given "starter" set of orthonormal basis vectors $\{\boldsymbol{u}_1, \ldots, \boldsymbol{u}_n\}$ in the tangent space, we first use the change-of-coordinate formulas for the wedge products to get correspondences between k-forms expressed as wedge products of the differentials dx and k-forms expressed as wedge products of the one-forms $\boldsymbol{v}_k$, starting from the two basic formulas $f(x^1, \ldots, x^n) = f(x^1, \ldots, x^n)$ for 0-forms and $dx^i = u^i_j \, \boldsymbol{v}_j$ for 1-forms, where u^i_j is such that

$$\boldsymbol{u}_j = u^i_j \frac{\partial}{\partial x^i}.$$

In that case, if (v^j_i) is the inverse of the matrix (u^i_j), we have

$$\boldsymbol{v}_j = v^j_i \, dx^i.$$

and $v^j_i u^k_j = \delta^k_i$. Because of this inverse relationship, we can get by with just the matrix (u^i_j).

Using the equations $\boldsymbol{v}_i(\boldsymbol{u}_j) = \delta^i_j = dx^i(\partial/\partial x^j)$, one can then set up a table of correspondences:

$$dx^{i_1} \wedge \cdots \wedge dx^{i_k} = \det \begin{pmatrix} u^{i_1}_{j_1} & \cdots & u^{i_1}_{j_k} \\ \vdots & \ddots & \vdots \\ u^{i_k}_{j_1} & \cdots & u^{i_k}_{j_k} \end{pmatrix} \boldsymbol{v}_{j_1} \wedge \cdots \wedge \boldsymbol{v}_{j_k},$$

where the right-hand side is to be summed over all k-tuples $(j_1, \ldots, j_k)$. (Only those for which no two of the indices are equal actually count, and all $k!$ permutations of a given *set* of indices $\{j_1, \ldots, j_k\}$ can be consolidated and written canonically as a single term with the indices in ascending order.)

Although this definition appears to be a bit messy, we really need it only for the k-forms involved in the Laplace–Beltrami operator on functions, that is, for 0-forms, 1-forms, $(n-1)$-forms, and n-forms.

For a 0-form $f(x^1, \ldots, x^n)$ we have $*f(x^1, \ldots, x^n) = f(x^1, \ldots, x^n)\, \boldsymbol{v}_1 \wedge \cdots \wedge \boldsymbol{v}_n$. Conversely, for an n-form we have $*f(x^1, \ldots, x^n)\, \boldsymbol{v}_1 \wedge \cdots \wedge \boldsymbol{v}_n = f(x^1, \ldots, x^n)$.

For a 1-form $\omega = P_1 \, \boldsymbol{v}_1 + P_2 \, \boldsymbol{v}_2 + \cdots + P_n \, \boldsymbol{v}_n$,

$$*\omega = P_1 \, \boldsymbol{v}_2 \wedge \cdots \wedge \boldsymbol{v}_n - P_2 \, \boldsymbol{v}_1 \wedge \boldsymbol{v}_3 \wedge \cdots \wedge \boldsymbol{v}_n + \cdots + (-1)^{n-1} P_n \, \boldsymbol{v}_1 \wedge \cdots \wedge \boldsymbol{v}_{n-1}.$$

Conversely, for an $(n-1)$-form $\omega = Q_1 \, \boldsymbol{v}_2 \wedge \cdots \wedge \boldsymbol{v}_n + Q_2 \, \boldsymbol{v}_1 \wedge \boldsymbol{v}_3 \wedge \cdots \wedge \boldsymbol{v}_n + \cdots + Q_n \, \boldsymbol{v}_1 \wedge \cdots \wedge \boldsymbol{v}_{n-1}$, we have

$$*\omega = (-1)^{n-1} Q_1 \, \boldsymbol{v}_1 + (-1)^{n-2} Q_2 \, \boldsymbol{v}_2 + \cdots + Q_n \, \boldsymbol{v}_n.$$

It is then easy to see that $*(*\omega) = (-1)^{n-1}\omega$ for 1-forms and $(n-1)$-forms. Since the basis $\{\boldsymbol{v}_1, \ldots, \boldsymbol{v}_n\}$ of the cotangent space was created artificially just in order to define the operation $*$, we need to see how the adjoint operator δ is expressed in terms of the standard basis $\{dx^1, \ldots, dx^n\}$ of the cotangent space.

Consider first a 1-form $\omega = P_i \, dx^i$. We define the operator δ by (1) expressing each dx^i in terms of $\boldsymbol{v}_1, \ldots, \boldsymbol{v}_n$, (2) applying the adjoint operation $*$ in the orthonormal basis $\{\boldsymbol{v}_1, \ldots, \boldsymbol{v}_n\}$ of the cotangent space, so as to get an $(n-1)$-form ω, (3) applying the differential operator d to obtain the n-form $d\omega$, and (4) applying the adjoint operation $*$ again to get a zero-form, whose expression depends only on the coordinate system and is independent of any bases in the tangent and cotangent spaces. We need the orthonormal bases $\{\boldsymbol{u}_1, \ldots, \boldsymbol{u}_n\}$ and $\{\boldsymbol{v}_1, \ldots, \boldsymbol{v}_n\}$

in order to apply the adjoint operation $*$, since its direct definition in terms of the basis $\{dx^i, \ldots, dx^n\}$ is too messy to write as a generic formula.

EXAMPLE 6.4. Consider polar coordinates on $\mathbb{R}^2$, for which $g_{11} = 1$, $g_{12} = g_{21} = 0$, and $g_{22} = r^2$. Thus $\langle \boldsymbol{u}, \boldsymbol{v} \rangle = u^1 v^1 + r^2 u^2 v^2$. (The superscript on r here is an exponent; those on u and v are simply superscripts.) We get an orthonormal basis of the tangent space easily: $\{\boldsymbol{u}_1, \boldsymbol{u}_2\} = \{\partial/\partial r, (1/r)\partial/\partial\theta\}$. The basis dual to this basis is $\boldsymbol{v}_1 = dr$, $\boldsymbol{v}_2 = r\,d\theta$. Since we are in two dimensions here, we see that $*\boldsymbol{v}_1 = \boldsymbol{v}_2$ and $*\boldsymbol{v}_2 = -\boldsymbol{v}_1$. Consequently, we have the following table of conversions:

$$
\begin{aligned}
f(r,\theta) &= f(r,\theta) \quad \text{for 0-forms } f(r,\theta)\,,\\
P(r,\theta)dr + Q(r,\theta)d\theta &= P(r,\theta)\boldsymbol{v}_1 + \frac{1}{r}Q(r,\theta)\,\boldsymbol{v}_2 \quad \text{for 1-forms}\,,\\
F(r,\theta)\,dr \wedge d\theta &= \frac{1}{r}F(r,\theta)\,\boldsymbol{v}_1 \wedge \boldsymbol{v}_2 \quad \text{for 2-forms}\,.
\end{aligned}
$$

From this table of conversions, we easily deduce that

$$
\begin{aligned}
*f(r,\theta) &= f(r,\theta)\,\boldsymbol{v}_1 \wedge \boldsymbol{v}_2\\
&= r\,f(r,\theta)\,dr \wedge d\theta\,,\\
*(P\,dr + Q\,d\theta) &= *\left(P\boldsymbol{v}_1 + \frac{Q}{r}\boldsymbol{v}_2\right)\\
&= P\boldsymbol{v}_2 - \frac{Q}{r}\boldsymbol{v}_1\\
&= -\frac{1}{r}Q\,dr + rP\,d\theta\,,\\
*F(r,\theta)\,dr \wedge d\theta &= *(F\,\boldsymbol{v}_1 \wedge \frac{1}{r}\boldsymbol{v}_2) = \frac{1}{r}F(r,\theta)\,.
\end{aligned}
$$

It is then easy to compute that

$$
\nabla^2 f = \delta(df) = \frac{1}{r}\frac{\partial}{\partial r}\left(r\frac{\partial f}{\partial r}\right) + \frac{1}{r^2}\frac{\partial^2 f}{\partial\theta^2}\,,
$$

which is exactly the standard expression for the two-variable Laplacian in polar coordinates.

To give an example that is less trivial than the case of polar coordinates in the plane, but still fairly simple, we shall look at the Laplace–Beltrami operator on a 2-sphere.

EXAMPLE 6.5. Consider longitude (θ) and latitude[16] (φ) coordinates on the sphere of radius r_0 in $\mathbb{R}^3$ with center at $(0,0,0)$, with the half-plane $y \leq 0$ removed, so that $-\pi < \theta < \pi$ and $-\pi/2 < \varphi < \pi/2$. The coordinate mapping is

$$
\boldsymbol{r}(\theta,\varphi) = (r_0\cos\theta\cos\varphi, r_0\sin\theta\cos\varphi, r_0\sin\varphi)\,.
$$

The matrix of metric coefficients (g_{ij}) is given by

$$
g_{11} = r_0^2\cos^2\varphi\,, \quad g_{12} = g_{21} = 0\,, \quad g_{22} = r_0^2\,.
$$

---

[16]Note that in this example, we are defying the preference of physics textbooks for co-latitude over latitude. Physicists also seem to prefer using φ for longitude and θ for co-latitude, again in direct opposition to the practice of mathematicians.

An orthonormal basis for the tangent space is given by

$$\boldsymbol{u}_1 = \frac{1}{r_0}\frac{\partial}{\partial\theta} + \frac{\sin\varphi}{r_0}\frac{\partial}{\partial\varphi},$$

$$\boldsymbol{u}_2 = \frac{\tan\varphi}{r_0}\frac{\partial}{\partial\theta} - \frac{\cos\varphi}{r_0}\frac{\partial}{\partial\varphi}.$$

That is to say

$$\langle \boldsymbol{u}_k, \boldsymbol{u}_l \rangle = g_{ij}u_k^i v_l^j = \delta_l^k.$$

The basis of the cotangent space dual to this orthonormal basis of the tangent space is

$$\boldsymbol{v}_1 = r_0\cos^2\varphi\,d\theta + r_0\sin\varphi\,d\varphi,$$

$$\boldsymbol{v}_2 = r_0\cos\varphi\sin\varphi\,d\theta - r_0\cos\varphi\,d\varphi.$$

In these terms we have the following relations:

$$d\theta = \frac{1}{r_0}\boldsymbol{v}_1 + \frac{\tan\varphi}{r_0}\boldsymbol{v}_2,$$

$$d\varphi = \frac{\sin\varphi}{r_0}\boldsymbol{v}_1 - \frac{\cos\varphi}{r_0}\boldsymbol{v}_2,$$

$$d\theta\wedge d\varphi = -\frac{\sec\varphi}{r_0^2}\boldsymbol{v}_1\wedge\boldsymbol{v}_2,$$

$$*f(\theta,\varphi) = -r_0^2\cos\varphi\,f(\theta,\varphi)\,d\theta\wedge d\varphi,$$

$$*(P\,d\theta + Q\,d\varphi) = Q\cos\varphi\,d\theta - P\sec\varphi\,d\varphi.$$

$$\delta(P\,d\theta + Q\,d\varphi) = \frac{\partial P}{\partial\theta}\frac{\sec^2\varphi}{r_0^2} + \frac{\partial Q}{\partial\varphi}\frac{1}{r_0^2} - Q\sin\varphi\frac{\sec\varphi}{r_0^2}.$$

As a result,

$$\nabla^2 f = \frac{1}{r_0^2}\left(\frac{\partial^2 f}{\partial\theta^2}\sec^2\varphi - \frac{\partial f}{\partial\varphi}\tan\varphi + \frac{\partial^2 f}{\partial\varphi^2}\right).$$

As this example shows, the process of constructing the Laplace–Beltrami operator is quite computable. Let us now look at the process in the abstract.

LEMMA 6.2. *A symmetric positive-definite square matrix M has a unique symmetric positive-definite square root, that is, a matrix N such that $N^2 = M$. Moreover, N is a polynomial in M and commutes with every matrix that commutes with M.*

PROOF. The matrix M and all its powers belong to a vector space of dimension n^2. The $n^2 + 1$ matrices $I, M, \ldots, M^{n^2}$ are therefore linearly dependent, and that means there is a monic polynomial $p(x) = x^k + c_1 x^{k-1} + \cdots + c_{k-1}x + c_k$ of degree $k \le n^2$ such that $p(M) = O$, where O is the matrix all of whose entries are zeros.[17] We thus have $M^k = -(c_1 M^{k-1} + \cdots + c_{k-1}M + c_k I)$, where I is the $n \times n$ identity matrix. By writing $M^j = \big(I + (M-I)\big)^j$, we see that there are constants $b_0, \ldots, b_{k-1}$ such that

$$(M-I)^k = b_0 I + b_1(M-I) + \cdots + b_{k-1}(M-I)^{k-1}.$$

---

[17]By the Cayley–Hamilton theorem, in fact, the minimal k is at most n. By the spectral theorem, the polynomial of minimal degree is $p(x) = (x - \lambda_1)\cdots(x - \lambda_k)$, where $\lambda_1, \ldots, \lambda_k$ are the distinct eigenvalues of M.

We can then find constants $b_{0i}, \ldots, b_{k-1\,i}$ such that

$$(M - I)^{k+i} = b_{0i}I + b_{1i}(M - I) + \cdots + b_{k-1\,i}(M - I)^{k-1}$$

for all $i = 1, 2, \ldots$.

Assume temporarily that M is a diagonal matrix and that all of its eigenvalues (diagonal entries) λ satisfy $0 < \lambda < 2$. In this case, it is quite trivial that there exists a nonnegative square root. If the distinct diagonal entries are, say $\lambda_1, \ldots, \lambda_k$, simply replace λ_j by $\sqrt{\lambda_j}$ and you have the required matrix. The Taylor series expansion is needed only to show that this square root is a polynomial in M. When the eigenvalues are in this range, the diagonal entries of the matrix $M - I$ lie in the interval $(-1, +1)$, and the series

$$\sqrt{M} = \left(I + (M - I)\right)^{1/2} = \sum_{n=0}^{\infty} \binom{\frac{1}{2}}{n}(M - I)^n$$

converges uniformly and absolutely, component by component. It can therefore be rearranged arbitrarily, and we find

$$
\begin{aligned}
\sqrt{M} &= \sum_{l=0}^{k-1} \left(b_l \binom{\frac{1}{2}}{l} + \sum_{i=1}^{\infty} b_{li} \binom{\frac{1}{2}}{k+i} \right)(M - I)^l \\
&= \sum_{l=0}^{k-1} c_l (M - I)^l.
\end{aligned}
$$

This equality shows that $\sqrt{M}$ is a polynomial in $M - I$, and therefore also a polynomial in M.

If M is diagonal, but has eigenvalues outside the range specified above, simply write $M = aN$, where $a > 0$ and N is a diagonal matrix with eigenvalues in the range $(0, 2)$. We then have

$$\sqrt{M} = \sqrt{a}\sqrt{N} = \sqrt{a}\sum_{l=0}^{k-1} c_l (N - I)^l = \sqrt{a}\sum_{l=0}^{k-1} c_l \left(\frac{1}{a}M - I\right)^l.$$

Finally, if M is not diagonal, the spectral theorem guarantees that there is an invertible (in fact, orthogonal) matrix P such that the matrix $N = P^{-1}MP$ is diagonal and positive-definite. Since

$$M = PNP^{-1} = P(\sqrt{N})^2 P^{-1} = \left(P(\sqrt{N})P^{-1}\right)\left(P(\sqrt{N})P^{-1}\right),$$

it follows that $\sqrt{M} = P(\sqrt{N})P^{-1}$, and

$$\sqrt{M} = P\sqrt{N}P^{-1} = P\left(\sum_{l=0}^{k-1} c_l(N-I)^l\right)P^{-1} = \sum_{l=0}^{k-1} c_l P(N-I)^l P^{-1} = \sum_{l=0}^{k-1} c_l(M-I)^l.$$

$\square$

This theorem is particularly useful on Riemannian manifolds, where the matrix of metric coefficients g and its inverse g^{-1} are symmetric and positive-definite. The latter is important in connection with the Laplace–Beltrami operator, since, as we are about to show, it furnishes an orthonormal basis of the tangent space that is naturally adapted to the metric, as opposed to more artificial ones, such as we might get by, for example, applying the Gram–Schmidt orthogonalization procedure to the standard basis $\{\partial/\partial x^1, \ldots, \partial/\partial x^n\}$.

EXAMPLE 6.6. Consider a sphere of radius r_0 in $\mathbb{R}^3$ with center at the origin. We can parameterize all of it except the "international date line" opposite the "prime meridian" by latitude and longitude coordinates $(\theta, \varphi) \mapsto \boldsymbol{r}(\theta, \varphi) = r_0 \cos\theta \cos\varphi\, \boldsymbol{i} + r_0 \sin\theta \cos\varphi\, \boldsymbol{j} + r_0 \sin\varphi\, \boldsymbol{k}$, $-\pi < \theta < \pi$, $-\pi/2 < \varphi < \pi/2$. When we do this, the matrix of metric coefficients is

$$g = \begin{pmatrix} r_0^2 \cos^2\varphi & 0 \\ 0 & r_0^2 \end{pmatrix}.$$

This matrix is positive-definite, and its inverse is

$$g^{-1} = \begin{pmatrix} r_0^{-2} \sec^2\varphi & 0 \\ 0 & r_0^{-2} \end{pmatrix}.$$

It is obvious that at a given point (θ_0, φ_0), we have

$$\sqrt{g^{-1}} = \begin{pmatrix} \frac{1}{r_0 \cos\varphi_0} & 0 \\ 0 & \frac{1}{r_0} \end{pmatrix}.$$

The theorem asserts, among other things, that $\sqrt{g^{-1}}$ is a polynomial in g^{-1}, and indeed it is quite trivial to verify that $\sqrt{g^{-1}} = aI + bg^{-1}$, where $a = 1/(r_0(1 + \cos\varphi_0))$ and $b = r_0 \cos\varphi_0/(1 + \cos\varphi_0)$. Consider now a new system of coordinates $y^1 = (r_0 \cos\varphi_0)\theta$ and $y^2 = r_0\varphi$ near the base point $(y_0^1, y_0^2) = ((r_0 \cos\varphi_0)\theta_0, r_0\varphi_0)$. We have

$$\begin{pmatrix} \theta \\ \varphi \end{pmatrix} = \sqrt{g^{-1}} \begin{pmatrix} y^1 \\ y^2 \end{pmatrix}.$$

The surface is then parameterized by the function

$$\boldsymbol{s}(y^1, y^2) = \boldsymbol{r}(\theta, \varphi) = r_0 \cos\left(\frac{y^1}{r_0 \cos\varphi_0}\right) \cos\left(\frac{y^2}{r_0}\right) \boldsymbol{i}$$
$$+ r_0 \sin\left(\frac{y^1}{r_0 \cos\varphi_0}\right) \cos\left(\frac{y^2}{r_0}\right) \boldsymbol{j} + r_0 \sin\left(\frac{y^2}{r_0}\right) \boldsymbol{k},$$

and the matrix of metric coefficients $\tilde{g}(y^1, y^2)$ satisfies $\tilde{g}(y_0^1, y_0^2) = I$. The vectors $\partial/\partial y^1 = (1/(r_0 \cos\varphi_0))\partial/\partial\theta$ and $\partial/\partial y^2 = (1/r_0)\partial/\partial\varphi$ form an orthonormal basis of the tangent space at the point $\boldsymbol{r}(\theta_0, \varphi_0) = \boldsymbol{s}(y_0^1, y_0^2)$.

THEOREM 6.10. For a 0-form $f(x^1, \ldots, x^n)$ in general coordinates where the matrix of metric coefficients is $g = (g_{ij})$ with inverse $g^{-1} = (g^{ij})$, the Laplace–Beltrami operator applied to f yields

(6.12) $$\nabla^2 f = \frac{1}{\sqrt{\det(g)}} \frac{\partial}{\partial x^l}\left(\sqrt{\det(g)}\, g^{il} \frac{\partial f}{\partial x^i}\right).$$

PROOF. By way of preliminaries we need a simple, well-known fact from linear algebra about the *classical adjoint matrix*[18] $*g$ of an $n \times n$ matrix $g = (g_{ij})$. It is the transposed matrix of co-factors of g. Specifically, $*g = (*g^{ij})$, where $*g^{ij}$ is $(-1)^{i+j}$ times the determinant of the $(n-1) \times (n-1)$ submatrix of g obtained by removing the row (row i) and column (column j) that contain the entry g_{ij}. (Actually, since we are dealing here with symmetric matrices, we need not worry

---

[18]The reader will notice that we are apparently abusing the star symbol by using it here for the classical adjoint of a matrix, when up to now we have used it for the adjoint of a k-form. The two uses are closely related, however, as we shall see very shortly.

about the distinction between rows and columns.) The fundamental fact is the well-known relation

$$g(*g) = \det(g)I\,,$$

where I is the $n \times n$ identity matrix. The important consequence of this relation is that

$$g^{-1} = \frac{1}{\det(g)}(*g)\,,$$

that is,

$$*g^{ij} = \det(g)g^{ij}\,,$$

where $g^{-1} = (g^{ij})$.

With these preliminaries out of the way, the proof is self-operating, just a matter of following the definitions.

The positive-definite matrix $g = (g_{ij})$ has a unique positive-definite square root $B = (b_{kl})$, whose inverse $A = (a^{kl})$ is the unique positive-definite square root of $g^{-1} = (g^{ij})$. We have the following relations among these matrices:

$$
\begin{aligned}
B^2 &= g\,, \\
A^2 &= g^{-1}\,, \\
AB &= BA = I\,, \\
Ag &= gA = B\,, \\
Bg^{-1} &= g^{-1}B = A\,, \\
\det(A) &= \frac{1}{\sqrt{\det(g)}}\,, \\
\det(B) &= \sqrt{\det(g)}\,, \\
*A &= \det(A)A^{-1} = \frac{1}{\sqrt{\det(g)}}B\,, \\
*B &= \det(B)B^{-1} = \sqrt{\det(g)}A\,.
\end{aligned}
$$

This last relation is the most important one, since it says $*b^{kl} = \sqrt{\det(g)}a^{kl}$.

The vectors $\boldsymbol{u}_j = a^{lj}\partial/\partial x^l$ form an orthonormal basis of the tangent space. The dual basis of the cotangent space is $\boldsymbol{v}_i = b_{ik}\,dx^k$, since

$$\boldsymbol{v}_i(\boldsymbol{u}_j) = b_{ik}a^{lj}dx^k\left(\frac{\partial}{\partial x^l}\right) = b_{ik}a^{kj} = \delta_i^j\,.$$

As a result, we have $dx^i = a^{ik}\boldsymbol{v}_k$. Now for 0-forms $f(x^1,\ldots,x^n)$, there is no difference between the expressions in terms of the bases $\{\boldsymbol{v}_1,\ldots,\boldsymbol{v}_n\}$ and $\{dx^1,\ldots,dx^n\}$. For n forms we have

$$
\begin{aligned}
F(x^1,\ldots,x^n)\,dx^1 \wedge \cdots \wedge dx^n &= F(x^1,\ldots,x^n)\det(A)\,\boldsymbol{v}_1 \wedge \cdots \wedge \boldsymbol{v}_n \\
&= F(x^1,\ldots,x^n)\frac{1}{\sqrt{\det(g)}}\,\boldsymbol{v}_1 \wedge \cdots \wedge \boldsymbol{v}_n\,.
\end{aligned}
$$

For 1-forms we find

$$
\begin{aligned}
*dx^i &= *(a^{ik}\boldsymbol{v}_k) \\
&= a^{ik}(*\boldsymbol{v}_k) \\
&= \sum_{k=1}^{n}(-1)^{k-1}a^{ik}\boldsymbol{v}_1 \wedge \cdots \wedge \boldsymbol{v}_{k-1} \wedge \boldsymbol{v}_{k+1} \wedge \cdots \wedge \boldsymbol{v}_n \\
&= \sum_{k=1}^{n}(-1)^{k-1}a^{ik}(b_{1j_1}\,dx^{j_1}) \wedge \cdots \\
&\qquad \cdots \wedge (b_{k-1\,j_{k-1}}\,dx^{j_{k-1}}) \wedge (b_{k+1\,j_{k+1}}\,dx^{j_{k+1}}) \wedge \cdots \wedge (b_{nj_n}\,dx^{j_n}) \\
&= \sum_{l=1}^{n}(-1)^{k-1}a^{ik}\det(P_{kl})\,dx^1 \wedge \cdots \wedge dx^{l-1} \wedge dx^{l+1} \wedge \cdots \wedge dx^n ,
\end{aligned}
$$

where

$$
P_{kl} = \begin{pmatrix}
b_{11} & \cdots & b_{1\,l-1} & b_{1\,l+1} & \cdots & b_{1n} \\
\vdots & \ddots & \vdots & \vdots & \ddots & \vdots \\
b_{k-1\,1} & \cdots & b_{k-1\,l-1} & b_{k-1\,l+1} & \cdots & b_{k-1\,n} \\
b_{k+1\,1} & \cdots & b_{k+1\,l-1} & b_{k+1\,l+1} & \cdots & b_{k+1\,n} \\
\vdots & \ddots & \vdots & \vdots & \ddots & \vdots \\
b_{n1} & \cdots & b_{n\,l-1} & b_{n\,l+1} & \cdots & b_{nn}
\end{pmatrix}.
$$

(Although the Einstein summation convention is generally in effect, we judged that it was necessary to include the summation signs here for the summations over k and l, since the "repeated" index over which summation takes place is the one whose associated subscript k or superscript l is *missing* in the wedge product.)

Thus we find

$$
\begin{aligned}
*dx^i &= \sum_{l=1}^{n}\sum_{k=1}^{n}(-1)^{k-1}a^{ik}\big((-1)^{k+l}*b^{kl}\big)\,dx^1 \wedge \cdots \wedge dx^{l-1} \wedge dx^{l+1} \wedge \cdots \wedge dx^n \\
&= \sum_{l=1}^{n}(-1)^{l-1}\sum_{k=1}^{n}a^{ik}(*b^{kl})\,dx^1 \wedge \cdots \wedge dx^{l-1} \wedge dx^{l+1} \wedge \cdots \wedge dx^n \\
&\quad - \sum_{l=1}^{n}(-1)^{l-1}\sqrt{\det(g)}\sum_{k=1}^{n}a^{ik}a^{kl}\,dx^1 \wedge \cdots \wedge dx^{l-1} \wedge dx^{l+1} \wedge \cdots \wedge dx^n \\
&= \sum_{l=1}^{n}(-1)^{l-1}\sqrt{\det(g)}g^{il}\,dx^1 \wedge \cdots \wedge dx^{l-1} \wedge dx^{l+1} \wedge \cdots \wedge dx^n .
\end{aligned}
$$

The heart of the proof is contained in this last set of equations. We have

$$
\begin{aligned}
\nabla^2 f &= *\big(d(*(df))\big) = *\Big(d\Big(*\Big(\frac{\partial f}{\partial x^i}\,dx^i\Big)\Big)\Big) \\
&= *\Big(d\Big(\sum_{l=1}^{n}(-1)^{l-1}\sqrt{\det(g)}g^{il}\frac{\partial f}{\partial x^i}\,dx^1 \wedge \cdots \wedge dx^{l-1} \wedge dx^{l+1} \wedge \cdots \wedge dx^n\Big)\Big) \\
&= *\Big(\frac{\partial}{\partial x^l}\Big(\sqrt{\det(g)}g^{il}\frac{\partial f}{\partial x^i}\Big)\,dx^1 \wedge \cdots \wedge dx^n\Big) \\
&= \frac{1}{\sqrt{\det(g)}}\frac{\partial}{\partial x^l}\Big(\sqrt{\det(g)}g^{il}\frac{\partial f}{\partial x^i}\Big).
\end{aligned}
$$

This is the formula we set out to prove.   □

REMARK 6.10. In expanded form, the Laplace–Beltrami operator is given by

$$(6.13) \qquad \nabla^2 f = g^{il} \frac{\partial^2 f}{\partial x^i \, \partial x^l} + \Big(\frac{g^{kl}}{2 \det(g)} \frac{\partial \big(\det(g) \big)}{\partial x^l} + \frac{\partial g^{kl}}{\partial x^l} \Big) \frac{\partial f}{\partial x^k} \, .$$

On $\mathbb{R}^3$, the Laplace–Beltrami operator (the ordinary Laplacian) is

$$\nabla \cdot \nabla f = \operatorname{div} \operatorname{grad} f \, .$$

The formula for the operator in generalized coordinates suggests that we define the gradient of a function $f(x^1, \dots, x^n)$ and the divergence of a vector field $\boldsymbol{u} = u^i \partial / \partial x^i$ on a general manifold as

$$
\begin{aligned}
\operatorname{grad} f \;&=\; g^{il} \frac{\partial f}{\partial x^l} \frac{\partial}{\partial x^i} \, , \\[2mm]
\operatorname{div} \boldsymbol{u} \;&=\; \frac{1}{\sqrt{\det(g)}} \frac{\partial \big(\sqrt{\det(g)} \, u^i \big)}{\partial x^i} \\[2mm]
&=\; \frac{\partial u^i}{\partial x^i} + \frac{u^i}{2 \det(g)} \frac{\partial(\det(g))}{\partial x^i} \\[2mm]
&=\; \frac{\partial u^i}{\partial x^i} + \frac{1}{2} g^{jk} u^i \frac{\partial g_{jk}}{\partial x^i} \, , \\[2mm]
&=\; \frac{\partial u^i}{\partial x^i} + u^i \Gamma_{ik}^k \, ,
\end{aligned}
$$

where the next-to-last equality follows from Eq. (6.16), proved below, and the last one by combining that result with the result of Problem 4.12. The tensor nature of the gradient is then easily verified.

The divergence is now seen to be connected with the covariant derivative via the relation

$$\operatorname{div} \boldsymbol{u} = dx^i \big(\nabla_{\frac{\partial}{\partial x^i}} \boldsymbol{u} \big) \, .$$

In other words, it is the contraction of the mixed tensor T of type $(1,1)$ whose component T_i^j is

$$T_i^j = \nabla_{\frac{\partial}{\partial x^i}} \Big(u^j \frac{\partial}{\partial x^j} \Big)$$

Warning! Habits of thought learned by constant use of flat spaces, such as in Newtonian mechanics, can be misleading. It is obvious that the gradient of a constant function is **0**. Conversely, if $\operatorname{grad} f$ vanishes over a region of parameter space, then its ordinary gradient, whose components are $(\partial f / \partial x^1, \dots, \partial f / \partial x^n)$ is orthogonal to every row of the matrix g^{-1}. Since g^{-1} is a nonsingular matrix, all of these partial derivatives must vanish, and hence the function is constant. On the other hand, it is *not* necessarily true that the divergence of a vector field whose local coordinates are constant is 0. As the formula for the divergence just given makes clear, this will be the case at points where all the Christoffel symbols vanish. But in general a vector field that has constant coefficients on a curved manifold may represent a nonzero flow of matter through a closed surface. (See Example 6.9 below for the Newtonian interpretation of a vanishing divergence.)

REMARK 6.11. Abusing the tilde-symbol once again, we note that with each vector $\boldsymbol{u}$, we can associate a tensor of type $(1,1)$, which we denote $\tilde{\boldsymbol{u}}$ and define as follows, for all vectors $\boldsymbol{v} = v^j \partial/\partial x^j$ and covectors $\boldsymbol{v} = v_k\,dx^k$:

$$\tilde{\boldsymbol{u}}(\boldsymbol{v},\boldsymbol{v}) = \boldsymbol{v}(\nabla_{\boldsymbol{v}}\boldsymbol{u}) = v_k v^j \left(\frac{\partial u^k}{\partial x^j} + u^i \Gamma_{ij}^k \right).$$

In the standard bases of the tangent and co-tangent spaces, the coordinates of $\tilde{\boldsymbol{u}}$ are

$$\tilde{u}_j^k = \frac{\partial u^k}{\partial x^j} + u^i \Gamma_{ij}^k,$$

and

$$\operatorname{div} \boldsymbol{u} = \tilde{u}_k^k.$$

Thus, the divergence of $\boldsymbol{u}$ is the contraction of the corresponding tensor $\tilde{\boldsymbol{u}}$, and consequently also a tensor. It follows that the Laplace–Beltrami operator is a tensor.

REMARK 6.12. We have made constant use of the assumption that the matrix $g = (g_{ij})$ is positive-definite, since we needed the square root of its determinant to produce the Laplace–Beltrami operator. The metric of relativistic space-time, however, is not positive-definite, and its determinant is negative. In his 1916 paper on general relativity, Einstein simply replaced the determinant by its negative before taking the square root. From our point of view, that was justifiable, since the imaginary quantity $\sqrt{-1}$ cancels out of the final expression anyway. Now that we have the representation in Eq. (6.13) for this operator, we can dispense with that assumption. An important example is the metric of flat space-time, which is not positive-definite, and for which

(6.14)
$$\nabla^2 f = \frac{\partial}{\partial x^l}\left(g^{il}\frac{\partial f}{\partial x^i} \right) = g^{il}\frac{\partial^2 f}{\partial x^i\,\partial x^l}.$$

Thus, on four-dimensional space-time, where

$$g = \begin{pmatrix} 1 & 0 & 0 & 0 \\ 0 & -1/c^2 & 0 & 0 \\ 0 & 0 & -1/c^2 & 0 \\ 0 & 0 & 0 & -1/c^2 \end{pmatrix},$$

we have

$$\nabla^2 f = \frac{\partial^2 f}{\partial t^2} - c^2\left(\frac{\partial^2 f}{\partial x^2} + \frac{\partial^2 f}{\partial y^2} + \frac{\partial^2 f}{\partial z^2} \right).$$

The Laplace–Beltrami operator for this case is called the *d'Alembertian*, after Jean Le Rond d'Alembert (1717–1783), and usually denoted $\square$. The differential equation $\square u = 0$ is the classical *wave equation*.[19] As a consequence of the Maxwell equations (Chapter 3), each component of an electric or magnetic field satisfies this equation. No wonder, then, that these equations have an intimate relation with the metric of space-time.

---

[19] One of the early definitive studies of the one-dimensional wave equation was made by d'Alembert, who derived this equation (in one spatial dimension) independently of earlier work by Brook Taylor (1685–1731).

6.5. Divergence of a tensor. We can define the divergence of any contravariant tensor of type $(n, 0)$ as a contravariant tensor of type $(n - 1, 0)$. We have need of this only for the case $n = 2$, but the generalization will be obvious. Let T be a contravariant tensor of type $(2, 0)$ whose coordinates are T^{ij}, that is,

$$T = T^{ij} \frac{\partial}{\partial x^i} \otimes \frac{\partial}{\partial x^j} \,.$$

Then div T is the contravariant vector

$$\left(\frac{\partial T^{ij}}{\partial x^j} + T^{iq} \Gamma^j_{qj} + T^{jq} \Gamma^i_{jq} \right) \frac{\partial}{\partial x^i} \,.$$

Thus, this divergence differs from the divergence of a tensor of type $(1, 0)$ by having two extra terms instead of one. Notice that in the case of the inverse-metric tensor, where $T^{ij} = g^{ij}$, Problem 4.13 implies that the additional two terms amount to $-\partial g^{ij}/\partial x^j$, so that the divergence of this tensor is $\mathbf{0}$.

To illustrate this concept, consider the stress-energy-momentum tensor associated with a mass of rest density ρ_0 and having a velocity three-vector $\mathbf{v} = (dx/dt, dy/dt, dz/dt)$, corresponding to the velocity four-vector

$$\mathbf{v}_4 = (\alpha; dx/ds, dy/ds, dz/ds) = (dx^1/ds; dx^2/ds, dx^3/ds, dx^4/ds) \,.$$

The tensor in question has coordinates (see Theorem 2.4)

$$T^{ij} = \rho_0 \frac{dx^i}{ds} \frac{dx^j}{ds} = \rho \frac{dx^i}{dt} \frac{dx^j}{dt} \,.$$

Since $dx^1/dt = 1$, it follows that, at a point where the Christoffel symbols vanish, such as the origin of a normal coordinate system, the first coordinate of the divergence of this tensor is

$$\frac{\partial T^{1j}}{\partial x^j} = \frac{\partial \rho}{\partial t} + \nabla \cdot (\rho \mathbf{v}) \,,$$

where this nabla-symbol is the classical divergence in $\mathbb{R}^3$.

This expression vanishes by virtue of the equation of continuity discussed in Section 10 of Chapter 2.

We can equally well define the divergence of a tensor of type $(0, 2)$ or of type $(1, 1)$. For example, for a tensor T of type $(0, 2)$ with components T_{ij}, we define the divergence as the tensor of type $(0, 1)$

$$\mathrm{div}\, T = \sum_j \left(\frac{\partial T_{ij}}{\partial x^j} - \left(\Gamma^l_{ij} T_{lj} - \Gamma^l_{jj} T_{il} \right) \right) dx^i \,.$$

(The summation sign is included here because the repeated index of summation j occurs only as a subscript.) Problem 4.12 shows that div g_{ij} is the zero one-form.

6.6. The Hessian. We can now explain what is meant by the Hessian of a function on a manifold.

DEFINITION 6.7. Let $f(x^1, \ldots, x^n)$ be a C^∞-function on a manifold with metric coefficients g_{ij} and inverse metric coefficients g^{ij}. The *Hessian* H_f is the tensor of type $(0, 2)$ given by

$$H_f(\mathbf{u}, \mathbf{v}) = \nabla^2_{\mathbf{u}, \mathbf{v}} f \,.$$

Notice that it is the second covariant derivative $\nabla^2_{\boldsymbol{u},\boldsymbol{v}}$ that appears here, not the Laplace–Beltrami operator ∇^2. In expanded form, we have

$$H_f(\boldsymbol{u}, \boldsymbol{v}) = \left(\frac{\partial^2 f}{\partial x^i \, \partial x^j} - \Gamma^k_{ij} \frac{\partial f}{\partial x^k} \right) u^i v^j \, .$$

Notice also that this expression differs from the ordinary Hessian previously defined for a function on the parameter space (a subset of $\mathbb{R}^n$) by the presence of the subtracted terms

$$\Gamma^k_{ij} \frac{\partial f}{\partial x^k} \, u^i v^j \, .$$

On the Euclidean space $\mathbb{R}^n$, the Christoffel symbols are identically zero, and therefore the Hessian as just defined coincides with the one we defined earlier. For more connections, see Problem 6.14 below.

Comparison of the expression just given for $H_f(\boldsymbol{u}, \boldsymbol{v})$ with Eq. (6.13) strongly suggests that there is a connection between the Hessian and the Laplace–Beltrami operator. And indeed there is. If we raise one index on the Hessian to get a mixed tensor $(H_f)^j_i = g^{jk}(H_f)_{ki}$, then contract, we get precisely the Laplace–Beltrami operator:

THEOREM 6.11.

(6.15) $$\nabla^2 f = g^{ij}(H_f)_{ij} = g^{ij} \frac{\partial^2 f}{\partial x^i \, \partial x^j} - g^{ij}\Gamma^k_{ij} \frac{\partial f}{\partial x^k} \, .$$

PROOF. Given Eq. (6.13), we see that we need to establish the relation

$$g^{ij}\Gamma^k_{ij} = -\frac{\partial g^{km}}{\partial x^m} - \frac{g^{km}}{2\det(g_{pq})} \frac{\partial\big(\det(g_{pq})\big)}{\partial x^m} = -\operatorname{div} \boldsymbol{g}^k \, ,$$

where $\boldsymbol{g}^k$ is the kth row (or column) of the matrix inverse to the matrix of metric coefficients.

Starting from the definition of the Christoffel symbol Γ^k_{ij}, we find

$$g^{ij}\Gamma^k_{ij} = \frac{1}{2}g^{ij}g^{km}\left(\frac{\partial g_{im}}{\partial x^j} + \frac{\partial g_{jm}}{\partial x^i} - \frac{g_{ij}}{\partial x^m} \right).$$

Since $g^{km}g_{im} = \delta^k_i$, it follows that

$$\frac{\partial(g^{km}g_{im})}{\partial x^j} = 0 \, ,$$

and therefore

$$g^{km}\frac{\partial g_{im}}{\partial x^j} = -\frac{\partial g^{km}}{\partial x^j}g_{im} \, .$$

Likewise,

$$g^{km}\frac{\partial g_{jm}}{\partial x^i} = -\frac{\partial g^{km}}{\partial x^i}g_{jm} \, .$$

Therefore,

$$\begin{aligned}
\frac{1}{2}g^{ij}g^{km}\left(\frac{\partial g_{im}}{\partial x^j} + \frac{\partial g_{im}}{\partial x^i} \right) &= -\frac{1}{2}\left(g^{ij}g_{im}\frac{\partial g^{km}}{\partial x^j} + g^{ij}g_{jm}\frac{\partial g^{km}}{\partial x^i} \right) \\
&= -\frac{1}{2}\left(\frac{\partial g^{km}}{\partial x^m} + \frac{\partial g^{km}}{\partial x^m} \right) \\
&= -\frac{\partial g^{km}}{\partial x^m} \, .
\end{aligned}$$

Thus, it now remains to be shown that

$$g^{km}g^{ij}\frac{\partial g_{ij}}{\partial x^m} = \frac{g^{km}}{\det(g_{pq})}\frac{\partial\big(\det(g_{pq})\big)}{\partial x^m}$$

for all k. We shall in fact show that

(6.16) $$g^{ij}\det(g_{pq})\frac{\partial g_{ij}}{\partial x^m} = \frac{\partial\big(\det(g_{pq})\big)}{\partial x^m}.$$

To that end, we observe that the coefficient of $\partial g_{ij}/\partial x^m$ on the left is $(-1)^{i+j}\big(*g_{ij}\big)$, where $*g_{ij}$ is the entry in row i and column j of the classical adjoint of the matrix $g = (g_{ij})$. In expanded form, this coefficient is

$$\sum_{j_1,\dots,j_{i-1},j_{i+1},\dots,j_n}(-1)^{i+j}\varepsilon(j_1,\dots,j_{i-1},j_{i+1},\dots,j_n)g_{1\,j_1}\cdots g_{i-1\,j_{i-1}}g_{i+1\,j_{i+1}}\cdots g_{n\,j_n}\,,$$

where the ordered $(n-1)$-tuple $(j_1,\dots,j_{i-1},j_{i+1},\dots,j_n)$ ranges over all $(n-1)!$ permutations of $(1,2,\dots,j-1,j+1,\dots,n)$.

Now $\det(g_{pq})$ has the expansion

$$\det(g_{pq}) = \varepsilon(j_1,\dots,j_n)g_{1\,j_1}\cdots g_{n\,j_n}\,,$$

where $(j_1,\dots,j_n)$ ranges over all $n!$ permutations of $(1,\dots,n)$. Thus, the coefficient of g_{ij} in this determinant is

$$\sum_{j_1,\dots,j_{i-1},j_{i+1},\dots,j_n}\varepsilon(j_1,\dots,j_{i-1},j,j_{i+1},\dots,j_n)g_{1\,j_1}\cdots g_{i-1\,j_{i-1}}g_{i+1\,j_{i+1}}\cdots g_{n\,j_n}\,,$$

where the range of summation is the same as above. This will be the coefficient of $\partial g_{ij}/\partial x^m$ when $\partial\big(\det(g_{pq})\big)/\partial x^m$ is expanded.

The proof thus comes down to establishing the simple equality

$$(-1)^{i+j}\varepsilon(j_1,\dots,j_{i-1},j_{i+1},\dots,j_n) = \varepsilon(j_1,\dots,j_{i-1},j,j_{i+1},\dots,j_n)\,.$$

This is very easy to do. Let p be the number of indices among $j_1,\dots,j_{i-1}$ that are less than j. Then, in the permutation $(j_1,\dots,j_{i-1},j,j_{i+1},\dots,j_n)$ there are $i-1-p$ ordered pairs (j_l,j) with $j_l > j$ and $j-1-p$ pairs (j,j_l) with $j_l < j$. Hence if N is the number of inversions in $(j_1,\dots,j_{i-1},j_{i+1},\dots,j_n)$, the number in $(j_1,\dots,j_{i-1},j,j_{i+1},\dots,j_n)$ is $N+i+j-2p-2$. The required equality follows immediately.   □

COROLLARY 6.3. *At a point P where all the Christoffel symbols vanish, the Laplace–Beltrami operator is given by*

$$\nabla^2 f = g^{pq}\frac{\partial^2 f}{\partial x^p\,\partial x^q}\,.$$

The corollary applies at the origin when we use normal coordinates. At a point where $g_{ij} = \delta_j^i$ and all the Christoffel symbols vanish, the Laplace–Beltrami operator is the ordinary Laplacian:

$$\nabla^2 f = \sum_{p=1}^n \frac{\partial^2 f}{\partial(x^p)^2}\,.$$

This formula holds in particular if the normal coordinates $u^1,\dots,u^n$ are taken in an orthonormal basis of the tangent space.

The equation in Corollary 6.3 is not a tensor relation, because it holds only at a particular point and only for parameterizations in which all the Christoffel symbols are zero at that point. Despite the restricted nature of its validity, it does provide for us a connection between the Laplace–Beltrami operator and the Riemann curvature tensor.

COROLLARY 6.4. *In normal coordinates* $(u^1, \ldots, u^n)$, *the following relation holds at the origin* $(0, \ldots, 0)$:

$$(6.17) \qquad \nabla^2 g_{pq} = g^{mr} \frac{\partial^2 g_{pq}}{\partial u^m \, \partial u^r} = -\frac{2}{3} \left(\frac{\partial \Gamma^r_{pq}}{\partial u^r} - \frac{\partial \Gamma^r_{rp}}{\partial u^q} \right) = -\frac{2}{3} R^r_{prq} ,$$

where R^i_{jkl} *is the coordinate of the Riemann curvature tensor at that point.*

PROOF. Since the Christoffel symbols vanish at $(0, \ldots, 0)$, we have

$$R^r_{prq} = \frac{\partial \Gamma^r_{pq}}{\partial u^r} - \frac{\partial \Gamma^r_{rp}}{\partial u^q} .$$

Applying relation (6.10) to the second term in this expression, taking account of the symmetry of the Christoffel symbols, we find

$$R^r_{prq} = 2 \frac{\partial \Gamma^r_{pq}}{\partial u^r} + \frac{\partial \Gamma^r_{rq}}{\partial u^p} .$$

Adding these two expressions and dividing the sum by 2, we find

$$R^r_{prq} = \frac{3}{2} \frac{\partial \Gamma^r_{pq}}{\partial u^r} + \frac{1}{2} \left(\frac{\partial \Gamma^r_{rq}}{\partial u^p} - \frac{\partial \Gamma^r_{rp}}{\partial u^q} \right) .$$

By writing out the definition of the Christoffel symbols and differentiating, bearing in mind that the partial derivatives of the metric coefficients vanish at $(0, \ldots, 0)$, we find that the second term here vanishes. (This is also a consequence of the fact that R^r_{prq} is symmetric in p and q. That fact, however, is not obvious.) Indeed we have

$$\frac{\partial \Gamma^r_{rq}}{\partial u^p} = \frac{1}{2} g^{rm} \left(\frac{\partial^2 g_{rm}}{\partial u^p \, \partial u^q} + \frac{\partial^2 g_{qm}}{\partial u^r \, \partial u^p} - \frac{\partial g_{rq}}{\partial u^p \, \partial u^m} \right) .$$

Since r and m are both merely dummy indices of summation here, they can be reversed in the last of these terms, and then the last two terms cancel each other. Thus we have

$$\frac{\partial \Gamma^r_{rq}}{\partial u^p} = \frac{1}{2} g^{rm} \frac{\partial^2 g_{rm}}{\partial u^p \, \partial u^q} .$$

Reversing p and q and subtracting demonstrates that

$$\frac{\partial \Gamma^r_{rq}}{\partial u^p} - \frac{\partial \Gamma^r_{rp}}{\partial u^q} = \frac{1}{2} g^{rm} \left(\frac{\partial^2 g_{rm}}{\partial u^p \, \partial u^q} - \frac{\partial^2 g_{rm}}{\partial u^q \, \partial u^p} \right) = 0 .$$

Thus, we find

$$R^r_{prq} = \frac{3}{2} \frac{\partial \Gamma^r_{pq}}{\partial u^r} .$$

At the same time, we find that

$$\frac{\partial \Gamma^r_{pq}}{\partial u^r} = \frac{1}{2} g^{rm} \left(\frac{\partial^2 g_{pm}}{\partial u^r \, \partial u^q} + \frac{\partial^2 g_{qm}}{\partial u^r \, \partial u^p} - \frac{\partial g_{pq}}{\partial u^r \, \partial u^m} \right) .$$

By Corollary 6.1, this equality yields

$$\frac{\partial \Gamma^r_{pq}}{\partial u^r} = -g^{rm} \frac{\partial g_{pq}}{\partial u^r \, \partial u^m} = -\nabla^2 g_{pq} .$$

The corollary is now proved. $\qquad \square$

To illustrate this corollary, it is not difficult to verify that in normal coordinates at the north pole of the sphere $\mathbb{S}^2_{r_0}$, we have

$$R^r_{prq} = \begin{cases} \frac{1}{r_0^2}, & \text{if } p = q \\ 0, & \text{if } p \neq q \end{cases}.$$

At the same point, in normal coordinates,

$$\nabla g_{pq} = \begin{cases} -\frac{2}{3r_0^2}, & \text{if } p = q \\ 0, & \text{if } p \neq q \end{cases}.$$

For the pseudo-hemisphere, which was our second example of normal coordinates, it turns out to be extremely complicated to compute all the tensors involved in this relation. However, by truncating the MacLaurin series of the metric coefficients at four terms, the whole problem becomes manageable, and it turns out that when we use rectangular coordinates in the tangent plane, taking the radius c to be 1 and the point whose projection into the equatorial plane has polar coordinates $r = 1/2$, $\theta = 0$,

$$R^r_{prq} = \begin{cases} -1 & \text{if } p = q \\ 0 & \text{if } p \neq q \end{cases} \quad ; \quad \nabla g_{pq} = \begin{cases} 2/3 & \text{if } p = q \\ 0 & \text{if } p \neq q \end{cases}.$$

REMARK 6.13. Corollary 6.4 is not a tensor relation, since it holds only in one particular set of coordinates and only at one point. You can easily see that no such relation holds in general between the two tensors involved. For example, in the space-time metric we used to compute the orbit of Mercury in Chapter 4, we deliberately chose the metric so that the Ricci tensor R^r_{prq} would vanish. But the Laplace–Beltrami operator, when applied to the four nonzero metric coefficients, yields respectively

$$-\frac{c^2 \rho_s^2}{\rho^4} , \; \frac{\rho_s^2}{(\rho - \rho_s)^2 \rho^2} , \; 6 - \frac{4\rho_s}{\rho} , 4\left(1 - \frac{\rho_s}{\rho} \sin^2 \varphi\right).$$

Theorem 6.11 provides the essential connection between the Laplace–Beltrami operator ∇^2 and the second covariant derivative $\nabla^2_{u,v}$, namely

$$\nabla^2 = g^{ij} \nabla^2_{\left(\frac{\partial}{\partial x^i}, \frac{\partial}{\partial x^j}\right)}.$$

By looking at the expressions given by Euler for the curvature of a surface in $\mathbb{R}^3$, we can see the analogy with the Hessian we have just defined, and the intimate connection between that Hessian, the matrix of second covariant derivatives, and the Laplace–Beltrami operator. Best of all, we now have a connection—at least in normal coordinates—between the Laplace–Beltrami operator whose action on the generalized potential functions g_{ij} should guide us in reformulating mechanics and the Ricci tensor obtained by contracting the Riemann curvature tensor. That, as mentioned in Chapter 4, was the essential element in producing Einstein's law of gravity. What remains for us to do at this point is just to explore the properties of the Ricci tensor. Once we do, we will be ready to sketch the transition from Newtonian mechanics to the mechanics of general relativity, which is our goal.

Before we take up that last task, we will give a brief review of the classical Laplacian that we have now generalized.

6.7. Geometric and physical interpretation. We shall give a number of examples of the use of the Laplacian from geometry and classical (nineteenth-century) physics.

EXAMPLE 6.7. Let $u(x, y, z)$ be a real-valued function defined on an open set U in $\mathbb{R}^3$.

Whatever physical quantity u represents, its differential du represents the change in that quantity in the direction perpendicular to the level surface S_r whose equation is $u(x, y, z) = u_0$. If, for example, $(x(t), y(t), z(t))$ are the coordinates of a particle at time t, then the velocity of the particle is

$$\boldsymbol{v}(t) = x'(t)\,\boldsymbol{i} + y'(t)\,\boldsymbol{j} + z'(t)\,\boldsymbol{k}\,,$$

and the infinitesimal change in $u(x(t), y(t), z(t))$ is

$$du = \frac{\partial u}{\partial x}\,dx + \frac{\partial u}{\partial y}\,dy + \frac{\partial u}{\partial z}\,dz = \left(\frac{\partial u}{\partial x}\,x'(t) + \frac{\partial u}{\partial y}\,y'(t) + \frac{\partial u}{\partial z}\,z'(t)\right) dt\,.$$

Now consider what the adjoint of du represents. It is a 2-form, namely

$$*du = \frac{\partial u}{\partial x}\,dy \wedge dz + \frac{\partial u}{\partial y}\,dz \wedge dx + \frac{\partial u}{\partial z}\,dx \wedge dy\,.$$

For any surface $S \subseteq U$ parameterized as $\boldsymbol{r}(s, t)$, we find that

$$*du = \left(\frac{\partial u}{\partial x}\det\begin{pmatrix}\frac{\partial y}{\partial s} & \frac{\partial y}{\partial t}\\ \frac{\partial z}{\partial s} & \frac{\partial z}{\partial t}\end{pmatrix} + \frac{\partial u}{\partial y}\det\begin{pmatrix}\frac{\partial z}{\partial s} & \frac{\partial z}{\partial t}\\ \frac{\partial x}{\partial s} & \frac{\partial x}{\partial t}\end{pmatrix} + \frac{\partial u}{\partial z}\det\begin{pmatrix}\frac{\partial x}{\partial s} & \frac{\partial x}{\partial t}\\ \frac{\partial y}{\partial s} & \frac{\partial y}{\partial t}\end{pmatrix}\right) ds \wedge dt\,.$$

In other words,

$$*du = \nabla u \cdot \frac{\partial \boldsymbol{r}}{\partial s} \times \frac{\partial \boldsymbol{r}}{\partial t}\,ds \wedge dt\,.$$

As a consequence, for any surface S contained in U,

$$\iint_S *du = \iint_S \nabla u \cdot d\boldsymbol{A}\,,$$

where $d\boldsymbol{A}$ is what we call the *vector element of surface area*. The integrand in this integral is intuitively the linear rate of change of u along each line through S perpendicular to S. The integral is called the *flux* of u (or, rather, ∇u) through S. If the surface S is closed and its interior is a region $B \subset U$, the divergence theorem says that this flux is given by the volume integral of the divergence of ∇u over B:

$$\iiint_B d(*du) = \iiint_B \nabla \cdot \nabla u\,dV = \iiint_B \nabla^2 u\,dV\,.$$

Interesting cases arise when u represents a density. For example, suppose $u(t; x, y, z)$ represents the temperature at a point (x, y, z) at time t. We think of this temperature as a "heat density," so that if $Q_B(t)$ is the total amount of thermal energy (kinetic molecular energy) in the region B at time t, then

$$Q_B(t) = \iiint_B u\,dV\,.$$

By Newton's law of cooling, the flow of heat at a given point in a given direction $\boldsymbol{n}$ (the direction being a unit vector) will be proportional to the component of the temperature gradient in the opposite direction, that is, it will be $-a\nabla u \cdot \boldsymbol{n}$, where a is a positive constant, and hence the total amount of heat flowing outward across an infinitesimal area $d\boldsymbol{A}$ in time dt will be $-a\nabla u \cdot d\boldsymbol{A}\,dt$. Thus, for the closed surface $S = \partial B$ that is the boundary of the region B, the rate at which heat is flowing across S and *into* B is

$$Q'_B(t) = a \iint\limits_{S} \nabla u \cdot d\boldsymbol{A}\,.$$

By the divergence theorem, we thus have

$$a \iiint\limits_{B} \nabla^2 u\,dV = Q'_B(t) = \iiint\limits_{B} \frac{\partial u}{\partial t}\,dV\,,$$

Since the region B is arbitrary, the two integrands must be equal, and we thus get the classical linearized *heat equation*

$$\frac{\partial u}{\partial t} = a\nabla^2 u\,.$$

In particular, when the temperature reaches a steady state, so that its time derivative is zero, there is no flow of heat, and the temperature satisfies Laplace's equation $\nabla^2 u = 0$. That means that a steady-state temperature is a *harmonic function*. A harmonic function represents a physical quantity that has no tendency to flow across any closed surface. That is, the integral of its gradient over a closed surface is zero, being equal to the integral of its Laplacian over the region enclosed by the surface. Whatever flows out of the surface at one place is counterbalanced by an inflow at some other point.

EXAMPLE 6.8. Most closely related to the historical roots of relativity is the wave equation, mentioned above, which is succinctly stated as "the d'Alembertian vanishes." In classical terms, it says

$$\frac{\partial^2 u}{\partial t^2} = c^2 \nabla^2 u\,,$$

where c is the speed with which the wave propagates. For a wave in a pair of interacting electric and magnetic fields, that speed is $1/\sqrt{\varepsilon\mu}$, where ε is the dielectric permittivity of the medium and μ its magnetic permeability. It was Maxwell's startling discovery in 1861 that in free space that speed happens to be the speed of light that led to the conclusion that light is an electromagnetic wave.

In this case, the Laplace equation $\nabla^2 z = 0$, which is the equation satisfied by a harmonic function, gives the shape of a standing wave.

EXAMPLE 6.9. Consider a fluid flowing through a region of space such that at time t the particle (molecule) at the point $\boldsymbol{x} = (x, y, z)$ has velocity $\boldsymbol{u}(t; x, y, z)$. If the fluid has density $\rho(t; x, y, z)$ at time t at point (x, y, z), the net amount of fluid dm flowing out of a region B bounded by a surface $\Sigma = \partial B$ during a brief time interval dt can be represented as the product

$$dm = -\Big(\int\limits_{\Sigma} \rho\boldsymbol{u} \cdot d\boldsymbol{A} \Big)\,dt\,.$$

(In portions of the surface where fluid is flowing *into* the region B, the dot product $\boldsymbol{u} \cdot d\boldsymbol{A}$ is negative.) By the divergence theorem, the mass of fluid m in the region satisfies

$$\frac{dm}{dt} = -\int\limits_{B} \operatorname{div}(\rho\boldsymbol{u})\,dV\,.$$

Over an infinitesimal region of volume dV we have $m = \rho\, dV$, and $dm/dt = \partial\rho/\partial t\, dV$. Since the integral is approximately $-\mathrm{div}\,(\rho\boldsymbol{u})\, dV$, we get the *equation of continuity*

$$\frac{\partial\rho}{\partial t} + \mathrm{div}\,(\rho\boldsymbol{u}) = 0\,.$$

In particular, the condition $\mathrm{div}\,(\rho\boldsymbol{u}) = 0$ at a point at all times means the fluid is not being compressed or expanded at that point.

The vanishing of the divergence of the mass-flow vector $\rho\boldsymbol{u}$ throughout a region is thus a sufficient condition for conservation of mass in that region.

7. Curvature, Phase 4: Ricci

In Corollary 6.4 above, we used the exponential mapping and normal coordinates to connect the Riemann curvature tensor with the Laplace–Beltrami operator at the origin, namely the formula

$$R^{r}_{prq} = -\frac{3}{2}\nabla^2 g_{pq}\,.$$

In fact, we established that—again in normal coordinates and at the origin—

$$R^{r}_{prq} = \frac{3}{2}\frac{\partial\Gamma^{r}_{pq}}{\partial u^{r}}\,.$$

We did not specify the basis to be used in the tangent space when constructing normal coordinates. The coordinates $(u^1, \ldots, u^n)$ could be taken in any basis. But, if we specify an orthonormal basis, as we are now going to do, we gain the additional simplification that $g_{pq} = \delta^{q}_{p}$ at the origin. Because the Christoffel symbols vanish at the origin, all the linear terms in the Maclaurin expansion of g_{pq} also vanish. We now wish to take up the question of the quadratic terms in that expansion. These will show up most clearly when we expand the determinant of the matrix g, and we recall that the square root of this determinant is the volume element on the manifold $\mathfrak{M}$ whose local parameterization is given by the coordinates $(u^1, \ldots, u^n)$: $dV = \sqrt{\det(g)}\, du^1, \ldots, du^n$.

THEOREM 6.12. *In normal coordinates formed using an orthonormal basis of the tangent space at a point P, the element of volume is given by*

$$dV = \left(1 - \frac{1}{6}R^{r}_{prq}\, u^p\, u^q + E\right) du^1 \cdots du^n\,,$$

where the error term E is of order $(O(|\boldsymbol{u}|^3)$.

PROOF. We use the notation

$$\Delta_{ijpq} = \frac{\partial g_{ij}}{\partial u^p\, \partial u^q}\,.$$

The matrix of metric coefficients is (assuming $n \geq 3$)

$$g = \begin{pmatrix} 1 - \frac{1}{2}\Delta_{11pq}u^p\, u^q & -\frac{1}{2}\Delta_{12pq}u^p\, u^q & \cdots & -\frac{1}{2}\Delta_{1npq}u^p\, u^q \\ -\frac{1}{2}\Delta_{21pq}u^p\, u^q & 1 - \frac{1}{2}\Delta_{22pq}u^p\, u^q & \cdots & -\frac{1}{2}\Delta_{2npq}u^p\, u^q \\ \vdots & \vdots & \ddots & \vdots \\ -\frac{1}{2}\Delta_{n1pq}u^p\, u^q & -\frac{1}{2}\Delta_{n2pq}u^p\, u^q & \cdots & 1 - \frac{1}{2}\Delta_{nnpq}u^p\, u^q \end{pmatrix} + E\,,$$

where the entries in the matrix E are all less than some absolute constant times $\left((u^1)^2 + \cdots + (u^n)^2\right)^{3/2}$. We are interested in the determinant of this matrix, which gives the square of the local volume element.

When $\det(g)$ is expanded, one of the terms is the product of the diagonal elements. Up to second order, that product is

$$1 - \frac{1}{2} \sum_{r=1}^{n} \Delta_{rrpq} \, u^p \, u^q \, .$$

All the other terms in the expansion of the determinant have at least two factors that are off-diagonal elements and are therefore less than some absolute constant times $\left((u^1)^2 + \cdots + (u^n)^2 \right)^2$. Thus the entire expansion of the determinant up to second order is just the preceding expression. Since

$$\Delta_{rrpq} = \frac{\partial \Gamma_{pq}^r}{\partial u^r} = \frac{2}{3} R_{prq}^r \, ,$$

it follows that

$$\det(g) = 1 - \frac{1}{3} R_{prq}^r \, u^p \, u^q + E \, ,$$

where the error term E is of the order $\left((u^1)^2 + \cdots + (u^n)^2 \right)^{3/2}$. Since $\sqrt{1-x} = 1 - \frac{1}{2}x + F$, where F is of the order x^2, we see finally that the volume element at points near P is

$$dV = \sqrt{\det(g)} \, du^1 \cdots du^n \approx \left(1 - \frac{1}{6} R_{prq}^r \, u^p \, u^q \right) du^1 \cdots du^n \, .$$

$\square$

We can now see that the expression $R_{prq}^r u^p u^q$ measures the departure of n-*dimensional volume* on the manifold from its Euclidean value in the parameter space. If it is constantly zero, then volume remains Euclidean (up to third order). When we reflect that gravitational forces distort objects and change their volume, we can dimly see a possible connection between this expression and gravitation. We introduced this object in Chapter 4 under the name *Ricci tensor* and set it equal to zero in order to define a metric on space-time in which Einstein's law of gravity is the simple statement that the world-line of a particle is a geodesic. We have now begun to put some foundation under what we did, and so it is time to develop the properties of this tensor.

7.1. The Bianchi identity. As background for what we are about to do, we need to prove the following lemma, known as the *first Bianchi identity*[20]

LEMMA 6.3. *The Riemann curvature tensor $R(\boldsymbol{u}, \boldsymbol{v})\boldsymbol{w}$ satisfies the relation*

$$R(\boldsymbol{u}, \boldsymbol{v})\boldsymbol{w} + R(\boldsymbol{w}, \boldsymbol{u})\boldsymbol{v} + R(\boldsymbol{v}, \boldsymbol{w})\boldsymbol{u} = \boldsymbol{0} \, .$$

PROOF. The very form of this relation suggests that it must follow from the Jacobi identity, and indeed it does. With each tangent vector $\boldsymbol{u}$, we associate an operator $L(\boldsymbol{u})$ on the tangent space called the *Lie derivative* with respect to $\boldsymbol{u}$ and defined by the relation $L(\boldsymbol{u})\boldsymbol{v} = [\boldsymbol{u}, \boldsymbol{v}] = \nabla_{\boldsymbol{u}} \boldsymbol{v} - \nabla_{\boldsymbol{v}} \boldsymbol{u}$. It is obvious from the definition that $L(\boldsymbol{x})\boldsymbol{y} = -L(\boldsymbol{y})\boldsymbol{x}$. The Jacobi identity implies that

$$L(\boldsymbol{u})\big(L(\boldsymbol{v})\boldsymbol{w}\big) + L(\boldsymbol{w})\big(L(\boldsymbol{u})\boldsymbol{v}\big) + L(\boldsymbol{v})\big(L(\boldsymbol{w})\boldsymbol{u}\big) = \boldsymbol{0} \, .$$

If we use the relation $L(\boldsymbol{x})\boldsymbol{y} = -L(\boldsymbol{y})\boldsymbol{x}$, we can rewrite this relation as

$$\boldsymbol{0} = L(\boldsymbol{u})\big(L(\boldsymbol{v})\boldsymbol{w}\big) - L\big(L(\boldsymbol{u})\boldsymbol{v}\big)\boldsymbol{w} - L(\boldsymbol{v})\big(L(\boldsymbol{u})\boldsymbol{w}\big) \, .$$

---

[20]Named after Luigi Bianchi (1856–1928), although it had been noticed two decades earlier by Ricci.

If we write out each of the terms on the right using the formulas $L(\boldsymbol{x})\boldsymbol{y} = \nabla_{\boldsymbol{x}}\boldsymbol{y} - \nabla_{\boldsymbol{y}}\boldsymbol{x}$ and $R(\boldsymbol{u},\boldsymbol{v})\boldsymbol{w} = \nabla_{\boldsymbol{u},\boldsymbol{v}}\boldsymbol{w} - \nabla_{\boldsymbol{v},\boldsymbol{u}}\boldsymbol{w}$, this relation becomes (when terms are rearranged) precisely the statement of the lemma. $\qquad\square$

7.2. The Ricci tensor. We begin by giving a formal definition of the Ricci tensor on a manifold of dimension n.

DEFINITION 6.8. The *Ricci tensor* $\mathrm{Ric}\,(\boldsymbol{u},\boldsymbol{v})$ is the tensor of type $(0,2)$ obtained from the Riemann curvature tensor whose coordinates are R^i_{jkl} by contracting on the indices i and k, that is, by setting $k = i$ and summing on this repeated index. As a bilinear functional, in a basis of the tangent space $\{\boldsymbol{u}_1,\ldots,\boldsymbol{u}_n\}$ in which $\boldsymbol{v} = v^j\boldsymbol{u}_j$ and $\boldsymbol{z} = z^l\boldsymbol{u}_l$,

$$\mathrm{Ric}\,(\boldsymbol{v},\boldsymbol{z}) = R^i_{jil}v^j z^l = \left(\frac{\partial\Gamma^i_{lj}}{\partial x^i} - \frac{\partial\Gamma^i_{ij}}{\partial x^l} + \left(\Gamma^i_{im}\Gamma^m_{lj} - \Gamma^i_{lm}\Gamma^m_{ij}\right)\right)v^j z^l\,.$$

Older texts, such as that of Eddington [**16**], define the Ricci tensor by contracting on the indices i and l rather than i and k, thereby obtaining the negative of what we are calling the Ricci tensor. Given that we mostly set it equal to zero, the difference in sign is not important. Eddington, Einstein, and others used the letter G to denote this tensor, a usage that conflicts with the notation for the universal gravitational constant. Modern presentations tend to use the letter R for this tensor, and that in turn conflicts with the notation for the Riemann curvature tensor $R(\boldsymbol{u},\boldsymbol{v})\boldsymbol{z}$. The notation we are using here is not standard, but at least has the virtue of suggesting exactly what it denotes.

Phrasing this definition another way, we can say that if $\{\boldsymbol{u}_1,\ldots,\boldsymbol{u}_n\}$ is a basis of the tangent space and $\{\boldsymbol{v}_1,\ldots,\boldsymbol{v}_n\}$ the dual basis of the cotangent space, then

$$\mathrm{Ric}\,(\boldsymbol{v},\boldsymbol{z}) = \sum_{i=1}^{n} \boldsymbol{v}_i\big(R(\boldsymbol{u}_i,\boldsymbol{z})\boldsymbol{v}\big)\,.$$

If $\boldsymbol{v}$ and $\boldsymbol{z}$ are fixed vectors, the mapping $\boldsymbol{w} \mapsto R(\boldsymbol{w},\boldsymbol{z})\boldsymbol{v}$ is a linear operator $T_{\boldsymbol{v},\boldsymbol{z}}$ on the tangent space. In those terms $\mathrm{Ric}\,(\boldsymbol{v},\boldsymbol{z})$ is the trace $\mathrm{Tr}\,(T_{\boldsymbol{v},\boldsymbol{z}})$ of the linear operator $T_{\boldsymbol{v},\boldsymbol{z}}$. The reader can easily verify this fact, since in the standard basis of the tangent space $\{\partial/\partial x^1,\ldots,\partial/\partial x^n\}$ we have

$$T_{\boldsymbol{v},\boldsymbol{z}}(\boldsymbol{u}) = R^i_{jkl}v^j u^k z^l \frac{\partial}{\partial x^i}\,,$$

so that the matrix of $T_{\boldsymbol{v},\boldsymbol{z}}$ in this standard basis is

$$T = \begin{pmatrix} t_{11} & \cdots & t_{1n} \\ \vdots & \ddots & \vdots \\ t_{n1} & \cdots & t_{nn} \end{pmatrix},$$

where

$$t_{ik} = R^i_{jkl}v^j z^l\,,$$

and so

$$\mathrm{Ric}\,(\boldsymbol{v},\boldsymbol{z}) = \sum_{i=1}^{n} t_{ii} = \mathrm{Tr}\,(T_{\boldsymbol{v},\boldsymbol{z}})\,.$$

Since the linear operator $T_{\boldsymbol{v},\boldsymbol{z}}$ is a tensor (transforms correctly under a change of coordinates), and the trace of a linear operator T is the same in any and all bases—it is the negative of the coefficient of λ^{n-1} in the characteristic polynomial of T—we see that $\mathrm{Ric}\,(\boldsymbol{u}, \boldsymbol{v})$ is a tensor. It also has the important property of symmetry, as we shall now prove.

THEOREM 6.13. *The Ricci tensor satisfies the relation*

$$\mathrm{Ric}\,(\boldsymbol{v}, \boldsymbol{z}) = \mathrm{Ric}\,(\boldsymbol{z}, \boldsymbol{v})$$

for all tangent vectors $\boldsymbol{v}$ and $\boldsymbol{z}$.

PROOF. We define a linear operator T on the tangent space to be the difference $T_{\boldsymbol{v},\boldsymbol{z}} - T_{\boldsymbol{z},\boldsymbol{v}}$. Since the trace is a linear function on the space of operators, it will suffice to show that the trace of T is zero. For any tangent vector $\boldsymbol{u}$ we have

$$T(\boldsymbol{u}) = R(\boldsymbol{u}, \boldsymbol{z})\boldsymbol{v} - R(\boldsymbol{u}, \boldsymbol{v})\boldsymbol{z} = R(\boldsymbol{u}, \boldsymbol{z})\boldsymbol{v} + R(\boldsymbol{v}, \boldsymbol{u})\boldsymbol{z}\,.$$

By the Bianchi identity (proved above as Lemma 6.3), we find

$$T(\boldsymbol{u}) = -R(\boldsymbol{z}, \boldsymbol{v})\boldsymbol{u} = R(\boldsymbol{v}, \boldsymbol{z})\boldsymbol{u}\,.$$

It then follows that in the standard inner product $\langle \cdot, \cdot \rangle$ on the tangent space we have

$$\langle \boldsymbol{u}, T(\boldsymbol{u}) \rangle = \langle \boldsymbol{u}, R(\boldsymbol{v}, \boldsymbol{z})\boldsymbol{u} \rangle = \mathrm{Rie}\,(\boldsymbol{u}, \boldsymbol{u}, \boldsymbol{v}, \boldsymbol{z})\,.$$

But, by the skew-symmetry of the covariant Riemann curvature tensor (Problem 5.7), this last quantity equals 0. Thus $\langle \boldsymbol{u}, T(\boldsymbol{u}) \rangle \equiv 0$ for all tangent vectors $\boldsymbol{u}$. In the language of linear algebra, the *numerical range* of the operator T consists of the number 0 alone. In particular, if we calculate the trace in a basis $\{\boldsymbol{u}_1, \dots, \boldsymbol{u}_n\}$ of the tangent space that is orthonormal with respect to the standard inner product, we find that

$$\mathrm{Tr}\,(T) = \sum_{k=1}^{n} \langle \boldsymbol{u}_k, T(\boldsymbol{u}_k) \rangle = 0\,.$$

$\square$

REMARK 6.14. It may be useful to say a word about the operation of contraction in general. If we are given a tensor of type (k, l), say a multilinear functional

$$T(\boldsymbol{v}_1, \dots, \boldsymbol{v}_k, \boldsymbol{u}_1 \dots, \boldsymbol{u}_l)$$

mapping k covectors $\boldsymbol{v}_1, \dots, \boldsymbol{v}_k$ and l vectors $\boldsymbol{u}_1, \dots, \boldsymbol{u}_l$ into the real numbers, where both k and l are positive, then holding $\boldsymbol{v}_i$ fixed for $i \neq a$ and $\boldsymbol{u}_j$ fixed for $j \neq b$ produces a bilinear functional $(\boldsymbol{v}_a, \boldsymbol{u}_b) \mapsto T(\boldsymbol{v}_1, \dots, \boldsymbol{v}_k, \boldsymbol{u}_1 \dots, \boldsymbol{u}_l)$. We can think of this bilinear functional as a transformation of $\boldsymbol{u}_b$ into a vector $L(\boldsymbol{u}_b)$ whose action on each covector $\boldsymbol{v}_a$ is given by $L(\boldsymbol{u}_b)(\boldsymbol{v}_a) = T(\boldsymbol{v}_1, \dots, \boldsymbol{v}_k, \boldsymbol{u}_1, \dots, \boldsymbol{u}_l)$. Contraction on the indices a and b produces a tensor of type $(k-1, l-1)$, which is the trace of the linear operator L. In contrast to contraction, the operation of lowering an index produces a tensor of type $(k-1, l+1)$. In the present case, starting from the Riemann tensor, which is of type $(1, 3)$, contraction produces the Ricci tensor of type $(0, 2)$. When we lowered the superscript in the Riemann curvature tensor, we got the covariant Riemann curvature tensor, which is of type $(0, 4)$.

7.3. Ricci curvature. At the beginning of this section, we showed that the element of volume dV near a point P of a manifold $\mathfrak{M}$ is given at a point Q having normal coordinates $(u^1, \ldots, u^n)$ based at P (that is, the coordinates of P are $(0, \ldots, 0)$) by the formula

$$(6.18) \qquad dV = du^1 \cdots du^n - \frac{1}{6}\mathrm{Ric}\,(\boldsymbol{u}, \boldsymbol{u})\,du^1 \cdots du^n + E\,,$$

where E is smaller than some absolute constant times $\left((u^1)^2 + \cdots + (u^n)^2\right)^{3/2}$. That motivates the following definition.

DEFINITION 6.9. The *Ricci curvature of* $\mathfrak{M}$ *at* P *in the direction of a tangent vector* $\boldsymbol{u}$, denoted $\mathrm{Ricc}\,(\boldsymbol{u})$, is $\mathrm{Ric}\,(\boldsymbol{u}, \boldsymbol{u})$.

We have at last managed to link this long winding journey through the concept of curvature to the discussion of planetary orbits in Chapter 4. Equation (6.18) shows that an infinitesimal n-dimensional piece of the tangent space having Euclidean volume dV in the tangent space maps to the same infinitesimal volume on the manifold $\mathfrak{M}$ at the point P. If $\mathrm{Ricc}\,(\boldsymbol{u})$ is positive at P, the ratio of the volume of the image of such an infinitesimal piece to the volume of its preimage in the parameter space will decrease in the direction of $\boldsymbol{u}$. Conversely, if the Ricci curvature in that direction is negative, that ratio will increase in the direction of $\boldsymbol{u}$. When the Ricci curvature is zero, as is the case in the metric chosen for space-time, volume will be stable in every direction.

For convenience, we use the symbol $\mathbb{R}^{n-1}_{\boldsymbol{v}^\perp}$ to denote the $(n-1)$-dimensional subspace of the tangent space at a point consisting of the vectors $\boldsymbol{u}$ for which $\langle \boldsymbol{u}, \boldsymbol{v}\rangle = 0$. In Theorems 6.14 and 6.15 below, as well as Definition 6.10 and Lemma 6.4, we assume a positive-definite matrix of metric coefficients. (The extension of these results to the general case is straightforward, but the general case of Theorem 6.15 requires the concept of a signed measure, and the extra space required to explain that is not justified by the additional understanding to be gained.)

THEOREM 6.14. Let $\{\boldsymbol{u}_1, \ldots, \boldsymbol{u}_{n-1}\}$ be an orthonormal basis of the subspace $\mathbb{R}^{n-1}_{\boldsymbol{v}^\perp}$ of the tangent space. Then the Ricci curvature $\mathrm{Ricc}\,(\boldsymbol{v})$ in the direction of a unit vector $\boldsymbol{v}$ is the sum of the sectional curvatures of the geodesic submanifolds tangent to the planes spanned by $\boldsymbol{v}$ and $\boldsymbol{u}_i$, $i = 1, 2, \ldots, n-1$. That is,

$$\mathrm{Ricc}\,(\boldsymbol{v}) = \sum_{i=1}^{n-1} \kappa(\boldsymbol{u}_i, \boldsymbol{v})\,.$$

PROOF. For any basis $\{\boldsymbol{u}_1, \ldots, \boldsymbol{u}_n\}$ whatever in the tangent space, we know that

$$\mathrm{Ric}\,(\boldsymbol{v}, \boldsymbol{v}) = \sum_{i=1}^{n} \langle \boldsymbol{u}_i, R(\boldsymbol{u}_i, \boldsymbol{v})\boldsymbol{v}\rangle = \kappa(\boldsymbol{u}_i, \boldsymbol{v})\left(\langle \boldsymbol{v}, \boldsymbol{v}\rangle\langle \boldsymbol{u}_i, \boldsymbol{u}_i\rangle - \langle \boldsymbol{v}, \boldsymbol{u}_i\rangle^2\right)\,.$$

By choosing the basis to be orthonormal and such that $\boldsymbol{u}_n = \boldsymbol{v}$, we get the last term of this sum to be zero and the others to be $\kappa(\boldsymbol{u}_i, \boldsymbol{v})$. Thus, as asserted,

$$\mathrm{Ricc}\,(\boldsymbol{v}) = \sum_{i=1}^{n-1} \kappa(\boldsymbol{u}_i, \boldsymbol{v})\,.$$

$\square$

It follows immediately from this result that the Ricci curvature of a two-dimensional surface, in any direction whatever, is its Gaussian curvature.

Since the right-hand side of this last equation is the same for all orthonormal bases of $\mathbb{R}^{n-1}_{\boldsymbol{v}^\perp}$, we make the following definition:

DEFINITION 6.10. The *average sectional curvature at P in the direction of $\boldsymbol{v}$* is the number $\chi(\boldsymbol{v})$ given by

$$\chi(\boldsymbol{v}) = \frac{1}{n-1} \sum_{i=1}^{n-1} \kappa(\boldsymbol{u}_i, \boldsymbol{v})$$

for some (any) orthonormal basis $\{\boldsymbol{u}_1, \ldots, \boldsymbol{u}_{n-1}\}$ of $\mathbb{R}^{n-1}_{\boldsymbol{v}^\perp}$.

In other words, the Ricci curvature Ricc $(\boldsymbol{v})$ is $n-1$ times the average sectional curvature $\chi(\boldsymbol{v})$. Again, this average sectional curvature reduces to the Gaussian curvature for two-dimensional manifolds.

7.4. The continuous average sectional curvature*. The average sectional curvature just defined is a discrete average. We are now going to prove that it is equal to a continuous average that has an even stronger claim to be called the average sectional curvature in the direction of $\boldsymbol{v}$.

To that end, we need to introduce some concepts involving polar coordinates on $\mathbb{R}^n$ that may be of independent interest. There is a measure $\sigma_{n-1}(E)$ defined on (Borel) subsets of the $(n-1)$-dimensional unit sphere in $\mathbb{R}^n$ that is orthogonally invariant, which means that $\sigma_{n-1}(T(E)) = \sigma_{n-1}(E)$ for all measurable sets E and all orthogonal linear transformations T. This measure, which is ordinary arc length on the unit circle in $\mathbb{R}^2$ and the usual surface area on the unit sphere in $\mathbb{R}^3$, makes it possible to compute n-fold integrals over $\mathbb{R}^n$ in polar coordinates, that is,

$$\int_{\mathbb{R}^n} f(\boldsymbol{x}) \, d\boldsymbol{x} = \int_0^\infty \int_{\mathbb{S}^{n-1}} f(\rho\boldsymbol{\xi}) \, \rho^{n-1} \, d\sigma_{n-1}(\boldsymbol{\xi}) \, d\rho \,.$$

The definition of this measure is straightforward: $\sigma_{n-1}(E) = n \, \mu_n(\widetilde{E})$, where μ_n is n-dimensional volume (Lebesgue measure) on $\mathbb{R}^n$, and for any subset $E \subseteq \mathbb{S}^{n-1}$, the set $\widetilde{E}$ is the "cone" whose base is E and whose apex is at the origin, that is,

$$\widetilde{E} = \{t\boldsymbol{x} : \boldsymbol{x} \in E, 0 < t \le 1\} \,.$$

The polar coordinate formula is proved by direct computation if the function $f(\boldsymbol{x})$ is the characteristic function of a set of the form

$$(a, b] \times E = \{t\boldsymbol{\xi} : \boldsymbol{\xi} \in E \subseteq \mathbb{S}^{n-1}, \, a < t \le b\},$$

a computation that requires only the fact that the mapping $\boldsymbol{x} \to t\boldsymbol{x}$ multiplies n-dimensional volumes by t^n. The formula is then extended to general integrable functions on $\mathbb{R}^n$ by the fact that linear combinations of such characteristic functions are dense in the space of integrable functions. Since we need the result only for continuous functions, we can rely on the fact that any continuous function of compact support on $\mathbb{R}^n$ can be uniformly approximated by a finite linear combination of such characteristic functions.

The formula

$$\int_0^\infty e^{-x^2} \, dx = \frac{1}{2}\sqrt{\pi}$$

is well-known and easy to establish:

$$\left(\int_0^\infty e^{-x^2} \, dx \right)^2 = \int_0^\infty \int_0^\infty e^{-x^2 - y^2} \, dx \, dy$$

$$= \int_0^{\pi/2} \int_0^\infty e^{-r^2} r \, dr \, d\theta = \frac{\pi}{4} \, .$$

This same technique of integrating the function $f(\boldsymbol{x}) = e^{-|\boldsymbol{x}|^2}$ in both polar and rectangular coordinates shows that, on the one hand

$$\int_{\mathbb{R}^n} e^{-|\boldsymbol{x}|^2} \, dx = \left(\int_{-\infty}^\infty e^{-x^2} \, dx \right)^n = \pi^{\frac{n}{2}} \, ,$$

and on the other hand

$$\int_{\mathbb{R}^n} e^{-|\boldsymbol{x}|^2} \, dx = \int_0^\infty \int_{\mathbb{S}^{n-1}} e^{-\rho|\boldsymbol{\xi}|^2} \, d\sigma_{n-1}(\boldsymbol{\xi}) \, \rho^{n-1} \, d\rho$$

$$= \sigma_{n-1}(\mathbb{S}^{n-1}) \int_0^\infty e^{-\rho^2} \rho^{n-1} \, d\rho \, .$$

It follows that the $(n-1)$-dimensional measure of $\mathbb{S}^{n-1}$, which we shall denote ω_{n-1}, is

$$\omega_{n-1} = \sigma_{n-1}(\mathbb{S}^{n-1}) = \omega_{n-1} = \frac{2\pi^{n/2}}{\Gamma(\frac{n}{2})} \, .$$

Here $\Gamma(z)$ is the Euler–Gauss Gamma function:

$$\Gamma(z) = \int_0^\infty t^{z-1} e^{-t} \, dt = 2 \int_0^\infty t^{2z-1} e^{-t^2} \, dt \, ,$$

The fundamental identity that we shall need is

$$\Gamma(z+1) = z\Gamma(z) \, ,$$

easily proved through integration by parts. We also need one consequence of this formula:

LEMMA 6.4.

$$\int_0^{\frac{\pi}{2}} \cos^{2\mu-1} \theta \, \sin^{2\nu-1} \theta \, d\theta = \frac{\Gamma(\mu + \nu)}{2\Gamma(\mu)\Gamma(\nu)} \, .$$

PROOF. We have

$$\Gamma(\mu)\Gamma(\nu) = 4 \int_0^\infty \int_0^\infty s^{2\mu-1} t^{2\nu-1} e^{-(s^2+t^2)} \, ds \, dt \, .$$

In polar coordinates, where $s = r\cos\theta$, $t = r\sin\theta$, this formula becomes

$$\Gamma(\mu)\Gamma(\nu) = 4 \int_0^\infty \int_0^{\frac{\pi}{2}} \cos^{2\mu-1} \theta \, \sin^{2\nu-1} \theta \, r^{2(\mu+\nu)-1} e^{-r^2} \, d\theta \, dr$$

$$= 2\Gamma(\mu + \nu) \int_0^{\pi/2} \cos^{2\mu-1} \theta \, \sin^{2\nu-1} \theta \, d\theta \, .$$

$\square$

Taking $\mu = \nu = 1/2$ and noting that $\Gamma(1) = 1$ (a trivial computation), we see that $\Gamma(1/2) = \sqrt{\pi}$. We remark that the number ω_{n-1} never involves any square root of π. When n is odd, the denominator $\Gamma(n/2)$ is a rational multiple of $\Gamma(1/2)$, which cancels that square root from the numerator. Thus, for example, $\Gamma(5/2) = (3/2)\Gamma(3/2) = (3/4)\Gamma(1/2) = 3\sqrt{\pi}/4$, and so $\omega_4 = 2\pi^{5/2}/(3\sqrt{\pi}/4) = 8\pi^2/3$.

Polar coordinates are theoretically applicable to any n-fold integral. In practice they are used most often for "radial" functions that depend only on the distance from the origin. About the only other time one ever finds them being used arises when the function being integrated depends on only one of the n variables. In that case, the following formula is useful.

LEMMA 6.5. *For a "zonal" function $f(\boldsymbol{x})$ of the form $f(\boldsymbol{x}) = \psi(\boldsymbol{x} \cdot \boldsymbol{y})$, where $\boldsymbol{y}$ is a fixed vector and ψ a function of a real variable, the integral over $\mathbb{S}^{n-1}$ is*

$$\int_{\mathbb{S}^{n-1}} f(\boldsymbol{\xi})\, d\sigma_{n-1}(\boldsymbol{\xi}) = \omega_{n-2} \int_{-\frac{\pi}{2}}^{\frac{\pi}{2}} \psi(|\boldsymbol{y}| \sin\theta)\, \cos^{n-2}\theta\, d\theta.$$

PROOF. There is no loss in generality in taking $\boldsymbol{y} = (0, 0, \ldots, 1)$, since the measure is rotation-invariant and we can replace $f(\boldsymbol{x})$ with $f(|\boldsymbol{y}|\boldsymbol{x})$ if we wish. Then $f(\boldsymbol{x})$ is a function of x^n only. By definition,

$$\int_{\mathbb{S}^{n-1}} f(\boldsymbol{\xi})\, d\sigma_{n-1}(\boldsymbol{\xi}) = n \int_{B^n} \tilde{f}(\boldsymbol{x})\, d\boldsymbol{x},$$

where B^n is the unit ball with the origin removed, that is, $\{\boldsymbol{x} : 0 < |\boldsymbol{x}| \le 1\}$, and

$$\tilde{f}(\boldsymbol{x}) = f\left(\frac{\boldsymbol{x}}{|\boldsymbol{x}|}\right).$$

In the present case, where f is a function of x^n only, we have

$$\tilde{f}(\boldsymbol{x}) = \psi\left(\frac{x^n}{\sqrt{|\boldsymbol{z}|^2 + (x^n)^2}}\right)\, dx,$$

where $\boldsymbol{z} = (x^1, \ldots, x^{n-1}) \in \mathbb{R}^{n-1}$.

Thus, we have

$$n \int_{B^n} \tilde{f}(\boldsymbol{x})\, dx = n \int_{-1}^{1} \int_{|\boldsymbol{z}|^2 \le 1 - (x^n)^2} \psi\left(\frac{x^n}{\sqrt{|\boldsymbol{z}|^2 + (x^n)^2}}\right)\, d\boldsymbol{z}\, dx^n$$

If we evaluate the inner integral here in polar coordinates on $\mathbb{R}^{n-1}$, we find that it is

$$\int_{0}^{\sqrt{1-(x^n)^2}} \int_{\mathbb{S}^{n-2}} \psi\left(\frac{x^n}{\sqrt{t^2 + (x^n)^2}}\right) t^{n-2}\, dt = \omega_{n-2} \int_{0}^{\sqrt{1-(x^n)^2}} \psi\left(\frac{x^n}{\sqrt{t^2 + (x^n)^2}}\right) t^{n-2}\, dt.$$

Thus we find that

$$\int_{\mathbb{S}^{n-1}} f(\boldsymbol{\xi})\, d\sigma_{n-1}(\boldsymbol{\xi}) = n\omega_{n-2} \int_{-1}^{1} \int_{0}^{\sqrt{1-(x^n)^2}} \psi\left(\frac{x^n}{\sqrt{t^2 + (x^n)^2}}\right) t^{n-2} dt\, dx^n.$$

This last integral is simply an integral over half of a unit disk, as we see by setting $t = r \cos \theta$, $x^n = r \sin \theta$:

$$\int_{\mathbb{S}^{n-1}} f(\boldsymbol{\xi}) = n\omega_{n-2} \int_{-\frac{\pi}{2}}^{\frac{\pi}{2}} \int_0^1 \psi(\sin\theta)\, r^{n-1} \cos^{n-2}\theta\, dr\, d\theta$$

$$= \omega_{n-2} \int_{-\frac{\pi}{2}}^{\frac{\pi}{2}} \psi(\sin\theta) \cos^{n-2}\theta\, d\theta.$$

$\square$

By invoking the $(n-2)$-dimensional sphere $\mathbb{S}^{n-2}_{\boldsymbol{v}^\perp}$ in $\mathbb{R}^{n-1}_{\boldsymbol{v}^\perp}$, we get the following theorem, which doubly justifies the terminology "average sectional curvature":

THEOREM 6.15. *If $\boldsymbol{v}$ is a unit tangent vector at P, the average sectional curvature at P in the direction of $\boldsymbol{v}$ is given by the formula*

$$\chi(\boldsymbol{v}) = \frac{1}{\omega_{n-2}} \int_{\mathbb{S}^{n-2}_{\boldsymbol{v}^\perp}} \kappa(\boldsymbol{\xi}, \boldsymbol{v})\, d\sigma_{n-2}(\boldsymbol{\xi}).$$

PROOF. Taking an orthonormal basis of the tangent space $\{\boldsymbol{u}_1, \ldots, \boldsymbol{u}_n\}$ in which $\boldsymbol{u}_n = \boldsymbol{v}$, we can write the average sectional curvature in the direction of $\boldsymbol{v}$, as defined above, in the form

$$\chi(\boldsymbol{v}) = \frac{a_{11} + \cdots + a_{n-1,n-1}}{n-1},$$

where

$$a_{ij} = g_{im} R^m_{jij}.$$

(There is no summation on i and j here, only on m.)

The continuous average defined in the statement of the theorem is

$$\chi(\boldsymbol{v}) = \frac{1}{\omega_{n-2}} \int_{\mathbb{S}^{n-2}_{\boldsymbol{v}^\perp}} a_{ij} \xi^i \xi^j\, d\sigma_{n-2}(\boldsymbol{\xi}).$$

Since the measure $d\sigma_{n-2}$ is invariant under the reflection

$$(\xi^1, \ldots, \xi^{i-1}, \xi^i, \xi^{i+1}, \ldots, \xi^{n-1}) \mapsto (\xi^1, \ldots, \xi^{i-1}, -\xi^i, \xi^{i+1}, \ldots, \xi^{n-1}),$$

which is an orthogonal transformation, while the term $a_{ij}\xi^i\xi^j$ reverses sign under this reflection, it follows that the integrals of all the terms containing a_{ij} with $i \ne j$ are zero. We claim now that

$$\int_{\mathbb{S}^{n-2}_{\boldsymbol{v}^\perp}} (\xi^i)^2\, d\sigma_{n-2}(\boldsymbol{\xi}) = \frac{\omega_{n-2}}{n-1}.$$

We now apply Lemma 6.5 with $\psi(t) = t^2$ and $\boldsymbol{y} = (\delta^1_i, \ldots, \delta^{n-1}_i)$, so that $|\boldsymbol{y}| = 1$. We thus have

$$\int_{\mathbb{S}^{n-2}_{\boldsymbol{v}^\perp}} (\xi^i)^2\, d\sigma_{n-2}(\boldsymbol{\xi}) = 2\omega_{n-3} \int_0^{\pi/2} \sin^2\theta \cos^{n-3}(\theta)\, d\theta = \omega_{n-3} \frac{\Gamma\left(\frac{n-2}{2}\right)\Gamma\left(\frac{3}{2}\right)}{\Gamma\left(\frac{n+1}{2}\right)}.$$

Sorting this out, we now have

$$\frac{1}{\omega_{n-2}} \int\limits_{\mathbb{S}^{n-2}_{v^\perp}} (\xi^i)^2 \, d\sigma_{n-2}(\boldsymbol{\xi}) = \frac{\Gamma\left(\frac{n-1}{2}\right)}{2\pi^{(n-1)/2}} \cdot \frac{2\pi^{(n-2)/2}}{\Gamma\left(\frac{n-2}{2}\right)} \cdot \frac{\Gamma\left(\frac{n-2}{2}\right)\Gamma\left(\frac{3}{2}\right)}{\Gamma\left(\frac{n+1}{2}\right)} = \frac{1}{n-1}\,,$$

as asserted.   □

REMARK 6.15. We recall that Einstein's law of gravity involved constructing a metric of a certain form in which the Ricci curvature vanished identically, then computing the geodesics in that metric. Because the Ricci tensor vanishes, the deviation of the "4-dimensional volume" of an infinitesimal region of space-time from its value at the origin is less than quadratically small in any coordinates $(t, \rho, \varphi, \theta)$. Because the matrix of metric coefficients in this case is diagonal with three negative entries and one positive entry, this "4-dimensional volume" is the square root of a negative number—an imaginary number. (As mentioned in Chapter 4, Einstein, Eddington, and others cut the Gordian knot of this difficulty by arbitrarily replacing the determinant with its negative, which is permissible, since the factor cancels out anyway.) As one can easily compute, it is

$$d\mu = \sqrt{-1}\frac{\rho^2 \sin^2 \varphi}{c^3} \, dt \, d\rho \, d\varphi \, d\theta\,,$$

and the density function here is of order $|\boldsymbol{u}|^4$ at points $(u^1, u^2, u^3, u^4) = (t, \rho, \varphi, \theta)$ near $(0,0,0,0)$. The physical dimension of this $d\mu$ is time raised to the fourth power. If $d\mu$ is scaled through multiplying each of the metric coefficients by c^2, its physical dimension becomes length raised to the fourth power; that is, it becomes

$$d\mu = \sqrt{-1}\rho^2 \sin^2 \varphi \, d\tau \, d\rho \, d\varphi \, d\theta\,,$$

where $\tau = ct$. (The coordinates τ and ρ are lengths, while the angles φ and θ are dimensionless.)

7.5. Scalar curvature. Finally, we introduce yet one more notion of curvature on a manifold.

DEFINITION 6.11. The *scalar curvature* R on a manifold is the contraction of the mixed Ricci tensor $\text{Ric}^j_i = g^{jk}\text{Ric}_{ik}$, that is,

$$R = g^{ij}\text{Ric}_{ij}\,.$$

(Although we have made strenuous efforts to avoid the abusive use of the symbol R for a large number of different entities, the use of this letter to denote the scalar curvature is so well established that we just have to warn the reader to watch out for the context when this symbol is encountered.)

The scalar curvature is obtained by first raising an index in the Ricci tensor to get the mixed Ricci tensor $\text{Ric}^j_i = g^{jk}\text{Ric}_{ik}$, then contracting on i and j (taking the trace of the matrix that represents the mixed tensor Ric^j_i).

The scalar curvature has the correct physical dimension to be a genuine curvature, namely $[d]^{-2}$, where $[d]^2$ is the dimension of the squared metric $ds^2 = g_{ij} \, dx^i \, dx^j$. The coordinates of the tensor Rg_{ij} have the same physical dimensions as the corresponding coordinates of the Ricci tensor, and hence we can consider linear combinations

$$\text{Ric}_{ij} + aRg_{ij}\,,$$

where a is any real number.

Expressions of this kind are the key to the relativistic reformulation of mechanics. Notice that R vanishes if the Ricci tensor is identically zero, as we assumed for the gravitational field of a particle. (In particular, the vanishing of the scalar curvature does not imply that a manifold is flat.) That fact meant that this term would not appear in the Einstein law of gravity in free space, which we discussed in Chapter 4. The task remaining to us is to insert this term back into the field equations and explore the relation between curvature and gravity in a wider context. That is the subject of the next chapter.

Meanwhile, we note that the succinct statement of Einstein's formulation of the gravitational field of a particle is that the Ricci tensor vanishes identically. As we have just recalled, the Ricci tensor may vanish on a curved space. We should point out, however, that the Ricci tensor *cannot* vanish identically on a curved surface in $\mathbb{R}^3$, since, as one can easily compute, the scalar curvature on such a surface is exactly twice the Gaussian curvature.

With that remark, we bring to a close this long and winding path through differential geometry. What remains—the subject matter of the next chapter—is to see how the use of this language affects the formulation of physical laws.

8. Problems

PROBLEM 6.1. A vector $\boldsymbol{a} = (a^1, a^2, a^3)$ in $\mathbb{R}^3$ can be naturally associated with a skew-symmetric 3×3 matrix

$$\boldsymbol{A} = \begin{pmatrix} 0 & a^3 & a^2 \\ -a^3 & 0 & a^1 \\ -a^2 & -a^1 & 0 \end{pmatrix}.$$

Show that, if $\boldsymbol{b} = (b^1, b^2, b^3)$ is associated in this way with the matrix $\boldsymbol{B}$, then the cross product $\boldsymbol{a} \times \boldsymbol{b}$ is associated with $[\boldsymbol{A}, \boldsymbol{B}] = \boldsymbol{A}\boldsymbol{B} - \boldsymbol{B}\boldsymbol{A}$. (Replacing an associative product with its commutator, that is, replacing AB with $[A, B]$ is a standard way of turning an associative algebra into a Lie algebra. If the associative algebra happens to be commutative, of course, the Lie algebra is trivial, since the Lie products are all equal to zero.)

PROBLEM 6.2. Consider a general surface in $\mathbb{R}^3$ parameterized by u and v, that is, $(u, v) \to \boldsymbol{r}(u, v)$, and a curve $\boldsymbol{\gamma}(s)$ on that surface:

$$\boldsymbol{\gamma}(s) = \boldsymbol{r}\big(u(s), v(s)\big).$$

For a fixed parameter value $s = s_0$, let $u_0 = u(s_0)$, $v_0 = v(s_0)$. Let $\boldsymbol{v}(u_0, v_0)$ be a vector at the point $P_0 = \boldsymbol{r}(u_0, v_0)$ on the surface:

$$\boldsymbol{v}(u_0, v_0) = a(u_0, v_0)\frac{\partial \boldsymbol{r}}{\partial u} + b(u_0, v_0)\frac{\partial \boldsymbol{r}}{\partial v}.$$

Show that, if $\boldsymbol{v}\big(u(s), v(s)\big)$ is the parallel transport of $\boldsymbol{v}(u_0, v_0)))$ from the point P_0 along the curve, that is

$$\boldsymbol{v}\big(u(s), v(s)\big) = a\big(u(s), v(s)\big)\frac{\partial \boldsymbol{r}}{\partial u} + b\big(u(s), v(s)\big)\frac{\partial \boldsymbol{r}}{\partial v},$$

then the squared length of the vector $\boldsymbol{v}\big(u(s), v(s)\big)$, which is

$$\big(a\big(u(s), v(s)\big)\big)^2 E\big(u(s), v(s)\big) + 2a\big(u(s), v(s)\big)b\big(u(s), v(s)\big)F\big(u(s), v(s)\big)$$
$$+ \big(b\big(u(s), v(s)\big)\big)^2 G\big(u(s)v(s)\big),$$

is constant. (That is, the derivative of this expression with respect to s is zero.)

PROBLEM 6.3. Finding geodesics is a task that can take bizarre twists. In Example 6.3, we found the complete family of geodesics on the pseudo-sphere through a given point, despite the fact that this surface has a complicated equation involving transcendental functions. To be sure, when cylindrical coordinates are used, z is independent of θ, and that makes the task much easier. Even so, compared to that example, one would expect it to be utterly trivial to find the geodesics on a surface as simple (algebraically) as the hyperbolic paraboloid whose equation is $z = xy/c$. In the natural parameterization $r(x, y) = x\,i + y\,j + (xy/c)\,k$, we have $g(0,0) = I$, and so all we need to do is find the geodesic $\gamma(t)$ with $\gamma(0) = \mathbf{0}$ whose tangent at $\mathbf{0}$ when arc length is the parameter is given by

$$\gamma'(0) = \frac{x}{\sqrt{x^2 + y^2}}\,i + \frac{y}{\sqrt{x^2 + y^2}}\,j\,.$$

for each pair x, y with $x^2 + y^2 = 1$. In polar coordinates, we need

$$\gamma(0) = \mathbf{0}\,,$$
$$\gamma'(0) = \cos\theta\,i + \sin\theta\,j\,.$$

Show that the Euler equations for this problem imply the equations

$$x' = v\,,$$
$$y' = w\,,$$
$$xv' - yw' = 0\,,$$
$$v^2 + w^2 + (xw + yv)^2/c^2 = 1\,.$$

(Attempt to solve these equations at your own risk!!)

PROBLEM 6.4. Show that Eq. (6.9) becomes $\theta = \theta_1$ when $a = c$. That is, it is a geodesic that is a meridian of longitude. For $a < c$, show by eliminating $\sinh\big((s_0 + s)/c\big)$ from Eqs. (6.8)–(6.9) that these equations imply Eq. (6.7) with $r_0 = \sqrt{c^2 - a^2}$ and $\theta_0 = \theta_1 + a/r_0$.

PROBLEM 6.5. Prove the formulas

$$[u, v] = \Big(u^l \frac{\partial v^i}{\partial x^l} - v^l \frac{\partial u^i}{\partial x^l}\Big)\frac{\partial}{\partial x^i} = \nabla_u v - \nabla_v u\,,$$

and

$$\nabla_{[u,v]} = \nabla_{\nabla_u v} - \nabla_{\nabla_v u}\,.$$

PROBLEM 6.6. Our derivation of the Bianchi identity suggests that it is possible to take the Lie derivative not only of a tangent vector but also of a linear mapping M from the tangent space to the set of linear operators on the space, producing in effect a trilinear mapping of triples of tangent vectors. The definition is

$$\big(L(u)M\big)(v, w) = L(u)\big(M(v)w\big) - \big(M(L(u)v)w - M(v)\big(L(u)w\big)\,.$$

Show that the Lie derivative of the covariant derivative operator ∇, regarded as the mapping from the tangent space to the space of linear operators on the tangent space given by the mapping $u \mapsto \nabla_u$, satisfies

$$\big(L(u)\nabla\big)(v, w) = R(u, v)w + \nabla^2_{v,w}u\,.$$

PROBLEM 6.7. Show that

$$\big(L(u)L\big)(v, w) = \big(L(u)\nabla\big)(v, w) - \big(L(u)\nabla\big)(w, v)\,.$$

and that

$$\big(L(\boldsymbol{u})\nabla\big)(\boldsymbol{w},\boldsymbol{v}) = \big(L(\boldsymbol{u})\nabla\big)(\boldsymbol{v},\boldsymbol{w})\,.$$

PROBLEM 6.8. Consider a coordinate system (for example, normal coordinates) in which the matrix of metric coefficients is the identity matrix at the point P. Prove that in these coordinates $R^{i}_{jkl} = -R^{j}_{ikl}$.

PROBLEM 6.9. Prove that the parity of the permutation $(i_1,\ldots,i_k,j_1,\ldots,j_{n-k})$ differs from that of $(j_1,\ldots,j_{n-k},i_1,\ldots,i_k)$ by the factor $(-1)^{k(n-k)}$.

PROBLEM 6.10. Show that in normal coordinates on the 2-sphere $\mathbb{S}^2$ in $\mathbb{R}^3$, the metric coefficients are

$$g_{11}(x,y) = \frac{y^2\left(\varphi\left(\frac{x^2+y^2}{r_0^2}\right)\right)^2 + x^2\left(\psi\left(\frac{x^2+y^2}{r_0^2}\right)\right)^2}{x^2+y^2} + \frac{x^2}{r_0^2}\left(\varphi\left(\frac{x^2+y^2}{r_0^2}\right)\right)^2,$$

$$g_{22}(x,y) = \frac{x^2\left(\varphi\left(\frac{x^2+y^2}{r_0^2}\right)\right)^2 + y^2\left(\psi\left(\frac{x^2+y^2}{r_0^2}\right)\right)}{x^2+y^2} + \frac{y^2}{r_0^2}\left(\varphi\left(\frac{x^2+y^2}{r_0^2}\right)\right)^2,$$

$$g_{12}(x,y) = \frac{\left(\left(\psi\left(\frac{x^2+y^2}{r_0^2}\right)\right)^2 - \left(\varphi\left(\frac{x^2+y^2}{r_0^2}\right)\right)^2\right)xy}{x^2+y^2} + \frac{xy}{r_0^2}\left(\varphi\left(\frac{x^2+y^2}{r_0^2}\right)\right)^2.$$

Here the functions $\varphi(t) = \sin(\sqrt{t})/\sqrt{t}$ and $\psi(t) = \cos(\sqrt{t})$ are needed only for nonnegative values of t, namely $t = (x^2+y^2)/r_0^2$. They satisfy the easily established relations $t\big(\varphi(t)\big)^2 + \big(\psi(t)\big)^2 = 1$, $\varphi'(t) = \frac{1}{2}\big(\psi(t) - \varphi(t)\big)$, and $\psi'(t) = -\frac{1}{2}\varphi(t)$. It is easy to see that $g_{11}(x,y) \to 1$, $g_{22}(x,y) \to 1$, and $g_{12}(x,y) \to 0$ as $(x,y) \to (0,0)$. Thus, suppressing the argument $(x^2+y^2)/r_0^2$ of φ and ψ, since $\varphi \to 1$ and $\psi \to 1$ as $(x,y) \to (0,0)$, we find

$$0 \le |g_{11}(x,y) - 1| = \left|\frac{y^2(\varphi^2-1) + x^2(\psi^2-1)}{x^2+y^2} + \frac{x^2}{r_0^2}\varphi^2\right|$$

$$\le \max\left(|\varphi^2-1|,|\psi^2-1|\right) + \frac{x^2}{r_0^2}\varphi^2 \to 0\,,$$

$$0 \le |g_{22}(x,y) - 1| = \left|\frac{x^2(\varphi^2-1) + y^2(\psi^2-1)}{x^2+y^2} + \frac{y^2}{r_0^2}\varphi^2\right|$$

$$\le \max\left(|\varphi^2-1|,|\psi^2-1|\right) + \frac{y^2}{r_0^2}\varphi^2 \to 0\,,$$

$$0 \le |g_{12}(x,y)| = \left|\frac{(\psi^2-\varphi^2)xy}{x^2+y^2} + \frac{xy}{r_0^2}\varphi^2\right| \le \frac{1}{2}|\psi^2-\varphi^2| + \frac{|xy|}{r_0^2}\varphi^2 \to 0\,.$$

Convert these expressions to those given in the text (Example 6.2).

Deduce as a consequence that the element of surface area in these coordinates is

$$\sqrt{g_{11}g_{22} - g_{12}^2} = \frac{r_0}{r}\sin\left(\frac{r}{r_0}\right) \approx 1 - \frac{1}{6}\frac{r^2}{r_0^2}\,.$$

Use this density function (not the approximation) to show that the area of the spherical cap centered at $(0,0,r_0)$ and whose boundary lies at geodesic distance s from this point has area $4\pi r_0^2 \sin^2(s/2r_0)$. In particular, the upper hemisphere, for which $s = \pi r_0/2$ has area $2\pi r_0^2$, as it ought to.

PROBLEM 6.11. Prove the following simple lemma:

Let $\mathfrak{M}_1$ and $\mathfrak{M}_2$ be manifolds of the same dimension n with coordinate systems $(x^1, \ldots, x^n)$ at point $P_1 \in \mathfrak{M}_1$ corresponding to parameter value $(0, \ldots, 0)$ and $P_2 \in \mathfrak{M}_2$ also corresponding to parameter value $(0, \ldots, 0)$. Let the metric coefficients of $\mathfrak{M}_1$ and $\mathfrak{M}_2$ be $g_{ij}(x^1, \ldots, x^n)$ and $h_{ij}(x^1, \ldots, x^n)$ respectively, $i, j = 1, \ldots, n$. If the Maclaurin series of g_{ij} and h_{ij} have the same coefficients up to order 2—that is, if $h_{ij}(x^1, \ldots, x^n) = g_{ij}(x^1, \ldots, x^n) + O\big(((x^1)^2 + \cdots + (x^n)^2)^{3/2}\big)$—then the curvature of $\mathfrak{M}_1$ at P_1 equals the curvature of $\mathfrak{M}_2$ at P_2.

As a corollary, when one is computing the curvature of a manifold at a given point from the metric coefficients, these coefficients can be replaced by their Taylor expansions through quadratic terms.

PROBLEM 6.12. Prove that the Laplace–Beltrami operator in cylindrical and spherical coordinates on $\mathbb{R}^3$ is just the form normally given for the Laplacian.

PROBLEM 6.13. Show that if $f(x^1, \ldots, x^n)$ is a C^2-function of compact support on $\mathbb{R}^n$, then

$$\int_{\mathbb{R}^n} f(\boldsymbol{x}) \nabla^2 f(\boldsymbol{x}) \, d\boldsymbol{x} = -\sum_{k=1}^{n} \int_{\mathbb{R}^n} \left(\frac{\partial f}{\partial x^k} \right)^2 d\boldsymbol{x} ,$$

and that this expression is negative unless $f(\boldsymbol{x}) \equiv 0$.

PROBLEM 6.14. Consider a surface in $\mathbb{R}^3$ consisting of the points $\big(x, y, z(x, y)\big)$. Let $x^1 = x$ and $x^2 = y$. First show that the Christoffel symbols for this surface have the simple expression

$$\Gamma_{ij}^k = \frac{\frac{\partial z}{\partial x^k} \frac{\partial^2 z}{\partial x^i \partial x^j}}{1 + \sum\limits_{l=1}^{2} \left(\frac{\partial z}{\partial x^l} \right)^2} .$$

Then show that the Hessian is given by

$$H_f(\boldsymbol{u}, \boldsymbol{v}) = \sum_{i,j=1}^{2} \left(\frac{\left(1 + \sum\limits_{l=1}^{2} \left(\frac{\partial z}{\partial x^l} \right)^2 \right) \frac{\partial^2 f}{\partial x^i \partial x^j} - \sum\limits_{l=1}^{2} \frac{\partial z}{\partial x^l} \frac{\partial f}{\partial x^l} \frac{\partial^2 z}{\partial x^i \partial x^j}}{1 + \sum\limits_{l=1}^{2} \left(\frac{\partial z}{\partial x^l} \right)^2} \right) u^i v^j .$$

Deduce as a corollary that

$$H_z(\boldsymbol{u}, \boldsymbol{v}) = \sum_{i,j=1}^{2} \frac{\frac{\partial^2 z}{\partial x^i \partial x_j} u^i v^j}{1 + \sum\limits_{l=1}^{2} \left(\frac{\partial z}{\partial x^l} \right)^2} .$$

Thus the Hessian of a function z regarded as a function on the surface that is its graph is just the ordinary Hessian of z as a function on $\mathbb{R}^2$ divided by the square of the area density of the surface. In particular, at a point where the surface is tangent to the xy-plane, it coincides with the ordinary Hessian of z.

PROBLEM 6.15. We proved Corollary 6.4 by applying Eq. (6.10) to the second term in the expression

$$R_{prq}^r = \frac{\partial \Gamma_{pq}^r}{\partial u^r} - \frac{\partial \Gamma_{rp}^r}{\partial u^q} .$$

Apply that equation instead to the first term and show that

$$R^r_{prq} = -\frac{3}{2}g^{rm}\frac{\partial g_{rm}}{\partial u^p\,\partial u^q}\,.$$

PROBLEM 6.16. When the 2-sphere $\mathbb{S}^2$ in $\mathbb{R}^3$ is parameterized by longitude and latitude coordinates, that is, by the mapping $(\theta,\varphi)\mapsto(\cos\theta\,\cos\varphi,\sin\theta\,\cos\varphi,\sin\varphi)$, the orthogonally invariant measure on it is $d\sigma_2(\theta,\varphi)=\cos\varphi\,d\theta\,d\varphi$. Prove that

$$\int_{\mathbb{S}^2}(\xi^1)^2\,d\sigma_2 = \int_{\mathbb{S}^2}(\xi^2)^2\,d\sigma_2 = \int_{\mathbb{S}^2}(\xi^3)^2\,d\sigma_2 = \frac{4\pi}{3}\,.$$

(This is the special case of Theorem 6.15 that occurs when $f(\boldsymbol{\xi})=(\boldsymbol{\xi}\cdot\boldsymbol{u})^2$ for the unit vectors $\boldsymbol{u}=(1,0,0)$, $\boldsymbol{u}=(0,1,0)$, and $\boldsymbol{u}=(0,0,1)$, since $\omega_2=4\pi$.)

PROBLEM 6.17. Let $\varepsilon>0$. Define two functions $f_\varepsilon(x)$ and $g_\varepsilon(x)$ on $[1,\infty)$ as follows:

$$f\varepsilon(x)=\begin{cases}\frac{n\varepsilon}{2}\sin^2\left(\pi n^3(x-n)\right), & \text{if } n\le x\le n+\frac{1}{n^3}, \quad n=1,2,3,\ldots,\\ 0, & \text{otherwise}\end{cases},$$

$$g_\varepsilon(x)=\frac{\pi^2\varepsilon}{24}+\int_1^x f_\varepsilon(s)\,ds\,.$$

Show that $f_\varepsilon(x)\ge 0$, so that and $f_\varepsilon(n+1/(2n^3)) = n\varepsilon/2$, so that $f_\varepsilon(n+1/(2n^3))\to\infty$ as $n\to\infty$. Thus, in particular, $f_\varepsilon(x)$ is not bounded. Then show that $g_\varepsilon(x)$ is an increasing function of x, and that for all x,

$$0<\frac{\pi^2\varepsilon}{24}\le g_\varepsilon(x)\le\frac{\pi^2\varepsilon}{12}<\varepsilon$$

By letting ε tend to zero, show that $g_\varepsilon(x)$ can be made arbitrarily small while its derivative $g'_\varepsilon(x)=f_\varepsilon(x)$ remains unbounded.

PROBLEM 6.18. Verify that the covariant derivative $\nabla_{\boldsymbol{v}}$ has the derivation property $\nabla_{\boldsymbol{v}}(f\boldsymbol{u})=\nabla_{\boldsymbol{v}}f\boldsymbol{u}+f\nabla_{\boldsymbol{v}}\boldsymbol{u}$.

PROBLEM 6.19. Show that the computed sectional curvature $\kappa(\boldsymbol{u},\boldsymbol{v})$ does not change if the vector $\boldsymbol{u}$ is replaced by $a\boldsymbol{u}+b\boldsymbol{v}$, for any nonzero a and any b. Thus, the sectional curvature depends only on the plane spanned by $\boldsymbol{u}$ and $\boldsymbol{v}$.

PROBLEM 6.20. Show that the length of the geodesic $\boldsymbol{\gamma_u}(t)$ used in defining the exponential mapping is $|\boldsymbol{u}|$.

PROBLEM 6.21. The three-sphere $\mathbb{S}^3$ of radius 1, whose sectional curvature we have computed, is an excellent example of a Lie group. It consists of the points in $\mathbb{R}^4$ that we may identify with points in space-time, calling them $T=(t^0;t^1,t^2,t^3)$. Then

$$\mathbb{S}^3=\{T:(t^0)^2+(t^1)^2+(t^2)^2+(t^3)^2=1\}\,.$$

What makes this manifold a Lie group is the group operation of *quaternion multiplication*. If we identify the quaternion T with the formal sum of a real number and a vector in $\mathbb{R}^3$, say $T=t^0+\boldsymbol{\tau}$, where t^0 is identified with the quaternion $(t^0,0,0,0)$ and $\boldsymbol{\tau}=t^1\boldsymbol{i}=t^2\boldsymbol{j}+t^3\boldsymbol{k}$ is identified with the quaternion $(0,t^1,t^2,t^3)$, the group operation is quaternion multiplication: If $S=s^0+\boldsymbol{\sigma}$, then $ST=(s^0t^0-\boldsymbol{\sigma}\cdot\boldsymbol{\tau})+(s\boldsymbol{\tau}+t\boldsymbol{\sigma}+\boldsymbol{\sigma}\times\boldsymbol{\tau})$.

Verify that $\mathbb{S}^3$ is a group under the operation of quaternion multiplication with identity $I=1+\boldsymbol{0}$.

PROBLEM 6.22. The exponential mapping at the point $(1; 0, 0, 0)$ in the group of unit quaternions is given by the analog of the mapping used in Example 6.2 on the sphere $\mathbb{S}^2$, namely

$$\exp(\boldsymbol{x}) = \psi(|\boldsymbol{x}|^2) + \varphi(|\boldsymbol{x}|^2)\boldsymbol{x}$$

when rectangular coordinates $\boldsymbol{x} = (x, y, z)$ are used on the tangent space $\mathbb{R}^3$. Here $\psi(t) = \cos(\sqrt{t})$ and $\varphi(t) = \sin(\sqrt{t})/\sqrt{t}$, as in Example 6.2. This expression shows that there is no singularity at the origin. In spherical coordinates (ρ, φ, θ), where there is a breakdown of the parameterization at that point, we have $\boldsymbol{x} = \left(\rho\cos\varphi\cos\theta, \rho\cos\varphi\sin\theta, \rho\sin\varphi\right)$, and

$$\exp(\boldsymbol{x}) = \cos(\rho) + \sin(\rho)\left(\cos\varphi\cos\theta\,\boldsymbol{i} + \cos\varphi\sin\theta\,\boldsymbol{j} + \sin\varphi\,\boldsymbol{k}\right).$$

Verify that for a unit vector $\boldsymbol{x}$ the mapping $s \mapsto \exp(s\boldsymbol{x})$ is a geodesic with s as arc length, and that

$$\exp(s\boldsymbol{x})\exp(t\boldsymbol{x}) = \exp\big((s+t)\boldsymbol{x}\big).$$

The product on the left is the quaternion product. This equation justifies the name *exponential mapping*.

PROBLEM 6.23. Let $\boldsymbol{\gamma}(t)$ be a smooth curve in a manifold, $0 \le t \le t_0$. For each tangent vector $\boldsymbol{w}$ at $\boldsymbol{\gamma}(0)$, let $\boldsymbol{w}(t)$ be the parallel transport of $\boldsymbol{w}$ along $\boldsymbol{\gamma}$ to $\boldsymbol{\gamma}(t)$. Show that the mapping $\boldsymbol{w} \mapsto \boldsymbol{w}(t)$ is a linear transformation and that it preserves the length of a vector. That is the expression $g_{ij}(\boldsymbol{\gamma}(t))w^i(t)w^j(t)$ is constant. (It follows that this mapping preserves the standard inner product on the tangent space, so that any two vectors map to two other vectors having the same lengths and forming the same angle.)

CHAPTER 7

The Geometrization of Gravity

But one thing is certain: I have never before in my life worked this hard, and I have acquired a great deal of respect for mathematics, whose finer points I had naively regarded as a pure luxury up to now. Compared with this problem, the original theory of relativity is child's play.

Einstein, letter to Arnold Sommerfeld, 29 October 1912. Klein, Kox, and Schulmann ([**47**], p. 505). My translation.

Now that the mathematical background of general relativity has been explored in Chapters 5 and 6, we can set the computational work of Chapter 4 in its proper context. Using the principle that mechanical laws should be expressed as tensor equations involving only at most second-order partial derivatives of the metric coefficients, Einstein was led to the Ricci tensor. Setting that tensor equal to zero in free space—when the attracting body is regarded as a massive particle—provided the simplest of all possible nontrivial tensor equations of this type and led to the very precise explanations of the precession of Mercury's perihelion and the deflection of light passing near the Sun. The closed-form Schwarzschild solution presented in Chapter 4 was a triumph of theoretical science and made a very satisfying connection between theory and observation.

If we were to leave off at that point, we would be presenting the reader with an oversimplified version of the story and a bit of a mystery to ponder. Two points in particular need to be addressed: (1) Why the Ricci tensor? Why not, for example, apply the Laplace–Beltrami operator to the metric coefficients instead? After all, we have perturbed the flat-space metric by adjoining terms connected with the potential energy. In Newtonian mechanics, it is the Laplacian of the potential energy that we work with. Why wouldn't the Hessian, which also describes the curvature, do equally well? (2) How are the equations of motion to be determined in other applications of relativistic mechanics? What replaces "force" and what replaces the famous "$F = ma$" in a situation where the gravitational field is produced by a continuous distribution of matter, and how do we formulate electromagnetic forces in this language?

We shall explore some of these questions in the present chapter. (Not all of them; we do not have space to discuss electromagnetism.) We shall introduce the stress-energy-momentum tensor promised in Chapter 2, along with the Einstein field equations that replace $F = ma$. We are entering very deep waters here, a full exploration of which would require another book the size of the present one. Accordingly, we shall keep safely close to shore, in water that is comparatively shallow. Our aim is to try to make the reformulation of mechanics using the Einstein field equations—which, it must be admitted, takes considerable getting used to for those trained in classical mechanics—seem less unnatural. We do this by stressing the similarities between the metric coefficients and potential functions, showing (for example) that the contravariant form of the Einstein tensor has zero divergence, and the like.

1. The Einstein Field Equations

Explaining the motion of a particle means giving its spatial coordinates as functions of time. A kinematic explanation, such as Ptolemy's model of planetary motion does only that. A dynamical explanation posits force, or something analogous to it (curvature) and states the relation between the spatial coordinates and time in the form of differential equations that are to be solved. Our present task is to see how such differential equations are to be expressed in the language of tensors, to build a bridge from the Newtonian scheme to that of general relativity. In both Newtonian and relativistic mechanics, the general scheme for giving a dynamic description of the motion of a particle is the same, consisting of two parts: (1) formulate an intuitive mathematical model that can be used to produce a set of differential equations; (2) solve the differential equations. Only the first of us concerns us in the present section.

In Newtonian mechanics, this part of the problem can be carried out in two ways that we have discussed. The first way—the Newtonian approach proper— is to formulate a mathematical expression for the forces acting on the particle, then use them as F in the equation $F = mr''$ that expresses Newton's second law. The second way, which we shall call Lagrangian, is available when the force is conservative. In that case we need to find a mathematical expression for the potential energy, use it to form the Lagrangian, and then get the equations of motion in the form of Euler's equations for a stationary time integral. The Newtonian approach requires us to use our intuition to write an expression for the relevant force, and Newton's second law of motion then provides the differential equations of the motion. The Lagrangian approach requires an equivalent intuition as to the form of the potential, after which the equations of motion are the Euler equations for a stationary value of the time integral of $T - V$.

In contrast, the method of general relativity requires an intuitive hypothesis of some suitable perturbation of the flat space-time metric adapted to the phenomenon being explained. Once that metric is found, the equations of motion are the Euler equations for a geodesic. The basic task, then, is to develop the appropriate intuition needed to imagine that perturbation of the space-time metric. Throughout this whole process, Einstein was guided by the solid foundation of Newtonian mechanics: It had been successful, and he knew that any improvement of it would have to yield the classical results as a limiting case. Two firm anchors from Newtonian mechanics were the speed of light c and the gravitational constant G. They formed the skeleton of the relativistic laws. Likewise, the reformulation of Newtonian mechanics in Lagrangian terms provided some important equations such as Laplace's equation and Poisson's equation, which again served as a guide to the formulation of the relativistic equations. In empty space, the Laplacian of the gravitational potential $\varphi = -GM/r$ vanishes (Laplace's equation, $\nabla^2 \varphi = 0$), while in space with a continuous distribution of matter of density ρ, this equation becomes Poisson's equation: $\nabla^2 \varphi = 4\pi\rho$. This equation will serve as a beacon to guide use as we navigate these deep waters.

Our aim is to start with the case where we have seen general relativity work well, the gravitational field around a massive particle in empty space, and then fill up that empty space with matter of a certain density. In empty space we got our metric coefficients by setting the Ricci tensor equal to zero. We discovered that we were essentially perturbing the flat-space metric by subtracting the ratio that the Schwarzschild radius ρ_s bears to the range ρ in the coefficient of dt^2 and subtracting the reciprocal of this ratio in the coefficient of $d\rho^2$. We noted that this ratio is the negative of twice the Newtonian potential per unit mass divided by the square of the speed of light ($\rho_s/\rho = -2V/c^2$, where $V = -GM/\rho$ is the Newtonian potential energy per unit mass of the orbiting particle). We now face the essential question: Why did Einstein think the Ricci tensor *ought* to vanish in this case? After we answer that question, we can make a guess as to how the metric should be perturbed in a more general case. Then, by looking at the Ricci tensor with the more general perturbation, we can arrive at a hypothetical set of field equations—the Einstein field equations—to replace Newton's second law. At that point, our work will be done.

1.1. Highlights of Newtonian/Lagrangian mechanics. Newton's equation of motion for a particle of mass m subject to a force $\boldsymbol{F}$ is given by his second law of motion

$$\boldsymbol{F} = \boldsymbol{p}' = m\boldsymbol{r}'' ,$$

where $\boldsymbol{r}(t)$ is the position of the particle at time t and the force $\boldsymbol{F} = \boldsymbol{F}(t; x^1, x^2, x^3)$ depends in general on both time and location. Here $\boldsymbol{p}$ is the momentum: $\boldsymbol{p} = m\boldsymbol{r}'$. This law can be viewed in two ways. The obvious way is to think of cause and effect. The force $\boldsymbol{F}$ on the left *causes* the acceleration $\boldsymbol{r}''$ on the right. From a second point of view, the equation furnishes a general model for solving mechanical problems. The right-hand side is a universal expression, applicable to any moving particle whose mass is known. The left-hand side is then hypothesized in a way specific to the individual motion; different versions of it are used in different situations. In the vibrating spring, for example, one assumes the force is directly proportional to the displacement and directed opposite to it. For the case of a vibrating string or membrane, the force is provided by tension and is proportional to the curvature. And, most famously, in Newton's law of gravity, the force is occult, but quantitatively easy to describe, being proportional to the product of the masses of the attracting body and the orbiting particle and inversely proportional to the distance between the particle and the center of mass of the attracting body. In this way, the Newtonian approach to mechanics can be used to model a large number of natural phenomena. The basic equation of motion combines the visual geometry implicit in the position of the particle, with the tactile notion of force, which we intuitively understand from our experience of the sensations we have in our muscles.

The Lagrangian formulation of mechanics introduces the potential energy V and kinetic energy T. The kinetic energy, like Newton's $m\boldsymbol{r}''$ is a universal expression: $T = m(\boldsymbol{r}' \cdot \boldsymbol{r}')/2$. The right-hand side of Newton's equation of motion is then simply described in terms of T. It is the time derivative of the "velocity-gradient" of the kinetic energy $\nabla_{\boldsymbol{r}'}T$:

$$m\boldsymbol{r}'' = \frac{d\boldsymbol{p}}{dt} = \frac{d}{dt}\left(\nabla_{\boldsymbol{r}'}T\right).$$

(See Chapter 2, p. 94 for an explanation of this notation.)

The potential energy V corresponds to Newton's force $\boldsymbol{F}$. When the force $\boldsymbol{F}$ is conservative (the work it does in moving a body over a path depends only on the endpoints of the path), such a potential exists, and the force $\boldsymbol{F}$ is the ordinary gradient of $-V$:

$$\boldsymbol{F} = \nabla_{\boldsymbol{r}}(-V).$$

In terms of the Lagrangian $L(\boldsymbol{r}, \boldsymbol{r}') = T(\boldsymbol{r}') - V(\boldsymbol{r})$, Newton's second law becomes the equation

$$\frac{d}{dt}\big(\nabla_{\boldsymbol{r}'}L(\boldsymbol{r}, \boldsymbol{r}')\big) = \nabla_{\boldsymbol{r}}L(\boldsymbol{r}, \boldsymbol{r}').$$

That vector equation is very familiar, being the set of Euler equations in the calculus of variations, used to find the stationary values of the time integral

$$I = \int_{t_0}^{t_1} L\big(\boldsymbol{r}(t), \boldsymbol{r}'(t)\big)\, dt.$$

As we remarked in Chapter 4, this approach does not give us any new equations; nor does it save us any trouble in solving the ones we already had in terms of Newton's formulation. But it is of great value in unifying our thinking about mechanics. For example, it was stated by Fermat that a ray of light always appears to follow a "path of least resistance," one that requires the minimal time to traverse compared with all nearby paths. In 1696, a generation after Fermat's death (which occurred in 1665), the same principle was introduced into mechanics by Johann Bernoulli (1667–1748) to solve a problem with which he then challenged other mathematicians: Finding a "sliding board" down which a frictionless particle would slide in least time from a given height—the *brachistochrone* (shortest-time) problem, which was later also solved by Newton.[1] In contrast to mechanics, optics had always been a purely geometric subject. But now an analogy with optics allowed the geometric point of view to supplant the concept of force, even in its stronghold of mechanics. Since the Euler equations are also the key to finding the shortest path between two points in a space with a general metric, the way was prepared early on for a further connection between geometry and mechanics. And in 1744, Bernoulli's protégé Leonhard Euler showed that a particle moving over a surface and subject to no tangential forces will move along a geodesic on that surface. (See Appendix 2 for a proof.) From that perspective, the classical Lagrangian approach can be thought of as saying that the particle moves along the path of "steepest descent," where *descent* means exchanging potential energy for kinetic energy. That is, given the states of these two energies at the initial and final positions, the particle gets from the initial state to the final state in minimal time.

1.2. Comparison with general relativity. Now let us compare the Lagrangian formulation of Newton's second law with the language of geodesics. The equations of a geodesic, with proper time as parameter, are

$$(x^i)'' + \Gamma^i_{jk}(\boldsymbol{x})(x^j)'\,(x^k)' = 0,$$

---

[1]Bernoulli imagined the particle was a ray of light moving in a medium whose index of refraction was inversely proportional to the square root of the distance fallen; that is because the speed of a particle falling with constant acceleration is proportional to the square root of the distance it has fallen.

where $\boldsymbol{x} = (x^1, x^2, x^3, x^4)$ are the coordinates of a point of space-time. The first term is readily recognizable as the time derivative of momentum per unit mass, that is, the time derivative of the "velocity-gradient" of kinetic energy.

A comparison of this equation with Newton's second law suggests that the Christoffel symbols Γ^i_{jk} are analogs of the components of the force, which is the gradient of the potential energy. Since the Christoffel symbols are constructed from the derivatives of the metric tensor, we conclude that potential energy corresponds to the whole set of metric coefficients g_{ij}. They are, as Eddington called them, relativistic potential functions. We should have expected this, since the term $\rho_s/\rho = 2GM/(\rho c^2)$ by which the metric coefficients are perturbed is equal to $-2V/c^2$, where $V = -GM/\rho$ is the gravitational potential per unit mass of an orbiting particle. But we also know that the Christoffel symbols determine the curvature of a manifold, and so we are once again confronted with the interchangeability of force and curvature.

That being said, inserting the potential energy into the metric of space time still amounts to a rather complex mathematical transformation, and it will not do simply to let the matter rest at this point. The detailed effects of this insertion cannot be immediately foreseen, and only a careful working out of the mathematical consequences will produce a usable geometric formulation of the laws of mechanics. With that goal in mind, let us begin with what we know to be usable: Einstein's law of gravity for empty space, which asserts that the Ricci tensor vanishes.

The first analogy that we notice is that, in empty space, the Newtonian potential per unit mass $V(r)/m$ is a harmonic function:

$$\nabla^2 V(r) = 0.$$

That equation (Laplace's equation for the potential per unit mass) thus appears to correspond to the equation Ric $= 0$; in other words, *in this special case*, the Ricci tensor of the metric coefficients corresponds to the Laplacian of the potential function per unit mass. As we saw in Chapter 6, there are at least four operators that correspond loosely to the Laplacian, in that they involve second-order derivatives of the metric coefficients. Besides the Ricci and Riemann tensors, there are the Laplace–Beltrami operator (seemingly the most obvious candidate) and the Hessian. We can put the Riemann tensor aside, since, as Einstein remarked, its vanishing makes the space flat. The number of connections we found among the other three in the previous chapter suggests that any one of them might do, and the choice among them would be mostly a matter of convenience.

The question we are trying to answer is: What is the relativistic form of the equations of motion when space is *not* empty? Suppose, for example, that a sphere with center at the origin is filled with very sparse matter having a constant density[2] ρ. In the Newtonian formulation, the gravitational potential per unit mass inside the sphere then becomes

$$V(r) = \frac{2\pi G\rho r^2}{3},$$

---

[2]With deep apologies, we must use the letter ρ, which previously stood for the distance between the center of attraction and the orbiting particle, for a mass density. Even worse, we are going to assume that that the matter in question is moving with constant speed v relative to an observer fixed in the frame of the attracting particle and has rest density ρ_0. To minimize confusion, we shall henceforth use r instead of ρ for the distance and replace the Schwarzschild radius ρ_s by $4r_s$, since we plan to use isotropic coordinates to simplify the algebra.

and Laplace's equation becomes *Poisson's equation*

$$\nabla^2 V = 4\pi G \rho .$$

It was this equation that Einstein and Grossmann attempted to generalize in their famous "first draft" [**24**] of the general theory of relativity, and we shall also use it to explore the relativistic reformulation of mechanics. Keep in mind that we need to reformulate both sides of the equation. One side, the analog of $m\boldsymbol{r}''$, needs to be a universal expression valid in all situations. The other side, the analog of $\boldsymbol{F}$, is hypothesized by any intuitive stroke of genius we can muster to explain a particular phenomenon. Bear in mind also that relativistic mechanics is one step more complicated than Newtonian. In Newtonian mechanics, the formulation of a hypothetical force $\boldsymbol{F}$ produces the differential equations of motion immediately. In relativistic mechanics, some intuitive physical picture of a perturbed space-time produces a set of differential equations for the metric coefficients g_{ij}. Only *after* those equations have been solved and the metric decided upon will the Euler equations for geodesics take a concrete form. The actual equations of motion are the Euler equations for these geodesics.

We are concentrating on the phenomenon of gravity and will confine our field equations to that particular case. Since we want a tensor equation, our quest is to find the tensor analog of $m\boldsymbol{r}''$. That analog is the *Einstein tensor* that we are about to introduce. Once we have it, we will need to formulate the tensor analog of $\boldsymbol{F}$, which is the stress-energy-momentum tensor. Formulating that tensor for a particular distribution of matter is a complicated problem, and we shall confine our exploration to the "shallow water" of the special case that occurs when the mass density ρ is constant. We shall assume that the space-time metric for this problem is simply the flat-space metric perturbed by the potential φ per unit mass given above. If that assumption is correct, computing the Einstein tensor for the given metric will tell us what the other side of the field equation is in this case. In this way, we get one concrete example of the stress-energy-momentum tensor that does not vanish identically. That is all the generality we have space to consider in the present work, and it will bring our narrative to a close. At that point, we shall look at just one more example, namely the Gödel metric that has become famous for its suggestion that time travel may be possible.

REMARK 7.1. To overcome the psychological difficulty involved in imagining a particle moving through space already occupied by matter with density ρ, we need to think of ρ as being very small, so that the term $2\pi G \rho r^2/3$ is of the same order as r_s/r, namely 10^{-7}. Given the value of G, this means that ρ has a value of about 0.003 kg per cubic meter. Sparse as that amount of matter is, it is still billions of times larger than the actual density of interplanetary space. Physicists tend to refer to such an interstellar distribution of matter as "dust." In ordinary language, that term rather suggests particulate matter, whereas the mathematical model is continuous. Perhaps it is better to picture this "dust" as a very thin atmosphere, so thin that the considerations of fluid dynamics do not apply to a particle moving through it, or—better yet—as "dark matter."

1.3. The Einstein tensor. If, following Eddington, we think of the metric coefficients g_{pq} as generalized potentials, we have a choice of various operators to play the role of the "Laplacian of the potential," namely the Laplace–Beltrami operator, the Hessian, and the Ricci tensor, all closely related to one another, as we have seen. Choosing the appropriate operator is the task of a physicist. Practical experience is helpful, and Einstein struggled with this problem for a considerable time.

We might be tempted to replace the Laplacian in Poisson's equation once again by the Ricci tensor, but then what do we do with the term $-4\pi G\rho$? We observe that, except for a coefficient of the form ar^2, that term is just the potential per unit mass. To be specific, it is $6\varphi/r^2$. Since we are using φ to perturb the metric coefficients g_{ij}, we are thus led to consider, as candidates for the left-hand side of the analog of Poisson's equation, combinations of the form

$$\mathrm{Ric} - C\varphi\,,$$

for a constant C of suitable dimension. The dimension of C, as already remarked, is the dimension of the scalar curvature R. The problem we face can thus be reduced to the choice of a dimensionless scalar a that will produce a suitable tensor $\mathrm{Ric}_{ij} - aRg_{ij}$. Notice that $R = 0$ when the Ricci tensor vanishes, and that explains why the extra term was absent in the free-space version of gravitation. Its existence would not have been suspected had we not tried to complicate the problem by replacing a particle with a continuous density.

What do we mean by *suitable* and how are we to choose a? Again, Newtonian mechanics is our guide. In addition to being conservative (meaning that the curl $\nabla \times \boldsymbol{F}$ is $\boldsymbol{0}$), the gravitational force $\boldsymbol{F} = -(GMmr^{-3})\boldsymbol{r}$ is also divergence-free: $\nabla \cdot \boldsymbol{F} = 0$. We thus attempt to choose a so that the tensor we are going to call the *Einstein tensor* and denote by the symbol Ein will have the form $\mathrm{Ein}_{ij} = \mathrm{Ric}_{ij} - aRg_{ij}$ and will have zero divergence. Since we have discussed only how to take the divergence of contravariant tensors, it will be necessary to raise the indices in this covariant tensor of type $(0, 2)$, getting a contravariant version of the Einstein tensor:[3]

$$(7.1) \qquad \mathrm{ContraEin}^{ij} = g^{ik}g^{jm}\big(\mathrm{Ric}_{km} - aRg_{km}\big) = \mathrm{Ric}^{ij} - aRg^{ij}\,.$$

Not to prolong the suspense, we shall reveal immediately that the magic coefficient is just $a = 1/2$.

THEOREM 7.1. *If the contravariant Einstein tensor ContraEin is defined by Eq (7.1) with $a = 1/2$, then*

$$\mathrm{div}\,(\mathrm{ContraEin}) = \begin{pmatrix} 0 \\ 0 \\ 0 \\ 0 \end{pmatrix}.$$

PROOF. Although this theorem is true in general, a full proof requires yet more combinatorial work in the area of the Bianchi identity, which we wish to spare the reader. Since we have need of this theorem only in the case of space-time coordinates in which the metric tensor is diagonal, we confine the proof to that case. We can then spare the reader all the computational torture by trusting *Mathematica* to compute the divergence of the contravariant Einstein tensor and verify that it is

---

[3]Because indices can be raised and lowered at will, any tensor equation stated in covariant form has an equivalent contravariant form. We choose to deal with the contravariant form in order to minimize the number of symbols we need. It is generally simpler, though, to state tensor equations in covariant form.

indeed the zero vector. *Mathematica* Notebook 11 in Volume 3 gives a proof for
a perfectly general *diagonal* space-time metric and requires only a minute or two
to run. Any attempt to get *Mathematica* to carry out this same labor with a full
4×4 matrix of metric coefficients will require considerable patience, since the two
minutes will expand to a day or more of computing, and the computer will need a
great deal of memory.

The output of this program is $\{0, 0, 0, 0\}$, and so we have assurance that the
contravariant form of the Einstein tensor is divergence-free, as desired.   □

REMARK 7.2. According to Wald ([**83**], p. 72), Einstein considered making
the equation Ein = 0 the fundamental equation of a gravitational field, but was
deterred by the fact that it would imply that ρ is constant throughout the universe.
This would make for a great deal of difficulty, since the Newtonian potential for
such a distribution would be identically 0. This curious step along the way to the
general field equations is an example of the trial-and-error process by which they
were discovered.

REMARK 7.3. Lovelock ([**57**]) has shown that the Einstein tensor is the only
divergence-free tensor that can be formed from the metric coefficients and their
first and second derivatives. Lovelock's theorem proves formally what Einstein had
asserted in 1916.

1.4. The field equation. We now have reason to believe that the Einstein
tensor represents one side of an "equation of motion" in space-time. Its contravariant form will have zero divergence, just like the Newtonian gravitational field, no
matter what the metric tensor is. In that sense, it is the universal part of the
fundamental equation of mechanics, just as $m\boldsymbol{r}''$ is in Newtonian mechanics. To
solve mechanical problems, we need the other side of the equation, analogous to
what is generally known as $\boldsymbol{F}$ in Newtonian mechanics, but should really be $k\boldsymbol{F}$,
where k is a constant of proportionality that reconciles the dimensions and sizes of
the two sides. Physical units, such as the MKS system are generally chosen so as
to make this constant equal to unity. Thus, the unit of force, the newton, is chosen
so that an acceleration of one meter per second-squared of one kilogram of mass
requires a force of one newton. Since we don't yet know what units will be most
convenient in relativity, we are going to state the fundamental field equations with
an unspecified constant of proportionality; after looking at one example, we shall
then give the constant the conventional value used by physicists.

We are now seeking the other side of the field equation, which is a constant
multiple of a tensor of type $(0, 2)$ called the *stress-energy-momentum* tensor, and
usually denoted T. We begin with a simple special case of the Einstein field equation
in the form

$$\text{Ein}_{pq} = \frac{8\pi G}{c^4} T_{pq} \,,$$

where the reason for the particular constant of proportionality will be seen in the
examples below. In standard notation, the metric ds^2 of special relativity is "spatialized," that is, $ds^2 = c^2\,dt^2 - dx^2 - dy^2 - dz^2$. In that case, $g_{11} = c^2$ has dimension
length^2/time^2, g_{1j} and g_{j1}, $j > 1$, all of which are zero, have dimension length/time,
and all the other g_{ij} are dimensionless. In that system, the dimensions of the components of the Ricci tensor (and hence also those of the Einstein tensor) are as follows:
Ric_{11} has dimension time^{-2}, Ric_{1j} has dimension (time × length)$^{-1}$, for $j > 1$, and

all other components have dimension length^{-2}. Now $G/c^4 = (2GM/c^2)(1/Mc^2) = r_s/(Mc^2)$, so that its dimension is time^2/(mass × length). It follows that the dimension of T_{11} must be mass × length/time^4, which is mass × velocity^4/volume. It will thus be a dimensionless constant times ρc^4, where ρ is a density In other words, the entry in the first row and column of Ein will be $8\pi G\rho$, where ρ has the physical dimension of density. When we look at examples below, we will see why $8\pi G$ is the most convenient value for the constant.

We will actually be working with the contravariant version of this tensor, and since g^{11} in special relativity is c^{-2}, that means the entry T^{11} will be simply ρ, which is $\rho c^2/c^2$, that is, an energy density ρc^2 divided by c^2. The entries T^{1j} and T^{j1}, $j > 1$, will represent components of momentum. The diagonal elements T^{jj}, $j > 1$ will represent longitudinal stresses, and the remaining six entries T^{ij}, $2 \le i,j \le 4$ will represent shear stresses. Our interest, however, will be confined to just T^{11}.

We shall look at the Einstein tensor only in the simple case when the gravitational field is due to a spherical mass distribution of constant density ρ, for which the Newtonian potential per unit mass is $V/m = 2\pi G\rho r^2/3$, as long as r is less than the radius of the sphere containing the mass distribution. We recall that in the case of attraction by a particle of mass M, we replaced the term dt^2 in the metric by $(1 - r_s/r)\,dt^2$, where r_s was the Schwarzschild radius. This, as we saw, was tantamount to introducing the coefficient $1 + 2V/(mc^2)$, where $V = -GMm/r$ is the Newtonian potential at distance r.

To keep the algebra simple, we are going to use isotropic spherical coordinates, in which the Schwarzschild radius r_s is four times its value in standard coordinates. In standard spherical coordinates we replace the ratio r_s/r by $-2V/(mc^2)$, so that in isotropic spherical coordinates we need to replace it by $-V/(2mc^2)$. But since we wish to confine our attention to distances r larger than r_s, where the coefficient of dt^2 should be smaller than 1, we are actually going to replace it with $V/2mc^2$. The metric becomes

$$ds^2 = \left(\frac{3c^2 - \pi G\rho r^2}{3c^2 + \pi G\rho r^2}\right)^2 dt^2 - \left(1 + \frac{\pi G\rho r^2}{3c^2}\right)^4 \left(dr^2 + r^2\,d\varphi^2 + r^2\sin^2\varphi\,d\theta^2\right).$$

With these coordinates, we find that

$$\text{Ein} = 8\pi G\rho \begin{pmatrix} \left(\frac{(1+\pi G\rho r^2/(3c^2))}{(1-\pi G\rho r^2/(3c^2))^7}\right) & 0 & 0 & 0 \\ 0 & \frac{\pi G\rho r^2}{3c^2}\frac{9c^2}{1-\left(\pi G\rho r^2/(3c^2)\right)^2} & 0 & 0 \\ 0 & 0 & 0 & 0 \\ 0 & 0 & 0 & 0 \end{pmatrix}.$$

Simple as it appears, this expression is still more complicated than we need for our purposes. We are assuming that the function $V/m = \pi G\rho r^2/3$ is "small." When we omit terms containing this factor, the Einstein tensor becomes

$$\text{Ein} = \begin{pmatrix} 8\pi G\rho & 0 & 0 & 0 \\ 0 & 0 & 0 & 0 \\ 0 & 0 & 0 & 0 \\ 0 & 0 & 0 & 0 \end{pmatrix}, \text{ that is, } (T_{pq}) = \begin{pmatrix} \rho c^4 & 0 & 0 & 0 \\ 0 & 0 & 0 & 0 \\ 0 & 0 & 0 & 0 \\ 0 & 0 & 0 & 0 \end{pmatrix},$$

which is not only of the same form that we indicated above, but even the same exact value!

Since, in first approximation, $g^{11} = 1/c^2$ for this case, it follows that the contravariant version of this tensor is the same thing multiplied by $1/c^4$. In other words, the contravariant stress-energy-momentum tensor in this case is

$$(T^{pq}) = \begin{pmatrix} \rho & 0 & 0 & 0 \\ 0 & 0 & 0 & 0 \\ 0 & 0 & 0 & 0 \\ 0 & 0 & 0 & 0 \end{pmatrix}.$$

This is exactly the stress-energy-momentum tensor we obtained in special relativity for a constant mass density ρ moving along a straight line at constant speed (see Chapter 2).

Although purists might prefer to think of this tensor equation as an approximation to the "true" equation, in reality the expression being approximated is also not "true" in the mathematical sense of infinite precision. The approximation is in every practical way better, since we can compute with it, and we have no elegant way of dealing with the "exact" Einstein tensor in this case. In the "Newtonian limit" as $c \to \infty$ and $r_s \to 0$, we find that $\pi G \rho r^2/(3c^2) \to 0$ also, and the entry in the first row and column of the "exact" Einstein tensor (the one before the approximation was made) becomes $2\nabla^2 V/m$, so that the equation becomes the classical Poisson equation, $\nabla^2(V/m) = 4\pi G\rho$.

REMARK 7.4. A digression on mathematical elegance and physics may be in order at this point. Both mathematicians and physicists care about elegance, despite the advice given to physicists by the great nineteenth-century physicist Ludwig Boltzmann (1844–1906) to "let elegance be the concern of shoemakers and tailors." ("Eleganz sei die Sache der Schuster und Schneider." Source: Wikiquotes.) This statement is often misattributed to Einstein, who did indeed quote it in 1916 in the preface to an exposition of relativity theory. Nevertheless, his attitude toward mathematical physics vehemently contradicts it. In fact, the context in which he made the quotation refers not to the theory itself but only to his exposition of it. What he said (my translation) was, "For the sake of clarity, I have found it necessary to repeat myself frequently, not paying the slightest attention to elegance of presentation; from the scientific point of view, I have followed the dictum of the brilliant theoretician L. Boltzmann, that one should let elegance be a concern of the tailors and shoemakers." Boltzmann's statement was rebutted by the mathematical physicist Franz von Krbek (1898–1984), a professor at the University of Greifswald, in his 1952 book *The Captive Infinite* (*Eingefangenes Unendlich*), p. 28.

Mathematicians and physicists seem to be pursuing the same goal. They diverge, however, in the matter of consistency. Physicists are generally willing to replace an exact expression by an approximate expression that uses simpler, more elementary functions. In both cases, to get a useful application to the physical world, we must resort to numerical computation of the resulting functions, and the numerical computations in this case do agree with each other and with observation. In that respect, Boltzmann's aphorism can be turned on its head. The process of taking a mathematical model and applying it to the world, as physicists do, is in many ways similar to what a tailor or shoemaker does with raw material taken from nature, trimming it here and there and sewing separate pieces together to fit the "client" (the physical world). And, just like the work of tailors and shoemakers, the end result is never a perfect fit. The universe, like the human body, is too complicated, and our tools can't make the raw material fit with infinite precision. But the aim is elegance, even if it amounts to elegance followed by ugly minor adjustments.

In any case, we should not be bothered by the need to make approximations that are closer than any observable difference. Mathematical physics has always accepted much coarser approximations. For example, the derivation of the classical vibrating string equation assumes that the restoring force on a stretched string at each point is proportional to the curvature at that point. If the equation of the instantaneous form of the string is $y = f(x,t)$ at time t, this means

$$\frac{\partial^2 y}{\partial t^2} = c^2 \frac{\frac{\partial^2 y}{\partial x^2}}{\left(1 + \left(\frac{\partial y}{\partial x}\right)^2\right)^{3/2}}.$$

But in fact, it is usually assumed that dy/dx is negligible, so that the equation becomes

$$\frac{\partial^2 y}{\partial t^2} = c^2 \frac{\partial^2 y}{\partial x^2},$$

which is the classical one-dimensional wave equation. If we didn't make such approximations, the Laplacian would arise far less often in Newtonian mechanics than it in fact does.

Let us now return to our main theme. The Einstein field equation we have so far is

$$\text{Ein} = \frac{8\pi G}{c^4} T,$$

where T is the stress-energy-momentum tensor.

In standard notation the Einstein tensor is written in terms of its coordinates, which are denoted $G_{\mu\nu}$, so that

$$G_{\mu\nu} = \text{Ric}_{\mu\nu} - \frac{1}{2} R g_{\mu\nu} = \frac{8\pi G}{c^4} T_{\mu\nu}.$$

That is a form more likely to be recognized by physicists, although they are accustomed to using just the letter R to denote the Ricci tensor. The constant still looks a bit messy, but physicists are accustomed to assuming units of measurement in which G and c are both numerically equal to 1, and thus the constant becomes simply 8π.

As we have already noted above, the contravariant version of this equation is

$$\text{ContraEin} = \frac{8\pi G}{c^4} \begin{pmatrix} \rho & 0 & 0 & 0 \\ 0 & 0 & 0 & 0 \\ 0 & 0 & 0 & 0 \\ 0 & 0 & 0 & 0 \end{pmatrix}.$$

There now remains only one point that needs to be mentioned in connection with the Einstein field equations.

1.5. The cosmological constant. The Einstein field equations were not arrived at over a cup of tea. Einstein proposed them only after many years of hard and deep thought. In the early years of general relativity, a century ago, definitions were still in a fluid state. Although gravitational effects are quite weak compared with electromagnetic effects, they still tend to cause a shrinkage in the size of the universe. Moreover, although the Newtonian gravitational interaction between two *particles* conserves all the usual things, such as angular momentum and total energy, such is not the case for solid bodies. Gravitational forces are responsible for the tides, and tidal friction produces heat. Consequently, the idealization of a planet as a particle, which we have used in Chapter 4, is not perfectly realistic, even though it does suffice to explain the precession of perihelion. In order to keep the universe from shrinking due to gravity, or expanding without limit due to an excess of energy, Einstein replaced the field equation written above by a slight modification, to obtain the now-standard *Einstein field equations*

$$(7.2) \qquad \mathrm{G}_{\mu\nu} + \Lambda g_{\mu\nu} = \frac{8\pi G}{c^4} T_{\mu\nu} \, .$$

The constant Λ was known to be small, and Eddington remarked at the time that its value was mostly theoretical. It allowed the universe to keep its size. As it happens, however, the solutions to Eq. (7.2) are not stable with respect to perturbations in the *cosmological constant* Λ, as Einstein called it. Einstein was therefore led to abandon this idea. As his friend George Gamow ([**32**], p. 44) recalled—inaccurately, some think—

> *Thus, Einstein's original gravity equation was correct, and changing it was a mistake. Much later, when I was discussing cosmological problems with Einstein, he remarked that the introduction of the cosmological term was the biggest blunder he ever made in his life. But this "blunder," rejected by Einstein is still used by cosmologists even today, and the cosmological constant denoted by the Greek letter Λ rears its ugly head again and again and again.*

It was accepted for many decades that the universe was, in fact, expanding, so that the constant Λ was not needed. As Gamow noted, however, it did not disappear from the literature, and recent evidence that the expansion is itself speeding up has brought it to prominence once again. Even Einstein's "blunder" turns out to have considerable merit.

2. Further Developments

Our discussion of general relativity has now gone up to the border of the Promised Land of general relativistic dynamics. We do not intend to cross that border, however. Our intention was only to get the reader mathematically closer to that step, while making various common-sense observations along the way about what we were doing. We have carried the story up to the year 1916, when Einstein published the exposition of his efforts to explain gravitation over a period of several years. His earlier work, now superseded, had attracted attention, and as early as 1915, Karl Schwarzschild had produced the exact solution of the field equations for the case of a tiny particle orbiting a heavy one in empty space. Also during that time, one of the giants of twentieth-century mathematics, David Hilbert (1862–1943) had begun to study the problem. Among other things, Hilbert studied the motion of a particle near a (nonrotating, stationary) black hole, as we shall briefly do in Section 4 below. Hilbert noticed certain peculiarities of this metric, notably that energy did not appear to be conserved, in contrast to special relativity and Newtonian mechanics. He asked his collaborators Felix Klein and Emmy Noether (1883–1934) to look into this problem. Noether did, and produced one of the most remarkable papers of twentieth-century mathematical physics. All that is beyond the scope of the present work.[4]

3. "Temporonautics" and the Gödel Rotating Universe

The mere fact that a solution of a set of differential equations exists in the mathematical sense does not endow that solution with any connection to physical reality, even when the equations were originally set up with a definite physical interpretation in mind for the functions they contain. A good example is furnished by a space-time metric introduced by Kurt Gödel (1906–1978) in 1949, in which a material particle can travel from any time and place to any other time and place, becoming a[5] "temporonaut."

Before introducing this metric, we need to say a few words about the meaning of the phrase *time travel*. All human experience leads us to believe that it is impossible, with one exception: our astronaut twin Mary from Chapter 1 really does find, when she gets back home, that more time has elapsed at home than on her clock. She has traveled into the future. Of course, we are all "traveling" into the future in one sense: we get old, and time "passes." But we have noticed that this journey has never brought us to any point in the past. Despite the feeling of *déjà vu* we all have occasionally, the natural world at any instant is sufficiently different from what it used to be that we don't actually believe we are in the past. The physical processes that mark the flow of time are oriented: water flows downhill, not uphill; radioactive elements transmute into lighter ones, not heavier ones; and—by "time's arrow," the second law of thermodynamics—the entropy of any isolated thermodynamic system (one that no external energy enters) increases. If we see water flowing uphill or rusted-out automobiles getting shiny, or lighter elements fusing into heavier ones,[6] it will not be difficult to find a source of energy producing these countercurrents to the general behavior of matter. The reason we recognize them as being unnatural at all is that our own sense of time passing retains its orientation as we observe them. This time orientation of physical processes does not follow from any space-time metric, but is an additional assumption. For an object at rest, the metric of special relativity states only that $ds^2 = dt^2$, so that proper time s and laboratory time t satisfy $s = C \pm t$. As far as a traveler knows on the basis of special relativity, laboratory time might be flowing in either direction relative to proper time. On a ray of light, s is frozen, since $ds = 0$, while $dt > 0$. We are going to construct an example where the reverse happens: dt assumes both positive and negative values, while $ds > 0$.

What do we mean when we think of traveling into the past? From the dozens of science-fiction movies and television series that we have all seen, the scenario is always the same: the characters in the drama are isolated in some traveling device,

---

[4]An anonymous reviewer pointed out that I have slighted Hilbert, who was the first to write down what is now called the Einstein–Hilbert action, and also probably discovered the Einstein equation, for which he gave priority to Einstein.

[5]This word is new as far as I know, although, considering the great linguistic fecundity of writers like Wells and Asimov, I wouldn't be surprised to learn that someone has already coined it. A story I read some 60 years ago contained the beautifully descriptive word *chronoclasm* to describe a catastrophe brought about by a time traveler meddling with the past. *Temporonaut* seems be the appropriate scientific-sounding term for a time traveler.

[6]In the only example of fusion outside a star up to now, the hydrogen bomb, a fission bomb provides the energy that causes the fusion of hydrogen into helium.

or in a confined space when some catastrophe occurs. When they step outside at
the end of their journey, or after the catastrophe, they find they are in the same
place where they started and not noticeably older than they were, but they are
either at a time before they left or many centuries later. We shall ignore the latter,
on the grounds that it appears to be possible on the basis of special relativity.
The science-fiction scenario suggests a way to state the problem of travel into the
past formally. To explain the problem, we consider the world-line of a particle in
space-time. Relativity provides two concepts of time for such a particle. There is
the laboratory time t of an observer—let that time correspond to the time passing
outside the particle where the temporonauts reside—and there is the proper time s
on the particle, which is the time shown on the clocks the temporonauts are carrying
with them. Our problem thus becomes to construct a path γ (not necessarily or
even usually a geodesic) that starts and ends at a single point of space-time, having
fixed laboratory coordinates.

With the usual interpretation of the coordinates in space-time, traveling into
the past cannot possibly mean simply standing in one place and commanding time
to flow backwards. When an object is standing in one place, its proper time differs
from laboratory time by a constant; and proper time, by definition, always increases.
Nor can it mean traversing a path parameterized by proper time s in such a way
that $dt/ds > 0$ at every point, yet t is smaller at the terminal value s_1 than at
the initial value s_0. If $dt/ds > 0$ at every point, as any calculus student knows,
t is bound to get larger as s does. What we actually do is exhibit a space-time
metric in which there exists a closed time-like curve, that is, one on which proper
time serves as the parameter, yet which returns periodically (in proper time) to the
same point in laboratory space-time. The passengers on a particle traveling along
this curve see their own watches apparently keeping normal time, but in relation
to the outside world that uses time t, they keep returning to the same point at the
same time t. To picture what this means, imagine passengers on the Circle Line
in London or the Ring Line in Moscow riding a train that never stops. It keeps
going past the same finite number of stations at a uniform speed according to their
measurements. Yet mysteriously, the time on every station clock that they pass
always shows exactly the same time it showed the last time they passed through it.
This would truly be an extreme manifestation of *déjà vu*! It would not be surprising
if some science-fiction writer used this scenario as the basis of a story.[7]

Such a loop, which is truly strange, fits exactly the definition introduced by
Douglas Hofstadter ([**42**], p. 10):

> The "Strange Loop" phenomenon occurs whenever, by moving up-
> wards (or downward) through the levels of some hierarchical sys-
> tem, we unexpectedly find ourselves right back where we started.

In the present case, *where* means a point in space-time, so that we find ourselves
not only *where*, but also *when* we started. Hofstadter illustrated the idea with the
1961 lithograph *Waterfall*, by Maurits Cornelis Escher (1898–1972). In this sketch,
the current is always directed downhill, as water should flow, and yet the stream
forms a loop, endlessly returning to its original height. If height replaces laboratory
time and current replaces proper time, this picture can be thought of as a closed

---

[7]Indeed, it may have happened already, as those who have seen the movie *Groundhog Day*
might think.

FIGURE 7.1. A strange loop: M. C. Escher's "Waterfall" © 2016
The M. C. Escher Company-The Netherlands. All rights reserved.
`www.mcescher.com`

time-like loop. Hofstadter later developed this concept in more detail in an entire
book [**43**] devoted to the topic. He did not, however, give this example, which
involves a creation of Kurt Gödel in an area remote from his work on logic and
language, which Hofstadter had expounded so masterfully earlier.

We have to make a rather complicated change of variable (due to Gödel) in
order to interpret this metric in normal space-time with a certain distribution of
rotating matter. Whether models of this type are physically realizable appears to
be an open question. The one we are defining is not very stable; you need the
cosmological constant to be very precisely related to a constant matter density ρ,
in fact by the equation $\Lambda = \ \ 2\pi G\rho/c^2$. We are not going to discuss the practicality
of this model, however. What we are about to embark on is purely an adventure
of the human mind.

3.1. The Gödel space-time metric. We introduce Gödel space-time as a
four-dimensional manifold $\mathfrak{M}$ whose differentiable structure (see Appendix 4) is
given by a single chart $\psi : \mathfrak{M} \to \mathbb{R}^4$. Except for trivial differences in notation, the Gödel manifold consists of space-time points that we deliberately "de-
dimensionalize", namely $\tau = \omega t$, $\xi = \omega x/c$, $\eta = \omega y/c$, and $\zeta = \omega z/c$, where c is the

speed of light and ω is a certain angular velocity introduced by Gödel, expressed in radians per unit time. If s is proper time in the dimensioned coordinates, then $\sigma = \omega s$ is proper time in the dimensionless coordinates. In dimensionless coordinates, the speed of light is the real number 1.

The metric Gödel assumed on this four-dimensional manifold is given by

$$d\sigma^2 = \frac{1}{2\omega^2}\left(d\tau^2 + 2e^\xi\, d\tau\, d\eta - d\xi^2 + \frac{e^{2\xi}}{2}\, d\eta^2 - d\zeta^2\right),$$

where the constant ω is an angular velocity that we shall bring into play below. The metric $d\sigma^2$ has the dimension of time-squared, consistent with what we have mostly done up to now.

If we "re-dimensionalize" this expression in the standard way, so that ds^2 has the physical dimension of length-squared, the result is

$$ds^2 = \frac{c^2}{2}\, dt^2 + ce^{\omega x/c}\, dt\, dy - \left(\frac{1}{2}dx^2 - \frac{e^{2\omega x/c}}{4}\, dy^2 + \frac{1}{2}dz^2\right).$$

These expressions show that the Gödel metric is a perturbation of the metric of special relativity. For the moment we find it more convenient to work in dimensionless variables and to ignore the factor $1/(2\omega^2)$, which we temporarily set equal to 1.

We intend to exhibit a path in this space on which proper time σ increases steadily, yet the path begins and ends at the same location and at the same laboratory time τ. To make the explanation clear, we need to reparameterize a portion of space-time with a highly artificial set of variables. Specifically, we introduce new variables t, u, v, w in terms of which the dimensionless time and space variables are given as follows by Momin ([**61**], p. 4):

$$\tau = 2\sqrt{2}\arctan\left(\tan\left(\frac{v}{2}\right)e^{-2u}\right) - \sqrt{2}v + 2t\,,$$

$$\xi = \ln\left(\cosh(2u) + \cos(v)\sinh(2u)\right),$$

$$\eta = \frac{\sqrt{2}\sin(v)\sinh(2u)}{\cosh(2u) + \cos(v)\sinh(2u)}\,,$$

$$\zeta = 2w\,.$$

(*Note added March 4, 2016:* Momin's article was posted on-line at the University of Toronto in 2015, but appears to have been removed since.)

Theoretically, the variable v should be restricted to the interval $(-\pi, \pi)$, since $\tan(v/2)$ is not defined at the endpoints of that interval. For any positive constant b, however, the function $2\arctan\left(b\tan(v/2)\right)$ approaches $-\pi$ as $v \downarrow -\pi$ and approaches π as $v \uparrow +\pi$. We can thus define this function, say for $\pi < v < 3\pi$ by letting it be $2\pi + 2\arctan\left(b\tan((v - 2\pi)/2)\right)$ on that interval. This function can, through such a procedure, be extended to a continuous function of v for all real values v and whose value equals the value of v at multiples of π. In fact, it is differentiable when so extended, since its derivative tends to $1/b$ as $v \to \pm\pi$. With this extension, τ becomes a continuous *periodic* function of v, having period 2π. Obviously, the same is true of ξ, η, and ζ when t, u, and w are held fixed.

We can now see that the straight line $t = 0$, $u = a$, $z = 0$, $-\infty < v < +\infty$, maps to a closed curve that is traversed infinitely many times. What we shall show is that, if a is suitably chosen, the variable v is the proper time σ on this path, so

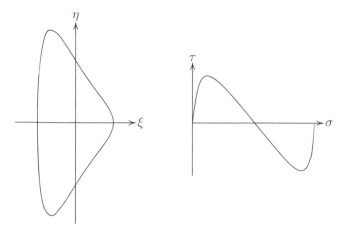

FIGURE 7.2. Left: A closed time-like curve in the $\xi\eta$-plane. Right: Laboratory time τ as a function of proper time σ over one traversal of the loop.

that it is timelike ($d\sigma^2 > 0$). In fact, as *Mathematica* will easily compute, in terms of the variables t, u, v, w, the metric is

$$d\sigma^2 = 4\left(dt^2 + 2\sqrt{2}\sinh^2(u)\, dt\, dv - du^2 + \left(\frac{7}{8} - \cosh(2u) + \frac{1}{8}\cosh(4u)\right) dv^2 - dz^2 \right).$$

Given that $dt = du = dz = 0$ along this path, we see that we will get $ds^2 = dv^2$ if $\cosh(2u) = 2 + \sqrt{2}$, that is, $u = a = \frac{1}{2}\ln\left(2 + \sqrt{2} + \sqrt{5 + 4\sqrt{2}}\right)$. The path returns to the point in space-time whose coordinates are $(0, 2a, 0, 0)$ whenever v is a multiple of 2π.

A geometric representation of this curve in the $\xi\eta$-plane is shown in Fig. 7.2, along with the value of the time τ on the "station clocks" in terms of the time σ shown on the wrist watches of the passengers on this "rapid-transit" line. If the train sets out from the leftmost point on the loop at noon, on both station clocks and clocks inside the cars, and travels counterclockwise around the loop, our passengers find that $\tau > \sigma$ early on, that is, the station clocks show a later time than their own clocks. But fairly soon, the station clocks show the latest time they will show and then begin to run backwards, as seen from inside the cars. By the time the right-most point of the loop is reached, the station clocks the passengers see will have moved back to noon, while the clocks inside the cars continue to move ahead at a uniform rate. Over the second half of the loop, the station clocks continue to retreat to a time before noon, but then begin to run forward again, getting back to noon exactly as the car arrives at its starting point.

REMARK 7.5. Any path along which t, u, and w are constant is a closed curve. The Euler equations show that for the fixed value $u = (1/4)\ln(4 + \sqrt{15})$, the corresponding closed curve is a geodesic. It is not, however, timelike. The passengers on such a "rapid transit loop" would have to have imaginary mass, but at least they wouldn't require any fuel!

3.2. Physical interpretation of this model. It is not surprising that one can invent a bizarre metric in which "time travel" is possible as a mere mathematical curiosity. What makes the Gödel metric more interesting is that it *might be* physically realizable in terms of physical space, provided we get just the right cosmological constant and just the right density of moving matter in space. To get a physical interpretation, we restore the physical dimensions of the coordinates, using Gödel's angular velocity variable ω and the speed of light c, as exhibited above. Although the metric does not have a diagonal matrix, so that the proof we gave for the vanishing of divergence does not apply, *Mathematica* will nevertheless verify instantly that the divergence of the contravariant Einstein tensor is zero in this case also.

If the cosmological constant Λ is taken equal to $-\omega^2/c^2$, we find that the contravariant version of the Einstein field equations says

$$\text{ContraEin} + (\omega^2/c^2)g^{-1} = \begin{pmatrix} \frac{4\omega^2}{c^4} & 0 & 0 & 0 \\ 0 & 0 & 0 & 0 \\ 0 & 0 & 0 & 0 \\ 0 & 0 & 0 & 0 \end{pmatrix}.$$

This means that the entry in the first row and column of the stress-energy-momentum tensor is $\omega^2/(2\pi G)$. Now this quantity, as one can easily see, has the physical dimension of a density ρ. Hence the covariant stress-energy-momentum tensor is exactly the one we obtained earlier with cosmological constant 0 for a mass density ρ equal to $\omega^2/(2\pi G)$. To picture a normal particle moving through such a mass density, it is best to think of the mass as a very rarified gas, such as in the stratosphere, or—even better—as dark matter, through which ordinary matter passes without any collisions, only gravitational interaction.

Thus, if the angular velocity ω is constant and the cosmological constant Λ is $-2\pi G\rho/c^2$, the Gödel universe with its peculiar metric behaves like ordinary space-time with the gravitational field given by this density. The exact sense in which the "dust" in this universe is rotating with angular velocity ω is a subject we leave the reader to explore independently.

4. Black Holes

We introduced the Schwarzschild radius $r_s = 2GM/c^2$ in Chapter 4 and noted that it is the parameter that determines the curvature of space-time in the gravitational field of a single particle. Although it is a tiny distance (only about 3 km in the case of the Sun) its effects are noticeable at distances tens of millions of times greater than itself. In the case of Mercury, the effect is a small amount of precession of the perihelion of the orbit, as seen by an observer at rest relative to the Sun. It needs to be emphasized that nothing in particular "happens" at this radius. It is a point where the *coordinates* have singularities, but there is no *physical* singularity there, and it is possible to change to coordinates that do not even have a mathematical singularity at the Schwarzschild radius. They are valid for all positive values of the distance to the origin.

But not at the origin itself; that is a true physical singularity. In the simplified model we have been using, in which the attracting body is regarded as a massive particle, that particle *is* a black hole. Having positive mass M, it possesses a positive Schwarzschild radius $r_s = 2GM/c^2$, while its own radius—the radius of a particle—is zero. Hence the entire mass is inside the Schwarzschild radius, and by definition, that makes it a black hole.

The processes by which black holes form involve nuclear physics beyond the scope of the present text, and so our inspection of some "macroscopic" physical phenomena associated with them will be cursory. Our main interest lies in the region just outside the Schwarzschild radius r_s, and we assume that the source of the gravitational field lies entirely inside that radius. The gravitational law of Chapter 4, which requires the vanishing of the Ricci tensor, holds at distances larger than r_s. The simplest way to study this region is to imagine a second particle or a light ray falling directly toward the gravitating particle. This model will reveal some bizarre aspects of relativistic gravitation, not at all what the Newtonian picture would lead us to expect. The stripped-down model we are going to study, which amounts to an object falling directly into a black hole, allows us to neglect the altitude and right-ascension coordinates φ and θ, so that we have only one spatial dimension to consider, for which we use the radial coordinate r.

4.1. Falling toward a center: the Newtonian case. We assume that the particle starts from rest at distance r_0 and falls directly to the origin. The initial-value problem to be solved is

$$r'' = -\frac{GM}{r^2}, \quad r(0) = r_0, \quad r'(0) = 0.$$

The usual trick of letting $p = r'$ in a second-order ordinary differential equation that does not contain r' works and shows that

$$r' = -\sqrt{2GM}\sqrt{\frac{1}{r} - \frac{1}{r_0}}.$$

(The negative sign is present because r is a decreasing function of time.)

This equation implies that r' tends to $-\infty$ as r decreases to zero, and hence the kinetic energy of the particle becomes arbitrarily large. Correspondingly, the potential energy that would be required to move it away from the origin is infinite. (The force holding it there would be infinitely large.)

This equation cannot be solved in closed form for r as a function of t, but the substitution $r = r_0 \cos^2 \theta$ allows us to solve it implicitly, getting the equation

$$t = \sqrt{\frac{r_0}{2GM}}\left(r_0 \arccos\left(\sqrt{\frac{r}{r_0}}\right) + \sqrt{r_0 r - r^2}\right).$$

From this relation, we see that the falling particle will reach the origin in a finite time t_0:[8]

$$t_0 = \frac{\pi}{2}\sqrt{\frac{r_0^3}{2GM}}.$$

By Kepler's third law, this is equal to the period it would have if orbiting in a circle at distance $\sqrt[3]{2}r_0/4$. If r_0 is the radius of the Earth's orbit, this corresponds to an orbit somewhere inside the orbit of Mercury (about 48 million kilometers from the Sun). More colorfully, if some catastrophe caused the Earth's forward motion in its orbit to cease, it would fall directly into the Sun, taking a little more than two months to do so.

---

[8]Notice the occurrence of the "geometric" constant π in this equation, which describes a physical phenomenon. The occurrence of π in this relation and in Kepler's third law may be considered a foreshadowing of the geometrization of physical laws.

4.2. Falling toward a center: the relativistic case. In the space-time coordinates $(t; r)$ of an observer fixed relative to the gravitating particle, the differential of proper time ds on the falling particle satisfies

$$(7.3) \qquad ds^2 = \left(1 - \frac{r_s}{r}\right) dt^2 - \frac{r}{c^2(r - r_s)}\, dr^2 \,,$$

We consider a particle falling into a black hole that, according to an external observer, is at distance r_0 at time $t = 0 = s$. To get the algebraically simplest case, we imagine the coordinates extrapolated into the past in such a way that, when its trajectory is written as an equation between r and t, we have $r \to \infty$ as $t \to -\infty$ and $r'(t) < 0$ at all times t. Since our space-time coordinates have a singularity at the distance $r = r_s$, we cannot extrapolate the path to distances shorter than r_s—our equations break down—but we shall take them to that limit and see what happens.

The simplest case is that of a photon (light ray), for which $ds \equiv 0$. We then have

$$\frac{dr}{dt} = -c\left(1 - \frac{r_s}{r}\right),$$

Thus, the speed of light decreases to zero at the event horizon. (We already knew this, from Section 8 of Chapter 4.) The sphere of radius r_s is such that light takes longer and longer to reach our external observer from points arbitrarily close to that horizon, no matter how close that observer may be. The event horizon itself is completely black, since no light leaves it.

Now consider a material particle moving at sublight speed $v = dr/dt$. It follows from Eq. (7.3) that $dt/ds \geq 1/\sqrt{1 - r_s/r}$, so that dt/ds tends to infinity as $r \downarrow r_s$. Also, since $v^2 = (dr/dt)^2 = c^2(1-r_s/r)\big((1-r_s/r)-(ds/dt)^2\big)$, the speed v measured by an external observer must tend to zero as $r \downarrow r_s$. Thus, the particle will appear to an outside observer to *slow down* as it gets nearer to the radius r_s. That is the very counterintuitive result of our equations, and there is no comparable effect in any model in which "force" is directed toward a center. Any force one imagines must appear to an outside observer to be a *repulsive* force at some point, slowing the fall. This phenomenon was first commented on by David Hilbert (1862–1943) in 1915. (See Loinger and Marsico [**53**]. As Loinger and Marsico point out, Einstein rediscovered this result later [**22**].) The same problem arises with Newton's law of gravity when we consider a particle falling toward a center: Time (t) can flow only until it reaches the value $(\pi/2)\sqrt{r_0^3/2GM}$, where r_0 is the value of r at time $t = 0$. At that point the particle will reach the origin where the actual physical body being idealized as a point is located, and as a result the mathematical model no longer applies.[9] This is obvious and not at all counterintuitive. But in relativity, the whole mass of the attracting body *may* be inside its Schwarzschild radius r_s, and we are forced to deal with the consequences of our equations, which imply this odd result.

To say any more, we need to look at the Euler equations, and for that purpose we now bring back our twins John and Mary from Chapter 1 to make their final

---

[9]This same singularity at the origin disturbed Gabriel Lamé (1795–1870) in his attempt [**49**] to explain light as a disturbance in an elastic medium. Knowing that he could not ignore the case of light emanating from a point source, he invoked a rather loosely-argued principle involving the ether, which essentially discarded all the Newtonian mechanics on which his own analysis had been based up to that point.

bow in, say the 24th century. Mary, who journeyed from London to Massachusetts in the eighteenth century and then to a nearby star and back in the 21st century, is now setting out on the ultimate travel adventure: into a black hole. All she has to do—provided we remove the rest of the universe—is to launch herself straight at it. Gravity will take care of everything else. We'll let her shrink to particle-size for the journey. We do this for her safety, since a body of any measurable size would be torn apart by tidal forces at the event horizon, the Schwarzschild radius r_s. The speed of information c with which the twins communicated back in the the 21st century cannot be improved on. At that time, we noted that, due to the large distances, messages would take a long time to get from one twin to the other. In this 24th-century exploit, it undergoes a further degradation, since $c \downarrow 0$ as Mary approaches the event horizon.

To start the discussion with a result that appears to come out of nowhere—its source will be explained below—we express the coordinates r and t that John uses to keep track of his sister in terms of Mary's proper time s as follows:

$$(7.4) \qquad r = r_0\left(1 - \frac{3cs}{2r_0}\right)^{\frac{2}{3}},$$

$$(7.5) \qquad t = t_0 + s - \frac{2\sqrt{r_s r}}{c} - \frac{r_s}{c}\ln\left(\frac{\sqrt{r} - \sqrt{r_s}}{\sqrt{r} + \sqrt{r_s}}\right),$$

where t_0 is chosen so that $r = r_0$ when $s = t = 0$, that is,

$$t_0 = \frac{2\sqrt{r_s r_0}}{c} + \frac{r_s}{c}\ln\left(\frac{\sqrt{r_0} - \sqrt{r_s}}{\sqrt{r_0} + \sqrt{r_s}}\right).$$

When the value of r from Eq. (7.4) is inserted into Eq. (7.5), the result is a very complicated expression, but one that is at least theoretically computable. These equations actually come from the Euler equations for geodesics in the metric given by Eq. (7.3). It is not difficult to verify that they satisfy these equations, that is,

$$\frac{d}{ds}\left(2\left(1 - \frac{r_s}{r}\right)\right)\frac{dt}{ds} = 0,$$

$$\frac{d}{ds}\left(2\left(\frac{r}{c^2(r - r_s)}\right)\right)\frac{dr}{ds} = \frac{r_s}{r^2}\left(\frac{dt}{ds}\right)^2 + \frac{r_s}{c^2(r - r_s)^2}\left(\frac{dr}{ds}\right)^2.$$

The second of these equations is messy, as it would have been back in Chapter 4, had we not replaced it with the "integrated" equation that, in the present situation becomes

$$1 = \left(1 - \frac{r_s}{r}\right)\left(\frac{dt}{ds}\right)^2 - \frac{r}{c^2(r - r_s)}\left(\frac{dr}{ds}\right)^2.$$

The solution of the first of the Euler equations is

$$\frac{dt}{ds} = \frac{\ell r}{r - r_s},$$

where ℓ is a positive constant. The integrated form of the second equation then becomes

$$\left(\frac{dr}{dt}\right)^2 = c^2\left(\left(1 - \frac{r_s}{r}\right)^2 - \frac{1}{\ell^2}\left(1 - \frac{r_s}{r}\right)^3\right).$$

This form of the equation was pointed out by Hilbert ([**40**], p. 289), who assumed units in which $c = 1$ and wrote α for r_s and A for the constant $-1/\ell^2$.

The solution divides into three cases according as $\ell < 1$, $\ell = 1$, and $\ell > 1$. We are confining ourselves to the case $\ell = 1$, which is the simplest. It is a routine though time-consuming computation to show that the variables t and r are related by the equation

(7.6)
$$\ln\left(\frac{\sqrt{r} + \sqrt{r_s}}{\sqrt{r} - \sqrt{r_s}}\right) - \frac{2}{3}\left(\frac{r}{r_s}\right)^{\frac{3}{2}} - 2\left(\frac{r}{r_s}\right)^{\frac{1}{2}} = K + \frac{ct}{r_s},$$

where again K is a constant such that $r = r_0$ at time $t = 0$:

$$K = \ln\left(\frac{\sqrt{r_0} + \sqrt{r_s}}{\sqrt{r_0} + \sqrt{r_s}}\right) - \frac{2}{3}\left(\frac{r_0}{r_s}\right)^{\frac{3}{2}} - 2\left(\frac{r_0}{r_s}\right)^{\frac{1}{2}}.$$

From the preceding equations we find the velocity v and acceleration a of the falling particle:

$$v = \frac{dr}{dt} = -c\left(1 - \frac{r_s}{r}\right)\sqrt{\frac{r_s}{r}},$$

$$a = \frac{d^2r}{dt^2} = -\frac{c^2 r_s}{2r^2}\left(1 - \frac{3r_s}{r}\right)\left(1 - \frac{r_s}{r}\right).$$

The acceleration is negative (toward the center) when $r > 3r_s$, but positive (away from the center) when $r_s < r < 3r_s$. Both velocity and acceleration tend to zero as $r \downarrow r_s$, showing that if Mary reached the event horizon (the Schwarzschild radius r_s) at a finite time t on John's clock, she would be "stuck there." There would be problems in communicating as Mary approached the event horizon, since the speed of light decreases to zero at that distance from the black hole. Messages would take longer and longer to send, eventually requiring an arbitrarily large amount of time to go from one sibling to the other, even though the distance between them remains bounded. From John's point of view, Mary would be falling forever, but never reaching the event horizon.

From Mary's point of view, on the other hand, her location r at time s satisfies the very simple differential equation

$$\frac{dr}{ds} = -\frac{c\sqrt{r_s}}{\sqrt{r}},$$

so that

$$r = \left(r_0^{3/2} - \frac{3c\sqrt{r_s}s}{2}\right)^{\frac{2}{3}}.$$

and the event horizon is reached at the finite proper time

$$s_0 = \frac{2(r_0^{3/2} - r_s^{3/2})}{3c\sqrt{r_s}}.$$

Proper time s continues to flow, from s_0 to $s_0 + 2r_s/(3c)$. At that point, Mary (shrunk to particle-size, remember) reaches the "pole" $r = 0$, which is a genuine space-time singularity, thereby becoming the Roald Amundsen of the twenty-fourth century.

As Eq. (7.6) shows, $t \to \infty$ as $r \downarrow r_s$. If we were to extrapolate Mary's journey into an imaginary remote past, we would see that, $r \to +\infty$ as $t \to -\infty$. Thus, if we imagine that she has been falling forever, then she was located at points arbitrarily remote from the black hole at times sufficiently long ago. Her trajectory, as seen in coordinates fixed with origin at the center of the black hole, is shown in Fig. 7.3.

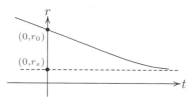

FIGURE 7.3. Trajectory of a particle falling toward a black hole with Schwarzschild radius $r_s = 2GM/c^2$, as seen by an observer.

REMARK 7.6. If we replace s and t by $s' = s$ and $t' = -t$ and $r(s)$ by $r(s')$, then $ds'/dt' = ds/dt$, and we get the same trajectory, only traversed in the opposite direction. That is, Mary has been moving away from the radius r_s from all eternity, and will continue in that direction, getting arbitrarily distant. Equation (7.6) becomes

$$(7.7) \qquad \ln\left(\frac{\sqrt{r}+\sqrt{r_s}}{\sqrt{r}-\sqrt{r_s}}\right) + \frac{2}{3}\left(\frac{r}{r_s}\right)^{\frac{3}{2}} + 2\left(\frac{r}{r_s}\right)^{\frac{1}{2}} = K + \frac{ct}{r_s},$$

where again K is chosen so that $r = r_0$ at time $t = 0 = s$.

REMARK 7.7. If $r_0 < r_s$, that is, if Mary began her journey *inside* the black hole—presumably, the twins were born there—she would, from John's perspective, never get out. Equation (7.6) would be replaced by

$$(7.8) \qquad \ln\left(\frac{\sqrt{r_s}+\sqrt{r}}{\sqrt{r_s}-\sqrt{r}}\right) - \frac{2}{3}\left(\frac{r}{r_s}\right)^{\frac{3}{2}} - 2\left(\frac{r}{r_s}\right)^{\frac{1}{2}} = K + \frac{ct}{r_s},$$

with K again chosen to make $r = r_0$ at time $t = 0 = s$. The equation is valid only as long as Mary stays away from any matter inside the black hole. Since we are regarding the black hole as being due to a particle, that requirement imposes no restriction. Depending on which direction time is flowing, her acceleration, as observed by John, is

$$a = \pm\frac{c^2 r_s}{2r^2}\left(1 - \frac{3r_s}{r}\right)\left(1 - \frac{r_s}{r}\right).$$

This quantity does not change sign when $r < r_s$. Hence, from John's point of view Mary approaches either 0 or r_s as time becomes infinite, never reaching either limit.

This brief and superficial excursion into the mysteries of a black hole is the last of the observations we intend to make on this subject. Our final chapter will be devoted to a chronology of important papers in the theory of relativity and to some commonsense metaphysical speculation about what it all means for our view of the natural world.

5. Problems

PROBLEM 7.1. Imagine a tunnel dug all the way through the Earth around one of its diameters. (Idealize the Earth as a perfect sphere and imagine its atmosphere has disappeared, so that the inside of the tunnel is a perfect vacuum.) What would happen to a particle dropped down the very center of the tunnel? (Use Newtonian reasoning.)

PROBLEM 7.2. Show that a plane whose first fundamental form is given by Eq. (7.3) has curvature $\kappa(r)$ depending only on r and given by

$$\kappa(r) = -\frac{c^2 r_s}{r^3}.$$

(Thus, somewhat surprisingly, given that the parameterization has a singularity at $r = r_s$, there is no singularity in the curvature at that point.)

PROBLEM 7.3. Show that the Laplace–Beltrami operator on the plane with first fundamental form (7.3) is

$$\nabla^2 f = \frac{r}{r - r_s}\frac{\partial^2 f}{\partial t^2} - \frac{c^2(r - r_s)}{r}\frac{\partial^2 f}{\partial r^2} - \frac{c^2 r_s}{r^2}\frac{\partial f}{\partial r}.$$

PROBLEM 7.4. Assuming $\ell = 1$, confirm Hilbert's statement that a body falling toward a black hole has acceleration toward the black hole if its speed $v = r'(t)$ satisfies $|v| < c(r - r_s)/(\sqrt{3}r)$ and away from it if $|v| > c(r - r_s)/(\sqrt{3}r)$.

PROBLEM 7.5. Show that the space-time given by the Gödel metric has no singularities. Also show that the Gaussian curvature of the section tangent to any plane containing the y-axis is 0, that of the sections tangent to the tx and tz planes is $-\omega^2$, and that of the section tangent to the xz-plane is $-3\omega^2$. Show finally that the scalar curvature is $-2\omega^2$.

PROBLEM 7.6. Prove that the manifold $\mathfrak{M}$ given by Gödel's original metric is a *homogeneous space*, in the sense that, for any two points P and Q in $\mathfrak{M}$, there is an isometry $T_{PQ} : \mathfrak{M} \to \mathfrak{M}$ (a one-to-one mapping of $\mathfrak{M}$ onto itself that preserves the metric) such that $T_{PQ}(P) = Q$. To do this, show that the mapping

$$\begin{aligned} T_{PQ}(t; x, y, z) &= (t + q^1 - p^1; x + q^2 - p^2, e^{p^2 - q^2}(y - p^3) + q^3, z + q^4 - p^4) \\ &= (\tau; \xi, \eta, \zeta). \end{aligned}$$

is a one-to-one diffeomorphic (infinitely differentiable, in fact, analytic) isometry of $\mathfrak{M}$ onto itself mapping $P = (p^1; p^2, p^3, p^4)$ to $Q = (q^1; q^2, q^3, q^4)$.

PROBLEM 7.7. Those who appreciate the power and beauty of the theory of analytic functions of a complex variable may yearn to see this theory extended to three-dimensional space $\mathbb{R}^3$. It is possible to create such an extension? A number of considerations come to mind, algebraic, geometric, and analytic.

The algebraic consideration is that $\mathbb{R}^3$ is *not* a field, but the plane $\mathbb{R}^2$ is. (The latter is the field of complex numbers.) Indeed, it was his attempt to find a suitable definition of multiplication for elements of $\mathbb{R}^3$ that led Hamilton to discover quaternions, which are a multiplication operation on $\mathbb{R}^4$. Thus, there is an algebraic barrier to such an extension.

The geometric barrier is even more formidable. Analytic function theory provides a plethora of conformal mappings. By the Riemann mapping theorem, any simply-connected subset of the plane that has at least two boundary points can be conformally mapped onto the unit disk. In contrast, only a very restricted class of mappings of open sets of $\mathbb{R}^3$ can be conformal.

Nevertheless, analysis can still forge ahead and define a mapping $\boldsymbol{F} : \mathbb{R}^3 \to \mathbb{R}^3$, given by $\boldsymbol{F}(x, y, z) = u(x, y, z)\boldsymbol{i} + v(x, y, z)\boldsymbol{j} + w(x, y, z)\boldsymbol{k}$, to be *conjugate-analytic* if $\nabla \times \boldsymbol{F} = \boldsymbol{0}$ and $\nabla \cdot \boldsymbol{F} = 0$. In particular, the Newtonian gravitational force is a conjugate-analytic function on $\mathbb{R}^3 \setminus \{\boldsymbol{0}\}$. Show that, when $w \equiv 0$ and u and v are independent of z (so that $\boldsymbol{F}$ maps the plane into itself), these equations reduce to the Cauchy–Riemann equations for the function $f(z) = f(x+iy) = u(x,y)-iv(x,y)$, namely $\partial u/\partial x = -\partial v/\partial y$ and $\partial u/\partial y = \partial v/\partial x$. Thus if $\boldsymbol{F} = u\boldsymbol{i} + v\boldsymbol{j}$ is identified with the complex function $f(x+iy) = u(x,y) + iv(x,y)$, it is the conjugate of an analytic function. Use these equations to show that u and v are harmonic functions, that is, $\nabla^2 u = 0 = \nabla^2 v$.

Also show that the components $u(x,y,z)$, $v(x,y,z)$, and $w(x,y,z)$ of a conjugate-analytic function $\boldsymbol{F} = u(x,y,z)\,\boldsymbol{i} + v(x,y,z)\,\boldsymbol{j} + w(x,y,z)\,\boldsymbol{k}$ are harmonic functions.

Part 3

Historical and Philosophical Context

CHAPTER 8

Experiments, Chronology, Metaphysics

Throughout the previous seven chapters, we have looked at a selection of topics taken from special and general theories of relativity, without aiming at a systematic exposition of either theory. The questions we have dealt with have all involved technical mathematics and physics. In this concluding chapter, we turn philosophical and speculate on what all this technical mathematics and physics means to an ordinary person hoping to understand the physical world. On this point, I have received very sensible advice from reviewers more knowledgeable than I, who have pointed out that many of the glib suggestions that I make in this chapter are actually areas of current research. The reader is hereby alerted that nothing I say in this chapter can be assumed to originate with me. The chapter consists of nine sections.

In the first section, we attempt to answer two questions: (1) To what extent is the theory of relativity confirmed by observation and experiment? (2) Are there rival theories that agree equally well or better with observation and experiment?

The second section attempts to put the whole subject into its social context, giving a list of some highlights from the history of physics and mathematics (not all of physics or mathematics, only a selection of important advances in the two subjects that eventually had some influence on the theory of relativity). This account of the way people formerly thought about physics is useful in judging the categories we now use in physical reasoning.

Sections 3 through 7 consist of metaphysical speculation of a very general sort on the relation between the physical universe and human thought, with an emphasis on the role played by measurement and mathematical models. Beginning with the earliest applied mathematics, which introduced the idea of precise measurement into the subject, we take note of the tendency toward abstraction, first in geometry, from the idealized shapes of geometric figures and finally to the abstract concept of space in general. That is where philosophers have disagreed with one another for centuries, but physics has compounded the difficulty and the interest of the subject by throwing up one abstract concept after another for our consideration: force, energy, electromagnetic and gravitational fields, subatomic particles, and the like, all of which have in common the fact that one cannot observe them directly through the senses. Their existence is believed in only because they enable us to account for what we do observe. Here we enter very deep waters, asking what is meant by physical reality and physical existence. In the modern world, we are accustomed to taking for granted the reality of such entities as radio waves, whose existence is not on the same plane of reality as the existence of, for example, air and water. Why then do we speak of radio waves with exactly the same sense of confidence in their reality that we ascribe to air and water? The physical quantities mentioned above are not directly seen or felt, and the instruments now used to detect these quantities are themselves the products of high technology; their functioning and, consequently, also the measurements they provide are interpreted using other physical theories. Nobody makes a *direct* measurement any more, in the sense that this phrase used to have. We take as our starting point the metaphysics of space and time and then extend it to look at other physical concepts, two of which are no longer regarded as viable entities by the physics community, and others that are. The main issue we intend to discuss is the appropriate language in which discussion of physical laws should take place, and what meaning is to be attached to statements that refer to these abstract physical entities. All of that presupposes people engaged in conversation on this subject. Unless one is attempting to communicate a picture of reality to someone else, there is no need to worry at all about the appropriate modes of thought for dealing with the physical world. At the end of this lengthy digression into many centuries of philosophy, we offer two rather commonplace suggestions for nonspecialists to consider: (1) Keep pondering the subject and revising your mental pictures of the occult objects that are used in physics, since new ways of thinking are desirable, but when discussing the subject, treat them as undefined terms; (2) in any discussion, concentrate on the mathematical functions (field intensity, for example) that can be measured, since the measurements encode what we actually know. This advice is reinforced with two analogies from mathematical practice, namely the use of axiom systems with undefined terms in a variety of areas and the particular use of various coordinate systems in differential geometry as a common language for discussing the geometry of a surface.

In the eighth section, we let two of the pioneers in general relativity, Albert Einstein and Sir Arthur Eddington, sum up their arguments in favor of (what was, when they wrote) the radical new approach to physics incorporated in the general theory of relativity.

In the ninth and final section, we sign off with a brief glance at the acceptance of relativity theory around the world and some reflections on the acceptance (or rejection) of scientific theories in general, leaving open the question whether the geometry that has played so large a role in relativity will retain its prominence in future physical theories.

1. Experimental Tests of General Relativity

The most important question to be asked is whether relativity theory is an accurate description of the universe. We discussed two of its early triumphs—the explanation of the precession of Mercury's perihelion, and the deflection of light around the Sun—in Chapter 4. Since alternative explanations of these two phenomena have been proposed, we now look at a few later experimental tests of the theory in competition with rival theories.

One point raised in Chapter 1 that we have not yet followed to its conclusion is the apparent absoluteness of rotational motion: the example given by Newton of the twirling bucket of water, in which the water does ride up the sides of the bucket, but only when the water itself is twirling, not when the bucket alone is doing so (until its rotation is communicated to the water). How does a relativistic view of the universe get reconciled to that commonplace observation? One way is through a principle known as *Mach's principle*, after Ernst Mach (1838–1916), according to which absolute space is replaced by the total distribution of matter in the universe. What appears to be absolute rotation is merely rotation relative to all the rest of the matter in the universe. That view makes two seemingly unverifiable predictions: (1) if the bucket could be made to remain still while all the rest of the matter rotated around it, the water would climb up the sides of the bucket; (2) if all the rest of the matter in the universe were removed and the water was rotated, it wouldn't rise. It has been suggested that the effect could be tested by putting water into a container encased in a heavy metal sphere and then spinning the sphere rapidly; but the engineering involved in doing so is formidable, since the sphere has to stand in for an immense quantity of matter—all the rest of the universe. It might appear that Mach's principle is the "only way out" if we wish to preserve relativity of motion, but it turns out that certain modifications of general relativity can do without it. In any case, the "total amount of matter in the universe" seems just as occult and unknowable a concept as absolute space. Indeed, what is meant by "all the matter in the universe"? Do we know that this quantity is constant? Are we sure matter isn't being created somewhere and destroyed elsewhere, as quantum theory allows? If it depends on time, are the fluctuations large enough to make any difference to observation? Whose clock is authorized to define the time on which it depends?

General relativity has always had to compete with rival theories. The most formidable and long-lasting of these, still "in the running" after half a century, (but just barely, now a minority point of view), is the Brans–Dicke theory, named for its creators Carl Brans (1935–) and Robert Henry Dicke (1916–1997)—see their paper [4]—in which a more elaborate version of the Einstein field equations is supplemented by a scalar equation involving the Laplace–Beltrami operator. The scalar equation lays hands on the most sacred principle of general relativity, the constancy of the gravitational "constant" G, allowing it to vary. According to Will ([86], pp. 152–153), Dicke was led to this theory by a (very) rough computation showing that G could be estimated as Tc^3/M, where T is the time since the origin of the universe and M the total amount of matter in the universe; in that respect, it dovetails nicely with Mach's principle. The Brans–Dicke theory and general relativity make closely similar quantitative predictions, and both are compatible with the observed precession of the perihelion of Mercury and the deflection of light around the Sun within the limits of observational error. The Brans–Dicke theory rules out fewer things because it contains a parameter that can be chosen to fit experimental data. As more and more precise experiments during the 1970s seemed to tell in favor of general relativity, that parameter had to be continually adjusted upward to remain compatible with observation. Because of it, the Brans–Dicke theory is harder to falsify than general relativity would be, and also more complicated. For those who take the principle popularly and inaccurately known as Occam's Razor seriously, those facts speak in favor of general relativity, which remains the dominant paradigm at the moment.

A large number of experimental tests of general relativity have been systematically catalogued and discussed by Clifford M. Will ([86], [87]). We have mentioned already (Chapter 4) the experiments of Baron Eötvös to detect a difference between gravitational and inertial mass and their negative results. This question arose again after the promulgation of general relativity, since it also posits the equality of these two masses.[1]

Here is a short list of the many experimental tests of general relativity, mostly taken from the book of Will [86]. Later results can be found in an on-line article [87] at the following url:

http://www.livingreviews.org/lrr-2006-3

- If we observe that the Schwarzschild radius $r_s = 2GM/c^2$ is directly proportional to the mass M of the fixed particle and that the Newtonian force is also, we can see that the strength of the gravitational field is directly proportional to r_s. In the metric of the gravitational field, the coefficient of dt^2 is $1 - r_s/r$. Thus, at a fixed distance r, the derivative $ds/dt = \sqrt{1 - r_s/r}$ decreases as r_s increases. In other words, a clock runs slower in a stronger gravitational field. By the principle of equivalence, we can increase the gravitational field on a clock by putting it into a fast-moving elevator—actually, a rocket ship—and making the elevator go up very fast. According to Will ([86], Chapter 3), that experiment was actually performed on 18 June 1976. A fine-tuned atomic clock was put aboard a Scout D rocket and shot off into space, experiencing $20g$'s of force. The effect is complicated by the fact that relativistic gravity affects not only the working of the clock, but also the frequency of the signal it transmits. Thus, two relativistic effects were being tested for at once. But the two effects were well separated, since the slowing of the clock occurred immediately after launch, while the frequency shift predominated at higher altitudes. The flight lasted two hours, and the results confirmed the predictions of general relativity up to 0.007%.

- The preceding example shows that the modern world would find it difficult to function without the general theory of relativity. Global positioning navigators use triangulation obtained from four or more satellites that transmit their proper time to a GPS navigator. These clocks have to be synchronized within 30 nanoseconds (3×10^{-8} sec). Because the receivers on the ground are in a stronger gravitational field, the atomic clocks they contain run slower than those in the satellites (an effect that overpowers the *slower* ticking of those in the sky predicted by special relativity). We can get some idea of how this works from a back-of-the-envelope computation based on what we already know. For a satellite in orbit above the Earth at height r, the clock it carries keeps proper time s, related to the "inertial time" t in a "laboratory" that is Earth-centered, but free of any gravitational field and not rotating with the Earth, by the equation

$$\left(\frac{ds}{dt}\right)^2 = \left(1 - \frac{r_s}{r}\right) - \frac{r^2}{c^2}\left(\frac{d\theta}{dt}\right)^2 .$$

---

[1] A headline in the Worcester, Massachusetts *Telegram* in 1919 screamed "EINSTEIN PROVEN TO BE WRONG BY AMERICAN SCIENTISTS." The accompanying article stated that physicists at Clark University in Worcester, among them Arthur Gordon Webster (1863–1923, Albert Michelson's successor as Professor of Physics at Clark University) had repeated Baron Eötvös's experiments and obtained a positive result. But, of course, one experiment doesn't "prove" anything unless it can be replicated. That one couldn't be. Webster eventually despaired of keeping up with the developments in modern physics and took his own life.

Kepler's third law can be stated by saying that the average angular velocity $(d\theta/dt)$ of a body orbiting in a circle of radius r is given by

$$\left(\frac{d\theta}{dt}\right)^2 = \frac{4\pi^2}{T^2} = \frac{GM}{r^3} = \frac{r_s c^2}{r^3}.$$

We therefore get $(r^2/c^2)(d\theta/dt)^2 = r_s/2r$, and it follows that for a satellite

$$\left(\frac{ds}{dt}\right)^2 = 1 - \frac{3r_s}{2r}.$$

The higher the satellite is, the closer its clock runs to the time in free space where there is no gravity. For a clock at the surface of the Earth, the value of $d\theta/dt$ is only half of what it is for a typical GPS satellite. There are about 30 of these satellites functioning at the moment, out of some 65 that have been launched; they revolve around the Earth twice a day with an orbital radius $r \approx 27,000$ km. If we somewhat unrealistically think of the clock on the ground as being in an orbit at height r_e equal to the radius of the Earth (about 6400 km), then the equation relating the proper time on that clock to free-space, gravity-free time is

$$\left(\frac{ds}{dt}\right)^2 = 1 - \frac{r_s}{r_e} - \frac{r_s r_e^2}{8r^3}.$$

Since the mass M of the Earth is 5.97×10^{24} kg and its Schwarzschild radius r_s is therefore about 9 millimeters (8.8×10^{-3} m), the discrepancy in ds/dt between the satellite and the ground clocks amounts to

$$\sqrt{1 - 3r_s/(2r)} - \sqrt{1 - r_s/r_e - r_s r_e^2/(8r^3)} \approx 4.442 \times 10^{-10},$$

that is, about 0.444 nanoseconds per second. As a result, the discrepancy between them would exceed the 30-nanosecond limit of tolerance in a little over one minute. If computations didn't correct for this error, the whole system would soon become unstable. We have here a good example of the practical usefulness of theory. No one would simply send up a bunch of satellites without knowing that this effect was to be expected. Relativistic effects are very small by traditional standards, but in the modern world of super-precise engineering, they have to be taken into account.

(The preceding computation is meant to have only heuristic value. We might equally well have considered a satellite in orbit at the radius of the Earth. Its angular velocity would be given by Kepler's third law, and the resulting discrepancy between the two values of ds/dt—that is $\sqrt{1 - 3r_s/(2r)} - \sqrt{1 - 3r_s/(2r_e)}$—would be 0.786 nanoseconds per second. The physical principle is what we are after rather than a precise analysis.)

- As early as 1916, Einstein developed the mathematics of gravitational waves, but they remained a curiosity for some 60 years, until the discovery of a binary pulsar some 1600 light years distant. If the two stars in this binary system change each other's shape, then they change each other's gravitational fields, and their oscillation about each other thereby produces a gravitational wave. It is too small to be detected from Earth of course, but it carries energy away from the system, causing the two stars to move closer together and thereby speeding up their period of revolution. The effect is extremely small, but with diligent observation, it was detected (see the book by Taylor and Weisberg [**80**]): gravitational waves do exist, just as theory predicts.

- The hypothesis that the speed of propagation of electromagnetic radiation in free space is independent of the wavelength has been recently tested by the measurement of a recent gamma-ray burst from a distant supernova. Even though the nova occurred billions of years ago, and the various frequencies of gamma rays have been traveling all that time to reach Earth, they all reached Earth within the span of a fraction of a second, as reported in *Nature Physics*, 16 March 2015. (Evidently, we may infer, this uniformity of speed applies only in Deep Space. Where matter is present, different frequencies must propagate at different speeds, else a prism would not be able to decompose light into a spectrum.)

`http://phys.org/news/2015-03-einstein-scientists-spacetime-foam.html`

2. Chronology

We list here some important events in the development of physics (mechanics and optics only), giving the approximate date when they occurred and their significance for the theory of relativity.[2] The development of physical theory is conveniently divided into three periods. The first consists of the earliest work on mechanics and optics, up to the end of the seventeenth century. The second period, from roughly 1700 through 1900, contains great advances in the application of calculus to both geometry and physics and the mathematization of both optics and electromagnetic theory; it is the "classical" period for mechanics, optics, and electromagnetism, all three of which were profoundly affected by relativity. The third period, from 1900 on, is the period of relativity theory and especially the geometrization of all three areas of physics, with forces being supplanted by fields producing a curved space-time.

2.1. Mechanics and optics. Some of the concepts of modern science have roots that are very ancient, although, as will be apparent from these descriptions, their original form differs considerably from the form they now have. This first period, two millennia long, saw the creation of all the concepts we associate with Newtonian mechanics.

ca. 330 BCE: Aristotle (384–322 BCE) or one of his students writes the treatise *Physics*, in which the concept of force (*dynamis*) is defined. The force applied to a "thing moved" is said to be directly proportional to its size and the distance moved and inversely proportional to the time required to move it that distance.

ca. 200 BCE: Diocles (ca. 240–180 BCE) writes a work *On Burning Mirrors*, in which he establishes the reflective property of parabolic mirrors (all rays of light parallel to the axis of a paraboloid of revolution are reflected to the focus of the generating parabola). Fragments of this work survived by accident, having been quoted by the early sixth-century commentator Eutocius, who was writing about Archimedes and Apollonius.

---

[2]I wish to warn the reader that developments in science tend to be more gradual than would appear from this list. Major breakthroughs are nearly always preceded by extensive preliminary work extending over considerable time, and ideas ascribed to one person are another are usually complex assemblages of ideas contributed by many people.

ca. 50: Heron of Alexandria (ca. 10?–ca. 75?) writes *Catoptrics* (reflection), in which he notes that the reflection property (angle of incidence equals angle of reflection) means that reflected light takes the shortest path from one point to another, given that it must first travel to the reflecting plane. He didn't think of this as a path of minimal *time* because he thought light traveled instantly, as did much later writers such as Fermat (see below).

ca. 150: Ptolemy (ca. 85–ca. 165) publishes five books of *Optics* containing results he (allegedly) obtained by experiment. The book contains a table of angles of incidence θ and refraction φ for light at the interface of water and air at increments of $10°$ from incidence angles of $10°$ through $80°$. This table is fairly close to the true values, but reveals itself to the mathematically trained eye to be a strict quadratic function, one we would write as

$$\varphi = \frac{33}{40}\theta - \frac{1}{400}\theta^2 .$$

It seems likely that it was this formula (or a finite-difference calculation from which it can be derived), rather than observation, that led to Ptolemy's table. If we take the relative velocities of light in air and water to have the ratio 4:3, the actual law is better represented by

$$\varphi = \frac{3}{4}\theta - \frac{1}{60,000}\theta^3 .$$

Still, Ptolemy was only off by 10% with the first coefficient, and his procedure (whatever it was) did show that the second term needs to be subtracted.

Ptolemy also wrote a treatise on astronomy under the title *Mathematike Syntaxis* (Mathematical Treatise), better known by its hybrid Greek–Arabic name *Almagest*, based on the hypothesis that the motions of the Sun, Moon, and planets can be described by superpositions of uniform circular motions known as epicycles, that is, circles turning on other circles. In that form, as Sternberg has pointed out [**77**], this hypothesis can be described in modern terms by saying that the coordinates of the planets are almost-periodic functions of time. That hypothesis remains true, even when the simple periodic elliptic orbits of Newtonian mechanics are replaced by their relativistic counterparts.

ca. 984: Ibn Sahl (ca. 940–1000) writes *On Burning Mirrors and Lenses*, which contains the law of refraction that we now call *Snell's law*: *The ratio $\sin\theta : \sin\varphi$ is the same for all angles of incidence θ*. Judging from the figure he drew, it seems that he connected refraction with a difference in the speed of light in the two media, measured by the ratio of distances traveled in equal times. This law was rediscovered seven centuries later in Europe and apparently stated by three people independently. It was first remarked on by Thomas Harriot (1560–1621) in 1602, then by Willebrord Snel van Royen (1580–1626) in 1621. Neither of them published it, however. It was finally published in 1637 by René Descartes (1596–1650), in his treatise *Dioptrics* (refraction). Descartes' argument was less than convincing; it was disputed in particular by Fermat.

ca. 1330: Scholars Thomas Bradwardine (1295–1349), William Heytesbury (1313–1372), and Richard Swineshead (dates uncertain) at Merton College, Oxford make the first systematic study of motion with variable velocity, considering the case of constant acceleration a. They produce the Merton rule, which asserts that the total distance covered over a given interval of time equals the distance covered by an object moving at a constant speed equal to the instantaneous speed at the midpoint of the time interval. In our terms, if the speed v is directly proportional to the time t traveled, say $v = at$, then the distance s is given by $s = (at/2)t = (a/2)t^2$. This work had important consequences for both mathematics and physics.

ca. 1375: Nicole d'Oresme (1323–1382), Bishop of Lisieux, writes the *Tractatus de configurationibus qualitatum et motuum* (Treatise on the shapes of qualities and motions), where "qualities" include speeds. In this treatise, he also studies non-uniform motion, arriving at the Merton rule, and illustrating it graphically, taking one axis to represent time and the other to represent the speed, so that, in our terms, the graph of the speed as a function of time, is a straight line through the origin of slope a. The distance d traveled up to any given time t (represented as the point on the horizontal axis at distance t from the origin) is then represented by the area between the line and the time axis. This treatise and a subsequent one, *Tractatus de latitudinibus formarum* (Treatise on the latitudes of forms), where "latitudes" are the ordinates (y-coordinates) of the "forms" (functions), anticipated the analytic geometry of Descartes and Fermat 250 years later.

1543: The Polish scholar Mikołaj Kopernik (Nicolaus Copernicus, 1573–1543) publishes *De revolutionibus orbium cœlestium* (On the revolutions of the celestial spheres), a copy of which was allegedly shown to him on his deathbed. In it, he proposes a heliocentric astronomy regarded as superior to the geocentric system of Ptolemy. This event is sometimes spoken of as the beginning of the Scientific Revolution, but of course it took account of a great deal of medieval work.

1609: After analyzing voluminous observational data collected by the Danish astronomer Tycho Brahe (1546–1601), Johannes Kepler (1571–1630) writes *Astronomia nova* (The new astronomy), in which he states what is now known as Kepler's second law, phrased by him as "The total sum of the distances is to the time of a full period as any part of the sum of the distances is to its time." In our terms, the "sum of the distances" is the area swept out by the radius vector from the Sun to a planet. The idea of regarding an area as a sum of lines was developed more fully by Bonaventura Cavalieri (1598–1647) in a treatise entitled *Geometria indivisibilibus continuorum nova quadam ratione promota* (Geometry advanced by a new type of reasoning using the indivisible parts of continua), which, however, was not published until 1635, by which time Kepler was already dead. Kepler's second law asserts that the line from the Sun to a planet sweeps out area at a constant rate, and this fact is now a pillar of mechanics known as conservation of angular momentum. In the same treatise, he published Kepler's first law, which he stated baldly after a rather difficult geometric argument, saying, "Therefore, the path of the planet is an ellipse." Kepler was the person who first used the word *focus* (Latin for fireplace) in relation to conic sections. He did so because the Sun was located at one of the foci of the elliptical orbit. This hypothesis, supplanting the circles of the Copernican theory, was an even greater simplification in describing the motions of the planets than heliocentrism had been compared with the geocentric system. Nine years later, in his *Harmonice mundi* (The harmony of the world, where "world" means the solar system), he published his third law, saying "The ratio of the periodic times of any two planets is precisely the sesquialteral ratio of the average distances...". That is, if two planets have periodic times T_1 and T_2, and their orbits are ellipses of semimajor axes a_1 and a_2, then $T_1/T_2 = a_1^{3/2}/a_2^{3/2}$. This last law was crucial to Newton in confirming the inverse-square law of gravitation.

1610: Galileo (1564–1642) publishes *Siderius nuncius* (The starry messenger), reporting his discovery of the four "Galilean" moons of Jupiter (Io, Callisto, Europa, Ganymede) and confirms that they are orbiting Jupiter, since they are periodically eclipsed by the planet. This was the first proven example of celestial bodies orbiting another celestial body and gave confirmation to the heliocentric system. It also provided, as Galileo noted, a potential "clock in the sky," since in principle it ought to be possible to tell the time by examining the configuration of the four moons relative to Jupiter. Such a clock was needed for navigation, since it is impossible to find one's longitude accurately without an accurate clock synchronized with solar time at a fixed location on Earth. The imprecision of such measurements made this proposed clock in the sky impractical, but the attempt by Ole Roemer to use it (see below) yielded a very important fact about the universe.

1638: Galileo writes the *Discorsi e dimostrazioni matematiche, intorno a due nuove scienze* (Dialogues and mathematical proofs related to two new sciences), the new sciences being Kepler's new astronomy and his own new mechanics. In it, he described the motion of falling bodies as uniform accelerations and derived again the Merton rule. In this treatise he also formulated the law of inertia (Newton's first law). In the same year, he described an experiment to measure the speed of light by having two people with shuttered lanterns send signals back and forth. Of course, the physical apparatus is nowhere near accurate enough to measure the speed in question, and the possibility that light is propagated with infinite speed remained open.

1644: Descartes publishes the *Principles of Philosophy*, in which he states that the total "quantity of motion" in the universe is constant (conservation of momentum), and that natural motion unaffected by any force, is in a straight line at constant speed (Newton's first law).

1657: In a letter to the physician-philosopher Marin Cureau de la Chambre (1594–1669), Pierre de Fermat (1603–1665) offers the opinion that refraction ought to be explained by a minimal principle. Speaking of the "resistance" of a medium to light rather than the speed of light passing through it—he didn't wish to challenge the view that light travels with infinite speed—he said that a light ray passing from point C in one medium

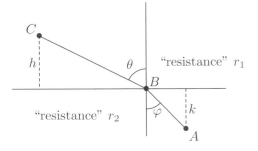

FIGURE 8.1. Fermat's principle in its original form: A ray of light that experiences "resistances" r_1 and r_2 respectively in two media with a planar interface will travel from point C in one medium to point A in the other, crossing the interface at a point B such that $CB + r\, BA$ is minimized, where $r = r_2/r_1$, that is, the total "effort" $r_1\, CB + r_2\, BA$ required to traverse the path is minimized.

to point A in another will cross a planar interface between the two media at a point B such that the sum $CB + rBA$ is minimized, r being the ratio of the two resistances. His line of reasoning was rather mystical, based on a faith that, as Fermat said, "Nature does nothing in vain." That is, Nature doesn't waste effort. Even though he allowed that light could be propagated with infinite speed, he thought Nature required some "effort" to do so, an effort that was directly proportional to the distance traveled multiplied by the resistance and hence within a medium of constant "resistance" directly proportional to the length of the path traversed. One can easily see that if effort is replaced by time, the mathematics of Fermat's reasoning remains the same, and now the ratio of distance to effort is replaced by the speed of light. Fermat noted, as Heron had done, that in a single medium where reflection takes place, minimizing time or effort results in the equality of the angles of incidence and reflection. He suggested applying the same reasoning to the case of refraction at the interface between two media. That leads to the law that ibn Sahl discovered. In his letter, which can be found in the work by Ross ([**69**], pp. 51–55), he considers only the case $r = 1/2$, which is a difficult enough problem. (See Fig. 8.1.) Since in that figure $CB = h \sec\theta$ and $CA = k \sec\varphi$, where h and k are the perpendicular distances from C and A respectively to the planar interface, Fermat is requiring that $h \sec\theta + (k/2) \sec\varphi$ be minimized subject to the constraint that $h \tan\theta + k \tan\varphi$ is constant. As calculus was not yet fully developed—Fermat was himself one of its pioneers, and was an expert at finding minima—that problem was, as he noted, a very difficult one. But he seems to have been the first to state the principle now known as Fermat's principle, as a minimization problem. Of course, from the practical point of view, he had no way to compute the ratio of the two "resistances" in advance. It could be computed from a table of known angles of incidence and refraction, but all that could have been verified empirically is that the ratio, which is the ratio of the sines of the two angles, is the same for all angles of incidence.

1676: In an effort to implement Galileo's "clock in the sky" by keeping careful records of the times when Jupiter eclipses its moon Io, the Danish astronomer Ole Roemer (1644–1710) notices that these eclipses tend to get closer together as Jupiter approaches opposition to the Sun and farther apart as Jupiter moves toward conjunction. The times when the eclipses begin fluctuate with an amplitude of about ±11 minutes from a regularly-spaced set of points in time, so that the deviation from most advanced to most retarded onset of an eclipse is about 22 minutes. Now Jupiter's conjunctions and oppositions are not intrinsic to the planet itself, but depend only on where the Earth is relative to Jupiter and the Sun. At opposition, the Earth is directly between the Sun and Jupiter; at conjunction, the Sun is directly between the Earth and Jupiter. That meant that there was an obvious explanation for the deviation of eclipse times. At conjunction, the light from Jupiter had farther to travel in order to reach the telescope than it had at opposition, the difference of the two distances being the diameter of the Earth's orbit. It followed that light was traveling at a *finite* speed, despite what Descartes and earlier scientists had believed. The Dutch scientist Christiaan Huygens (1629–1695) used Roemer's data and the current best estimate of the diameter of the Earth's orbit to conclude that the speed of light must be around 131,000 miles per second.

1687: Isaac Newton (1643–1727) publishes *Philosophiæ naturalis principia mathematica* (Mathematical principles of natural philosophy), a systematic treatise synthesizing all the mechanics and astronomy that had gone before. In order to complete his proofs, Newton had to resort in some cases to infinitesimal methods, which he called the *method of first and last ratios* (differential calculus). Although many of its parts were discovered piecemeal by the aforementioned Cavalieri and by certain French mathematicians of the first half of the seventeenth century, including Fermat and Descartes, along with Blaise Pascal (1623–1665) and Gilles Personne de Roberval (1602–1675), it was primarily Newton and Gottfried Wilhelm von Leibniz (1646–1711) who organized calculus into the comprehensive, systematic mathematical theory that we now have. Newton no doubt did this a little earlier, but Leibniz did it much more elegantly, with a differential and integral notation that is still used today.

1696: Johann Bernoulli (1667–1748) uses Leibniz' journal *Acta eruditorum* to propose the first modern problem that we now recognize as a variational problem, that is, a problem whose solution is a curve rather than a point or a number. The only earlier problem of the same type is the isoperimetric problem of finding the curve of fixed length enclosing the largest area; a restricted formulation of that problem had been solved by the Athenian geometer Zenodorus (ca. 200–ca. 140 BCE) and his solution of it was reported by the commentator Pappus of Alexandria (probably fourth century CE). The problem posed by Bernoulli is known as the *brachistochrone* (shortest-time) problem. It asks the shape of a curve down which a frictionless particle will slide under the influence of gravity in least time from a point A to a point B not directly below A. Bernoulli himself solved the problem using calculus and imagining that gravity is tantamount to an index of refraction proportional to the square root of the distance fallen. That gave a relation between the slope of the tangent line to the curve at each instant and the vertical distance fallen at that instant—in other words, a differential equation for the path of a ray of light. He argued from Fermat's principle that a particle moving as a ray of light would move must travel from A to B in minimal time. Here we have the first extension of a minimal principle from optics to mechanics. As optics is a very geometric subject, and the index of refraction allows the geometry to adopt a rich variety of metrics, by the year 1700, the seeds were planted for the geometrization of mechanics and astronomy that was to be complete 250 years later.

2.2. Two centuries of consolidation and discovery. The Continental mathematicians and physicists extended Newton's basic work and reformulated it into a much more sophisticated and elegant system. A key part of this reshaping was the concept of energy and the law of conservation of energy. These concepts were not named in connection with mechanics, however. A concept that is in effect kinetic energy, the product of mass and the square of velocity mv^2 was introduced by Leibniz under the name *vis viva* (living force),[3] and this concept was extensively used throughout the eighteenth century, as we shall see below. It was the development of thermodynamics during the early nineteenth century that really brought the concepts of work and kinetic, potential, and thermal energy into focus, along with the word *energy* itself. That part of the history lies outside our direct narrative.

Throughout the two centuries that followed the development of calculus, geometers were applying this new mathematics to very general surfaces in three-dimensional space, and physicists were finding new worlds to conquer in the phenomena associated with elasticity, electromagnetism, heat conduction, acoustics, and many other areas. We shall list the works on geometry first, through 1855, then separately list the advances in physics, starting in 1851.

> **1744:** Leonhard Euler (1707–1783) publishes *Methodus inveniendi lineas curvas* (A method of finding curves [**25**]), which contained the fundamental equation of the calculus of variations, the natural analog of setting the derivative (or differential) equal to zero. The procedure for finding the value of x at which a numerical-valued function $y = f(x)$ achieves a maximum or minimum is to set its differential $dy = f'(x)\,dx$ equal to zero, that is, to solve the numerical equation $f'(x) = 0$. The procedure for finding a function $y(x)$ such that an integral $I = \int F(x, y, y')\,dx$ has a minimum or maximum is to set its variation δI equal to zero. This analog of the numerical equation $f'(x) = 0$ is the second-order differential

---

[3]In a note published in the March 1686 *Acta eruditorum*, Leibniz gave what he called a short demonstration of an error made by "Descartes and others" (meaning, of course, Newton). These people, he said, misuse the law of conservation of quantity of motion. It was, he said, mv^2, not mv, that was conserved. Of course, *both* are conserved in elastic collisions, but Leibniz wanted the amount of work put into an object to be the same as what could be got out of it. In that respect, he was off by a factor of 2, since if you lift a mass m to height h, the work you have done is mgh. If you then drop it, it will reach ground level at speed $v = \sqrt{2gh}$, so that $mv^2 = 2mgh$. That is why we now write kinetic energy as $mv^2/2$. What Leibniz, Fermat, Descartes, and Newton all agreed on was that conservation laws were important.

equation known as the *Euler equation*:

$$\delta I = \frac{d}{dx}\left(\frac{\partial F}{\partial y'}\right) - \frac{\partial F}{\partial y} = 0 \,.$$

Immediate applications of this principle, provided one could solve the corresponding Euler equation, were in isoperimetric problems and finding geodesics on surfaces in $\mathbb{R}^3$, as well as the least-time paths followed by light rays. In what may be seen as an anticipation of the coming geometrization of mechanics, Euler used his method to prove that a particle moving on a surface and subject to no tangential acceleration would trace a geodesic at constant linear speed. Like Fermat, Euler had a strong faith in the efficient design of the universe. As he wrote in this treatise (p. 245), "Since the material of the universe is the most perfect and proceeds from a supremely wise Creator, nothing at all is found in the world that does not illustrate some maximal or minimal principle."

1755: Joseph-Louis (Giuseppe-Lodovico) Lagrange (1736–1811) writes to Euler introducing the δI notation, which he calls the *variation*, analogous to Leibniz' dy. Euler thereafter referred to the subject as "calculus of variations." Lagrange concentrated on algebraizing physics, believing that algebra provided protection against slipshod reasoning. A generation later, he publishes *Méchanique analitique* (Analytical mechanics [**48**]), in which he replaces Newton's $F = ma$ by a very general expression

$$\frac{d}{dt}\left(\frac{\partial T}{\partial \dot{q}_r}\right) - \frac{\partial T}{\partial q_r} = \frac{\partial U}{\partial q_r} \,,$$

where T is the kinetic energy of the system whose position coordinates are q_r, $\dot{q}_r$ is the (traditional) notation for the time derivative, and U is a generalized potential energy. (The modern concept of a potential function was introduced by George Green (1793–1841) in 1827.) The Lagrangian, nowadays the difference of kinetic and potential energy, has the property that the Euler equations for an extremal of its time integral yield the equations of Newton's fundamental second law of motion.

1767: Euler publishes "Recherches sur la courbure des surfaces" (Investigations on the curvature of surfaces [**26**]), a paper he had actually written in 1760, in which he introduces the concept of curvature of a surface. He observed that there were two planes passing through a given point of the surface for which the *linear* curvature of the curves of intersection with the surface were maximal and minimal. He defined the *curvature of the surface* at that point as the product of those extreme curvatures. A decade later he introduced the parametric description of a surface, the basis of the later work of Gauss and a prefiguration of the concept of a manifold.

1825: While engaged in supervising the mapping of the region around Göttingen, Germany, Gauss considers the problems of conformal mapping and writes *Disquisitiones generales circa superficies curvas* (General reflections on curved surfaces), in which he introduces the metric coefficients that determine the length of an infinitesimal piece of a curve on a surface. He then proves that a concept of curvature equivalent to the one introduced by Euler is found by projecting a small area around a point to the unit sphere at the origin, taking the ratio of the area of the projection to the area on the surface, and letting the area on the surface then shrink to a point. By a rather arduous route, this approach led him to an explicit formula for curvature in terms of the metric coefficients and their first and second partial derivatives in terms of the parameters. (He did not verify that this number was independent of the parameterization, a crucial point in modern mathematics.) Knowing that the sum of the angles of a triangle is two right angles only if Euclid's parallel postulate holds, he sent crews up three mountains with lanterns, instructing each of them to measure the angle subtended by the other two. As happened with Galileo's attempt to measure the speed of light, however, the instruments and method used were not sensitive enough to make the measurement: He found that the sum was two right angles within the limits of probable error. Any curvature of physical space that *might* exist was too small to be detected at that scale.

1826: Nikolai Ivanovich Lobachevskii (1792–1856) develops "imaginary geometry," which we now call hyperbolic geometry. The basic parts of this subject had been developed in the eighteenth and early nineteenth century by a number of scholars, some of whom thought they were proving the parallel postulate. The first of them in Europe was Girolamo Saccheri (1667–1733), who developed the subject very far and allowed his strict rigor to lapse in a very small point, just enough to motivate him to publish *Euclides ab omne nævo vindicatus* (Euclid acquitted of every blemish). Another who worked on the problem was Johann Heinrich Lambert (1728–1777), who made the prescient remark that what we call hyperbolic plane geometry had the trigonometry of a sphere of imaginary radius. (Actually, what we now call inaccurately *Saccheri quadrilaterals* and *Lambert quadrilaterals* had been studied centuries earlier by yet another mathematician trying to establish the parallel postulate: Thabit ibn-Qurra (836–901).) Gauss himself realized early on that it was impossible to prove the parallel postulate, but he published nothing on the subject. In 1816, however, he wrote to his student Christian Ludwig Gerling (1788–1864) suggesting that if it were true, it would be sensible to take as a universal unit of distance the side of an equilateral triangle whose angles were extremely close to 60°. Two years later, he was surprised to receive from Gerling a paper the latter had received from a Marburg lawyer named Karl Schweikart (1780–1859), developing what Schweikart called *astral geometry* and was in essence hyperbolic geometry.

The ideas of hyperbolic geometry were slowly being recognized, but Lobachevskii deserves the credit for being the first to publish on the subject, albeit only in the very obscure *Proceedings of the Kazan' Physico-mathematical Society*. Five years later, in 1831, János Bólyai (1802–1860) published an excessively condensed version of the theory as an appendix to a book written by his father Farkas Bólyai (1775–1856), a former classmate of Gauss. Both Lobachevskii and János Bólyai developed the trigonometry of the hyperbolic plane, confirming what Lambert had said. It took some time for these ideas to gain acceptance, and mathematical cranks continued to dispute them, far into the twentieth century. (See the

1931 book *Euclid or Einstein?*, by Rev. Jeremiah J. Callahan (1878–1969), president of Duquesne University.)

1854: Georg Friedrich Bernhard Riemann (1826–1866) gives his inaugural lecture at the University of Göttingen, with the aged Gauss in the audience. The lecture, which was published in 1867, the year after Riemann's death, bears the title "Ueber die Hypothesen welche der Geometrie zu Grunde liegen" (On the assumptions that underlie geometry [**68**]). In it, he took Gauss' metric coefficients on a surface, which were always those induced by the ambient space $\mathbb{R}^3$ in which the surface was embedded, and replaced them with arbitrary coefficients on an object parameterized by a set of n variables, which he called an n-fold extended quantity. His metric is the one we still use today: $ds = \sqrt{\Sigma a_{ij}(x^1, \ldots, x^n)\, dx^i\, dx^j}$. This lecture was made before a general audience, and the mathematical details were omitted, to be filled in later. Riemann assumed that the quantity under the radical was positive, that is, the metric was defined by a positive-definite matrix (a_{ij}). Such a manifold is nowadays called Riemannian. Riemann had many friends in Italy, where he frequently went for his health—he died (of tuberculosis) and is buried at Selasca, Italy—and it was mainly the Italian geometers Eugenio Beltrami (1835–1900), Gregorio Ricci-Curbastro (1853–1925), Tullio Levi-Civita (1873–1941), and Luigi Bianchi (1873–1928) who developed the multilinear algebra needed for manifolds. In 1901 Ricci-Curbastro and Levi-Civita published the paper "Méthodes de calcul différentiel absolu et leurs applications" (Absolute differential calculus methods and their applications [**67**]). This paper contained the basic ideas of tensor calculus. Beltrami is noted for producing the pseudo-sphere model of hyperbolic geometry and for another model that can be pictured on a Euclidean disk. In an 1868 paper on spaces of constant curvature, he introduced the Laplace–Beltrami operator discussed in Chapter 6 above.

Among the many profound contributions that Riemann made to physics, one that is mentioned less often than his work in geometry and complex analysis is a paper he sent to the Royal Society of Göttingen in 1858, containing, as he said, "a remark that brings the theory of electricity and magnetism into a close connection with the theory of light and heat radiation. I have found that the electrodynamic effects of galvanic currents can be understood by assuming that the effect of one quantity of electricity on others is not instantaneous but propagates to them with a velocity that is constant (equal to that of light within observational error)." This remark was stunningly accurate, and anticipated by three years Maxwell's realization that electromagnetic radiation is propagated at the speed of light. Riemann did not have Maxwell's laws at his disposal, and it was Maxwell who drew the converse conclusion that light is in fact electromagnetic radiation.

The mathematical background of relativity now being laid out, we turn to the physics behind it. By 1850, Newton's proposal that light consists of a stream of particles had been out of favor for some time, as the interference experiments of Thomas Young (1773–1829) showed that it behaved like a wave. The wave theory had been pioneered by Huygens, who provided the important principle that each point of the "wave surface" acted as the source of a new wave. Complicated phenomena such as double refraction, seen in crystals like Iceland spar, were considered by Huygens, who finally solved the problem by picturing the wave surface as a sphere internally tangential to an ellipsoid of revolution. Later, in 1816, Augustin-Jean Fresnel (1788–1827) considered a more complicated type of double refraction, finding the wave surface to be a fourth-degree polynomial equation in spatial variables x, y, and z (actually, it is quadratic in x^2, y^2, and z^2). In parallel with this work on optics, the theory of electricity and magnetism was developing rapidly, and in 1846, Gauss and his collaborator Wilhelm Weber (1804–1891) produced a formula now known as Weber's law giving the mutual force of repulsion between two like charges in relative motion. That formula involved a velocity that, in 1855, Weber determined to be 4.3945×10^{10} cm/sec. The following year, Gustave Kirchhoff (1824–1887) remarked that Weber's velocity was apparently the velocity of light multiplied by[4] $\sqrt{2}$.

1851: Armand Hippolyte Louis Fizeau (1819–1896) publishes "Sur les hypothèses relatives à l'éther lumineux," (On the hypotheses concerning the ether [**29**]), describing an experiment in which he caused light to pass through flowing water to determine if its speed is affected by the motion of the medium in which it propagates. He found that its velocity did appear to increase when the medium moved in the direction of propagation, but that the increase was not as much as would be predicted if it were an elastic disturbance in the medium itself. Einstein later cited this result as an important "retrodiction" of his special theory of relativity.

1861: James Clerk Maxwell (1831–1879) writes to William Thomson (Lord Kelvin, 1824–1907) to announce his computation of the speed with which a self-sustaining electromagnetic wave must propagate, given the known values of dielectric permittivity ε and magnetic permeability μ. (The velocity is $1/\sqrt{\varepsilon\mu}$.) The known values of ε and μ gave Maxwell a value of 193,088 miles per second, close enough to the speed of light to be "more than a coincidence." He wrote that the magnetic and luminiferous media must be the same and that "Weber's number is really, as it appears to be, one-half of the velocity of light in millimeters per second." (According to Siegel ([**75**], p. 139), Weber's system of measurements differed from Maxwell's by a factor of $\sqrt{2}$. Thus when Maxwell did the computation, he came out with a different value for the constant. I am grateful to Adrian Rice for pointing out this reference.) He thus duplicated unknowingly what Riemann had concluded three years earlier.

1887: Edward Williams Morley (1838–1923) and Albert Abraham Michelson (1852–1931) publish the paper "On the relative motion of the Earth and the luminiferous ether" [**59**], describing an experiment they conducted at Case Western Reserve University in Cleveland, Ohio. This was the the famous Michelson–Morley experiment designed to detect the motion of the Earth through the presumed medium (ether) that was the conductor of light waves. The experiment showed no evidence of any such motion.

---

[4]We commented in Chapter 3 on the variety of units used in electrical computations. Maxwell, apparently repeating Weber's computation in his own units, arrived at a value equal to one-half of the velocity of light, as will be seen below.

1889: George Francis FitzGerald (1851–1901) professor of physics at Trinity College in Dublin, writes a letter to *Science* [**28**], suggesting a way of accounting for the negative results of the Michelson–Morley experiment. His idea was to assume that an object in motion is contracted along the line of its motion in just the amount now accounted for by the theory of special relativity. His description of this idea was rather laconic, the whole of it being contained in just one sentence: "I would suggest that almost the only hypothesis that can reconcile this opposition is that the length of material bodies changes according as they are moving through the ether or across it, by an amount depending on the square of the ratio of their velocities to that of light."

1892: Hendrik Antoon Lorentz (1853–1928) professor of physics at the University of Leiden, publishes "La Théorie electromagnétique de Maxwell et son application aux corps mouvants" (Maxwell's electromagnetic theory and its application to moving bodies [**54**]). The subject of the paper was "electrons" (a name suggested by FitzGerald) but it contained a conjecture that the length of a moving body decreases along its line of motion, by exactly the same amount that FitzGerald had conjectured. Lorentz worked out the details of what is now called the Lorentz transformation. He is also one of the creators of the Lorentz law giving the force on a charged particle moving in electric and magnetic fields in the form $F = q(E + (1/c)v \times B)$, where q is the charge, v its velocity relative to the observer, E the electric field intensity, and B the magnetic field intensity. This law had been stated in 1889 and should probably be called the Heaviside law, after its creator Oliver Heaviside (1850–1925). When the force is interpreted as in Newtonian mechanics, this law means that two observers in uniform relative motion will agree on the total force F (as any two observers in uniform relative motion must, given Newton's second law) and also on the magnetic field, even though the velocity would be different for them. Consequently, they had to disagree about the value of the electric field intensity. This was a peculiar asymmetry in the laws of electromagnetism, and one that cried out for explanation. When the force is interpreted as in special relativity, however, the observers disagree symmetrically about both fields; moreover, the transformation between the two sets of fields makes it possible to reduce Maxwell's four laws to just two independent ones, the two divergence equations. (See Chapter 3 for details.) That was a striking theoretical triumph of relativity theory.

1895: From the mid-1880s Lorentz had been interested in the problem of the motion of Earth through the hypothetical ether and had published several papers on it. One very important one, in 1895, was "Versuch einer Theorie der electrischen und optischen Erscheinungen in bewegten Körpern" (An attempt to construct a theory of electrical and optic phenomena in moving bodies [**55**]), which contained a sort of "first draft" of the Lorentz transformation. In it, he introduced what he called "local time," what we now call "proper time." This concept deeply impressed Henri Poincaré, who engaged in correspondence with Lorentz.

1905: Henri Poincaré (1854–1912) publishes "Sur la dynamique de l'électron" (On the dynamics of the electron [**64**]), writing that "the main point proved by Lorentz, is that the equations of the electromagnetic field are invariant under a certain transformation, which I shall call the Lorentz transformation. This transformation has the form $x' = k\ell(x + \varepsilon t)$, $t' = k\ell(t + \varepsilon x)$, $y' = \ell y$, $z' = \ell z$, where $k = 1/\sqrt{1 - \varepsilon^2}$." Poincaré of course was "editing" what Lorentz had done. He often did that, generously ascribing his own contributions to the original authors who had given him the idea. (This practice had led to some friction with Felix Klein (1849–1925)—soon resolved, however—when Poincaré gave the name *Fuchsian* to some groups that Klein had studied but Fuchs hadn't. Poincaré later gave the name *Kleinian* to some groups with which Klein had a rather tenuous connection.)

2.3. The development of relativity. After these preliminaries, we can give a short synopsis of the development of the theory of relativity proper in the first half of the twentieth century. The list that follows is only a minuscule portion of the literature in general relativity and is confined to the special topics that were discussed in the previous chapters of the book.

1905: Albert Einstein (1879–1955) publishes "Zur Elektrodynamik bewegter Körper" (On the electrodynamics of moving bodies [**17**]), in which he discusses the whole problem of the speed of light and the asymmetries in the laws of electrodynamics. He derives the Lorentz transformation and gives the relativistic formula for addition of velocities and the relativistic conversion of Maxwell's laws.

1909: Hermann Minkowski (1864–1909) publishes "Raum und Zeit" (Space and time [**60**]), in which he formalizes what Einstein had done in the special theory of relativity, introducing four-vectors, the term *proper time* (*Eigenzeit*), and the "mystical formula" $3 \cdot 10^5 \, \text{km} = \sqrt{-1} \, \text{sec}$.

1911: The Dutch physicist Willem de Sitter (1872–1934) publishes "On the bearing of the principle of relativity on gravitational astronomy" [**10**], which contains an exploration of the application of the theory of relativity (as it existed at the time) to gravitational theory.

1913: Vladimir Varićak (1865–1942) publishes "Anwendung der Lobatschefskijschen Geometrie in der Relativtheorie," (Application of Lobachevskian geometry in the theory of relativity [**82**]), a paper he had written three years earlier, pointing out that velocities in special relativity can be vector-added to form triangles whose trigonometry is that of the hyperbolic plane of curvature $-1/c^2$. This idea attracted some attention after it was rediscovered by the distinguished French mathematician Emile Borel (1871–1956). (See Chapter 1 and Appendix 1 for details.)

1913: Einstein and Marcel Grossman (1878–1936) publish a paper bearing the title "Entwurf einer verallgemeinerten Relativitätsheorie und einer Theorie der Gravitation," (Draft of a generalized theory of relativity and gravitation [**24**]), the first part of which, the mathematical part, was Grossman's and the second part, the physical part, Einstein's. At this point, they had not yet succeeded in getting a law of gravity that was invariant across observers; some frames of reference were still privileged over others.

1915: Einstein publishes four papers on general relativity, the third of which was "Erklärung der Perihelbewegung des Merkur aus der allgemeinen Relativitätstheorie," (Explanation of the movement of the perihelion of Mercury from the general theory of relativity [**19**]) and the last of which was "Feldgleichungen der Gravitation," (Gravitational field equations [**20**]). This last paper contained the final form of these field equations.

1916: Karl Schwarzschild (1873–1916) publishes two papers on general relativity, the first of which was "Über das Gravitationsfeld eines Massenpunktes nach der Einsteinschen Theorie," (On the gravitational field of a point mass according to the Einstein theory [**74**]) and contained a more complicated version of what we called the Schwarzschild solution in Chapter 4. The second was "Über das Gravitationsfeld einer Kugel aus incompressibler Flüssigkeit," (On the gravitational field of a ball of incompressible fluid [**73**]), and it considered the more general case of a continuous distribution of matter.

1916: Einstein publishes an extended, systematic exposition of relativity as "Die Grundlage der allgemeinen Relativitätstheorie," (Foundations of the general theory of relativity, *Annalen der Physik*, Vierte Folge, **49**, 769–822).

1916: David Hilbert (1862–1943) gives a course in foundations of physics, published in 1924 as *Grundlagen der Physik* ([**40**], [**41**]), a series of papers in *Mathematische Annalen*. In these papers, he attempted to axiomatize physics (an activity he was fond of) and gave a thorough discussion of the dynamics of a particle or light ray falling into a black hole.

1916: De Sitter publishes "Einstein's theory of gravitation and its astronomical consequences" [**11**], a thorough analysis of the free-space gravitational equations of general relativity.

1922: Alexander Friedmann (1888–1925), publishes "Über die Krümmung des Raumes" (On the curvature of space [**30**]), which was exactly what its title indicated.

1923: Elie Cartan (1869–1951) publishes "Sur les variétés à connexion affine et la théorie de la relativité généralisée (première partie)" (On manifolds with an affine connection and the general theory of relativity (part 1)" [**6**]).

1924: Cartan publishes an extension of the previous work [**7**].

1924: Friedmann publishes "Über die Möglichkeit einer Welt mit konstanter negativer Krümmung des Raumes" (On the possibility of a universe in which space has constant negative curvature [**31**]), which explores the possibility that the geometry of space could be hyperbolic rather than elliptic.

1927: Georges Lemaître (1894–1966) publishes "Un univers homogène de masse constante et de rayon croissant rendant compte de la vitesse radiale des nébuleuses extra-galactiques" (A homogeneous universe of constant mass and increasing radius taking account of the radial velocity of extra-galactic nebulae [**52**]), anticipating the work of Hubble, whose name is attached to the idea of an expanding universe.

1929: Edwin Hubble (1889–1953), apparently unaware of the earlier work of Lemaître, publishes "A relation between distance and radial velocity among extra-galactic nebulae" [**45**].

1949: Kurt Gödel (1906–1978) publishes "An example of a new type of cosmological solution of Einstein's field equations of gravitation" [**34**], in which he describes a (barely) conceivable distribution of matter and a cosmological constant that lead to a universe in which time travel is possible, as described above in Section 3 of Chapter 7.

3. Space and Time

Space is not an empirical concept derived from an external perception. For the mental picture of space must already be present as background in order for certain perceptions to be deduced from anything external to me... Accordingly, the mental picture of space cannot be derived from the relations of external appearance through perception. Rather, this external experience itself is only possible in the first place by means of a mental picture. Space is a necessary a priori *mental picture underlying all other intuitions. One can never picture space absent, although one can perfectly well believe that no objects are to be found in it. Space is thus to be viewed as the precondition for the possibility of phenomena rather than as a condition dependent on them. Space is an* a priori *mental picture that necessarily underlies external phenomena. The logically certain knowledge of all geometric propositions and the* a priori *possibility of constructing them is based on this* a priori *necessity.*

Time is not an empirical concept that can be deduced from perception. For simultaneity or sequentiality would not form part of a perception unless the mental picture of time lay at its base. It is only under the assumption that time exists that one can imagine that two things happen at the same time (simultaneously) or at different times (sequentially). Time is a necessary mental picture underlying all other intuitions. One cannot by any means eliminate time itself from the perception of phenomena, although one can perfectly well remove the phenomena from time. Time is thus given a priori. *It is only in time that all the reality of phenomena is possible. The latter can be removed, but the universal precondition for their possibility cannot.*

Immanuel Kant (1724–1804), *Critique of Pure Reason.*

Throughout the early period of my life almost all my serious working time was devoted to mathematics. I supposed in those days that I was more interested in the application of mathematics to the explanation of natural phenomena than in pure mathematics for its own sake. This emphasis changed with time, and it was the purest of pure mathematics that finally claimed me. This change had various causes but one of the most important was a desire to refute Kant, whose theory of space and time as a priori *intuitions seemed to me horrid.*

Bertrand Russell ([**71**], p. 42).

These quotations from Kant mark the summit of influence of Euclidean geometry on science and philosophy. They came at the end of a long process of development of human thought, beginning with the earliest purely practical needs of commerce and surveying, becoming "philosophized" under the influence of Plato and Aristotle, and culminating in the abstract concept of absolute space and time used by Newton and analyzed by Kant. As the contrasting quote from Russell shows, Kant's view of absolute space and time fell into disfavor during the nineteenth and twentieth centuries and was generally rejected. We are now going to make a survey of this rise and fall with the aim of providing a wider view of the theory that was expounded in detail in the previous chapters.

The attempt to understand the physical world is one of the oldest intellectual pursuits of the human race, and mathematics has played an important role in it from the beginning. The crux of the matter is precisely what is meant by "understanding" the physical world. A modest amount of reading through the records of several millennia of metaphysical speculation has convinced this author that the important issue in understanding the universe is the question, "How shall we talk about the physical world?" Philosophers have debated endlessly such questions as the difference between appearance and reality, essence and existence, and the like. These questions have produced some interesting literature, but very little that arises in the conversations of generally educated people in daily life. Because these mysteries have been with the human race for so many centuries, the literature about them is vast, much more than anyone could possibly read in a decade. In view of that fact, the reader will perhaps indulge one whose experience is confined to the short span of a single lifetime, and moreover a life not devoted to metaphysical questions, for speculating on the meaning of all this human effort from the limited perspective provided by that experience. The point I hope to make is that modern mathematics, including set theory, provides a model of human knowledge in general, one that allows certain unknowable entities to remain unknowable in their essence, but at the same time allows us to say that we know certain propositions in which they are mentioned. Even though we abandon as fruitless the effort to know (experience) what these entities are, however, we each must have some mental picture of them to think about, and that need justifies the attempts of philosophers to express their own picture of them in words. The transition from thought to language is by no means easy or unimportant.

The basic concepts that have both geometric and mechanical significance are mass, space (distance, area, volume), and time. These are the materials from which mechanical theories are constructed. By combining these materials, physics provides us with a variety of ways of imagining the world. By examining how this process occurs and the ways people have thought about these concepts over the centuries, we may arrive at a useful guide to thinking about physical laws. Of the three abstractions just mentioned, mass has altered very little from earliest times. The Greeks spoke of *bodies* rather than mass, but, as Archimedes' work *On Floating Bodies* demonstrates, they had the concept of density, from which the concept of mass arises via the equation mass = density × volume. Our perception of mass or weight is tactile rather than visual, and hence geometry enters into its analysis indirectly, in such laws as Archimedes' law of the lever. Moreover, the concepts of mass and density did not change at all from the earliest times until the advent of special relativity (see Chapter 2). For that reason, we shall confine our discussion mostly to the concepts of time and space, which are both inherently geometrical. After we look at some of what philosophers have said on these subjects, we can examine the larger stock of objects that constitute modern physical theory, such as gravitational, electric, and magnetic fields and compare them with now-obsolete entities such as phlogiston and ether to see the extent to which we are justified in regarding them as real.

Before we begin, we make one disclaimer: The sections that follow are aimed at non-specialists, people who just want a way of thinking about informally described physical theories in a way that does not ignore certain obvious conceptual difficulties that would arise if they had to write an essay explaining what they know about the subject. We do not presume to tell professional physicists or philosophers how they ought to think about their subjects.

3.1. Measurement. Mass, length, area, and volume are the most intuitive of the mathematical abstractions that play a role in physics. Standard weights and lengths have been around as long as human civilization, and were crucial, along with elementary geometry, for allocating land and trading in the produce of the land. The basic geometric problem in surveying, for example, was to assign a number (area) to a plot of land. Since plots of land come in different shapes, it was necessary to have some means of assigning this number that would be intuitively fair. That made the problem of comparing the *sizes* of plots of different *shapes* an important one. This problem shows how uniform human intuition is across many cultures. The same shapes (triangles, rectangles, circles) occur repeatedly in mathematical texts from China, India, Mesopotamia, and Egypt. The presence of rectangles among them, with the constant use of the formula area = width × length, shows that human intuition is essentially Euclidean. (Rectangles do not exist in non-Euclidean geometry.)

This "numerical" geometry, which focused on the size of geometric figures, became "philosophized" in one civilization, that of ancient Greece. The result was an emphasis on the figure itself, its shape, and the proportions between the parts of different figures, rather than mere quantity (length, area, volume, or weight). Philosophers such as Aristotle and mathematical physicists such as Archimedes spoke of particular shapes rather than space and of bodies rather than matter or mass. Geometry, in addition to acquiring a logical structure, became the study of proportion.

But the older, practical approach, in which quantity was the main issue (the amount of land, or the weight or volume of grain, and the like) persisted and reappeared in mathematical treatises that have survived from 2000 years ago. The mathematicians Zenodorus (second century BCE) and Heron (first century CE) wrote quantitatively about *area* ($\dot{\epsilon}\mu\beta\alpha\delta\acute{o}\nu$), where Archimedes had written about a *surface* ($\dot{\epsilon}\pi\iota\varphi\acute{\alpha}\nu\epsilon\iota\alpha$). Heron gave the area of a figure as an absolute number, whereas Archimedes spoke only of the ratio of two surfaces. He said, for example, that a sphere is four times as large as its equatorial disk. In Heron's language and ours this theorem says that the *area* of the surface of a sphere—thought of as a *number* of square units—is four times the area of its equatorial disk. Greek geometry, influenced by Platonism, represents a very beautiful, but ultimately self-limiting, departure from the quantitative numerical geometry that came both before and after it. Modern geometry takes something from both of these, but adds the crucial element of symbolic algebra. We are not confined, as Heron was, to computing particular areas in order to convey our ideas through examples; we can say simply and generally that the area of a sphere of radius r is $4\pi r^2$, which is much more efficient. Modern geometry also mixes in the methods of calculus to produce the differential geometry that has made general relativity possible.

This may be a good place to mention the property of *continuity* that we ascribe to all three of the basic mechanical concepts.[5] Being continuous, they present certain challenges to those who would measure them. There are some interesting philosophical subtleties involved in the progression from *counting* discrete collections of things, which is not problematic at all, to *measuring* continuous things. The latter process involves choosing a unit of measurement for each type of thing measured (kilogram, meter, second), and it always involves approximations and usually the use of fractions. Units of measurement of continuous quantities are pure conventions, which differ from one time and place to another. Most people would consider themselves hard-used if they had to deal with lengths given in old Russian *arshins* or ancient Greek *parasangs*, or even furlongs. Counting, in contrast, is universal. "Five sheep" will translate perfectly into its equivalent in any other language. In the higher-level abstraction of Greek mathematics, it becomes provable that it is not possible to choose, for example, a unit of length that will exactly measure both the side and diagonal of a square. But we pass over all that and take the measurement of continuous quantities as a given. Measurement is an important element of science and is involved in the both of two philosophical questions we are going to discuss: (1) What exists? (What is real?) (2) How do two observers interpret each other's language so as to know when they are talking about the same object and whether they are in agreement as to what they are saying about it?

3.2. The views of Immanuel Kant. One consequence of the Greek abstraction in geometry was the emergence of a still higher-order abstraction embracing all the already-abstract geometric figures. That is the concept of *space*, in which these shapes "live." The concept was fully formed by modern times, as Newton's discussion of it (quoted in Chapter 1) attests. All modern languages have a word for it. For a thorough discussion of the evolution of spatial concepts, see the monograph by Jeremy Gray [**36**]. As mentioned above, Euclidean geometry reached its high point in the seventeenth and eighteenth centuries, as shown by the works of people such as Newton and Kant. We shall accordingly use their words, and the words of their critics, as a springboard for launching our own discussion.

As the epigrams quoted at the beginning of this section show, the philosopher Immanuel Kant laid down some cogent ideas on the subject of time, space, and our knowledge of the external world, in his *Critique of Pure Reason*. While Kant's views have clearly visible defects from the modern point of view, they were stated with great eloquence, and deserve to be taken seriously. Kant distinguished between two kinds of propositions. *Analytic* propositions are those that are true or false "by definition," such as the statement that every uncle has either a niece or a nephew (true) and that herbivores live by eating field mice (false). *Synthetic*

---

[5]We are going to ignore the contradictory fact that some modern physical theories actually regard all three as being "made up" of discrete "atoms." We have enough problems to deal with already, and can't afford to make things yet more complicated.

propositions are those in which the truth of the relation asserted between the subject and the predicate is not determined by the definitions of those two things.

Examples are the statement that Julius Caesar was murdered (true) and that Genghis Khan invaded Russia in the nineteenth century (false). Crossing elegantly with the analytic-synthetic distinction was Kant's distinction between two kinds of knowledge. *A priori* (literally "from before") knowledge is knowledge for which the logical grounds are innate to the human mind and independent of any experience. We not only know it, we also know how to prove it logically. All analytic statements that people know to be true belong to this class. *A posteriori* (literally "from afterward") knowledge is derived from perception or experience; examples are the facts established by experimental science and history. All *a posteriori* knowledge is synthetic. From those principles, the question naturally arises: Are *analytic* and *a priori* propositions co-extensive, so that, by inference, *synthetic* and *a posteriori* propositions are also? Those who can imagine the appropriate Venn diagram will see that there remains one doubtful area: Is there such a thing as synthetic *a priori* knowledge? Kant believed that there was, and he found examples of it in mathematics, at least the arithmetic and geometry of his day. (He had nothing to say about algebra, even though it had undergone an amazing development in the seventeenth century.)

In the case of arithmetic, he thought that the equality $5 + 7 = 12$ (his example) was an assertion that is *a priori*, that is, we are born knowing that it must be true, yet synthetic in that the concepts of 5, 7, and addition exist independently of the number 12. Nowadays, this claim appears doubtful. Given just the empty set $\varnothing$ and the successor relation that expands a set E by placing the set $\{E\}$ among its elements, resulting in the set E^+, one can construct the ordinal numbers, define the natural numbers and addition, and prove that $5 + 7 = 12$. We define $1 = \varnothing^+ = \{\varnothing\}$, $2 = 1^+ = \{\varnothing, \{\varnothing\}\}$, $3 = 2^+ = \{\varnothing, \{\varnothing\}, \{\{\varnothing, \{\varnothing\}\}\}\}$, and so on. Thus, mathematics appears to clear up this matter from a purely logical point of view. We need the undefined terms that come from set theory, and these must be left undefined. On that basis, we can actually prove the propositions of elementary arithmetic. We then have logical *grounds* for our belief, with an irreducible residue of undefined terms that we have to live with. The logical problem is solved, to the extent that it can be. This logical solution, however, does not account for psychology. The *cause* of our belief that $5 + 7 = 12$ is not the very non-intuitive definition we have indicated here for the positive integers. This fact is one that we learned in school. But how did it get into the arithmetic curriculum? That question shows that there is still some virtue in Kant's approach: How did it happen that people the world over discovered the same fact, namely that $5 + 7 = 12$? The discovery was surely not made in accordance with the proof of its correctness; it involved, as Kant correctly said, intuition. This logical/psychological seesaw will be seen in all the examples we discuss below. Our knowledge can be asserted with confidence except for certain undefined terms; but physics requires a physical interpretation of the undefined terms. We cannot help attempting to express our intuitive picture of those terms in language, a task that remains as difficult as ever. We get around the difficulty, in the end, by positing physical laws that express numerical relations among measurable variables. The variables and their numerical values suffice for theoretical physics, and we relegate the question of the intrinsic nature of the basic concepts to the subject of metaphysics.

In geometry, likewise, Kant thought that the proposition that there exists a triangle with sides of lengths x, y, and z whenever the sum of the smaller two exceeds the largest of these three quantities was *a priori*: Everybody "just knows" that it is true, and this knowledge has not come from examining physical triangles. It is synthetic *a priori* knowledge. In stating his belief that mathematical concepts (along with certain other metaphysical concepts such as causation) are synthetic *a priori* knowledge, Kant took it for granted that we "just know" the parallel postulate, which is to say, we know that geometry must be Euclidean. By the time of his death in 1804, however, mathematicians had already shown that we do not "know" this at all. It has become commonplace to claim that non-Euclidean geometry discredits this neat Kantian scheme. Gauss, who discovered hyperbolic geometry, thought so, and most mathematicians accepted the judgment of Gauss. This opinion was eloquently stated by the physicist Ludwig Boltzmann, who was disturbed by Kant's attempts to limit the use of entities that go beyond experience and took rather an evolutionary view of human thought.[6] Still, as the modern philosopher Philip Kitcher has pointed out to me, it is in the nature of synthetic propositions to admit logically conceivable alternatives: If they didn't, they would be analytic rather than synthetic. The validity of Kant's view is not refuted by the construction of a logically consistent non-Euclidean geometry.

Kant regarded space and time as mental pictures—innate knowledge—underlying all perception rather than as physical objects, and this point of view seems useful: We attach coordinates to physical space by referring to physical bodies; the coordinates themselves are not "things" in the same physical sense as the bodies that occupy portions of space. Some of them are thought of as points occupied by matter or energy; others are not, but are pure abstractions. The view seems sound to me. It is nevertheless vulnerable to attack, as the quotation above from Bertrand Russell and the ones below from Ernst Mach attest.

Kant's claim that we can imagine pure empty space without any bodies in it raises a number of questions. Is this empty space "made up" of points? If so, how is one point of it to be distinguished from any other point? If not, how can a physical body occupy only part of a thing that has no parts? Most people find it difficult or impossible to imagine empty space without some physical boundary. As architects know very well, space isn't fully appreciated as space unless it is bounded by something physical. To those questions, one supposes, Kant might reply that he only said space was a form of intuition and wasn't talking about space as understood by physicists and astronomers. But if Kant's space is divorced from the physical applications of geometry, of what value is it? The distances between geographical points and celestial bodies have been measured using geometry and are known. If these bodies are not located in the space Kant has in mind, then what is added to our knowledge by the statement that space is a form of intuition? It seems clear that he did mean his space to be physical space, since he spoke of it containing bodies. But, after all, we perceive the bodies directly. There is a relation between two physical bodies, possibly changing over time, called the distance between them. It does not depend on an abstraction called space. The distance between two

---

[6]Boltzmann replaces Kant's *a priori* with the notion of *innate knowledge*, which is not quite the same thing. Perhaps the difference is that innate knowledge is not necessarily accompanied by knowledge of the grounds for it. (In the case of synthetic propositions, those grounds are intuitive rather than logical.)

bodies is a fact that we can know, even though we cannot know what "space" in the abstract is. As we shall see in more detail below, what we know arises from measurement. The bodies and their mutual distances at different times are the "stuff" of mechanics. What more do we need? How does it help our understanding to talk about abstract space? I think the help is psychological, and psychological needs are not to be despised.

Kant's claim that we cannot imagine space being absent, although we can imagine that there are no bodies in it, appears to be both logically absurd and at the same time psychologically necessary. To model the situation, let us accept that we use a three-dimensional system of orthogonal coordinates to talk about the points of space. Mathematically, that is fine; these are pure ideas, and there is no need to connect them with physical reality. Yet do want to connect them with reality; we need (or imagine we need) the concept of physical space in order to discuss mechanics. In order to interpret our coordinates in physical terms, we need at least four noncoplanar bodies whose mutual distances are constant to set up an origin and three coordinate axes. That much of space has to be occupied by something physical, as Mach was later to point out. The rest of the points, apparently, do not need to be occupied by anything physical. We can go to a physical location whose coordinates are (x, y, z) and verify that these are our coordinates by measuring the distances to our four reference points. But given a point that is not occupied by anything physical and where no event ever occurs, what *is* that point? How can it be described except by giving its three real-number coordinates? We can't help picturing it in our minds as a thing, but that thing is not physical.

Kant believed that this intuitive idea of space was logically necessary for the development of geometry, that propositions could not be stated without invoking it at least implicitly. In that, modern geometry contradicts him. In Hilbert's axiomatization of geometry, it is not necessary to mention "space," only to take as starting points certain terms left undefined, such as points, lines, and planes and to make certain assumptions, classified as axioms of incidence, axioms of order, axioms of congruence, and the axiom of continuity. The subject is then a purely verbal creation that can be applied just as well as ever in physics, but does not *logically* require any intuitive interpretation such as Kant believed necessary. That being said, Hilbert's axiomatic version of geometry is rather a cold, bloodless thing, lacking the poetic beauty of Euclid's original work. Those who love geometry will probably always prefer Euclid's and Kant's intuitive visual version of it. As a last defense of Kant, we could still argue that his "form of intuition" is a *psychological* necessity for thinking about geometry. Hilbert, after all, did not formulate his axiom system in a vacuum. Any useful system of axioms has to be created with some interpretation in mind. And if human intuition leads people the world over to think the same way, it facilitates the conversation when people come to communicate with one another, even if it is superfluous from a purely logical point of view.

Even granting that Kant was wrong in his belief that space is Euclidean, we can still find some good in his philosophy if we accept that geometry is based on intuition and that everybody's intuition is Euclidean. The universality of the Euclidean version of the Pythagorean theorem shows that such is indeed the case. That we can go beyond our intuition and make effective use of non-Euclidean geometry is an inspiring statement about human minds, but our point of departure is still our Euclidean intuition. What is needed, it seems to me, is an amendment to Kant's philosophy: We need to refrain from using the word *knowledge* when speaking about synthetic *a priori* propositions, and we need to recognize that experience and perception are the main sources of intuition, which has been described as our stock of inherited prejudices.

The sources that *cause* us to believe in what Kant regarded as *a priori* geometric knowledge, can be seen through psychology. Psychologists have studied the development of geometric intuition in children and find that their understanding of (for example) the relative sizes of things is acquired gradually, though at a very early age. Setting aside the question of *causes* of our geometric beliefs, we suggest a modification in Kant's view of the *grounds* for them. The propositions that Kant laid down as *a priori knowledge* should rather be regarded as *working hypotheses*, useful starting points for theory, in which one has some degree of confidence, which some would call faith.[7] By accepting that modification, we come to recognize that our knowledge is tentative and incomplete, only as secure as the assumptions with which we began. We must abandon the absolute and follow where experiment and theory lead us.

There is no assurance that we will ever attain perfect knowledge, and indeed it seems most likely that we never shall. But even imperfect knowledge, subject to revision and/or rejection, nevertheless gives us the satisfaction of an ever-improving understanding of the world. It has been said that to travel hopefully is better than to arrive. If that aphorism appears to negate any reasonable motive for going anywhere, we can observe that at most points, there is a *direction* that appears to lead to something better—a *proximate* goal. One can travel (hopefully) in that direction, without necessarily expecting to arrive and without needing any *ultimate* goal, while always—it must be admitted—running the risk of winding up hopelessly lost. I consider this view to be an optimistic one. It shows that, while we must begin with our inherited, prejudiced ways of looking at the physical world, we can, through logic and mathematics, revise our working hypotheses and get a new picture that works better.

3.3. Measuring time. Since this book, as its title shows, is to a large degree about time, we need to discuss the metaphysics of time, which is the most abstract of the three mechanical concepts, not perceived by vision or touch, and yet part of the consciousness of all people. Every language has a word for it, and the picture of it in the mind of nearly everyone has changed very little over the centuries. Almost universally, it is regarded as something that flows, and the metaphor of a river is very commonly used to describe it.[8] The reconciliation of time between observers in relative motion lies at the heart of the entire theory of relativity, where the

---

[7]Whether science requires faith to practice is a question that can be explored from both an epistemological and a psychological point of view. There is a difference between *entertaining a working hypothesis and exploring its consequences* and the absolute assertion of the truth of that hypothesis. One usually has a degree of confidence that the working hypothesis is approximately true; else no one would devote time and effort to exploring its consequences. Ideally, the degree of confidence should be tentative. That being said, it is well known that what science has rendered probable is often asserted as if it were a faith. One should have more confidence in the results of science than in the proclamations of the many oracles that are available, but not absolute confidence in it. The difference is that, when contrary evidence comes to light, it needs to be seriously considered, and if it cannot be refuted, then one's view should change.

[8]"Time is but the stream I go a-fishing in... Its thin current slides away, but eternity remains." – Henry Thoreau (1817–1863).

distinction between proper time and laboratory time is crucial. Poincaré thought that the invention of the concept of proper time by Lorentz (who called it *local time*) was a brilliant insight.

Given that we now have two varieties of the thing, we pose a natural question: What *is* time? We can use the word *time* in sentences that any person of normal intellect will find to be both meaningful and true. But does time "exist"? And how do we perceive it? Certainly, reality has temporal aspects, but is time a *thing*, in the same sense in which physical bodies are things? Even the measurement of time is a non-trivial task, much harder than measuring distance or mass. As a great philosopher of the late fourth and early fifth centuries, Augustine of Hippo, wrote, in a quotation that can be found at the website

www.gutenberg.org/files/3296/3296-h/3296-h.htm,

> *For what is time? Who can readily and briefly explain this? Who can even in thought comprehend it, so as to utter a word about it? But what in discourse do we mention more familiarly and knowingly, than time? And we understand when we speak of it; we understand also when we hear it spoken of by another. What then is time? If no one asks me, I know: if I wish to explain it to one that asketh, I know not: yet I say boldly that I know, that if nothing passed away, time past were not; and if nothing were coming, a time to come were not; and if nothing were, time present were not. Those two times then, past and to come, how are they, seeing the past now is not, and that to come is not yet? But the present, should it always be present, and never pass into time past, verily it should not be time, but eternity. If time present (if it is to be time) only cometh into existence, because it passeth into time past, how can we say that either this is, whose cause of being is, that it shall not be; so, namely, that we cannot truly say that time is, but because it is tending not to be?* [Translation by Edward Bouverie Pusey].

Everyone would say that time can be spoken of meaningfully. That is, the word "time" can be used as a noun in a sentence that will convey accurate information about the world. We can know many propositions that involve time—for example, that two events must either be simultaneous, or in a definite sequential order (since relativity we need to add the qualification "for each individual observer"), and if event B comes after event A and event C comes after event B, then event C comes after event A (again, for each individual observer)—but we do not know what Aristotle would have called the essence of time. That is precisely what Augustine is saying in this passage. It does not appear to add anything to our understanding when we say that time is "real" or that it "exists."

The second quotation from Immanuel Kant at beginning of this section represents an eighteenth-century attempt to come to grips with the mystery of time. The problem was stated by Augustine in the passage just quoted. Kant's solution to it, declaring time to be an *a priori* form of intuition, reflects the tension described by Augustine: on the one hand, our inability to know the *essence* of time and on the other hand, our ability to know *facts* about it. His claim that we can imagine time without any events is doubtful. To think of time, we need to think of things

3. SPACE AND TIME

happening in sequence; that is a psychological necessity. The objections to this claim of Kant's are the same objections that we made in connection with his views on space.

The point that time isn't a *thing* was made by the physicist Ernst Mach (1838–1916), who commented, after quoting Newton on absolute time (the same passage we quoted in Chapter 1):

> *It is utterly beyond our power to measure the changes of things by time. Quite the contrary, time is an abstraction, at which we arrive by means of the changes of things. A motion is termed uniform in which equal increments of space traversed correspond to equal increments of space traversed in some motion with which we form a comparison, as the rotation of the Earth. A motion may, with respect to another motion, be uniform. But the question whether a motion is in itself uniform, is senseless. With just as little justice, also, may we speak of an "absolute time," of a time independent of change. This absolute time can be measured by comparison with no motion; it has therefore neither a practical nor a scientific value, and no one is justified in saying that he knows anything about it. It is an idle metaphysical conception ([**58**], p. 209, my translation).*

In defense of Kant, we must point out the hidden assumption in the process Mach has described here. He says we determine whether motion is uniform by singling out a particular motion as a standard (like the motion of the hands on a clock) and comparing other motions with it. What he doesn't say is that the comparison requires us to note the positions in the two motions at the beginning and end of the comparison, and hence implicitly requires a concept of simultaneity between the two processes and sequentiality between the beginning and end of each of them. Kant would not have thought that Mach had refuted his view of time, and perhaps Mach didn't intend to. His aim was to point out that it is impossible to prove in any absolute sense that, for example, the time interval between 10:15 AM and 11:05 AM on a given day is *equal to* the time interval between 12:10 PM and 1:00 PM on any other given day. The assumption that the hands of a clock move uniformly (within limits) remains always an assumption.

Perhaps we might call that assumption synthetic *a priori* knowledge, since we are all thoroughly committed to it (bearing in mind that we now agree to interpret this knowledge as a working hypothesis). We are wise to make that commitment, and Mach has overstated his case if he means to cast doubt on it. In fact, any competent musician can detect the difference between a uniform tempo and a non-uniform one. True, if one were to analyze the brain activity of, say, Itzhak Perlman as he performs a solo, then the person doing the analysis would indeed be comparing two physical processes (the stressed beats in the music and the firing of certain neurons in Perlman's brain keeping time through an innate sense of rhythm) in order to determine that they were or were not uniform relative to each other, just as Mach stated. But for Perlman himself, there would be only one process going on, namely the music, and he would perceive its uniformity or non-uniformity directly. Galileo is said to have used his own sense of a uniform tempo to measure the distances traversed by a ball rolling downhill over equal time intervals. We don't have an absolutely precise sense of time passing, but we can subjectively estimate it with a small error, even in a dark silent room.

Moreover, the two time intervals just mentioned can be measured by a variety of clocks, all working on different physical principles: the unwinding of a coiled spring against a balance wheel, the oscillations of a pendulum, the dripping of sand through an hourglass, the vibrations of certain crystals, and the position of the stars. If the substantial agreement of the measurements by all these instruments does not correspond to some physically real time in which the measured intervals are congruent, we are faced with a major mystery. Surely the reasonable conclusion is that they actually do correspond to congruent time intervals. All the laws of mechanics and electromagnetism that involve time become incomprehensible on any other assumption. They need to be, as mathematicians say, invariant under a translation of the time axis.

3.4. Absolute space. Mach was equally scornful of Newton's view on absolute space, which he also quoted at length (again, the same passage we quoted in Chapter 1):

> It is scarcely necessary to remark that in the reflections here presented Newton has again acted contrary to his expressed intention only to investigate actual facts. No one is competent to predicate things about absolute space and absolute motion; they are pure objects of thought, pure mental constructs, that cannot be produced in experience. All our principles of mechanics are, as we have shown in detail, experimental knowledge concerning the relative positions and motions of bodies. Even in the provinces in which they are now recognized as valid, they could not be, and were not, admitted without previously being subjected to experimental tests. No one is warranted in extending these principles beyond the boundaries of experience. In fact, such an extension is meaningless, as no one possesses the requisite knowledge to make use of it ([**58**], pp. 213–214, my translation).

He then states the view that, when elaborated by Einstein, has come to be known as *Mach's principle*:

> When we say that a body K alters its direction and velocity solely through the influence of another body J, we have asserted a conception that is impossible to arrive at unless other bodies A, B, C ... are present with reference to which the motion of the body K has been measured. In reality, therefore, we are merely aware of a relation of the body K to A, B, C .... If now we suddenly neglect A, B, C ... and attempt to speak of the behavior of the body K in absolute space, we implicate ourselves in a twofold error. In the first place, we cannot know how K would act in the absence of A, B, C,...; and in the second place, every means would be wanting of forming a judgment of the behavior of K and of putting to the test what we had predicated, which latter therefore would be bereft of all scientific significance ([**58**], p. 214, my translation).

As now understood, *Mach's principle* replaces Newton's absolute space with the whole body of physical objects in the universe. This replacement goes beyond what Mach actually said, and one may doubt if he would agree with it. It gives rise to difficulties connected with the fact that different observers in relative motion do not agree about the relative locations of other physical objects. To make it work without assuming the absolute Newtonian time and space that Mach rejected, we would need to assume some value for the average density of mass in every direction for all observers, and we have no reliable way of estimating what that density is. In the last analysis, we do need mathematical models of things. Each observer in special relativity has an absolute Newtonian space and time, made real by a set of four coordinates. The space and time of one observer are reconciled with those of another observer via the Lorentz transformation. The reconciliation is between the measured intervals of time and space and expressed by the equations of the Lorentz transformation. Once again, we see that the physical knowledge involved is about the results of measurement. Underlying it, however, are two different Newtonian measurements of the events whose coordinates are being communicated back and forth. Mach's belief that individuals have no justification for picturing space and time as Newton did is simply wrong.

Even if the absolute time and space that Newton and Kant believed in do not correspond to anything observable, such models are often useful ways of thinking about the physical world. It would be very difficult to write about physics without making constant use of nouns such as force, time, magnetic induction, and so on, nouns that the reader's mind processes as if they were palpable objects. We should rather celebrate the fact that through algebra and calculus we have been able to produce more and more concepts of this type that modify the mental pictures of the universe that were once common. An excellent example is the replacement of time and space by space-time. One can argue that relativistic space-time is also an absolute geometric object, not all that different psychologically from Newton's absolute space and time. This point has been discussed by Lehmkuhl [**51**]. The individual picture each physicist has is important for that individual's understanding, even if it is not directly relevant to physics as a community enterprise. What is relevant in that enterprise is the communicable set of facts that come from measurement of these hypothetical structures. For all those reasons, we declare that Mach was wrong to assert that absolute space is "bereft of all scientific significance." It may not appear in a formal treatise on physics, but it almost certainly plays a role in the thought of individual scientists.

4. The Reality of Physical Concepts

Given space, time, and mass as primitive concepts, physicists define more complex objects, such as velocity, acceleration, momentum, force, angular momentum, energy, and others. As the mathematization of physics progressed, it threw up higher and higher-level abstract entities, not directly perceivable, but pictured as real objects, just as one thinks about rocks and trees: light, gravity, electric and magnetic fields, protons, quarks, and the like. All of these concepts can be named and discussed meaningfully, but no one directly and immediately *perceives*, for example, a radio wave. A radio wave cannot be seen, heard, smelled, tasted, or felt. In the mind of any individual physicist, there is probably a mental picture of each of them. That is all there need be until that physicist attempts to communicate with other physicists.

The conversion of a mental picture into a physical principle stated in human language gives rise to the two problems that we are attempting to analyze in the present chapter. To take the first of the two problems, the meaning and/or reality of intuitive concepts (such as gravitational fields) posited in order to explain observations, the accepted procedure in mathematics is to begin with undefined terms. Concepts such as gravity appear to be exactly that. Yet they are not quite that, since physicists need to interpret their primitive concepts and mathematicians do not. When mathematicians create an axiomatic system, the undefined terms may remain forever undefined. The minute one tries to apply the system to the real world, however, the undefined terms have to be given an interpretation. In physics, by way of contrast, interpretation is always needed. That is where intuition enters the story. Just as mathematicians can deduce theorems from their axioms and undefined terms, physicists can produce testable predictions by positing physical laws (usually mathematical in nature) involving such concepts as gravity and magnetic fields. The interpretation, as we have been saying, comes from measurement and involves quantitative relations satisfied by the hypothetical objects. These relations, if confirmed by observation and experiment, become knowledge. The underlying issues of what objects the primitive terms refer to, what they "essentially are" and the extent to which we can know them, belong to metaphysics. The measurements are facts, and they are the subject matter of physics proper.

Let us take an example from everyday life. Radio waves are important in physics and engineering, and who would venture to deny that they "exist"? By reasoning about them as if they were palpable objects, mathematical physicists constructed theories in which they played a role, implying that certain events that can be observed would occur.[9] Radio waves cannot be observed directly, but the assumption that they exist and have certain properties leads to a correct explanation of what can be observed. Because we are so familiar with radio broadcasts, we have no difficulty picturing a radio wave as being like a wave in water. But in fact, the radio waves of theory are electromagnetic waves, periodic variations in the intensity of mathematical objects called electric and magnetic fields that are coupled in a precise way in accordance with Maxwell's laws. Their existence is of a very rarified type, even though we picture them mentally as being just like things we do observe.

There are other mathematical objects of importance in physics that seem even less like the physical objects we can observe directly, and it is difficult to picture

---

[9] In the case of radio waves, the theory explained a phenomenon that had been noticed earlier: When an electrical circuit is opened or closed, a spark appears in the gap of a nearby broken metal ring that is not part of the circuit. Heinrich Hertz (1857–1894) tried the experiment in 1887 and found that it worked. In the mid-1890s, Guglielmo Marconi (1874–1937) applied this principle to create a practical wireless telegraphy, starting with home experiments in 1894. Since the Italian government refused to support him and in some quarters regarded him as a lunatic, he finally moved to England, where he found support. A radio receiver had been invented earlier, at Tufts University in 1882, by Amos Dolbear (1837–1910), who obtained a patent for it in 1886, a patent that Marconi was eventually forced to buy. In fairness to the Russians, it should be noted that Aleksandr Stepanovich Popov (1859–1905) demonstrated a radio receiver in a paper delivered in May 1895. He doesn't get credit for it in the West because he was a patriotic Russian who refused to license it abroad, even though the Russian government refused to support him. In that respect, his experience resembled that of Marconi.

them as "things." A good example of such an object is energy, particularly potential energy. If you carry a stone to the top of a high hill, you will have done work on it and thereby increased its potential energy. But no physical examination of the stone itself would ever reveal any mysterious component part of it that could be called its potential energy. That is because potential energy is a mathematical relation determined by the position of the stone relative to a gravitational field. Energy isn't a "thing," and yet certain forms of energy can be detected and *measured*. We do so constantly, every time our electric meter is read. The reading tells us how many kilowatt hours of energy we have consumed. We say that potential energy can be exchanged for kinetic energy, and kinetic energy (of falling water, for example) can be used to produce electrical energy. The question thus arises, to what extent is potential energy real, and how should we picture it?

We have now brought out on stage a variety of concepts from classical (mass, length, area, volume, time) to modern (energy, gravitation, magnetism, electricity) that we wish to examine from a philosophical point of view. These concepts provide us with enough examples to inquire about the relation between the physical world and our mathematical models of it.

4.1. Knowledge of the physical world*. The present section and the one that follows wander into areas that are even more abstract and remote from everyday experience than the physics we have been discussing. They are included in order to provide the broadest possible framework for the more concrete topics that are the core of the present book. In particular, we are going to discuss what the words *exist* and *real* mean, and what it means to *know* what exists and what is real.

Philosophers have long attempted to arrive at an absolutely true and accurate description of the physical world, and they have done so by inventing abstract concepts, just as physicists do. Plato of Athens (ca. 425–347 BCE) is the earliest philosopher who wrote systematic treatises on the subject that are still extant. He may have been led to formulate his metaphysics in an attempt to reconcile two principles proclaimed by the philosophers Heraclitus (fl. ca. 500 BCE) of Ephesus and Parmenides (ca. 515–460 BCE) of Elea. From the fragments of quotations of these philosophers that have survived, it appears that Heraclitus argued through many examples that there is nothing permanent in the world: all is in flux. Parmenides, on the other hand, argued that there can be no true knowledge of anything that changes. (One may suppose that he reasoned as follows: Since knowledge of an object must be phrased as a sentence with a subject, if the meaning of the subject changes, the sentence can no longer be held to be true, and hence does not represent knowledge.) Plato's search for the absolute that would be the exception to Heraclitus's world of flux and the foundation of a theory of knowledge satisfying the requirements of Parmenides while still being applicable to the observable world, led him to imagine a world of pure "forms," whose properties were unchanging, and which, when mixed together in the physical world that we inhabit, accounted for the phenomena that we actually do observe. He emphasized the importance of mathematics, which to him was "Euclidean" geometry—this was a century before Euclid lived—and number theory, apparently seeing in its subject matter (lines, points, numbers, circles, and the like) the kind of timeless pure entities that resembled the forms he believed were the fundamental elements of the physical world. In his later years, he seems to have realized the defects of this approach, but did not have enough time left on Earth to revise it and present an alternative approach to epistemology.

Plato was not the last natural philosopher to look to mathematics for absolute concepts that would give a full explanation of the physical world. His student Aristotle was much less mystical, more practical, and more scientific than Plato, but he too was looking for an absolute reality that would rise above the changing world we observe. Instead of Plato's timeless forms, Aristotle made what he regarded as an important distinction, that between *substance* and *accidents*, that is, between what we would regard as *things* and *properties* of things. This distinction is mirrored by the grammatical distinction between subjects and predicates: A substance can be the subject of many predicates (accidents). Some things can be both subjects and predicates; for example, the United States has 50 states that are members of it; and it has membership in the United Nations. In the statement "Iowa is one of the 50 United States" membership in the United States is a predicate of Iowa, whereas in the statement "The United States is a member of the United Nations," membership in the United Nations is a predicate of the United States. Thus the United States is both a subject and a predicate. At the rock-bottom level, however, some things, Aristotle thought, are pure predicates, for example beauty and virtue, while others (substances) are pure subjects, such as individual people (souls). The work we have done in Chapter 5 seems to fit this model, with a surface in $\mathbb{R}^3$ being an example of a subject, and the various parameters used to describe it playing the role of predicates. The distinction is hard to apply in most cases, because, while predicates appear to modify subjects, we could not articulate a predicate without using some information about the subjects it applies to.

Set theory mirrors this way of viewing the world, at least to the extent of denying the possibility of an infinite regression from predicates to subjects. The elements ("subjects") of a set ("predicate") may be other sets ("predicates"), but if one looks at their elements ("subjects"), which may also be sets ("predicates"), and continues "digging," at some finite stage, one arrives at the empty set. It has no elements, and so is not a predicate of anything at all. That is, it is always false to say that an element belongs to the empty set.[10] The empty set itself, however, does belong to other sets, and thus is, in Aristotelian terms, a pure subject, the only one in the entire universe of set theory. It has properties, but its elements do not, because they don't exist.

4.2. Reality*. Since one of the main topics of the present chapter is the question of what exists (is real), we note that Kant made another useful contribution to the language of philosophy in his *Critique*, in a context that is also relevant to the empty set. He pointed out that the verb *exist* is not a predicate. It is

---

[10]This regression terminates as described only because it is explicitly postulated in the standard axiomatization of set theory, in which an axiom known as the axiom of regularity is laid down. This axiom is one way of avoiding Russell's paradox. Formally, it says that for any set A there at least one element $a \in A$ such that $a \cap A = \varnothing$, that is, *as a set*, a is disjoint from the set A that it is an element of. This axiom is so non-intuitive that, to common sense, it thwarts the original purpose of set theory, which was to provide a clear and secure (non-contradictory) foundation for mathematics. This axiom is not clear to common sense. Mathematicians not concerned with foundations don't think about it much, if at all. They use set theory because it is a convenient common language of discourse. As for the security and non-contradictory nature of mathematics, even set theorists are sometimes reduced to saying *we have faith* that our axioms are consistent.

part of the syntactical structure of a sentence. This point becomes clear when we examine the two kinds of statements, universal and particular. If I say that all swans swim (a universal statement), that is a meaningful statement, and its negation is also meaningful: there exists a swan that does not swim. Exactly one of the two statements is true, but both are at least meaningful. In standard logic, the universal statement is by convention true if there are no swans, since a statement asserted about nothing makes no claim that can be falsified. Of course, such a statement is useless in any discussion of the physical world, since it conveys no information about anything real. In the other direction, if I say that some swans swim, that is, there exists a swan that swims (a particular assertion), I am making a claim that cannot be true unless there actually are swans. Once that verification is made, the negation of this statement is also meaningful, though easy to refute empirically: No swans swim. The verb *swim* can be used as a predicate in both kinds of statements.

If, in contrast, I say that all swans exist, which is *empirically* a true and trivial statement, I am talking logical nonsense, because the negation of that statement would be, "There exists a swan that does not exist." In this case, the positive statement remains true by convention if there are no swans, but says nothing about the physical world. The existence of swans has to be established empirically *before* one can know that this claim is meaningful; otherwise, it is a statement about the members of the empty set, and therefore, though true, devoid of significance. Likewise, the particular statement "Some swans exist," equivalent to "There exists a swan that exists," which seems to be a mere tautology, is a true statement about the world, but only because the existence of at least one swan has been empirically verified. The statement could in that case be shortened by eliminating the last two words. Attempts to get around these considerations by replacing the verb *exist* with the adjective *real* merely make things complicated without changing any of the essential principles.

The concept of *existence* is a predicate only in an indirect sense. If I say, "There exists an entity x having property P," I am making an assertion, not about hypothetical entities x having property P, but rather about the *set* $E = \{x : x \text{ has property } P\}$. I am asserting that E is non-empty. That fact reveals the fallacy in some traditional attempts to conjure things into existence by defining them as existing. As one example, take the classical "being whose non-existence is inconceivable." Consider the set $E = \{x : \text{the non-existence of } x \text{ is inconceivable}\}$. As the argument runs, the definition has a clear meaning; that is, we understand what it means to say that the non-existence of a thing is inconceivable. Thus the concept is meaningful. And since it is, such a being necessarily does exist.

Now, in the statement that such a being necessarily exists, the descriptive phrase *such a being* is equivalent to *a being that belongs to the set E*. In other words, the argument says that if a being belongs to E, then it exists. We may grant that that statement is true, but as a proof that some being *does* belong to E, this statement assumes what is to be proved. One would need to exhibit the being and prove it belonged to E before the argument would apply. And if the being were exhibited, there would no longer be any need for the verbiage involved in the definition of the set E in order to demonstrate that it exists. We have here the old familiar logical vicious circle. If the set E is in fact empty, then any assertions about its elements are true, but have no application in either the physical world or the world of pure thought.

We are in some very empyrean heights here, going beyond the boundaries of even theoretical physics. Fortunately, the entities whose reality or unreality is relevant to physics are not mere words. Proving that they exist is not a mere matter of word-spinning, as it is in the case just considered. Instead, we associate them with mathematical functions and interpret the values of those functions as predictions of the readings on various measuring instruments. In that way, we obtain an interpretation of the undefined terms that can be communicated from one person to another.

As a final remark, we add that whether an object "exists" in the mathematical sense is a purely formal and verbal matter. Certain axioms of set theory begin with the symbol $\exists$, and mathematical objects exist (in the mathematical sense) only in cases where a statement beginning with this symbol can be derived from those axioms. What those objects are therefore depends on the strictness with which axioms beginning with this symbol are chosen. One somewhat controversial axiom is the *axiom of choice*, which asserts that for any set $\mathfrak{A}$, all of whose elements are non-empty sets, there exists a function f whose domain is $\mathfrak{A}$ and such that $f(A) \in A$ for each $A \in \mathfrak{A}$. If this axiom is adopted, there exist (again, in the mathematical sense) sets of real numbers that are not measurable in the sense of Henri Lebesgue (1875–1941). Without this axiom, such sets cannot be proved to exist. Not all mathematicians do accept this axiom, and hence what does or does not exist in the world of mathematics is a party question. In any case, statements asserting existence in the mathematical sense do not say anything about the physical world. They dwell in the same empyrean heights as the philosophical arguments just discussed, providing models for a real-world interpretation of what are undefined terms in a purely formal and axiomatic version of physics.

We shall return to these esoteric questions after a digression to reflect on the role that mathematics plays both in physics and in answering the questions we have posed here.

5. The Harmony Between Mathematics and the Physical World

In the modern era, the absolute spheres that Ptolemy and other Greek and medieval astronomers believed made up the cosmos was displaced by the Copernican theory, and the cosmos came to be thought of as an infinite Euclidean space, although its infinity was a matter of dispute, since the human mind finds it equally difficult to grasp both an unbounded universe and a bounded one. Here again, mathematics eventually provided a model, in the form of a compact manifold without boundary, examples of which are the spheres originally believed in by the Greeks.

It was natural that space should be conceived of as Euclidean. There is something in the human mind that readily seizes on this simplest of all forms of geometry. We have already mentioned that the concept of a rectangle, which exists only in Euclidean geometry, is universal across cultures. Likewise, the Pythagorean theorem for a right triangle with legs x and y and hypotenuse z, expressed nowadays as the equation $x^2 + y^2 = z^2$, was discovered independently in many places between four thousand and two thousand years ago, including China, India, and Mesopotamia. But this is the Euclidean form of the theorem; its use implies that space is Euclidean.

On a sphere of radius r (elliptic geometry), the corresponding theorem takes the form $\cos(z/r) = \cos(x/r)\cos(y/r)$ for a right triangle having great-circle sides of lengths x, y, and z; but since a sphere can be isometrically embedded in three-dimensional Euclidean space,[11] this result was not seen as the two-dimensional version of the metric relations in a non-Euclidean three-dimensional space. Differential geometry, however, and a persistent tradition inherited from the Greeks of axiomatizing geometry, led mathematicians eventually to consider seriously the possibility of a more radically different kind of geometry—the hyperbolic plane—only part of which can be embedded isometrically in three-dimensional Euclidean space. In this geometry the Pythagorean theorem has the form $\cosh(z/r) = \cosh(x/r)\cosh(y/r)$, or, using complex numbers $\cos(z/ir) = \cos(x/ir)\cos(y/ir)$. Thus it is the geometry of a sphere of imaginary radius ir.

The Euclidean model appears to be perfectly suited to Newtonian mechanics. For example, suppose three observers A, B, and C are in relative motion, all being in the same place at one instant of time, and that A observes B moving at speed u and C moving at speed v, along straight lines making angle θ with each other (A is of course, located at the intersection of those two lines). If you want to know the relative speed w that B and C have—in Newtonian mechanics, A, B, and C will all agree on the value of that speed—and what angles each of them observes between the directions of motion of the other two observers—again, all three observers will agree about those angles—all you have to do is draw two lines whose lengths are proportional to u and v making an angle θ at a point P, then join the other ends Q and R of the two lines to form the third side w of a triangle and solve the resulting triangle PQR by the Euclidean law of cosines: $w^2 = u^2 + v^2 - 2uv\cos(\theta)$. That works perfectly in Newtonian mechanics.

In special relativity, however, as we have seen, if A wants to know the speed that B and C observe each other to have (which will not be the relative speed he observes them to have) and the angles that each finds between the other two observers, it will be necessary to use the hyperbolic law of cosines: $\cosh(W/k) = \cosh(U/k)\cosh(V/k) - \sinh(U/k)\sinh(V/k)\cos(\theta)$, where c is the speed of light, $k\sqrt{-1}$ is the radius of curvature of the hyperbolic space, and $U = k\operatorname{arctanh}(u/c)$, $V = k\operatorname{arctanh}(v/c)$, $W = k\operatorname{arctanh}(w/c)$. Wherever the two forms of mechanics disagree in a way that can be measured (only for a minority of observable phenomena), the relativistic model turns out to be the correct one.

The conclusion we draw from all of this is twofold: (1) mathematical concepts provide an uncannily accurate "map" of the physical world, and (2) that map, when it is most accurate, is at variance with our intuition. There is no surprise in this for those who have looked at the implications of the theory of evolution. Genes survive through a random selection process that optimizes a function of a large number of variables. Only one of those variables is precision in perceiving the physical world. An organism genetically equipped to deal with a very small curvature of space would represent an enormous investment of biological energy and would not be more fit than one that deals with the world through simpler approximations. Euclidean geometry may be only an approximation to the physical world; but it is

---

[11] Technically, the geometry of the sphere is called *doubly elliptic* since two lines (great circles) intersect in two points rather than one. If each pair of antipodal points is regarded as a single point, the geometry becomes singly elliptic and is the geometry of the projective plane, which cannot be embedded in three-dimensional Euclidean space.

simple, and using it requires far less energy—meaning energy invested in creating the nerve connections and reflexes of an animal—than would be the case if the organism were to deal with a hyperbolic world. The wonder is that the human mind is able to refine this intuition and replace it with a more complicated set of concepts that can deal with aspects of the physical world that our ancient ancestors did not have any need to think about. Moreover, the new mathematical creations of the human mind are sometimes stunningly in harmony with the physical world. That fact requires some separate commentary.

Since this book has been written from the point of view of a mathematician rather than that of a physicist, it aims at discovering why it is that there appears to be what Felix Klein called a "pre-established harmony" between certain areas of mathematics and certain physical problems. This harmony was also expressed by Eugene Wigner in his famous essay "The unreasonable effectiveness of mathematics" [**85**]. Why is it that vector analysis encodes so perfectly the laws of mechanics and electromagnetism? Why is the real part of the square of a quaternion equal to the space-time interval between two events in special relativity? Why do self-adjoint operators in Hilbert space represent in such wonderful detail the properties of "observables" in quantum mechanics? Why does the mathematical principle that the more concentrated a function is, the more dispersed its Fourier transform will be express precisely the Heisenberg uncertainty principle? In keeping with the general level of this chapter, we are going to provide a short list of observations that amount to nothing more than common sense. From them, at least a partial explanation may be derived.

(1) *Deliberate invention.* To paraphrase a remark of Voltaire, we might rather wonder why our ears and noses are so well adapted for supporting eyeglasses. The explanation in that case is obvious. A similar explanation will do for some areas of applied mathematics, where mathematically trained scientists deliberately set out to describe physical phenomena mathematically—in seventeenth-century mechanics (calculus), eighteenth-century theory of heat conduction (Fourier analysis), and nineteenth-century mechanics and electromagnetism (vector analysis). The descriptive mathematical model was, like eyeglasses, invented rather than discovered. Afterward, those inventions continued to be developed by pure mathematicians, and in most cases the latest results are totally divorced from applications. In other cases, the explanation is more indirect. The fact that the space-time metric of special relativity is the real part of the square of a quaternion can be made to seem natural. Special relativity was designed for the purpose of treating Maxwell's laws more efficiently, and those laws involve the vector curl and dot products, exactly the two operations that can be used to define quaternion operations. It may be a coincidence that quaternions seem to appear in relativity, but "it's a small world," where basic algebra is concerned.

The use of self-adjoint operators to represent position and momentum in quantum mechanics is another example of this small world. Mathematicians developed this theory because it was natural for people working in functional analysis to regard the derivative as a linear operator on an infinite-dimensional space of functions and to find some way of compensating for the fact that it wasn't a continuous operation. The compensation was the fact that it is self-adjoint, or rather, that the operator $f \mapsto -2\pi i f'$ is self-adjoint. But, as everyone knows, the time derivative of position is velocity, that is, momentum per unit mass. Likewise, the mapping $f(x) \mapsto xf(x)$ can be thought of as representing position if you take $f(x)$ to be the probability density function for the position of a particle. This mapping also is self-adjoint but not continuous. The "unreasonable" part of the effectiveness in this case occurs when the Fourier transform is invoked; applying the momentum operator to a function $f(x)$ corresponds to applying the position operator to its Fourier transform, and then, very surprisingly, the Heisenberg uncertainty principle falls out in a neat quantitative form.

(2) *Abstraction and oversimplification.* It sometimes happens that an elaborate mathematical structure turns out to be a ready-made fit to the needs of physics. The classical example is the conic sections, invented by the Greeks for the purest of reasons: to solve the problems of trisecting the angle and duplicating the cube. In the seventeenth century, Galileo found that projectiles near the earth move in parabolas, and Kepler found that planets move in ellipses. In general, as we know, Newtonian orbits in the two-body problem are conic sections. (See Chapter 4.) But that ready-made structure soon becomes inadequate even in oversimplified models such as the motion of a pendulum neglecting the weight of the pendulum bob, air resistance, friction, and elasticity considerations. An exact analysis under these assumptions still requires elliptic functions. When the importance of Laplace's equation was recognized, its solution in cylindrical and spherical coordinates led to a need for still more elaborate mathematical structures, which had to be invented (Bessel functions and spherical harmonics). As we saw in Chapter 4, the simplest relativistic model of the two-body problem also leads to elliptic functions. Similarly, even the simplest cases of rigid-body motion, such as those studied by Euler and Lagrange (see Appendix 5 for details) lead to a need for elliptic functions, and a more general case of rigid-body motion studied by Sonya Kovalevskaya requires hyperelliptic functions (integrals involving the square root of a polynomial of degree five). In summary, as Grattan-Guinness ([**35**]) points out, the effectiveness of mathematics in physics is not always unreasonable, but it is always limited. Of course, mathematicians are constantly developing theories far more complicated than those that fit physical models; most of them have, so far, not found any physical application.

Mathematicians from a large number of fields are busy producing aesthetically beautiful structures, and occasionally (rarely) those structures turn out to describe something useful to physicists. The tensor calculus is an outstanding example. Given the prevalence of differential equations in physics, the mathematical areas that have most prominently figured in this cooperative enterprise are those that grew out of algebra and calculus: real and complex analysis. The most powerful tool in the analyst's arsenal, arguably, is the class of theta functions. Through them, theoretically, one can get exact expressions for the integral of any algebraic function. There are, it turns out, just enough independent algebraic integrals for the problem of a rotating rigid body to make the cases studied by Euler, Lagrange, and Kovalevskaya exactly solvable. But the general case of this problem suffers from a dearth of algebraic integrals (see Appendix 5) and is not solvable in this way. An even more daunting challenge is the very natural three-body problem of Newtonian mechanics. It has only ten algebraic integrals, whereas eighteen would be needed to express the solutions in terms of theta functions. For that reason, the problem is more often studied geometrically, one of the most outstanding such studies having been made by Poincaré in the 1880s. In this problem, the simple conic sections are no longer adequate, and some very wild orbits, such as the Lissajous orbits,[12] become possible. Despite their fecundity, mathematicians have not yet produced a structure that will solve this problem.

(3) *Logical necessity.* We describe below the role played by symbolic algebra and analysis in physics. On that basis, we can give a partial explanation of the "unreasonable effectiveness" of mathematics. Pretending we are physicists, we idealize the physical universe—for example, perhaps replacing bodies by point masses—and then visualize quantitative relations such as the inverse-square law of gravitation. Using symbols to represent force, distance, and mass, we get a familiar algebraic relation, known as Newton's law of gravity. The important feature of mathematical relations is that they take the form of implications. From an assumed law, we can deduce consequences in the form of differential equations of motion. By solving those equations, we can further deduce the actual trajectories of bodies. These mathematical relations are not themselves "about" anything at all. If the premises hold for any interpretation of the symbols in them, then the conclusions must also hold in that interpretation. Essentially, we take a look at the universe, use it to set up a mathematical system, then see where the mathematical system leads. If it leads to something we can observe, we have been successful. *How* it works is easy to explain.

Yet, even when we see all the logical connections, the aesthetic beauty of the result is still awe-inspiring. To take the uncertainty principle as an example, a simple wave is expressed in complex form as $e^{2\pi i \omega t}$, where t is the time and ω the frequency in cycles per unit time. (The product ωt is thus physically dimensionless.) We can analyze a function of time $f(t)$ by creating its Fourier transform

$$\hat{f}(\omega) = \int_{-\infty}^{\infty} f(t)e^{-2\pi i \omega t} \, dt \, .$$

We can then synthesize the original function from the set of frequencies $\hat{f}(\omega)$ by the inverse Fourier transform

$$f(t) = \int_{-\infty}^{\infty} \hat{f}(\omega)e^{2\pi i \omega t} \, d\omega \, .$$

---

[12]Named for Jules-Antoine Lissajous (1822–1880).

As mentioned above, applying the position operator to $f(t)$ amounts to applying the momentum operator to $\hat{f}(\omega)$, and vice versa (at least for even functions). Simple, well-known inequalities for the standard deviations of random variables corresponding to density functions then imply that the product of the two standard deviations cannot be less than a certain positive minimum. This is the Heisenberg uncertainty principle. The whole picture can be understood, and yet even when it is grasped in its entirety, it remains mysterious that this logically necessary consequence of the behavior of abstract mathematical concepts describes a real physical constraint that restricts our ability to measure the position and momentum of a small particle. The miracle doesn't disappear when it is understood.

(4) *Coincidence and opportunism.* Only rarely does an already-existing mathematical object like the Fibonacci series, created as an amusing puzzle, turn out to have real-world applications such as in the phyllotaxis models of the brothers Bravais, Auguste (1811–1863) and Louis François (1801–1843). When this does happen, it can be described as the sudden realization of an opportunity that only a few inventive people are capable of seeing. These rare cases remain mysterious, if they are not merely lucky coincidences. We saw one such example in Chapter 1, the case of hyperbolic geometry. It was not inevitable that mathematicians would invent this form of geometry. Indeed, given that human intuition is essentially Euclidean, the possibility of such a geometry would have been unlikely to occur to anyone, except that the Greek invention of axiom systems called attention to the fact that one needed to assume something equivalent to the parallel postulate in order to prove theorems formally. Outside the culture of ancient Greece and its heirs (the medieval Muslim and Christian geometers), mathematicians solved problems but generally did not prove theorems. When hyperbolic geometry was invented, there was not the slightest reason to think it would ever apply to anything in physical science. And yet, as we saw in Chapter 1, it meshes perfectly with the Lorentz transformation to express the relative velocities of bodies in motion.

We have not space to analyze any more instances of this phenomenon. The interested reader might wish to investigate the case of RSA codes,[13] for which the necessary number theory was developed more than half a century before the codes were invented.

(5) *Physical structure of the brain.* We have noted that there is a strong uniformity in intuitive human thinking the world over, shown most clearly in the universal use of certain simple geometric shapes. No doubt, this uniformity is a product of evolution. Our minds are *to some extent*, shaped by the physical world. In that broad perspective, perhaps what our minds create in mathematics and mathematical physics can't help reflecting the physical world. Our brains, after all, certainly rely on chemical and electrical processes to do what they do. (While it cannot be said that there is nothing more to human thought than physical and chemical processes,

---

[13]Named after Ron Rivest, Adi Shamir, and Leonard Adleman.

what more there may be is not yet known and obviously would be unknowable through physics and chemistry.) At the very least, then, physical and
chemical processes play a role in shaping our thought. On that broad basis, a very vague connection can be made suggesting that those processes
might naturally cause us to create imaginary structures that resemble the
operation of the processes that produce our thinking. In other words,
the mathematical models we think up are the natural consequence of the
processes that produce the thinking.

That is all we intend to say on this subject. We now return to the epistemological problems we have posed, abandoning the incompletely solved problem to the
philosophers.

5.1. The role of algebra. Much of the answer to our two questions about
the origin and understanding of our knowledge of physics comes from the subject
of algebra, and glance at the gradual penetration of this subject into physics is
enlightening. Mathematical physics in a form that we can recognize began with
Archimedes' law of the lever. This law was stated geometrically and lacked the all-
important component provided by algebra. The translation of the concept of direct
proportion into the concept of a linear transformation had not yet been made. As
another example of the need for algebra, Ptolemy's table of angles of incidence and
refraction fits a simple quadratic model, and this is not quite right, though it is
a good approximation. Likewise, Heron's attempt to explain the principle of the
inclined plane geometrically doesn't work; the error in it was corrected by Jordanus
Nemorarius in the thirteenth century.

As is well known, a great revolution in science occurred in the seventeenth
century. In the early years of that century we find Galileo and Kepler still using
only Euclidean geometry and numerical observations. But a few decades later,
Descartes and Fermat took the radical step of incorporating the algebra that had
been developed largely in Italy in the sixteenth century and applying it to geometry.
That was a momentous step. Even more important, given that the sixteenth-
century Italian algebra was still being expounded in the geometric language of
Euclid, Descartes and Fermat adopted the simple symbolic notation of François
Viète (1540–1603)—Fermat completely, Descartes with modifications that made his
notation standard right down to the present day. The introduction of symbols to
represent unspecified or unknown numbers was, I believe, more important than just
algebraic *reasoning* about such numbers. Such symbols are called *variables*,[14] and
they are one of the significant qualitative differences between modern and ancient
mathematics. The symbols can be combined into formulas that are manipulated
according to a definite set of rules.

It is precisely here that symbolic algebra and its offspring the calculus become
a crucial part of the transition from unknowable primitive terms to knowable facts
about those primitive terms. Each primitive term—a magnetic field, for instance—
is labeled with a symbol representing a variable that depends on other variables,
usually time and space coordinates, which themselves are symbols for the elusive

---

[14]Modern computer science has kept the human imagination well grounded and forced mathematicians to be precise. When defining a variable, one does not think of it as representing any
specific quantity. Yet it is still necessary to specify what *type* of variable it is, which is to say, what
objects can be substituted for it in formulas: real numbers, integers, rational numbers, complex
numbers, matrices, and the like.

points of space-time that we have discussed above. Physical laws can then be stated as algebraic or differential equations relating these variables, such as the Lorentz law for the force on a charge moving in a magnetic field. If we mix in the other crucial element—that the values of both dependent and independent variables can be measured by processes that can be communicated from one person to another— we can convert these algebraic or differential equations into equalities between the numbers that result from the measurement, thereby replacing the equations with equalities between numbers. If those equalities hold, then the physical law is verified to some extent, and we may, if we wish, take that as an indication that the original primitive term denotes something real, even if it transcends our senses.

In Chapters 5 and 6 of the present book, we have seen the power of analytic geometry and calculus in the analysis of geometric figures that would have seemed impossibly complicated to the ancient Greeks. Euler used these techniques to solve a variety of problems in geometry and physics. In his hands, the use of parameters to represent curves and surfaces in $\mathbb{R}^3$ led to the results on curvature discussed at the beginning of Chapter 5. Gauss took the further step of showing how to compute the curvature of a surface directly from its metric coefficients, thus potentially eliminating any need for the absolute Euclidean space $\mathbb{R}^3$ in which the surface was embedded. That step was subsequently built upon by Riemann and a galaxy of brilliant Italian geometers to produce the tensor calculus and the general notion of a manifold. This was the mathematical theory presented to the world in more or less finished form at the end of the nineteenth century, just in time for Einstein to use it in the general theory of relativity.

The algebra involved in general relativity is not difficult in itself, although it is often tedious. The difficulties that come with it are twofold: first, convincing oneself that the algebra does indeed express the geometric or physical concept it is being used to represent, especially when that concept involves more than three geometric dimensions; second, verifying that, where parameters are used, the concept being defined is independent of any particular choice of parameters. The first of these accounts for the lengthiness of Chapters 5 and 6. We needed some assurance that the relativistic formulation of physics using tensors really does connect with differential geometry through the concept of curvature; as a result, we spent many pages looking at surfaces in $\mathbb{R}^3$, which are intuitive. The second difficulty is addressed in Appendix 6, where it is shown that the value obtained for the curvature of a surface is independent of the parameters used.

5.2. The role of calculus. It was stated in the preface that this book is aimed at generality and breadth, making connections among the diverse parts of its subject rather than exploring all the important parts of it in depth. The main connection of this sort that I wish to make is one that, in retrospect, seems to be an inevitable consequence of the work of the seventeenth-century scientists and mathematicians. By inventing the calculus, they showed how, in first approximation, all smooth functions could be linearized using first-order derivatives, thereby making geometry applicable to all forms of motion. At the same time, they revolutionized mechanics by making acceleration rather than velocity the key concept that makes sense of the notion of force. Acceleration, in turn, is expressed by the second-order derivative. In geometry, that second-order derivative is associated with curvature. When the facts are arranged in this way, it appears that curvature was bound to have an important role to play in the application of geometry to physics. And so it has turned out. Curvature and force are both variants of the second derivative and amount to different ways of looking at the same physical phenomena.

5.3. The role of modern analysis. The successive terms of a Maclaurin series mirror the way physical models develop: We begin with an oversimplified model in which everything is linear, a good example being the eighteenth-century analysis of the vibrating string, which is summarized in Appendix 2. Once that oversimplified model is understood, if it appears to have explanatory value, we refine it by introducing complications that make it more realistic. This process mimics the Maclaurin series, in which the first two coefficients determine the best approximation of a curve by a straight line and the third coefficient determines the extent to which the curve deviates locally from that straight line. The power of the calculus and its outgrowths is shown by the fact that mathematical physics was able to give precise, quantitative descriptions of dozens of physical phenomena, including the motion of particles and planets, the propagation of light and sound, and the action of electricity and magnetism, all using only second-order differential equations, which is to say, just these first three terms of the Maclaurin series. It was not until the nineteenth century that a physical phenomenon led to a third-order differential equation. The best-known of these phenomena—the propagation of waves in a shallow channel—led to what is called the Korteweg–de Vries equation.[15]

Of course, the order of a derivative depends on the function you start with. In terms of the metric coefficients, the curvature is defined using only second-order derivatives; but often the metric coefficients come from a parameterization and are themselves first-order derivatives.

In several places in this book, we have called upon elliptic functions to provide us with exact solutions of important differential equations. There is a seeming paradox here, in that these functions can be fully understood only from the point of view of analytic functions of a complex variable, even though the physical quantities that those variables represent are interpreted as assuming only real numbers as values. The two giant areas known as real analysis and complex analysis could hardly be more different. Real analysis is a vast, chaotic jungle of counterintuitive pathological functions that exhibit little stability, functions that are often wildly discontinuous. Complex analysis, on the other hand, deals only with the smoothest, most regular functions imaginable beyond those that are utterly trivial (the constants). In fact, the requirements made of an analytic function of a complex variable are so restrictive that it seems a miracle that any non-trivial ones at all exist. That they nevertheless do exist in such abundance that one can be found mapping any simply connected open set in the plane of complex numbers (except the plane itself) onto a disk is an amazing fact (the Riemann mapping theorem).

One of the surprises buried within mathematical physics is that complex function theory is often more successful at solving physical problems than real-variable theory. This fact is especially surprising, given that in mechanics the independent variable often represents time. What can it mean to use a complex variable to represent time? The eighteenth- and nineteenth-century mathematicians do not appear to have written much on this subject. The only extended remark I have found is the following, made by Weierstrass in 1885 (see his *Werke*, Bd. 3, S. 24):

---

[15]Named after Diederik Johannes Korteweg (1848–1941) and Gustav de Vries (1866–1934), who introduced it in 1895. It had actually been written about nearly two decades earlier in an 1877 work by Joseph Valentin Boussinesq (1842–1929).

> *It is very remarkable that in a problem of mathematical physics where one seeks an unknown function of two variables that, in terms of their physical meaning, can have only real values and is such that for a particular value of one of the variables the function must equal a prescribed function of the other, an expression often results that is an analytic function of the variable and hence also has a meaning for complex values of the latter.*

The interpretation of physical variables as complex numbers can lead to useful insight into a physical phenomenon. For example, in the 1897 lectures on the spinning top in which he introduced what are now called the Cayley–Klein parameters, Felix Klein remarked that if the physical variables are regarded as complex, a rotating rigid body can be treated either as a motion in hyperbolic space or as a motion in Euclidean space accompanied by a strain. Perhaps, since they had seen that complex numbers were needed to produce the three real roots of a cubic equation with real coefficients, it may not have seemed strange to these mathematicians that the complex-variable properties of solutions of differential equations might be relevant in the study of mechanical problems.

Time is sometimes represented as a two-dimensional quantity in the study of Gibbs random fields. In recent years, the physicist Stephen Hawking [38] has proposed using what he calls *imaginary time* to solve certain problems. This is not quite the same as regarding time as a complex variable. It is rather only a number along the imaginary axis in the complex plane. Just as we "spatialized" time by replacing t with $\tau = ct$, resulting in a metric $x^2 + y^2 + z^2 - \tau^2$, if we use Hawking's idea, which amounts to employing Minkowski's "mystical formula" (see p. 5), we can replace t with $\tau = ict$, and then our metric is $x^2 + y^2 + z^2 + \tau^2$, which is the ordinary Euclidean metric on four-dimensional space. Hawking cautions, however that this is a purely mathematical device, saying ([38], p. 135) "...we may regard our use of imaginary time and Euclidean space-time as merely a mathematical device (or trick) to calculate answers about real space-time." The views that Hawking espoused in [38] do not seem to have attracted many adherents among physicists up to the present time. Nevertheless we have here an example of a harmony between a creation of pure thought (imaginary numbers) and a theory that explains physical phenomena.

Lagrange, who invented the term *analytic function*, thought physics ought to be able to get by with these functions alone, since they are completely determined at every point by their values in a neighborhood (no matter how small) of any one point. That property, he thought, corresponded to the determinacy of nature, so that knowing the state of a physical system over any finite interval of time would determine its state for all time. Physicists are no longer committed to Lagrange's deterministic view of the natural world, although Einstein famously resisted the uncertainty principle. There is, however, a second reason for favoring the use of analytic functions, so that mathematicians can justify the modes of thought that physicists need for their work. Analytic functions seem to be required in the approximations used by physicists, such as those used by Eddington (see the last section of Chapter 6). Eddington approximates the metric coefficients of a hypothetical space-time, then proceeds to compute an approximate Ricci tensor on that basis. There is no theorem in real analysis that can justify such an approximation. Two functions of a real variable can be very close together, yet have derivatives that are very far apart. But for analytic functions, due to what is called Morera's theorem and Cauchy's formulas for the derivatives of an analytic function, there is such a theorem.

6. Knowledge of Hypothetical Objects: An Example

To illustrate how we can know facts about hypothetical objects that we do not directly experience, we use the example of a magnetic field—those magnetic "lines of force" talked about in high-school physics courses and illustrated with magnets and iron filings. The reality of a magnetic field is convincingly demonstrated by its effects not only on compasses, but even on living things, as the following example attests.

> *To test the hypothesis that lobsters derive positional information from the Earth's magnetic field, lobsters were exposed to fields replicating those that exist at specific locations in their environment. Lobsters tested in a field north of the capture site oriented themselves southwards, whereas those tested in a field south of the capture site oriented themselves northwards.* [**2**]

The phenomenon described in this quotation was reported some 40 years earlier by Lancelot Hogben ([**44**], p. 81), including a drawing of a shrimp swimming in a magnetic field. Hogben's explanation was that the shrimp has an "inertial guidance system" in the form of a liquid-filled sac lined with sensitive cilia and containing a ball of solid matter that orients the shrimp by pressing against the side of the sac opposite to the direction of motion. According to Hogben, when this solid ball is replaced by ferromagnetic material and the shrimp is placed in a magnetic field, it will move along the magnetic lines of force. (I have been unable to determine who performed this delicate operation and experiment on the shrimp.) The topic has been well-studied over the past few decades (see, for example, [**5**]). Given the work of Boles and Lohmann [**2**], the ball in the inertial guidance system mentioned by Hogben *already* contains ferromagnetic material, so that the poor shrimp might have been spared the ordeal of the operation.

To illustrate the passage from primitive undefined concepts to facts about those concepts, we turn to algebra, introducing a variable to represent the magnetic field and formulating an algebraic equation relating it to other primitive terms such as charge, velocity, and force. Electromagnetic theory provides a vector-valued function $\boldsymbol{B}(t; x, y, z)$ to represent the magnetic field. Its numerical value at a given point and time can be measured by observing the motion of a charged particle or by measuring the current in a moving electrical conductor. It is this combination of algebra and measurement that provides the vocabulary for physicists to communicate and reconcile theory with observation. Physics can talk meaningfully about the function $\boldsymbol{B}(t; x, y, z)$ in a particular situation, and they can verify their assertions about it by measuring its value. Their knowledge consists of certain propositions expressed as vector equations. For example, given a charge q moving with velocity $\boldsymbol{v}$, the force on the charge is given by the Lorentz law $\boldsymbol{F} = q((\boldsymbol{v}/c) \times \boldsymbol{B})$. For that reason, the physicists need not be concerned with the "essence" of a magnetic field or any of the other metaphysical questions we are considering in this chapter, any more than mathematics books bother to discuss the desirability of the axiom of choice. From the point of view of most professional physicists and mathematicians, there is no need to delve into this philosophy. The great conversation goes on at professional meetings without any need to worry about it. We are discussing it here only to offer a few suggestions to non-specialists who may be inclined to ponder such matters. For the non-specialist, the fact that the Lorentz law is consistent with observation provides sufficient grounds to say that the magnetic field represented by the symbol $\boldsymbol{B}(t; x, y, z)$ is a real thing. As long as the present theory continues to be used by physicists, non-specialists can talk confidently about its terms, taking for granted that they are real. If it becomes necessary to revise or replace the present theory, that is a job for the specialist, and the non-specialist can await the results.

There remains another philosophical problem connected with this process that we need to address: Since a physical quantity can be measured in different ways, how can two people using different instruments, whose functioning may even depend on different physical theories, be sure they are talking about the same quantity and agree on its value? This problem is no problem for practicing physicists. The needed reconciliation has already been performed, and its invocation is immediate. We have already mentioned the widespread agreement exhibited by terrestrial clocks operating on a wide variety of physical principles. Similarly, no one questions whether a CAT scan and an MRI of the same patient will give consistent results or worries that an optical telescope and a radio telescope may contradict each other. If the readings from different measuring devices agree, then they are measuring the same quantity and no one needs to ask what the "intrinsic nature" of that quantity is. To elaborate on this point, we are going to suggest another analogy from mathematics, this time from the theory of manifolds.

6.1. Manifolds as a model for communication. Granting the need for undefined, primitive concepts in any theoretical discussion, and granting that we can know facts involving those concepts without knowing the concepts themselves, there remains the problem of communicating those facts to other people. How do we know if two observers are talking about the same object? And given some quantitative statement about it, how can we determine whether those two observers agree or disagree about the numerical values involved? We took that process for granted when we derived the Lorentz transformation in Chapter 1, by which two observers reconcile their observations of a single space time point. The reconciliation is easy in some cases. If one observer says that the temperature of a body is 40° Celsius and another says it is 104° Fahrenheit, then they agree about the temperature. But there are many devices for measuring temperature. The ordinary thermometer, based on the physical principal that heat normally causes materials to expand, is suitable for many terrestrial purposes. But digital thermometers, based on electrical conductivity are also in common use. For measuring temperatures at more remote locations such as Mars, astronomers once used a device known as a thermocouple. For measuring the temperature of stars, the spectrum of a star comes into play. If, by chance, two different methods of measuring temperature are used, what set of physical conditions in each of the two measuring instrument will cause them to provide the same reading, so that the two observers will agree they are measuring the same temperature? According to theory, the reading on each thermometer is a mathematical function of the states of a finite set of physical variables associated with the thermometer. Those physical variables amount to a sort of parameterization of the object whose temperature is being measured, and the observers know they are measuring the same physical quantity if the two readings are identical. The process resembles the use of different coordinate systems on a manifold.

Each manifold is perceived in a unique way by people using a particular set of parameters to describe it. The price of that individual freedom is the restriction of discourse to objects (tensors and related concepts) for which agreed-upon translations from one parameter language to another exist. To take a simple example, consider the ordinary plane with a closed half-line removed. One person may coordinatize that object by regarding it as the xy-plane of Euclidean plane geometry, with the non-positive portion of the x-axis removed. It then becomes, for that person, the set of all ordered pairs (x, y) satisfying the inequality $x + |x| + |y| > 0$. Another person may prefer polar coordinates and regard it as the set of ordered pairs (r, θ) with $r > 0$, $-\pi < \theta < \pi$. The two can agree that they are talking about "the same point P" if the coordinates that they assign to P are convertible via the relations

$$x = r \cos \theta,$$

$$y = r \sin \theta,$$

$$r = \sqrt{x^2 + y^2},$$

$$\theta = 2 \arctan \left(\frac{y}{x + \sqrt{x^2 + y^2}} \right).$$

Given that two people use these sets of coordinates in the indicated domains and agree to identify the pairs (x, y) and (r, θ) when these equations are satisfied, we may ask whether the original object that motivated our choice of these transformations is needed at all. Any discussion they have is going to involve symbolic expressions using these variables? What need is there for the intuitive object? Why do we need the point P at all, if we have these coordinate transformations? The object containing P may or may not be real in some unknowable metaphysical sense, but its reality or unreality is *irrelevant to the conversation between the two observers*. They can both define additional concepts just from their parameters, such as tangent vectors and covectors, and they can use tensor principles to determine whether a law stated by one of them translates into the same law as understood by the other. What one regards as a function $f(x, y)$, the other regards as $\hat{f}(r, \theta) = f(r \cos \theta, r \sin \theta)$, and conversely, what the latter regards as a function $g(r, \theta)$, the former regards as $\tilde{g}(x, y) = g(\sqrt{x^2 + y^2}, 2 \arctan(y/(x + \sqrt{x^2 + y^2})))$. We obviously have $\hat{\tilde{g}} = g$ and $\tilde{\hat{f}} = f$ for all functions $f(x, y)$ and $g(r, \theta)$. The differentials are also easily translated: $dx = \cos \theta \, dr - r \sin \theta \, d\theta$, and so forth. In that way, the geometry imposed on this space by any metric ds^2 can be expressed in either language, and the relevant Christoffel symbols, Riemann curvature tensor, and the like computed in both languages. What they know, they can talk about, and they need not worry about what the underlying manifold "essentially is."

As long as the mathematical language in which physicists communicate with one another continues to produce consistent results, as it has in the past, the *working hypothesis* of an objective physical world and the physical concepts used to describe it are all we need. If that working hypothesis should ever fail, future scientists will have to deal with that problem. But that event has not yet occurred.[16]

Thus, as we have now amply illustrated, measurement plays the essential role in making knowledge possible. Without it, we would be floundering in futility like many of the ancient and medieval philosophers, trying to make sense of unknowable entities. A hypothetical law phrased mathematically suggests measurements that can be made. If they confirm the law, it can then be used to make further predictions. To illustrate this process, consider Newton's law of gravity $F = GMm/r^2$. We can check it against observation only *after* we have determined the universal gravitational constant G. To do that, as Henry Cavendish (1731–1810) did in 1797, we have to *assume* that the quantity Fr^2/Mm is constant, where M and m are the masses of two test bodies, r the distance between them, and F the force of mutual attraction measured by experiment. (Newton's law of gravity actually asserts only that this expression is constant over all masses and radii.) If the attractive force measured in different experiments with different masses and distances reveals that this quantity is always the same, within the limits of accuracy, then we can conclude simultaneously that the law is valid and that we have determined the constant G. As a consequence, it really does not make sense to talk about, say, the fiftieth decimal digit of G, since G is not a mathematical constant like π. If two experimenters happened to get different values for that digit—and no one has come even close to this level of precision—we need not conclude that one is right and the other wrong. There is always the possibility that the quantity Fr^2/Mm varies slightly with distance or with mass.

We can now respond to the seemingly natural question that is often asked: What *is* a field of force? Our response is that we do not know and we do not have to care. Mathematically, the field corresponds to a *vector-valued function* $\boldsymbol{F}(t; x, y, z)$ on a region of space-time. We can determine the numerical value of this function experimentally (assuming the validity of the laws of gravity and electromagnetism) by observing the way test particles move in these fields. The fact that they do move as predicted is taken as evidence that the field really is there, although it gives no information as to what the field is "made of." We accept the reality of the field on the basis of these observations and give up trying to imagine a "substance" spread over space that "is" the field.

Giving up on the search for the substance that constitutes a force field leaves us with another important question, however. Maxwell and Hertz established the existence of electromagnetic waves, of which light is an example. Now our experience of waves, based on observations of bodies of water and the transmission of sound

---

[16]This possibility—that the future might not resemble the past—raises yet another longstanding metaphysical problem, that of induction. Philosophers have tried to justify some principle of induction so as to put a foundation under the way that we all think. Kant's contemporary, the philosopher David Hume (1711–1776) delivered a devastating critique of the principle. But why is a justification needed? Induction is a very useful working hypothesis—we expect water to boil when placed on a hot burner, not freeze. What need do we have to prove that inductive thinking is valid? What *causes* us to think inductively is obvious on the basis of evolution: if we didn't think that way, we wouldn't learn from experience. Why philosophers have felt a need to provide logical *grounds* for induction is a mystery. If it ever fails on a large scale, we'll abandon the hypothesis. We don't expect the Sun to rise tomorrow merely because it has done so every day in human history; we have other (also inductively established!) physical principles that imply it probably will. We don't have space at this late point in our narrative to say any more about this subject.

through vibrations, has given us an intuition in which a wave is a periodic, undulatory motion of something material. If electric and magnetic fields are not material, just what is it that is undulating when an electromagnetic wave propagates? And what do we mean by the speed of propagation? One answer is that the *intensity* or *strength* of the field—that is, the numerical value of the function—is varying periodically at each point of space between two extreme values. That variation *is* the wave, and the picture of something material that is moving is not needed. What is traveling is the set of coordinates of points where the field has a given intensity, say a maximum (crest) or minimum (trough). The points where these intensities occur change continuously with time, so that a particle that happened to be "riding a crest" really would travel through space. The picture of a surfer riding waves into shore here seems to fit well here. The actual molecules of water at the seashore do not move—at least not very far—as waves travel along. Wave crests and troughs do cause them to move in a roughly circular manner, but the motion of the individual water molecules does not resemble the linear motion of the wave crests. This explanation rests on the fact that field intensities can be measured. It is measurement that provides the crucial link between the hypothetical objects that we cannot know and the facts that we can nevertheless know about those hypothetical objects. A radio wave cannot be *experienced* in the way a chemical substance can be seen and felt, but both can be measured, and modern physics and chemistry are all about the measurements. Two scientists can easily communicate what they know about measurements, whereas if they were in the situation of the ancient natural philosophers, attempting to get to the "essence" of the physical world, they could spend years discussing what the universe is made of.

7. Knowledge of the Physical World

To summarize our main points: (1) We have knowledge of the physical world, but it is based on an irreducible minimum of unknowable entities; (2) nevertheless we can know *facts* about these unknowable entities; (3) our knowledge arises because we invent symbols to denote the unknowable objects and formulate physical laws expressed as equations whose variables correspond to measurements that can be made. If the variables are replaced by the numerical values of the measured quantities, the equations become equalities between numbers and thereby confirm or refute the law, depending on whether they are observed to be true or false. One question now remains: What do we mean by knowing a fact or thing?

The present discussion is less precise in English than it would be in German or French because of the relative poverty of English words for expressing what we know. French and German have one word used in sentences that assert knowledge of a fact (*savoir* and *wissen* respectively) and a different word for asserting knowledge of (familiarity with, acquaintance with, experience of) a person or thing (*connaître* and *kennen* respectively). We do not *know* time, space, magnetic fields, and the like in the latter sense. To know something in that sense is rather like knowing what Aristotelian philosophers used to call its *essence* (which Bertrand Russell said was "a hopelessly muddle-headed action, incapable of precision"). Yet we do *know* in the factual sense many propositions about these unknowable entities, and that knowledge comes from our mathematical or linguistic descriptions of their properties. Mathematics furnishes a model here. The abstract things defined by physics play a role similar to that of the undefined terms with which an axiomatic mathematical theory begins. We know the axioms and the theorems we deduce about these undefined terms in the "savoir/wissen" sense. We do not know the undefined terms themselves in the "connaître/kennen" sense. Even so, as remarked above about the axiomatization of Euclidean geometry, we generally have in mind some visualizable *interpretation* of the undefined terms.

Each person has a description of the world, analogous to a set of parameters for describing a manifold. As long as that person is content to think and contemplate the world in isolation, no difficulties arise with that description. But science is a social enterprise, and scientists must talk to one another. We use language to communicate our descriptions, and the language of physics—like the language of tensors in differential geometry and indeed, *including* that language—enables us to communicate with one another and determine whether our descriptions agree or not. We repeat that the crucial link in this chain from unknowable undefined concepts to known facts is measurement.[17] The measurements are what we know, and they enable physics to function as an intellectual and applied, *social* enterprise. Such entities as quarks and force fields are concepts that can be communicated among physicists and lead to quantitative predictions that have been consistently confirmed by experiment. Anyone who desires a stronger assertion of their reality than that must assume the burden of stating what that stronger form of reality means, and how we could know about it. Thus, we take the modest position that *one should talk and reason as if these objects were real, until such time as better explanations of what we observe are found.* We assert, for example, that chemists before the eighteenth century were right in speaking *as if* phlogiston was a real thing; but we do not assert that it *was* a real thing. Similarly, nineteenth-century mathematical physicists were correct in regarding "luminiferous ether" as a real thing; but we do not assert that it *was* a real thing. It is conceivable that some day better theories will be found, and radio waves will no longer be a necessary part of physics.[18] But in the meantime, we are correct in speaking *as if* they exist.

Finally, we note another important aspect of scientific theories that affects their plausibility. When a number of physical principles interact fruitfully—for example, the synergy between the Lorentz transformation and Maxwell's laws, as discussed in Chapter 3—each of them gains plausibility. Theories have to be consistent with observation and experiment, it is true. But beyond that testability, our confidence in the correctness of a theory depends very much on the way it "fits" with other physical theories. That interlocking reinforcement prevents the automatic rejection of a principle on the basis of a single observation or experiment. This interlocking of theories shows up especially well in the precise measurements made in modern experimental science. We accept such things as radio waves and regard them as having a real existence because that is the most efficient way of talking about phenomena that we perceive directly, such as the computer screen I am looking at as

---

[17]To express this transition in terms used in mathematical logic, the contrast between the two corresponds to the difference between the syntax of a formal language and the semantics expressed in its metalanguage. The theory (syntax) is purely formal and intellectual. Measurement (semantics) gives it meaning and applies it to the physical world.

[18]In my opinion, radio waves are far less likely to become obsolete as an explanation than phlogiston and the ether, first because they arose when much more was known about the physical universe, and second because they are more closely linked to measurement. I expect them at the very least to have a longer life than these earlier, now obsolete concepts, but perhaps not everlasting life.

I type these words. The common-sense distinction between theory and observation becomes murky in the light of modern physics, and describing reality is a matter of getting the language just right. Even people who scornfully dismiss what is "only a theory" will usually admit that an observation made with a mass spectrometer is a valid observation, thereby ignoring the fact that the functioning of a mass spectrometer is interpreted and understood only by means of very complicated physical *theories*. What appears to be agreement between theory and observation is more often nowadays agreement among a number of theories. Again, we need to emphasize that the interlocking is often a matter of measuring a quantity using instruments based on different physical theories. Agreement among several different methods of measurement increases our confidence that each individual theory is more or less right.

Those who are seeking absolute certainty and are disquieted by the thought that future generations may look back at us and pity us for speaking about such things as bosons and fermions as if they were real will not like this conclusion. But our aim here is a practical one: to provide a way of looking at physical propositions that can be confidently used, but is always subject to refutation at any time. The human race is not likely ever to reach absolute truth. Each new physical paradigm is like a vehicle that (we hope) is carrying us closer to some ultimate truth, but breaks down after a time and has to be repaired or replaced. We have no choice but to live our lives midway on this journey, riding in the vehicle that is currently running, traveling hopefully rather than arriving, as already discussed. To make ourselves more content with our lot, we shall close this section with a look at some of the abandoned wreckage of past vehicles.

7.1. Obsolete concept # 1: phlogiston. The seemingly simple task of explaining combustion scientifically took an extraordinarily long time. We now know that the delay in understanding was caused by ignorance of the way chemical elements combine and that combustion is a form of oxidation. But when an intuitive, qualitative explanation of combustion seemed to be needed, a substance (or, at least, the name of a substance) was invented to meet the need. A word related to our word *flame*, the ancient Greek word *phlogiston* (φλογιστόν), meaning *burned-up*, was called into service. (These obscure Greek and Latin names for unknowable things call to mind Molière's send-up of medical education in *Le malade imaginaire*, in which the aspiring doctor "explains" that opium causes sleep because it has "a dormitive virtue.") The theory was that the flame one could see when a combustible substance burned was the release of a mysterious substance that it contained called *phlogiston*. Thus, losing phlogiston was what we now describe as combining with oxygen. Pure oxygen itself was referred to as *dephlogisticated air*. Peculiar as it now seems, the theory did explain a few things. For example, granting that phlogiston had the property of levity rather than gravity (the flame moved upward in the air), it explained why the residue after something was burned might sometimes weigh more than the original material.

7.2. Obsolete concept # 2: ether. When the explanation of light using the model of a wave became successful, explaining such phenomena as interference and refraction, certain logical problems arose. The main one concerns our intuitive picture of a wave. We see waves mostly in water, where the wave itself is invisible, but manifests itself in an undulatory motion of the physical water. The vibrations of a tuning fork are another example illustrating that a wave requires a *periodic* motion of some substance. This is certainly true of sound waves, which cannot travel in a vacuum. But light *does* travel through interstellar space. If it is a wave, and something is vibrating as either its cause or its effect, just what *is* the medium that is transmitting the wave?

In accordance with the precedents set by gravity and phlogiston, a suitable name—*ether*, another ancient Greek word ($\alpha i \theta \acute{\eta} \rho$), meaning *bright, clear air*—was invented as a patch to cover our ignorance. It seemed necessary to do so, since certain phenomena associated with light were seemingly impossible of explanation in mechanical terms. In the nineteenth century, the accepted theory was that light was a disturbance propagating through an elastic medium, and that was one of the principal applications of elasticity theory in the book of Gabriel Lamé (1795–1870). Lamé was sure that elasticity theory could explain even double refraction, and he did manage to obtain Fresnel's equation for the wave surface. In his treatise [**49**], Lamé was able to explain the propagation of a *planar* light wave in purely mechanical terms involving the vibrations of molecules in an elastic solid. Lamé had great ambitions for his elegant theory of elasticity. But one important case stumped him: propagation of light from a point source. As he showed, the wave surface propagates normally at every point except the point of origin. There, he said,

> In order to obey the laws found above [the origin] must undergo vibrations of infinite amplitude in all directions at once. Should one therefore conclude that the assumption of a continuous stream of waves produced by a point source is impossible? I do not think so. . .

In this section (§ 129 of Lecture 24, pp. 325–326) under the heading "Necessité d'admettre l'éther," he continued, demonstrating the amazing ability of a first-rate mind to resist drawing uncomfortable conclusions. To Lamé, this anomaly showed the need for the medium of ether, and, in very obscure language, he sought to save his beloved elasticity theory from this disaster by appealing to its protection:[19]

> It thus follows necessarily that the central system, and then all the doubly-refracting space, must contain another type of matter which is the actual medium vibrating under the influence of the light. Thus physical mass plays only a passive role, modifying by a sort of resistance the directions of the vibrations and the velocities of propagation. . .

This passage, at the very end of Lamé's treatise, is vague and comforting and uses the ether as a trump card when nothing else will take the trick. It also renders irrelevant all the intricate mathematical theory he had developed in the first 23 of his lectures. As Bertrand Russell later wrote ([**72**], pp. 19–20):

---

[19]Karl Weierstrass (1815–1897) thought he could find another way out. He thought Lamé hadn't found all the solutions to his differential equations and tried to find other solutions by a new method involving what we now call the divergence theorem (Gauss's theorem). He gave his paper on the subject to his pupil Sonya Kovalevskaya (1850–1891), who needed a publication at the time to get a position at the University of Stockholm. Using the method, she produced a formula for a solution. Unfortunately, as Vito Volterra (1860–1940) discovered a few weeks after her death, the solutions didn't actually satisfy the equations, since she had overlooked the multivaluedness of the functions involved.

> *...the æther was never so comfortably material as "gross" matter. It could vibrate, but it did not seem to consist of little bits each with its own individuality, or to be subject to any discoverable molar motions. No one knew whether it was a jelly or a gas. Its properties could not be inferred from those of billiard balls, but were merely those demanded by its functions. In fact, like a painfully good boy, it only did what it was told, and might therefore be expected to die young.*

Lamé's words here may evoke in the reader the same feeling as the now-current phrase *dark matter*. At the very least, the latter—still elusive—concept, about which little more is known at present than was known about the ether two centuries ago, may incline us to be lenient in our judgment of those who used the ether in their theories.

8. A Few Words from the Discoverers

To end this narrative, we recall the famous Allegory of the Cave in Book VII of the *Republic*. Plato compared the philosopher to a person from a race of people confined to the half-light of a cave who emerges from the cave into the bright sunshine and then is faced with the task of returning to the cave to explain to the others who have not left that there is a brilliant world beyond the one they know, that they know only appearances and the reality is much greater. Relativity did something like that for the physicists of the world, who had been confined to the cave of three-dimensional space. String theory now proposes to regard even space-time as a sort of cave and comes to tell us about a still larger and more marvelous world.

All this heavy use of mathematics—a pure creation of the human mind—inevitably provokes a sense of mystery. You can't help wondering *why* mathematics appears to be so effective in explaining the external world. On the practical level, it raises an even more urgent question: Have we "got it right"? What grounds can there be for believing that general relativity reveals actual information about the universe that makes a difference to anybody's life? To put the matter more succinctly, how plausible is the general relativistic formulation of mechanics? These are the questions we have been, if not answering, at least discussing in the present chapter. To bring the discussion to a close, we turn to two of the intellectual giants who created the theory a century ago, and we let them have the last word.

8.1. Einstein's justification. The explanation that Einstein gave in his 1916 paper[20] on general relativity stressed the following points. In this list, we have drawn out the implications of his reasoning and put them in brackets:

- The law of gravity was stated by Newton as a set of second-order ordinary differential equations in the spatial coordinates. [Therefore the simplest refinement of the Newtonian model that will improve it should be a set of differential equations of at most third-order in the space-time coordinates.]
- The differential equations that express the law of gravity should have the same form in any coordinate system. [Therefore they should be tensor equations.]

---

[20]This paper, which Einstein submitted in November of 1915, represents a breakthrough "Eureka!" moment in his thought, when he finally, after many years of striving that we have not described here, found a set of generally covariant equations for the gravitational field.

- Many examples show that there is an intimate connection between force and curvature. [Therefore the Gaussian metric coefficients should be used to express what Newton stated in terms of force.]
- Curvature is quantified in differential geometry by means of the Riemann–Christoffel (Riemann curvature) tensor.
- The Riemann–Christoffel tensor of the metric of special relativity vanishes identically, and conversely, a space where this tensor vanishes has a flat metric. [Therefore the requirement that this tensor vanish would be too strong.]
- The only tensor involving at most second-order partial derivatives of the metric coefficients that is symmetric and is linear in the second-order partial derivatives is the Ricci tensor. [Therefore, let us explore the consequences of assuming a metric in which the Ricci tensor vanishes.]

After stating these considerations (the first of them being a tacit assumption), Einstein presented his readers with the result and argued that it should be accepted as physically correct since it (1) reduced to the Newtonian law in first approximation and (2) explained the precession of the perihelion of Mercury. (See Chapter 4 for the elaboration of both of these statements.)

8.2. Eddington's justification. Although there is no arguing with the success of this approach, the grounds on which the equations were formulated seem perhaps less satisfying to us than they would have seemed to Einstein, who had just spent several years searching for a set of equations that would fulfill all these conditions. We would like something more, something to give us an intuitive feeling that the vanishing of the Ricci tensor ought to impose a metric that "feels like" gravity. What we need is an expression in terms of tensors that corresponds to Newton's second law $F = ma$ and, if possible, includes the three great conservation laws (conservation of energy, conservation of momentum, conservation of angular momentum). In the previous chapter, we attempted to make the Einstein tensor seem intuitive, the natural analog of mass times acceleration. The centerpiece of our heuristic argument was Theorem 7.1, which established that the divergence of the contravariant form of the Einstein tensor vanishes. That was the main point urged by Eddington [16], and we shall let him have the final word. He first posed the problem of justifying the heuristic reformulation of mechanics as follows:

> We may put to the experiments three questions in crescendo. Do they verify? Do they suggest? Do they (within certain limitations) compel the laws we adopt? It is when the last question is put that the difficulty arises for there are always limitations which will embarrass the mathematician who wishes to keep strictly to rigorous inference ([16], p. 105).

He then stressed the importance of the fact that the divergence of the Einstein tensor vanishes:

> In three dimensions the vanishing of the divergence is the condition of continuity of flux, e.g., in hydrodynamics $du/dx + dv/dy + dw/dz = 0$. Adding a time-coordinate, this becomes the condition of conservation or permanence, as will be shown in detail later. It will be realised how important for a theory of the material world

is the discovery of a world-tensor which is inherently permanent ([**16**], p. 115).

Finally, after establishing that the divergence of the (mixed) Einstein tensor vanishes,[21] he reformulated the whole theory of relativistic gravitation for a continuous density of matter, and concluded as follows:

> *We thus arrive at the law of gravitation for continuous matter...*
> *but with a different justification. Appeal is now made to a Principle of Identification. Our deductive theory starts with the interval...from which the tensor $g_{\mu\nu}$ is immediately obtained. By pure mathematics we derive other tensors $G_{\mu\nu}$, $B_{\mu\nu\sigma\rho}$ [the Ricci tensor that we denoted* Ric *and covariant Riemann curvature tensor that we denoted* Rie*], and if necessary more complicated tensors. These constitute our world-building material; and the aim of the deductive theory is to construct from this a world which functions in the same way as the known physical world. If we succeed, mass, momentum, stress, etc. must be the vulgar names for certain analytical quantities in the deductive theory; and it is this stage of naming the analytical tensors which is reached in [the preceding section]. If the theory provides a tensor $G_\mu^\nu - g_\mu^\nu G$ [the mixed Einstein tensor] which behaves in exactly the same way as the tensor summarising the mass, momentum and stress of matter is observed to behave, it is difficult to see how anything more could be required of it* ([**16**], pp. 119–120).

9. Epilogue: The Reception of Relativity

Scientific debates have never aroused the passion that religious and political differences have inspired and continue to inspire. Science is by comparison a peaceful and rational activity of the mind. Nevertheless the bottom layer of irrationality and emotion in every human being distorts reality, even in the questions of objective fact that natural science deals with. Cynics have claimed that scientific theories are never really proved. Rather, it is claimed, what often happens when a new theory supplants an old one is that the adherents of the old theory gradually retire, fade from the picture and die. This view sometimes appears to fit the facts. Ernst Mach, for example, is quoted as saying, "I can no more accept the theory of relativity than I can accept the existence of atoms and other such dogmas." (He had a famous dispute with Ludwig Boltzmann over the existence of atoms.) He made this statement in the preface to his posthumously published book on the principles of physical optics. Ironically, Einstein approved of Mach's rejection of absolute space, wrote that Mach might very likely have discovered the theory of relativity himself, and gave Mach's principle both the formulation and the name it now has.

---

[21] Eddington's argument for this relation seems to me fallacious. He switches to a particular system of parameters (which may, for example, be normal coordinates) in which the Christoffel symbols vanish at the origin, then proves the relation only at the origin. But Corollaries 6.3 and 6.4 are examples of equations that hold at the origin in normal coordinates, but are nevertheless not tensor equations. Although it is not clear what restrictions he imposes on the metric coefficients, it is not at all difficult to exhibit metric coefficients in which this divergence is not zero, for example, by taking $g_{11} = 1 + t^2 x + y^2$, $g_{22} = 1 - tz^2 + x^2 y^2$, $g_{33} = 1 + x^2 y - t^3 z$, $g_{44} = 1 - t^2 - x^2 - y^2 - z^4$, $g_{ij} = 0$ if $i \neq j$.

That view, however, does not take into account the rules of scientific evidence that have been established over the past five centuries. The majority of scientists, especially in the fields of physics, chemistry, biology, and geology, may propose hypotheses that challenge an accepted view—indeed, it is their duty to do so—but they almost never make these proposals dogmatically or refuse to be convinced by contrary evidence. (The Brans–Dicke theory is an excellent example of this rational process.)

For a non-scientist (such as the present author), whose opinion carries no scientific weight but who wants to understand the world in terms of the most reliable theories available, the best guide is something like a principle of "majority rules." In a true open market of ideas, individual biases tend to cancel one another out. Any particular scientist may have a crazy pet theory that, meeting with no support among peers, never wins the day and gets into textbooks. The overwhelming majority of scientists in a particular field do not share the irrational thinking that leads an individual to advocate silly theories. There are nearly always contrarians, even in well established areas of science: people with doctorates in physics who do not believe in the reality of electric currents, or people with doctorates in some biological or geological area who do not accept the reality of evolution. Einstein himself, as is well-known, tried to find ways of getting around the uncertainty principle; but he never asserted this personal preference for a deterministic physics as a fact. All too frequently, the contrarians—unlike Einstein—make dogmatic claims that they are right, in the teeth of the evidence. Such people represent the irreducible minimum of irrationality that is inherent in any area of human endeavor. Unfortunately, their academic credentials are often exploited by special interests pursuing a political or commercial end and cited as proof that there is still room for doubt or (even worse) that the scientific consensus is simply a fraud and a hoax. In the abstract, it is true, a scientific consensus is not the same thing as certain proof. But in the world of public policy absolute proof is never available. Decisions have to be made on the basis of incomplete information. Under those circumstances, it is reckless to trust the contrarians, even though the non-expert cannot refute them.

Any account of the history of a scientific theory must take account of many human factors like these, which are not themselves scientific. The history of the theory of relativity fits that pattern to some extent. The gradual acceptance of the special and general theories of relativity, and the various routes to that end in several countries was the subject of a collection of essays edited by Thomas Glick [**33**]. These essays are very nuanced, and we do not have space to summarize any of them. Suffice it to say that old habits of thought die hard, and that new ideas often have to run an obstacle course through the Academy.

9.1. The future of geometry in physics. One of the main purposes of the present book has been to describe how, through relativity, the geometric concept of curvature gradually displaced the older physical notion of force. To be sure, the geometry involved was highly algebraized, so much so that one hardly needed figures to describe it. One could get by perfectly well just manipulating formulas. That paradigm shift was fully developed during the century that followed Einstein's 1915 paper on general relativity and became one of the dominant features of mathematical physics. Whether it will remain so in the future is uncertain. Already there are signs that the search for a Grand Unified Theory (GUT) or a Theory Of Everything (TOE) will require an entirely different form of mathematics. If that happens, then general relativity will represent the high-water mark of the penetration of geometry into physics. But that is far too broad and deep a topic to enter into at this point, and so we end our narrative here.

Bibliography

[1] Ali R. Amir-Móez, *transl.* (1919–2007). Discussion of difficulties in Euclid by Omar ibn-Abrahim al-Khayyami (Omar Khayyam). *Scripta Mathematica*, 24(4):275–303, 1959.

[2] Larry C. Boles and Kenneth J. Lohmann. True navigation and magnetic maps in spiny lobsters. *Nature*, 421 (2 January):60–63, 2003.

[3] Emile Borel (1871–1956). *Introduction géométrique à quelques théories physiques*. Gauthier-Villars, Paris, 1914.

[4] Carl H. Brans and Robert Henry Dicke (1916–1997). Mach's principle and a relativistic theory of gravitation. *Physical Review*, 124:925–935, 1961.

[5] Ruth E. Buskirk and William P. O'Brien. *Magnetic remanence and response to magnetic fields in crustacea*, volume 5 of *Topics in Geobiology*, chapter 17, pages 365–383. Springer-Verlag, Berlin, 1985.

[6] Elie Cartan (1869–1952). Sur les variétés à connexion affine et la théorie de la relativité généralisée (première partie). *Annales de l'Ecole Normale Supérieure*, 40:325–412, 1923.

[7] Elie Cartan (1869–1952). Sur les variétés à connexion affine et la théorie de la relativité généralisée (première partie). *Annales de l'Ecole Normale Supérieure*, 41:1–25, 1923.

[8] Ronald W. Clark (1918–1987). *Einstein: The Life and Times*. World Publishing Company, New York and Cleveland, 1971.

[9] P. C. W. Davies. *About Time: Einstein's Unfinished Revolution*. Simon & Schuster, New York, 1995.

[10] Willem de Sitter (1872–1934). On the bearing of the principle of relativity on gravitational astronomy. *Monthly Notices of the Royal Astronomical Society*, 71:388–415, 1911.

[11] Willem de Sitter (1872–1934). On Einstein's theory of gravitation and its astronomical consequences. *Monthly Notices of the Royal Astronomical Society*, 77:155–184, 1916.

[12] René Descartes (1596–1650). *Principia philosophiæ (1647)*, volume 3 of *Œuvres*. Levrault, Paris, 1824.

[13] Tevian Dray. *The Geometry of Special Relativity*. CRC Press, Boca Raton, 2012.

[14] Tevian Dray. *Differential Forms and the Geometry of General Relativity*. CRC Press, Boca Raton, 2015.

[15] Johannes Droste (1886–1963). The field of a single centre in Einstein's theory of gravitation, and the motion of a particle in that field. *Proceedings of the Royal Netherlands Academy of Arts and Sciences*, 19, Part 1:197–215, 1916.

[16] Arthur Stanley Eddington (1882–1944). *The Mathematical Theory of Relativity*. Cambridge University Press, 1923.

[17] Albert Einstein (1879–1955). Zur Elektrodynamik bewegter Körper. *Annalen der Physik, vierte Folge*, 17:891–921, 1905.

[18] Albert Einstein (1879–1955). Über den Einfluß der Schwerkraft auf die Ausbreitung des Lichtes. *Annalen der Physik, vierte Folge*, 35:898–908, 1911.

[19] Albert Einstein (1879–1955). Erklärung der Perihelbewegung des Merkur aus der allgemeinen Relativitätstheorie. *Sitzungsberichte der preussischen Akademie der Wissenschaften*, pages 831–839, 1915.

[20] Albert Einstein (1879–1955). Feldgleichungen der Gravitation. *Sitzungsberichte der preussischen Akademie der Wissenschaften*, pages 844–847, 1915.

[21] Albert Einstein (1879–1955). Die Grundlage der allgemeinen Relativitätstheorie. *Annalen der Physik*, pages 769–822, 1916.

[22] Albert Einstein (1879–1955). On a stationary system with spherical symmetry consisting of many gravitating masses. *Annals of Mathematics, Second series*, 40:922–936, 1939.

[23] Albert Einstein (1879–1955). *Relativity: The Special and the General Theory.* Bonanza Books, New York, 1961.

[24] Albert Einstein (1879–1955) and Marcel Grossmann (1878–1936). Entwurf einer verallgemeinerten Relativitätstheorie und einer Theorie der Gravitation. *Zeitschrift für Mathematik und Physik*, 62:225–261, 1913.

[25] Leonhard Euler (1707–1783). *Methodus inveniendi lineas curvas maximi minimive proprietate gaudentes.* Bousquet & Cie., Lausanne, 1744.

[26] Leonhard Euler (1707–1783). Recherches sur la courbure des surfaces. *Mémoires de l'Académie des Sciences de Berlin*, 16, 1767.

[27] Richard Faber (1940–2011). *Differential Geometry and Relativity Theory.* Marcel Dekker, New York, 1983.

[28] George Francis FitzGerald (1851–1901). The ether and the earth's atmosphere. *Science*, 13:349, 1889.

[29] Armand Hipployte Louis Fizeau (1819–1896). Sur les hypothèses relatives à l'éther lumineux. *Comptes rendus*, 33:349–355, 1851.

[30] Alexander Friedmann (1888–1925). Über die Krümmung des Raumes. *Zeitschrift für Physik*, 10:377–386, 1922.

[31] Alexander Friedmann (1888–1925). Über die Möglichkeit einer Welt mit konstanter negativer Krümmung des Raumes. *Zeitschrift für Physik*, 21:326–332, 1924.

[32] George Gamow (1904–1968). *My World Line.* Viking Press, New York, 1970.

[33] Thomas Glick, editor. *The Comparative Reception of Relativity.* D. Reidel, Boston, 1987.

[34] Kurt Gödel (1906–1978). An example of a new type of cosmological solution of Einstein's field equations of gravitation. *Review of Modern Physics*, 21:447–450, 1949.

[35] Ivor Grattan-Guinness (1941–2014). Solving Wigner's mystery: the reasonable (though perhaps limited) effectiveness of mathematics in the natural sciences. *The Mathematical Intelligencer*, 30:7–17, 2008.

[36] Jeremy Gray. *Ideas of Space: Euclidean, Non-Euclidean, and Relativistic.* Clarendon Press, Oxford, 1989.

[37] G. M. Harvey. Gravitational deflection of light. *The Observatory*, 99:195–198, 1979.

[38] Stephen Hawking. *A Brief History of Time.* Bantam Books, New York, 1988.

[39] Heinrich Hertz (1857–1894). *Die Prinzipien der Mechanik in neuem Zusammenhange dargestellt*, volume III of *Gesammelte Werke.* Johann Ambrosius Barth, Leipzig, 1894.

[40] David Hilbert (1862–1943). Die Grundlagen der Physik. *Mathematische Annalen*, 92:1–32, 1924.

[41] David Hilbert (1862–1943). *Die Grundlagen der Physik*, pages 258–289. Chelsea, New York, 1965.

[42] Douglas R. Hofstadter. *Gödel, Escher, Bach: An Eternal Golden Braid.* Basic Books, New York, 1979.

[43] Douglas R. Hofstadter. *I Am a Strange Loop.* Basic Books, New York, 2007.

[44] Lancelot Hogben (1895–1975). *Mathematics in the Making.* Rathbone Books, London, 1960.

[45] Edwin Hubble (1889–1953). A relation between distance and radial velocity among extra-galactic nebulae. *Proceedings of the National Academy of Sciences*, 15:168–173, 1929.

[46] Felix Klein (1849–1925). *Vorlesungen über die Entwicklung der Mathematik im 19. Jahrhundert, Part II.* Springer-Verlag, Berlin, 1927.

[47] Martin J. Klein (1924–2009), A. J. Kox, and Robert Schulmann, editors. *The Collected Papers of Albert Einstein, Vol. 5. The Swiss Years: Correspondence 1902–1914.* Princeton University Press, 1993.

[48] Joseph-Louis Lagrange (1736–1811). *Méchanique analitique.* Desaint, Paris, 1788.

[49] Gabriel Lamé (1795–1870). *Leçons sur la théorie de l'élasticité des corps solides, deuxième édition.* Gauthier–Villars, Paris, 1866.

[50] Urbain Le Verrier (1811–1877). Lettre à M. Faye sur la théorie de Mercure et sur le mouvement du périhélie de cette planète. *Comptes rendus*, 49:379–383, 1859.

[51] Dennis Lehmkuhl. Mass-energy-momentum in general relativity. Only there because of space-time? *British Journal for the Philosophy of Science*, 62:453–488, 2011.

[52] Georges Lemaître (1894–1966). Un univers homogène de masse constante et de rayon croissant rendant compte de la vitesse radiale des nébuleuses extra-galactiques. *Annales de la Société Scientifique de Bruxelles*, 47:49–59, 1927.

[53] Angelo Loinger and Tiziana Marsico. On Hilbert's gravitational repulsion (*a historical note*). *ArXiv:0904.1578v1*, pages 1–7, 2009.

[54] Hendrik Antoon Lorentz (1853–1928). Théorie électromagnétique de Maxwell et son application aux corps mouvants. *Archives néerlandaises des sciences exactes et naturelles*, 25:363–552, 1892.

[55] Hendrik Antoon Lorentz (1853–1928). *Versuch einer Theorie der electrischen und optischen Erscheinungen in bewegten Körpern*. E. J. Brill, Leiden, 1895.

[56] Hendrik Antoon Lorentz (1853–1928). The Michelson–Morley experiment and the dimensions of moving bodies. *Nature*, 106:793–795, 1921.

[57] David Lovelock. The uniqueness of the Einstein field equations in a four-dimensional space. *Archive for Rational Mechanics and Analysis*, 33:54–70, 1969.

[58] Ernst Mach (1838–1916). *Die Mechanik in ihrer Entwicklung historisch-kritisch dargestellt*. F. A. Brockhaus, Leipzig, 1883.

[59] Albert Abraham Michelson (1852–1931) and Edward Williams Morley (1838–1923). On the relative motion of the earth and the luminiferous ether. *American Journal of Science*, 34:333–345, 1887.

[60] Hermann Minkowski (1864–1909). Raum und Zeit. *Jahresbericht der deutschen Mathematiker-Vereinigung*, pages 75–88, 1909.

[61] Al Momin. The Gödel solution to the Einstein field equations. (Posted online), 2008.

[62] Jayant V. Narlikar. *An Introduction to Relativity*. Cambridge University Press, 2010.

[63] Isaac Newton (1643–1727). *Philosophiæ naturalis principia mathematica (1687)*, translated as *The Mathematical Principles of Natural Philosophy*, volume 34 of *Great Books of the Western World*. Encyclopedia Britannica, 1952.

[64] Henri Poincaré (1854–1912). Sur la dynamique de l'électron. *Comptes rendus*, 140:1504–1508, 1905.

[65] Henri Poincaré (1854–1912). *The Value of Science*. Modern Library Science Series. Modern Library, New York, 2001.

[66] Henri Poincaré (1854–1912). *Science and Method (1908)*, translated by Francis Maitland (reprint). Barnes and Noble Books, 2004.

[67] Grigorio Ricci-Curbastro (1853–1952) and Tullio Levi-Civita (1873–1941). Méthodes de calcul différentiel absolu et leurs applications. *Mathematische Annalen*, 54:125–201, 1900.

[68] Georg Friedrich Bernhard Riemann (1826–1866). Ueber die hypothesen welche der Geometric zu Grunde liegen. *Abhandlungen der königlichen Gesellschaft der Wissenschaften zu Göttingen*, 13:133–152, 1867.

[69] Jason Ross, editor. *Light: A History. Fermat's Complete Correspondence on Light*. ΔYNAMIΣ, The Journal of the La Rouche–Riemann Method of Physical Economics, 2008.

[70] Bertrand Arthur William Russell (1872–1970). *History of Western Philosophy*. Simon and Schuster, New York, 1945.

[71] Bertrand Arthur William Russell (1872–1970). *The Pursuit of Truth*. Fact and Fiction. Simon and Schuster, New York, 1962.

[72] Bertrand Arthur William Russell (1872–1970). *The Analysis of Matter* (reprint of the 1927 edition). Routledge, London, 1992.

[73] Karl Schwarzschild (1873–1916). Über das Gravitationsfeld einer Kugel aus incompressibler Flüssigkeit. *Sitzungsberichte der preussischen Akademie der Wissenschaften*, pages 424–434, 1916.

[74] Karl Schwarzschild (1873–1916). Über das Gravitationsfeld eines Massenpunktes nach der Einsteinschen Theorie. *Sitzungsberichte der preussischen Akademie der Wissenschaften*, pages 189–196, 1916.

[75] Daniel M. Siegel. *Innovation in Maxwell's Electromagnetic Theory: Molecular Vortices, Displacement Current, and Light*. Cambridge University Press, 2002.

[76] Johann Georg Soldner (1776–1833). Ueber die Ablenkung eines Lichtstrals von seiner geradlinigen Bewegung, durch die Attraktion eines Weltkörpers, an welchem er nahe vorbei geht. *Berliner astronomisches Jahrbuch*, pages 161–172, 1804.

[77] Shlomo Sternberg. *Celestial Mechanics*. Benjamin, New York, 1969.

[78] Shlomo Sternberg. *Curvature in Mathematics and Physics*. Dover, New York, 2012.

[79] Peter Guthrie Tait (1831–1901). On the importance of quaternions in physics. *Philosophical Magazine and Journal of Science, Fifth series*, pages 84–97, January 1890.

[80] J. H. Taylor and J. M. Weisberg. A new test of general relativity – Gravitational radiation and the binary pulsar PSR 1913+16. *Astrophysical Journal*, 253:908–920, 1982.

[81] Roberto Torretti. *Relativity and Geometry*. Foundations and Philosophy of Science and Technology. Pergamon Press, Oxford, 1983.

[82] Vladimir Varićak (1865–1942). Anwendung der Lobatschefskijschen Geometrie in der Relativtheorie. *Physicalische Zeitschrift*, 11:93–96, 1910.

[83] Robert M. Wald. *General Relativity*. University of Chicago Press, 1984.

[84] Richard Westfall (1924–1996). Newton's marvelous years of discovery and their aftermath: myth versus manuscript. *Isis*, 71:109–121, 1980.

[85] Eugene Wigner (1902–1995). The unreasonable effectiveness of mathematics. *Communications in Pure and Applied Mathematics*, 13:1–14, 1960.

[86] Clifford M. Will. *Was Einstein Right? Putting General Relativity to the Test*. Second edition. Basic Books, 1986.

[87] Clifford M. Will. The confrontation between general relativity and experiment. (Posted online at http://www.livingreviews.org/lrr-2006-3), 2006.

[88] A. Zee. *Einstein Gravity in a Nutshell*. Princeton University Press, Princeton, 2013.

Subject Index

Name Index